ADVANCES IN CIVIL, ARCHITECTURAL, STRUCTURAL AND CONSTRUCTIONAL ENGINEERING

PROCEEDINGS OF THE INTERNATIONAL CONFERENCE ON CIVIL, ARCHITECTURAL, STRUCTURAL AND CONSTRUCTIONAL ENGINEERING, DONG-A UNIVERSITY, BUSAN, SOUTH KOREA, 21–23 AUGUST, 2015

Advances in Civil, Architectural, Structural and Constructional Engineering

Editors

Dong-Keon Kim
Department of Architectural Engineering, Dong-A University, Busan, South Korea

Jongwon Jung
Department of Civil and Environmental Engineering, Louisiana State University, Banton, LA, USA

Junwon Seo
Department of Civil and Environmental Engineering, South Dakota State University, Brookings, USA

CRC Press
Taylor & Francis Group
Boca Raton London New York

CRC Press is an imprint of the
Taylor & Francis Group, an **informa** business
A BALKEMA BOOK

Published by:
CRC Press/Balkema
P.O. Box 447, 2300 AK Leiden, The Netherlands
e-mail: Pub.NL@taylorandfrancis.com
www.crcpress.com – www.taylorandfrancis.com

First issued in paperback 2020

© 2016 by Taylor & Francis Group, LLC
CRC Press/Balkema is an imprint of the Taylor & Francis Group, an informa business

No claim to original U.S. Government works

Typeset by V Publishing Solutions Pvt Ltd., Chennai, India

ISBN 13: 978-0-367-73730-6 (pbk)
ISBN 13: 978-1-138-02849-4 (hbk)

Visit the Taylor & Francis Web site at
http://www.taylorandfrancis.com

and the CRC Press Web site at
http://www.crcpress.com

Table of contents

Structures, building performance study and power energy

Preface

The ICCASCE 2015 is an annual international conference on Civil, Architectural, Structural and Constructional Engineering. ICCASCE 2015 took place in Busan, at Dong-A University, South-Korea, on August 21–23, 2015. This conference was sponsored by the Research Institute of Construction Technology and Planning (RICTP) at the Dong-A University. The conference program covered invited, oral, and poster presentations from scientists working in similar areas with a view to establish platforms for collaborative research projects in this field. This conference has brought together leaders from industry and academia to exchange and share their experiences, present research results, explore collaborations and to spark new ideas, with the aim of developing new projects and exploiting new technology in this field.

The book is a collection of accepted papers. All these accepted papers were subjected to strict peer-reviewing by 2–3 expert referees, including a preliminary review process conducted by conference editors and committee members before their publication by CRC Press. This book is divided into four main chapters including 1. Advanced material properties and applications, 2. Mechanical and management engineering, 3. Structures, building performance study and power energy, 4. Water, irrigation and architectural engineering application. The committee of ICCASCE 2015 would like to express their sincere thanks to all authors for their high-quality research papers and careful presentations. All reviewers are also thanked for their careful comments and advices. Thanks are finally given to CRC Press as well for producing this volume.

The Organizing Committee of ICCASCE 2015
Committee Chair Prof. Dong-keon Kim,
Dong-A University, South Korea

Advances in Civil, Architectural, Structural and Constructional Engineering – Kim, Jung & Seo (Eds)
© 2016 Taylor & Francis Group, London, ISBN 978-1-138-02849-4

Organization

RESEARCH INSTITUTE OF CONSTRUCTION TECHNOLOGY AND PLANNING (DEPARTMENT OF ARCHITECTURAL ENGINEERING) AT THE DONG-A UNIVERSITY

In these days it is vital to be prepared for disasters and calamities and give proper attention to prevention measures. As a result of this understanding we have initiated the Research Institute of Construction Technology and Planning that was founded in 1965 at the Dong-A University with the aim to contribute to the development of the construction industry.

Accordingly, the Research Institute of Construction Technology and Planning at the Dong-A University has advanced research on construction technology and planning through the various seminars, conferences and lectures. This research could be conducted, in cooperation with the Dong-A University, in various fields such as structures, soil, hydraulic engineering and the environment. The Research Institute of Construction Technology and Planning aims to be a leader in the construction industry through stimulating various research activities, technical exchanges and organizing global conferences.

Advances material properties and applications

Advances in Civil, Architectural, Structural and Constructional Engineering – Kim, Jung & Seo (Eds)
© *2016 Taylor & Francis Group, London, ISBN 978-1-138-02849-4*

Hardened properties of sawdust-crete containing pre-coated sawdust with nano-silica

Bashar S. Mohammed, M.F. Nuruddin & N.H. Ishak
Department of Civil and Environmental Engineering, Universiti Teknologi PETRONAS, Bandar Seri Iskandar, Perak, Malaysia

ABSTRACT: Inclusion of sawdust as a partial replacement to fine aggregate in Portland cement concrete has many advantages such as safe waste-disposal and improving some hardened concrete properties. Sawdust-crete is lighter in weight and has lower thermal conductivity in comparison with normal concrete. However, the major drawback is the reduction in the strength of the sawdust. Therefore, nano-silica is used to pre-coat the sawdust before mixing with other sawdust-crete ingredients. The test results show that the use of nano-silica will lead to the decrease in Mercury Intrusion Porosimetry (MIP) and Interfacial Transition Zone (ITZ) and the increase in the compressive strength of the hardened sawdust-crete.

1 INTRODUCTION

Construction industry is facing a challenge of integrating sustainability in its production development, by either preserving the natural raw materials or reducing CO_2 emission (Torkaman et al. 2014). One of the possible ways to achieve this goal is to include waste materials in concrete mixture production (Mohammed et al. 2012 and Mohammed & Fang 2011). Sawdust is an industrial by-product waste from sawmill and continues to increase due to the growing demand on wood products. To reduce pollution due to sawdust combustion, sawdust has been used in concrete mixtures as a partial replacement to fine aggregate by volume (Mohammed et al. 2014). It has been reported that concrete containing sawdust is lighter in weight and exhibits lower thermal conductivity in comparison with normal concrete (Mohammed et al. 2014). However, the adverse effect of the inclusion of sawdust in concrete as a partial replacement to fine aggregate is the reduction in strengths. Therefore, researchers have suggested the treatment of wood waste before inclusion in concrete to enhance bonding between wood waste surfaces and the cement matrix (Coatanlem et al. 2006). Coatanlem et al. (2006) reported an improvement in the strength of concrete containing wood chippings by soaking these chippings in sodium silicate solution prior to mixing in concrete. This led to the improvement of the bond between the wood waste and the cement matrix due to the development of ettringite needles at surfaces of the wood waste and near surface of the cement matrix. Nano-silica (SiO_2) has been widely used to improve the concrete properties due to its high reactivity, accessibility and effective final cost (Mukharjee & Barai 2014). The nano-silica enhances the hardened properties of the concrete through chemical and physical mechanisms. The chemical mechanism involves the conversion of calcium hydroxide into C-S-H gel, while the physical mechanism involves filling the nano-voids in the nano-phase of cement paste. Both mechanisms lead to the densification of the microstructure of the concrete and crushing of the Interfacial Transition Zone (ITZ) between the hardened cement pastes and the aggregate that subsequently improve the strength and durability of the concrete. Therefore, the main objective of this work was to investigate the effects of the coating of sawdust with nano-silica on the properties of the dry concrete containing sawdust as a partial replacement to fine aggregate.

2 MIXTURE PROPORTIONS AND SAMPLE PREPARATION

A total of 21 dry concrete mixtures were prepared for testing in this investigation. Two variables were selected, namely sawdust and nano-silica as a partial replacement by volume to fine aggregate and cement, respectively. Seven replacement percentages of sawdust (0%, 5%, 10%, 15%, 20%, 25% and 30%) and three replacement percentages of nano-silica (0%, 5% and 10%) were considered to study the effects of coating sawdust particles with nano-silica on the hardened properties of the sawdust concrete or sawdust-crete. Table 1 presents the mixture ratio proportions by volume. The main mixture ratio is 1 (cement):1 (coarse aggregate):2 (fine aggregate), while the water content required

Table 1. Mixture ratio proportions by volume for dry concrete mixtures.

Mixture		Cementitious Materials			Coarse Aggregate	Fine Aggregate	
NS	SD%	Cement	Fly ash	Nano silica		Sand	Saw-dust
0	0	0.85	0.15	0	1	2	0
	5	0.85	0.15	0	1	1.9	0.1
	10	0.85	0.15	0	1	1.8	0.2
	15	0.85	0.15	0	1	1.7	0.3
	20	0.85	0.15	0	1	1.6	0.4
	25	0.85	0.15	0	1	1.5	0.5
	30	0.85	0.15	0	1	1.4	0.6
5	0	0.8	0.15	0.05	1	2	0
	5	0.8	0.15	0.05	1	1.9	0.1
	10	0.8	0.15	0.05	1	1.8	0.2
	15	0.8	0.15	0.05	1	1.7	0.3
	20	0.8	0.15	0.05	1	1.6	0.4
	25	0.8	0.15	0.05	1	1.5	0.5
	30	0.8	0.15	0.05	1	1.4	0.6
10	0	0.75	0.15	0.1	1	2	0
	5	0.75	0.15	0.1	1	1.9	0.1
	10	0.75	0.15	0.1	1	1.8	0.2
	15	0.75	0.15	0.1	1	1.7	0.3
	20	0.75	0.15	0.1	1	1.6	0.4
	25	0.75	0.15	0.1	1	1.5	0.5
	30	0.75	0.15	0.1	1	1.4	0.6

for each dry mixture is 8% of the total batch dry weight. Four tests were performed on the samples from selected mixtures for 28 days, which include: compressive strength, Scanning Electron Microscopy (SEM), thermal conductivity test and Mercury Intrusion Porosimetry (MIP). In the mixing process, 2% of water in the total mixture was added to sawdust and mixed with nano-silica for about 5 minutes, and then cement, flay ash, fine aggregate, coarse aggregate and the remaining amount of water were added and the mixing was continued for another 5 minutes. All tests specimens were prepared under compaction pressure, whereas the materials were cast inside the mold in three equal depth layers with 25 blows for each layer and cured in the laboratory environment for 28 days before testing.

3 TEST RESULTS AND ANALYSIS

Compressive strength test in accordance with the requirements of ASTM C140 was conducted on three samples from each mixture for 28 days. Figure 1 shows the compressive strength for each mixture; and the values, reported in Figure 2, are average of three samples. As shown in Figure 1, the

compressive strength decreases with the increasing sawdust percentage from 0% to 30% and increases with the increasing amount of nano-silica percentage from 0% to 10%. This reduction in the compressive strength of the hardened sawdust-crete might be due to the following reasons: water demand, less bonding between the cement matrix and sawdust particles, and hardness of the sawdust particles. For water demand, the water absorption of sawdust is much higher than that of the sand and also its size of the particles is finer than that of the sand; therefore, increasing sawdust percentage in the concrete mixture will lead to reduction in the amount of water available for cement hydration, which in turn will result in concrete with lower compressive strength. Because the sawdust particles are finer than sand, they have a surface area larger than that of the replaced sand so that

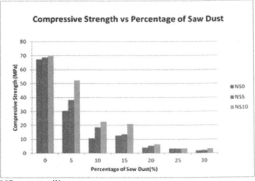

NS = nano-silica

Figure 1. Compressive strength versus percentage of sawdust.

a) SEM image for NS0SD15

b) SEM image for NS5SD15.

C) SEM image for NS10SD15.

Figure 2. SEM images of sawdust-crete mixtures.

more cement is required to cover all these particles, which also lead to the decrease in the compressive strength of the hardened sawdust-crete. Another reason is that the sawdust increases the air content of sawdust-crete, which leads to the development of weaker points inside the mixture (sawdust particles), resulting in stress concentration points in the mixture leading to premature failure.

The Mercury Intrusion Porosimetry (MIP) test is the most popular test used for determining the pore characteristics in concrete. It is conducted using Thermo Finnigan Pascal 240 with a high pressure station up to 200 MPa and pore radius analysis ranging from 7.5 to 0.0037 μm. It is used to measure the volume and size of the pores. Samples from five mixtures were tested, and these selected mixtures showed the variation effects of nano-silica percentages (0%, 5% and 10%) with constant sawdust containing 15% and also variation effects of sawdust percentages (10%, 15% and 20%) with constant nano-silica containing 5%. As shown in Table 2, the accessible porosity is increased as the percentage of the sawdust is increased, and it decreased as the nano-silica percentage increased.

Scanning Electron Microscope (SEM) with ultra-high-resolution imaging is designed to fulfill the requirements of analysis up to nano-scale surface structure and morphology of solids. Samples from three mixtures were tested to establish the variation effects of nano-silica percentages (0%, 5% and 10%) on the thickness of the Interfacial Transition Zone (ITZ) between the cement paste and the surface of the coarse aggregate. Figure 2 and Table 3 summarize the results of ITZ, where

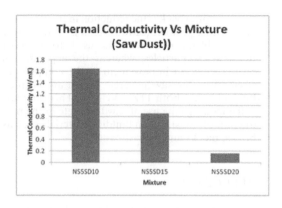

Figure 3. Thermal conductivity versus sawdust contents.

the thickness of ITZ decreases as the nano-silica percentage increases from 0% to 10%.

The decrease in the MIP and ITZ of sawdust-crete mixtures with the increasing nano-silica contents is due to the physical and chemical effects of nano-silica. The physical effect is due to the filling of nano-particles into the nano-voids in the hardened cement paste, which leads to the reduction in the porosity of the sawdust-crete. In contrast, the chemical effect is due to the high reactivity of the nano-silica with hydroxide calcium to produce C-S-H gel, which densifies the ITZ between the aggregate and the cement paste. Both effects of nano-silica (physical and chemical) lead to the enhancement of the bond strength between the cement paste and the aggregates' surface and in turn increase the compressive strength and decrease the permeability.

Thermal conductivity test was conducted to determine how the material can conduct heat and how it behaves in the concrete. Three mixtures were selected to measure the variable effects of sawdust percentages (10%, 15% and 20%) on the heat conductivity of sawdust-crete mixtures. As shown in Figure 3, increasing sawdust leads to decreasing thermal conductivity of the mixtures. This is due to the fact that increasing sawdust percentage in the sawdust-crete mixtures leads to increasing air contents and voids inside the mixtures. As the air has lower thermal conductivity than cement paste, increasing air contents will lead to lower final thermal conductivity of the mixture.

Table 2. Total pore volume and accessible porosity for sawdust-crete mixtures.

Sample	Total pore volume (mm^3/g)	Accessible porosity (%)
NS0SD15	128.43	28.13
NS5SD10	117.63	25.42
NS5SD15	120.45	26.35
NS5SD20	133.69	28.58
NS10SD15	110.19	25.45

NS: nano-silica, SD: sawdust.

Table 3. ITZ for sawdust-crete mixtures.

Samples	Interfacial Transition Zone, ITZ (μm)
NS0SD15	7.12
NS5SD15	3.28
NS10SD15	2.97

NS: nano-silica, SD: sawdust.

4 CONCLUSIONS

Many advantages have been reported on the inclusion of sawdust as a partial replacement to fine aggregate in Portland cement concrete. However, the main drawback is the reduction in the strengths of the final product (sawdust-crete). To overcome this problem, nano-silica is used to pre-coat the sawdust particles, whereas the experimental results show an improvement in the compressive strength of the sawdust-crete, which are justified by decreasing both ITZ and MIP. This outcome will encourage the use of sawdust-crete in the production of construction building materials such as bricks, blocks and precast panels.

REFERENCES

Coatanlem, P., Jauberthie, R. & Rendell, F. 2006. Lightweight wood chipping concrete durability. *Construction and Building Materials*, 20: 776–781.

Mohammed, B.S., Abdullahi, M. & Hoong, C.K. 2014. Statistical models for concrete containing wood chipping as partial replacement to fine aggregate. *Construction and Building Materials*, 55: 13–3.

Mohammed, B.S. & Fang, O.C. 2011. Mechanical and durability properties of concretes containing paper-mill residuals and fly ash. *Construction and Building Materials*, 25:717–725

Mohammed, B.S., Fang, O.C., Hossain, K.M.A. & Lachemi, M. 2012. Mix proportioning of concrete containing paper mill residuals using response surface methodology. *Construction and Building Materials*, 35: 63–68.

Mukharjee, B.B. & Barai, S.V. 2014. Influence of incorporation of nano-silica and recycled aggregates on compressive strength and microstructure of concrete. *Construction and Building Materials*, 71: 570–578.

Torkaman, J., Ashori, A. & Momtazi, A.S. 2014. Using wood fiber waste, rice husk ash, and limestone powder waste as cement replacement materials for lightweight concrete blocks. *Construction and Building Materials*, 50: 432–436.

Advances in Civil, Architectural, Structural and Constructional Engineering – Kim, Jung & Seo (Eds)
© *2016 Taylor & Francis Group, London, ISBN 978-1-138-02849-4*

Chloride resistance of PVA cement-based materials

J.F. Shao, Y.F. Fan & S.Y. Zhang
Dalian Maritime University, Dalian, China

ABSTRACT: The performance of PVA cement-based materials on chloride resistance at various curing ages is systematically discussed in this paper. The chloride resistance of fly ash at different dosages and different fiber contents is investigated by the Rapid Chloride Migration (RCM) method. The test results indicate that the chloride diffusion coefficient of PVA cement-based materials when mixed with 60% fly ash decreases by 13.96% and 2.59% when compared with 50% and 70% fly ash. Chloride penetration can be effectively restrained by the addition of fiber. The 28-day samples can inhibit the penetration of chloride ions prominently. The chloride diffusion coefficient of PVA cement-based material samples is smallest when 2% fiber is added, which is 24.1% lower than the specimens without fiber.

1 INTRODUCTION

In recent years, many structures were damaged much earlier than their target service lives, especially in the marine environment. Chloride ions penetrating into the cement-based materials will cause the steel to corrode and finally lead to the breaking of the materials directly. Developing new materials with high chloride penetration resistance would be an efficient and fundamental way. In order to improve the poor tensile properties and ductility properties of cement-based materials, fiber-reinforced materials have been used recently. Steel fiber, polypropylene fiber and carbon fiber-reinforced materials are most widely used in the market. Steel fiber is hard to extend because it is hard to mix and disperse and because of its heavy weight. The poor performance of the bond matrix and low tensile strength limits the use of polypropylene fiber to be widespread in the market. Carbon fiber is expensive. PVA has a brilliant development because it has excellent characteristics.

The study on the properties of PVA (Poly Vinyl Alcohol) cement-based materials has obtained a lot of achievements. A large number of research results show that it has good ductility and energy absorption capacity. The characteristics of strain hardening and multiple cracking have been shown by the direct tensile test. The tensile strength is improved obviously. However, the durability of PVA cement-based materials has been relatively failed to achieve in a wide range of research. Some research results show that it has good crack resistance and frost resistance capacity (Li et al. 2011, Xu & Li 2009, Li & Xu 2009, Şahmaran et al. 2008). But there are not many results when adding a large number of fibers, especially for resistance to chloride penetration.

The objective of this research is to obtain the relationships between the chloride diffusion coefficient of PVA cement-based materials and fly ash and fiber content. Then, it aims to obtain the permeability of PVA cement-based materials. Therefore, it provides a reference for the design of the mix proportions of PVA cement-based materials.

2 MATERIALS AND METHODS

2.1 Experimental materials

Ordinary Portland cement of Type 42.5R was used in all mixes. Its chemical composition is presented in Table 1. The fly ash used is of first grade and its density is 2548 kg/m³. Fine aggregates are river sand (0.6–1.18 m) and quartz sand (110 μm–220 μm).

Polycarboxylate superplasticizer was selected, with a solid content of 20%. The fiber used was obtained from Japan Kuraray Company's production of high-strength PVA fiber. The basic parameters are listed in Table 2.

Table 1. Chemical composition of cement.

Chemical composition	CaO	SiO_2	Al_2O_3	Fe_2O_3	MgO	SO_3	LOI
Content/%	59.30	21.91	6.27	3.78	1.64	2.41	4.69

Table 2. The basic parameters of PVA fiber.

Density (g/cm³)	Diameter (μm)	Length (mm)	Strength (MPa)	Elastic Modulus (GPa)	Elongation (%)
1.3	40	12	1560	41	6.5

2.2 Specimen fabrication

The experiment chose quartz sand in the chloride resistance of fly ash at different dosages of 50%, 60% and 70%, and the water-to-cement ratio was 0.35. Mixing proportions are presented in Table 3. In PVA cement-based materials for using river sand, the water-to-cement ratio was 0.35, the fly ash content was 60%, and the fiber content used was 0%, 0.8%, 1%, 1.2%, 1.4%, 1.8% and 2%. Mixing proportions are presented in Table 4.

First, sand, cement, plasticizers and fly ash were added into a mixing pot, dry mixed for 2mins, and then added to the mixture of water and water reducer wet mix for 3mins. Finally, the fiber was added and then mixed for 5mins.

The test specimens were divided into 30 groups, with three specimens per group. The specimen size was Φ100 mm × 50 mm. Conserving the test specimens to the specified test age, tests were performed on the specimens.

2.3 Methods

To evaluate the chloride permeability of cement based materials, a series of testing methods were employed. The Rapid Chloride Migration (RCM)

Table 3. Mix proportions of fly ash in the test.

Cement	Quartz sand	Fly ash	Fiber	Water	Water reducer
624.82	388.29	624.82 (50%)	19.5	437.37	3.12
499.85	388.29	749.78 (60%)	19.5	437.37	3.12
374.89	388.29	874.74 (70%)	19.5	437.37	3.12

Table 4. Mix proportions of fiber in the test.

Sand	Cement	Fly ash	Water	Water reducer	Fiber	Cellulose
394.2	507.5	761.2	440	5.1	0 (0%)	0.333
391	503.4	755.1	432.4	10.1	10.4 (0.8%)	0.333
390.2	502.4	753.6	439.6	11.3	13 (1%)	0.333
389.5	501.4	752.1	438.7	12.5	15.6 (1.2%)	0.333
388.7	500.3	750.5	423.8	17.5	18.2 (1.4%)	0.333
387.1	498.3	747.5	422.1	17.4	23.4 (1.8%)	0.333
386.3	497.3	746	421.2	17.4	26 (2%)	0.333

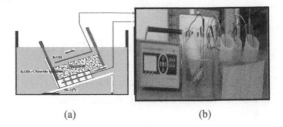

(a)　　　　　　　　　(b)

Figure 1. RCM-DAL chloride ion diffusion coefficient tester. a. Sketch b. test device.

Table 5. Initial current and testing time.

Initial current I$_0$/mA	I$_0$ < 5	5 ≤ I$_0$ < 10	10 ≤ I$_0$ < 30
Testing time/h	168	96	48
Initial current I0/mA	30 ≤ I0 < 60	60 ≤ I0 < 120	120 ≤ I0
Testing time/h	24	8	4

method proposed by Tang and Nilsson was applied in this study. The principle of this method is to generate chloride penetration through the sample by a solution concentration gradient and accelerate the movement of chloride using an electrical field (Fan & Zhang 2014, Stanish et al. 1997, Spiesz et al. 2012, Tang & Nilsson 1992).

The testing setup used in this paper is shown in Figure 1. The relationship between the applied initial current and the testing time is presented in Table 5. The depth was used to determine the diffusion coefficient through the Nemst-Einstein equation, which is described as follows:

$$D_{RCM} = 2.872 \times m^2/s \frac{Th\left(x_d - \alpha\sqrt{x_d}\right)}{t}$$

$$\alpha = 3.338 \times 10^{-3}\sqrt{Th}$$

(1)

where D_{RCM} is the chloride diffusion coefficient tested by the RCM method (m²/s); T is the average temperature of the initial and final temperatures of the anode solution (K); h is the height of the testing specimen (m); x_d is the diffusion depth of chloride anion (m); t is the electricity test time (s); and a is a dimensionless constant.

3 RESULTS AND DISCUSSION

3.1 Effect of fly ash on the chloride permeability of PVA cement-based materials

Conserving the test specimen to a predetermined age, the chloride permeability tests were performed on the specimens. Table 6 presents the chloride dif-

Table 6. Chloride diffusion coefficient values of the specimens at different ages (m²/s).

Fly ash addition/%	Age/d		
	7	14	28
50	6.86E-11	7.536E-12	1.479E-12
60	5.57E-11	6.472E-12	1.273E-12
70	3.77E-11	7.871E-12	1.307E-12

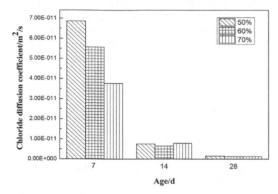

Figure 2. Chloride diffusion coefficient values of the specimens at different ages.

fusion coefficient values of the specimens at different ages. The development of the chloride diffusion coefficients with the addition of fly ash is shown in Figure 2.

As shown in Figure 2, the chloride diffusion coefficient of cement-based material concrete decreased obviously with the addition of fly ash at the age of 7 days. Thus, the chloride resistance increased. At the age of 14 days, the chloride diffusion coefficient of the specimens with increasing fly ash content first decreased and then increased. The chloride diffusion coefficient was minimum when mixed with 60% fly ash.

The trend of the chloride diffusion coefficient at the age of 28 days was the same as that at the age of 14 days. The chloride diffusion coefficients of PVA cement-based materials when mixed with 60% fly ash decreased by 13.96% and 2.59% when compared with 50% and 70% fly ash.

According to the study by Liu (Liu 2009), the effect of fly ash on the chloride permeability of PVA cement-based materials is obvious. This is due to the large changes in the micro-structure when mixed with fly ash. The harmful holes are reduced and the porosity is decreased. Therefore, the chloride permeability resistance of cement-based materials is enhanced.

With the increase in the content of fly ash, cement content is reduced relatively. The ductility of the cement-based materials is improved. But the strength is relatively weakened. The generation and development of cracks are not well prevented. This explains why the chloride diffusion coefficient increases with the addition of fly ash.

3.2 Effect of fiber on the chloride permeability of PVA cement-based materials

Conserving the test specimen to a predetermined age, the chloride permeability tests were performed on the specimens. Table 7 presents the chloride diffusion coefficient values of the specimens at different ages. The development of the chloride diffusion coefficients with the addition of fiber is shown in Figure 3.

Table 7. Chloride diffusion coefficient values of the specimens at different ages (m²/s).

Fiber addition/%	Age/d		
	3	7	28
0	1.22E-10	6.43E-11	5.34E-12
0.8	1.19E-10	6.02E-11	5.88E-12
1	1.24E-10	6.17E-11	4.59E-12
1.2	1.16E-10	5.82E-11	5.04E-12
1.4	1.13E-10	6.05E-11	4.22E-12
1.8	1.15E-10	6.05E-11	4.08E-12
2	1.14E-10	5.98E-11	4.05E-12

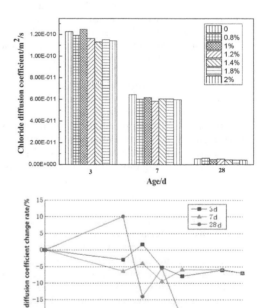

Figure 3. Chloride diffusion coefficient values of different fiber contents.

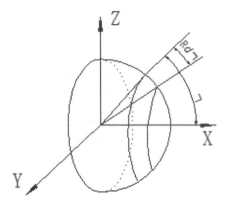

Figure 4. Hemisphere model.

As shown in Table 7 and Figure 4, the chloride diffusion coefficient decreases obviously with curing ages, indicating that all curves have similar trends, curves preliminarily change rapidly and then level off. Thus, the chloride resistance increased. The addition of fiber into cement-based materials can promote the ability of resistance to chloride penetration. Particularly at the age of 28 days, the chloride diffusion coefficient of the cement-based materials decreased significantly. The chloride diffusion coefficient for the fiber content of about 1.3% was smaller than that for other contents at 3 and 7 curing days. Decreasing by 7.9% and 5.9%, with the increasing fiber content, the chloride resistance increased. The 28-day samples can inhibit the penetration of chloride prominently. The chloride diffusion coefficient of PVA cement-based material samples is smallest when 2% fiber is added, which is 24.1% lower than those without the fiber.

According to the study by Wang (2009), fiber can change the structure and porosity of concrete internal pores. The appropriate amount of fiber can effectively improve the internal pores' structure and improve the density of cement-based materials. However, adding more fiber into cement-based materials will reduce the density and increase the porosity, making it difficult to disperse. There will be a lot of bubbles produced in the materials. Chloride penetration is accelerated due to the addition of the inside interface. The shrinkage stress will appear during the capillary water evaporation in the hydration process. If tensile stress is relatively large, cement-based material will produce plastic cracking. The distribution of fiber will become complex, preventing cracking and crack extension. On the other hand, fiber improves the capillary structure, even blocking it. This is the main factor responsible in reducing the chloride diffusion coefficient. However, at early ages, fiber and mortar still cannot form a reliable bonding. Thus, it cannot effectively prevent cracking.

It is clear that increasing the fiber content will reduce the cross-sectional area of the mortar. Our test is to force chloride ions to move by means of the electric field. The resistance value of the material is inversely proportional to the cross-sectional area, leading to the increase in the resistance values of cement-based materials. Theoretical calculations show that this effect is much larger than that using the fiber content. We chose the hemisphere model and the secondary projection model (Romualdi & Mandel 1964), and considered that each fiber has an independent identical distribution. Figure 4 shows the hemisphere model. Its probability density is $sin\alpha\, d\alpha$. The cross-sectional enhancement coefficient can be obtained by the following formula:

$$C_1 = \int_0^{\frac{\pi}{2}} \sec\alpha \sin\alpha\, d\alpha \tag{2}$$

However, this is an abnormal integral, so let us assume α = arctan (12/0.04) = 89.81° as the integral's upper limit. It can be understood that the probability of the fiber perpendicular to the horizontal plane is 0. Therefore, Equation (2) is equal to 5.71.

Figure 5 shows the secondary projection model. Its probability density is shown in the following formula:

$$f(\alpha,\beta) = \frac{1}{2\pi}\cos\alpha\, d\alpha\, d\beta \tag{3}$$

The cross-sectional enhancement coefficient can be obtained by the following formula:

$$C_2 = \left(\frac{1}{2\pi}\right)\iint \sec\alpha \sec\beta \cos\alpha\, d\alpha\, d\beta = 6.4 \tag{4}$$

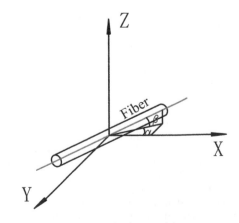

Figure 5. Secondary projection model.

There is little difference between the two coefficients. In the hemisphere model, the resistance increases by 10.6% when the fiber content is 2%. So the impact of fiber content on the resistance cannot be ignored. But at early ages, the impact can be ignored because of much water inside the cement-based materials.

For these reasons, PVA cement-based materials exhibited early resistance to chloride penetration as the fiber content first increased and then decreased. However, the 28-day samples can inhibit the penetration of chloride prominently.

4 CONCLUSIONS

The experiment of the Rapid Chloride Migration (RCM) method was conducted to study the chloride penetration of PVA cement-based materials. The chloride diffusion coefficient of PVA cement-based materials was investigated at different curing ages, different dosages of fly ash and different fiber contents. Based on the results, the following conclusions can be drawn:

1. The chloride diffusion coefficients of PVA cement-based materials when mixed with 60% fly ash decreased by 13.96% and 2.59% when compared with 50% and 70% fly ash. That is, when mixed with 60% fly ash, the penetration resistance of chloride ions was better than that mixed with 50% and 70% fly ash.
2. In the cement-based materials, fiber can promote the growth of penetration resistance of chloride ions. The chloride diffusion coefficient for the fiber content of about 1.3% was smaller than that for other contents at 3 and 7 curing days. Decreasing by 7.9% and 5.9%, with the increasing fiber content, the chloride resistance decreased. The 28-day samples could inhibit the penetration of chloride prominently. The chloride diffusion coefficient of PVA cement-based material samples was smallest when 2% fiber was added, which was 24.1% lower than the specimens without fiber.

ACKNOWLEDGMENT

This paper was financially supported by a National Natural Science Foundation of PR China (Grant No. 51178069), Program for New Century Excellent Talents in University (Grant No. NCET-11-0860).

REFERENCES

Fan, Y.F. & Zhang, S.Y. 2014. Influence of kaolinite clay on the chloride diffusion property of cement-based materials. *Cement & Concrete Composites*, 45: 117–124.
Li, Q.H. & Xu, S.L. 2009. Performance and application of ultra-high toughness cementitious composite: a review. *Engineering Mechanics*, 26(2): 23–60.
Li, Y., Liang, X.W. & Liu, Z.J. 2011. Experimental research on preparation of ecological cementitious composites PVA-ECC. *Industrial construction magazine agency*, 41(4): 97–102.
Liu, Z.Y. 2009. Experimental study on the shrinkage properties of fly ash concrete and reference concrete under different temperature and humidity environments. *China civil engineering journal*, 42(5): 69–73.
Romualdi, J.P. & Mandel, J.A. 1964. Tensile strength of concrete affected by uniformly distributed and closely spaced short length of wire reinforcement. *ACI Journal Proceedings*, 657–670.
Şahmaran, M., Li, V.C. & Andrade, C. 2008. Corrosion resistance performance steel-reinforced engineered cementitious composite beams. *ACI Materials Journal*, 105(3): 243–250.
Spiesz, P., Ballari, M.M. & Brouwers, H.J.H. 2012. RCM: A new model accounting for the non-linear chloride binding isotherm and the non-equilibrium conditions between the free and bound-chloride concentrations. *Construction and Building Materials*, 27: 293–304.
Stanish, K.D., Hooton, R.D. & Thomas, M.D.A. 1997. *Testing the chloride penetration resistance of concrete: a literature review*. FHWA Contract DTFH61-97-R-00022 "Prediction of Chloride Penetration in Concrete". Ontario (Canada): University of Toronto.
Tang, L. & Nilsson, L.O. 1992. Rapid determination of chloride diffusivity of concrete by applying an electric field. *ACI Materials Journal*, 49(1): 49–53.
Wang, C.F. 2009. *Study on durability of polypropylene fiber concrete in chloride environment*, Xian: Xian University of Architecture and Technology, 33–48.
Xu, S.L. & Li, H.D. 2009. Uniaxial tensile experiments of ultra-high toughness cementitious composites. *China civil engineering journal*, 42(9): 33–40.

Advances in Civil, Architectural, Structural and Constructional Engineering – Kim, Jung & Seo (Eds)
© *2016 Taylor & Francis Group, London, ISBN 978-1-138-02849-4*

The interpretation of Norman Forster's ecological works

Peijun Yu
Department of Mechanical Engineering, Binghamton University-State University of New York, NY, USA

ABSTRACT: This article presents the ecological technological strategies used in Norman Forster's building works and the effects achieved by them. By analyzing the skills to grasp and handle the relationships of modern architecture and ecological environment in his architectural design, the enlightening and creative thinking is presented and a new train of thought for China future architecture design is provided. Nowadays, China is in the stage of rapid urbanization, due to the pursuit of "high-end", such as heating, air-conditioning, lighting, ventilation, office appliances and other aspects, and the energy consumption is greatly increased. Building energy consumption in China accounts for one-third of the total energy consumption, so the progress on ecological energy-saving low-carbon road is urgent for them to realize sustainable development.

1 INTRODUCTION

The ecological architecture, also known as "sustainable architecture" or "green architecture", is the production of architectural ecology research. Ecological architecture is devoted to the study of the relationship between the whole ecosphere and the architecture, which is considered as a creature of the ecosphere. According to its definition, in the narrow sense, it is a living space environment which is energy-efficient, healthy and harmless, recyclable and in harmony with the nature. Broadly defined, it is a system which is built by the principle of guarantee, the ecological system's virtuous circle, the basic of ecological economy, the connotation of ecological society, the support of ecological technology and the goal of the ecological environment.

Architecture ecological technology strategy is a total name of all kinds of technological or non-technological methods to realize the design target of the ecological architecture (Yang et al. 2014a). It is the technology application part of architecture ecology and its content appears to diversify. Irrespective of the kinds of ecological technology strategy, we should adjust 'suitable technology' to local conditions, such as local environment, climate, economy and culture. Here, 'suitable technology' is not an absolute concept; it may be traditional technology strategy, high technology strategy or low technology strategy between them. However, to one place, choose one strategy, or two or all, it must be suitable for the local environment. All the measures to approach to the ecological architecture design target can be called the "suitable technology" (Bassiouny & Koura 2008).

Although architecture ecological technology strategy is just one branch of many academic ecological technology strategies in the process of human ecological society, it links to other subjects closely. The strategy is not permanent. With the improvement of human knowledge, the advancement of other related subjects and the development of ecological architecture will also change and advance. We can say that there is no end for the research on the architecture ecological technology strategy. So any research on the architecture ecological technology strategy must be based on the background of the times. Any research will be conducive to the architecture's ecological progress if separated from reality (Yang 2010).

Norman Foster, the representative of 'high-tech', is one of the most successful architects in the world. When he is mentioned, he will be associated with the steel and high-tech materials. Meanwhile, he is the leader of ecological architecture, who uses the principle of ecology and the concept of high technology, low energy consumption, recyclable and green ecological design to create an ecological energy-saving system and changes traditional design concept by "technical thinking". Foster has many well-known works such as Headquarters Building of Frankfurt Commerzbank, Government Building of Berlin, Gary Art Center of France, Tokyo Millennium Tower of Japan, London City Hall of England and Hong Kong International Airport of China (Ong 2003). He is familiar with new technology, especially for the full application of ecological technology that adds a great temptation for his architectural creation. In this paper, we would like to start from Foster's works to expound the application of ecological technology strategy in architecture design (Mochida & Lun 2008).

2 LONDON CITY HALL OF ENGLAND

London City Hall is one of the most important new buildings in London, the capital of England (Figure 1). Its design aims not only to express the openness of the national democratic system in the process of implementation, but also to show the overall sustainability and the potential of protecting the environment from pollution as a public building (Figure 2) (Lin 2007).

There are a variety of considerations on the new site and construction scheme at that time. In 1998, the original eight designs were exhibited for public in London. The visitors from different communities gave their comments. Finally, the Tower Bridge was selected as the address and the design of Norman Foster's office was adopted by the mayor of London in 1999. The London City Hall is located in the south bank of the River Thames, a prominent position, adjacent to many famous historic landmarks such as London Bridge and Tower Bridge (Qian 2014). It takes only 5~7 minutes to reach London Bridge, the railway station of Tower Bridge and the subway station. The building is a steel structure that was in construction for 15 months before its opening to the public in the summer of 2002 (Figure 3).

The London City Hall is located in the banks of the River Thames, besides the Tower Bridge. Its floor space and building area are 5.25 ha and 17000 m^2, respectively. The main function rooms include: meeting hall, the mayor's office, senators' office and public service occupancy that can accommodate 440 persons to work and an auditorium that can accommodate 250 persons. Instead of a strict government office, the Hall is a facility for public services, so the whole outer walls of the building are decorated with high transparency of glass curtain wall (Zhang et al. 2004). The intention of its construction is to emphasize the transparency of government work, to let the citizens know the internal situation of the office and to reflect the office transparency and public participation (Figure 4). At the same time, at the building's top, there is a flexible space, 'London Parlour', which can be an exhibition hall open to the public. It can accommodate 200 persons. There is a café at the bottom of the building (Yang 2013a).

Figure 1. Full view of London City hall.

Figure 2. Transverse section of London City hall.

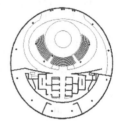

Figure 3. Second plane of London City hall.

Figure 4. The inner of London City hall.

Sitting in the café, one can appreciate the beautiful scenery of the River Thames through the glass. The vertical transportation is by the elevator and the gentle slope (Gottfried 2003). It is efficient and convenient for people to use all the facilities in this building. Meanwhile, when people are walking on the spiral rising slope, they can see the situation of government workers—as the executor of the democratic system—the government buildings, show the political ideology of "democracy".

Concerning the shape, the building uses a unique shape-irregular sphere to show a sense of movement and closed to the natural shape. It is obviously different from Foster's other works of high-tech. The shape is a distorted sphere that is not optional, but by calculation and verification the area exposed to solar radiation is minimized to reach the goal of reducing the absorption of solar heat in summer and internal heat loss in winter (Figure 5) (Derek & Clements 1997). Therefore, the optimization of energy efficiency and Foster's ecological concept are achieved. In the process of design, the experiment model is used and the heat distribution of building's surface is achieved by the analysis of whole year's sunshine regulation. The sunny slope picks out to form a beautiful cambered surface step by step, so that the facades directly exposed to the sun are reduced to a minimum (Ma & Zhou 2005). The arc curvature, the most important basis of building outside surface decoration engineering design, is worked out by the analysis of annual sunlight. The decrease in the outside area can promote the energy efficiency maximization. Through the calculation we learn that compared with the same volume of the cuboid, the superficial area of these spheres similar to the sphere can reduce by 25%.

On the building's top, there is an open place for various activities, "London Parlour", which can be used in exhibitions and all kinds of activities. It can accommodate 200 persons. It is a very good viewing platform on which people can see the whole scenery of London. There are many public facilities on the bottom of the building, such as restaurants, cafe, exhibition area and library (Yang et al. 2014b). In these areas, people can see the London Bridge and Tower Bridge clearly, which are full of historical meanings.

The natural draft is used in the building, and all the office windows can be opened. The heating system is controlled by a computer system in which key points' temperature data will be collected by a sensor, and then coordinates the heating; the inner heat will be in center together to recycle (Kubota et al. 2008). By using these measures, unnecessary energy consumption can be minimized.

In addition, a series of active and passive shading devices are used. The building opens to the south obliquely. This orientation can ensure the floors skillfully become one of the most important

shading devices with the internal natural ventilation and aeration. The tilted building is towards south, with each layer picking out step by step whose hovering distance is calculated, and just can barrier the strongest direct sunlight naturally in summer. Cooling system makes full use of the low-temperature groundwater to reduce energy consumption (Han & Guo 2008). The building contains an equipment room, extracting the groundwater from depths through the pipes to the cooling system. After cooling the building, part of them will be set to the bathroom, kitchen, gardens for washing and irrigation, while others are once again set into the underground by natural cooling. In this way, large amounts of electricity consumption can be avoided in summer.

By the comprehensive application of all kinds of energy saving technology discussed above, conventional air conditioning equipment is not needed and even additional heating system would not be required in colder seasons (Li 2015). It has proved that these measures are effective. The building heating and cooling systems energy consumption is only quarter of the same scale office building equipped with the typical central air conditioning system. The level has reached British energy-saving building evaluation gold grade standard so it is a real sense of "green building".

3 SWISS REINSURANCE BUILDING

Swiss Reinsurance building, located in the "Financial City" of London, is the first high-level ecological building in London. Its peculiar shape "gherkin", "bullet" makes it the unique scenery line in London (Figure 6). Meanwhile, it is also a well-designed green intelligent building. With research on air dynamics, Foster used the spiral design to meet the goal of natural lighting and ventilation as far as possible to reduce the building energy consumption to minimize. Swiss Reinsurance building won the "RIBA Stirling Prize" as "the most innovative design architecture" in 2004 (Calkins 2005).

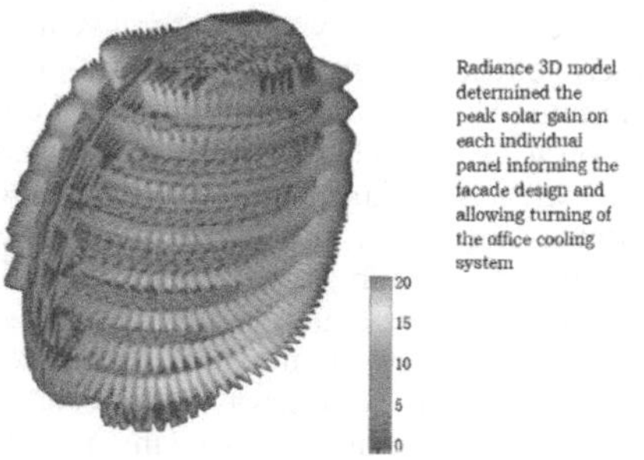

Radiance 3D model determined the peak solar gain on each individual panel informing the facade design and allowing turning of the office cooling system

20
15
10
5
0

Figure 5.   Solar radiation analysis of London City hall.

Figure 6.   Swiss reinsurance building.

The whole facade of the building is completely smooth surface, using the circular surface shape and breaking the traditional office buildings designed to a square box structure. Therefore, the contour line of building can incorporate into the surrounding environment maximally and let the underlying plaza get full of sunshine to reduce the heat loss in winter. At the same time, the smooth surface design of the building also conforms to the principle of aerodynamics (Qiu et al. 2007). Compared with the traditional square box buildings, its smooth surface can guide the air flow through its surrounding smoothly (see Figure 12) without producing the obvious adverse effects such as "wind shadow area", "strong wind" and "funneling". Thus, the influence of the microclimate changes caused by the building to the underlying and surrounding walkers is reduced (Yang 2013b).

On the surface of the building, there are six rising spiral lines that correspond respectively to the guiding air ventilation chambers inside the building. The spiral chambers play a dual role of light well and ventilation shaft and the streamline spiral atrium brings good natural ventilation into the building (Figure 7). Therefore, dependence of the internal air conditioning system is reduced effectively (Wu & Huang 2002). The design of atrium and smooth-like appearance gives an opportunity to gain the maximum natural light and to reduce the inside lighting energy consumption. Meanwhile, staff in the building also can get enough vision space. The design of the balconies on the atrium edge between each floor and the placement of public office equipments creates a strong overall visual effect. The interior space gets expressed through the unique spiral gray curtain wall of the building's facade (Qian 2015). Whether in appearance or in the layout of the building interior, annual energy consumption can be reduced by 40% through both the natural ventilation and natural lighting methods (Figure 8).

The whole support system is a cylinder, the core tube section, located in the center of the building. The office space is located around the cylinder to ensure each office has three-face lighting. Thus,

Figure 7. Simulate the air flow through the building by STAR-CD software.

Figure 8. Simulate the air flow through the building by STAR-CD software.

the building has a greater contact with nature, so the sun acquisition is enough to meet the demand of artificial lighting, namely the lighting energy consumption is reduced dramatically. At the same time, illumination and motion sensor equipped in the chamber makes the building more intelligent, avoiding all unnecessary lighting and reducing the energy consumption of the building further (Simiu & Scanlan 1996).

Although the building envelope structure uses the full glass curtain wall consisting of 5500 pieces of flat triangles and diamond-shaped glasses in total, the processing of the glass curtain wall is distinguished into many areas according to the different internal function space (Zhao & Hu 2007). Their several areas constitute a very complex curtain wall system. This system provides a set of breathed peripheral protection structure according to the different requirements of lighting and ventilation in different functional areas. Meanwhile, the appearance is distinguished to ensure the logic of building throughout both the inside and outside of the building and the process of the design. The whole glass curtain wall is divided into two parts: office area wall and chamber area wall. The former consists of the outer wall with double glasses and the inner wall with single glass (Jeong & Bienkiewicz 1997). Equipped with computer controlled shutters and weather sensing system, the walls have realized the constant monitoring of temperature, wind velocity and light intensity; the windows can be opened automatically for fresh air when necessary. This type of natural ventilation can largely reduce the energy consumption of air conditioning. The open double glass panel constitutes the spiral rising wall of the chamber area, which is tinted with gray glasses, and the high-performance coatings can effectively reduce the sunlight. The curtain wall is equipped with open fans. When the surrounding air is caught by it, the air will spiral up along the screw chambers if it is driven by the fluctuation floor room leeway (Yang et al. 2014c). A variety of appropriate energy saving measures make the building energy consumption, which is only equivalent to 50% of the same scale. According to the LEED green building rating standards, evaluated from five aspects of the sustainability of the site plan, saving water and electricity, energy efficiency and renewable energy, conserving materials and resources, and indoor environmental quality, the Swiss Reinsurance Building reaches the "green gold" level.

4 T3 TERMINAL OF BEIJING CAPITAL INTERNATIONAL AIRPORT

The T3 Terminal of Beijing Capital International Airport, located in the east of Beijing Capital International Airport, with a total building area of 986,000 m², is the world's largest single terminal at present and composed by the main building, international terminal gallery and domestic terminal gallery (Figure 9). Norman Foster's architecture was involved in the design of the T3 Terminal and his outstanding creativity of architectural design laid a solid foundation for the success of the T3 Terminal's design (Xin & Zhao 2004). Foster put the concepts of eco-energy saving and sustainable development into the design, which became a feature of the scheme. As a super-hub airport terminal, nearly one million square meters area means daily operation will consume huge energy. Thus, it has great potential for development with the aspect of green energy and can get great economic benefits and environmental benefits. Initially, Foster made efforts to explore the eco-energy saving and sustainable development strategy on various aspects of the building. In the whole process of building the T3 Terminal, several kinds of ecological energy saving strategy were used as follows.

4.1 Glass curtain walls and roof shading technology

The design of T3 Terminal external glass curtain wall gives people a streamline feeling (Figure 10). It not only gives a person with the aesthetic feeling of flow and architecture form aesthetic feeling, but also reflects the concept of ecological energy-saving. In the design of the external glass curtain wall, "hanging wall system" was used creatively according to

Figure 9.   The night bird's eye view of the T3 Terminal of Beijing.

Figure 10.   Glass wall of T3 Terminal of Beijing.

Figure 11.   Interior hanging shade system of the T3 Terminal of Beijing.

the past experiences of similar projects. The vertical support of the glass unit is canceled and the bearing member wire rope is hidden in the glass plate instead (Li 2012). Then, it is suspended and hidden in the steel beam of the roof truss. The horizontal members fixed on the suspension wire play a double role of a horizontal loading and shading effect. Finally, the glass is inlaid on the horizontal visor. This new concept of the external glass curtain wall system weakens the impact of the vertical component and allows the roof become transparent and the visitors can see the outside to achieve the purpose of decompression for passengers (Zhang 2000). The horizontal visor is the structural member for bearing and the shading facilities for shading illuminative effect (Figure 11). This idea uniting the bearing with the shading not only makes the whole building more transparent and more artistic expressive, but also can effectively reduce the influence of outdoor solar radiation on the indoor energy consumption.

4.2 Making full use of natural light to reduce lighting energy consumption

In the design of T3 Terminal, Foster tries to minimize the building's depth as much as possible and to maintain the external skin transparent (Figure 12). Therefore, most of the interior space in normal circumstances keeps illumination by the outdoor lighting. Meanwhile, to keep the transparency of the interior space, top walls and electrical facilities are eliminated. The overall roof used the unified model. By using the unified model and the unified design from the vision to the geometric structure of terminal, the whole roof is dynamic with a flow and full of vitality (Figure 13). The roof is a continuous hyperboloid whose height and thickness must meet the requirements of building interior space height,

Figure 12. Transparent outer epidermis of the T3 Terminal of Beijing.

Figure 13. Roof skylights of the T3 Terminal.

Figure 14. APM system of the T3 Terminal.

form and structure design (Krishna 1995). While the natural light directly reaching the interior public area through three pyramid skylights toward southeast would be the most distinctive architectural element. This skylight's setting is not random, but it was well calculated depending on the changes of sun angle over time. In the morning, sunshine can be directed to the ticket hall and noon it can be reflected and spread to the hall through the masking device and hot afternoon it can be cut off. The width of the triangular skylights set on the roof is about 9 m. It consists of a slope roof component and two glazed side elevations. Wherever these skylights are located, their orientation is consistent. The angle between two glazed facades faces towards southeast to minimize the energy consumption. Another example of using natural light is the open passenger MRT tunnel space (Watson et al. 2003). In order to facilitate the sliding aircraft on the airfield, MRT trains must operate below the ground. Similar to the subway, the train track layout is set in a closed underground space, while the T3 layout is set in an open gully. Thus, the passengers will not feel oppressed, and what is important is that the cost of the system's daily operation is low.

4.3 Intelligent building environmental control system—i-bus system

T3 was used in the current world's most advanced i-bus intelligent building control system when it was constructed. Through the system, the following intelligent control functions are realized: curtain wall equipment control, lighting control,

intelligent panel manual control, timing control, meteorological center (sensor) automatic control, the light sensor control, i-bus sun latitude and longitude automatic tracking control, computer graphics software centralized control, fire linkage control and BA system linkage control function. I-bus system can adjust indoor lighting, temperature and shade environment to reduce the energy consumption as much as possible according to personnel activity status, work routines and natural light condition (Okada & Ha 1992). By the application of the i-bus system, reduction using the fresh air system could be achieved in 5 months annually, which not only realizes the comfortable office environment, but also maximizes to reduce the energy (Figure 14).

4.4 Intelligent building environmental control system—i-bus system

To cut the cost of the entire terminal building, Foster always kept his usual appropriate techniques and means to perform the buildings. Through the reasonable grid, standard modulus design and the prefabrication of building components, create the lowest difficulties of site operation are created. Problems of the interface and coordination were solved through the accurate design options. In the deepening process, the staff sought low-cost building materials and equipments in the local market as far as possible to reduce the impact on the environment and project costs. What is important is that by all means, it reflected the ecological concept.

5 CONCLUSIONS

With respect to the living environment, humans originally lived in empty houses and later in tree houses, and then in the vernacular architecture. All the buildings contain the nuclide ecological concept and can effectively combine with the specific natural environment. Afterwards, with the development of science and technology and the appearance of modern architecture, the tendency of heavy technology and light nature generated a lot of "anti-ecology" as well as anti-nature architecture. This is also the most important element of global environment deterioration. Foster pointed out through his works that there is no contradiction between "high-tech" and "original ecology". They can combine together organically. The application of architectural ecological energy saving strategy makes the buildings not only contain the harmonious characteristics between local architecture and nature, but also include the modern scientific technology means. Norman Foster showed the world his creative thinking. Today, China is in the stage of high-speed urbanization. People are overly pursuing the "upscale" heating, air condition, lighting, ventilation and office appliances that consume a lot of energy, such as large room and large areas of glass curtain wall and too much artificial lighting. Building energy consumption accounted for one-third of the total energy consumption in China. In the future, to realize the sustainable development, the progress on the ecological energy-saving low-carbon road is urgent.

ACKNOWLEDGMENT

This work was financially supported by grants 201320138568.0 and 201320138582.0.

REFERENCES

Bassiouny, R. & Koura, N.S.A. 2008. An analytical and numerical study of solar chimney use for room natural ventilation. *Energy and buildings*, 40(5): 865–873.

Calkins, M. 2005. Strategy use and challenges of ecological design in landscape architecture. *Landscape and Urban planning*, 73(1): 29–48.

Derek, T & Clements, C.J. 1997. What do we mean by intelligent buildings?, *Automation in Construction*, 6(5): 395–400.

Gładyszewska, F.K. & Gajewski, A. 2012. Effect of wind on stack ventilation performance. *Energy and Buildings*, 51: 242–247.

Gottfried, D.A. 2003. blueprint for green building economics. *Industry and Environment*, 26(2): 20–21.

Han, S. & Guo, B. 2008. Building natural ventilation the temperature effect. *Building Energy*, 18–22.

Jeong, S.H. & Bienkiewicz, B. 1997. Application of autoregressive modeling in proper orthogonal decomposition of building wind pressure. *Journal of wind engineering and industrial aerodynamics*, 69: 685–695.

Krishna, P. 1995. Wind loads on low rise buildings—A review. *Journal of wind engineering and industrial aerodynamics*, 54: 383–396.

Kubota, T. Miura, M. & Tominaga, Y. 2008. Wind tunnel tests on the relationship between building density and pedestrian-level wind velocity: Development of guidelines for realizing acceptable wind environment in residential neighborhoods. *Building and Environment*, 43(10): 1699–1708.

Li, Y. 2015. Interpreting solar house—using International Solar Decathlon works as example, *Advanced Materials Research*, 1092–1093: 567–572.

Li, Y. 2012. Analysis of Planning of Neighborhood Communication Space in the Livable Community, *Applied mechanics and Materials*, 174–177: 3018–3022.

Lin, X.D. 2007. *Green Building*. Beijing: China Building Industry Press.

Ma, S.Y. & Zhou, D.H. 2005. On Norman·Foster and the school of High—Tech. Journal of HeFei University of Technology, 2, 85–88.

Mochida, A. & Lun, I.Y.F. 2008. Prediction of wind environment and thermal comfort at pedestrian level in urban area. *Journal of Wind Engineering and Industrial Aerodynamics*, 96(10): 1498–1527.

Okada, H. & Ha, Y.C. 1992. Comparison of wind tunnel and full-scale pressure measurement tests on the Texas Texh Building. *Journal of Wind Engineering and Industrial Aerodynamics*, 43(1): 1601–1612.

Ong, B.L. 2003. Green plot ratio: an ecological measure for architecture and urban planning. *Landscape and urban planning*, 63(4): 197–211.

Qian, F. 2014. Insulation and Energy-saving Technology for the External Wall of Residential Building, *Advanced Materials Research*,1073–1076(2): 1263–1270.

Qian, F. 2015. Analysis of Energy Saving Design of Solar Building-Take Tongji University solar decathlon works for example, *Applied Mechanics and Materials*, 737: 139–144.

Qiu, N.I. Yang, Q. & Yan, C. 2007. Research of the Ecological and Livable Residential Estates in Dong Guan. *Guangdong Landscape*.

Simiu, E. & Scanlan, R.H. 1996. *Wind effects on structures: fundamentals and applications to design*. John Wiley.

Watson, D. Plattus, A.J. & Shibley, R.G. 2003. *Time-saver standards for urban design*. New York: McGraw-Hill.

Wu, D. & Huang, Q. 2002. Virtual Reality Technology and Research Status of the Development Process. *Ocean Mapping*, 22(6): 15–17.

Xin, J.G. & Zhao, J.L. 2004. *Application Analysis of the Ecological Concept on Architectural Design*. China Technology Information.

Yang, L. 2010. Computational Fluid Dynamics Technology and Its Application in Wind Environment Analysis. *Journal of Urban Technology*, 17(3): 53–67.

Yang, L. 2013a. Research of Urban Thermal Environment Based on Digital Technologies, *Nature Environment and Pollution Technology*, 12(4): 645–650.

Yang L. 2013b. Research on Building Wind Environment Based on the Compare of Wind Tunnel Experiments and Numerical Simulations, *Nature Environment and Pollution Technology*, 12(3): 375–382.

Yang, L. He, B.J. & Ye, M. 2014a. CFD Simulation Research on Residential Indoor Air Quality, *Science of the Total Environment*, 472: 1137–1144.

Yang, L. He, B.J. & Ye, M. 2014b. Application Research of ECOTECT in Residential Estate Planning, *Energy and buildings*, 72: 195–202.

Yang, L. He, B.J. & Ye, M. 2014c. The application of solar technologies in building energy efficiency: BISE design in solar-powered residential buildings, *Technology in Society*, 1–8.

Zhang, M. 2000. Virtual Reality Applications in Architectural Design. *Central Building*, 18(1): 51–52.

Zhang, Z.Q. Wang, Z.J. & Lian, L.M. 2004. Indoor thermal environment of residential buildings Numerical Simulation. *Building heat ventilation air-conditioning*, 23(5): 88–92.

Zhao, J.L. & Hu, Y.J. 2007. The regional expression of Norman Foster. *Shanxi architecture*, 33(29): 63–65.

Advances in Civil, Architectural, Structural and Constructional Engineering – Kim, Jung & Seo (Eds)
© 2016 Taylor & Francis Group, London, ISBN 978-1-138-02849-4

Wettability and surface tension of liquid aluminium-copper alloys on graphite at 1273 K

W.J. Mao & N. Shinozaki

Graduate School of Life Science and Systems Engineering, Kyushu Institute of Technology, Fukuoka, Japan

ABSTRACT: The wettability and surface tension of liquid aluminium-copper alloys on graphite substrates with Cu content of 10 mass%, 20 mass%, 30 mass% and 40 mass% were studied at 1273 K by sessile drop method under vacuum. The results showed that the wetting slightly decreased with increasing the content of Cu, which the equilibrium contact angles were 71°, 76°, 81° and 85° for the Al-10 mass%Cu/C, Al-20 mass%Cu/C, Al-30 mass%Cu/C and Al-40 mass%Cu/C systems, respectively. An Al_4C_3 layer was formed at the interface of Al-Cu alloys/graphite in all cases, and its thickness decreased with the increase of the content of Cu. The surface tension of liquid Al-10 mass%Cu, Al-20 mass%Cu, Al-30 mass%Cu and Al-40 mass%Cu were 852, 938, 964 and 990 mN/m, respectively, and it showed a monotonic increase with the increase of Cu content.

1 INTRODUCTION

Liquid aluminium-copper alloys play an important role in many lightweight constructions and transport applications due to its high strength and ductility, superior castability, the presence of low temperature eutectic compositions and low fabrication cost (Zhang et al. 2007, Lokker et al. 2001). On the other hand, graphite is well-known as a solid lubricant and its presence makes the aluminium alloys self-lubricating. Recently, Al-Cu composites reinforced by graphite with excellent mechanical properties was not only considered as an important structural material for automobile and aeronautical manufacturing fields, but also have attracted particular attention to the exploration of their possible applicability. Therefore, the understanding of interfacial phenomena and wettability of the Al-Cu alloy/graphite systems are of crucial importance.

The surface tension of liquid metals is not only an intrinsic property, but also a key parameter of technological importance. Especially, accurate information on the surface tension of liquid Al and its alloys is beneficial to many materials-related processes, such as casting, welding, brazing, and the fabrication of matrix composites. In the past, the measurement of the surface tension of liquid Al-Cu alloys have been performed by many researchers (Poirier et al. 1987, Schmitz et al. 2009, Keene et al. 1993) using the sessile drop method and the maximum bubble pressure method, and all of them found that the surface tension increased monotonically with increasing the content of Cu

at temperature above 1073 K. Noteworthily, due to the oxidation of aluminium alloys was unavoidable under even a ultra-high vacuum system, the value of the surface tension of liquid Al alloys should be associated with the oxidized surface. In order to obtain a relatively reliable value of the surface tension of liquid Al-Cu alloys, it is very necessary to improve the measurement method and adopt the surface tension value which the extent of the surface oxidation is reduced to the minimum as far as possible.

In the present work, the wettability and surface tension of liquid Al-Cu alloys on graphite substrates with Cu content of 10 mass%, 20 mass%, 30 mass% and 40 mass% were studied at 1273 K by sessile drop method under vacuum. Such a study aims to contribute to further understanding of wetting behavior of the Al-Cu alloy/graphite systems, and the measurement of the surface tension of liquid Al-Cu alloys helps providing useful and reliable data for any metallurgical and materials-related processes.

2 EXPERIMENTAL PROCEDURE

High purity aluminium (99.99 mass%) cubes and copper (99.99 mass%) plates were employed in this study and the drop of Al-Cu alloys were prepared in situ by direct melting of Cu on Al themselves placed on the graphite substrate (we demonstrated that the wetting behavior of Al-10 mass% Cu/graphite in this placement state was the same as the placement state in which Al and Cu metals were pre-alloyed by arc-melting and placed on the graphite substrate. Therefore, the former placement state that was easier to carry out was selected for all systems in this work). The metals of Al and Cu were weighted in designed proportion which the mass ratio of Al/Cu were 90:10, 80:20, 70:30 and 60:40, respectively, and the total mass was approximately 0.15 g in all cases. Before the wetting experiment, the metals of Al and Cu were polished by emery papers from 400 to 1500 grit and cleaned with ethanol ultrasonically in order to prevent further oxidation.

The substrates used were high purity (99.99 mass%) graphite plates with a density of 1.88 Mg/m³ and an ash content less than 5 ppm. These substrates were 12 mm in square and 2 mm in thickness. Due to both sides of substrate were different in surface roughness, the relatively smooth surface was chosen for this experiment and the average roughness Ra of this surface obtained from 3D-SEM was 0.9 μm ± 0.02 μm. Before experiment, the substrates were ultrasonically cleaned in ethanol, then dried in air drying oven.

Sessile drop experiments were performed in a quartz tube furnace under vacuum conditions. The metal sample was placed on the center of the graphite substrate and then adjusted horizontally. Two ceramic boats containing sponge titanium were used as the deoxidizing agent and placed at the left and right sides, about 5 cm away from the sample respectively. The furnace was evacuated to a vacuum approximately 1.5 Pa at the room temperature and then heated to the temperature of 1273 K in about 35 minutes. The wetting behavior of the sample was recorded using a digital camera connected with a macro-lens by an extension bellows set placed in front of the observation window. The pictures of the sample were recorded at intervals of 5 minutes at 1273 K. After experiments, the contact angle was measured from the drop profile using image analysis software, and in order to obtain an average of the contact angle, both sides of droplet were measured. The surface tension of liquid Al-Cu alloys was calculated from the profile of the molten drop using Rotenberg's method (Rotenberg et al. 1983). Further detail on the measurement has been reported in a previous literature (Mukai et al. 2000). The density (g/cm³) of the molten Al-Cu alloys reported by Y. Plevachuk et al. (Plevachuk et al. 2008) was used for calculating the surface tension.

After the sessile drop experiments, the selected samples were embedded in resin, sectioned perpendicular to the interface and polished to examine the interfacial microstructures using an electron probe micro-analyzer (Jeol, JXA-8530F, Japan). And A X-ray diffraction analyzer (Jeol, JDX-3500K, Japan) was used to determine the phases of the interface.

3 RESULTS AND DISCUSSION

3.1 *Wettability of the Al-Cu/graphite systems*

Figure 1 shows the time dependence of the contact angle measured at 1273 K for molten Al-Cu alloys on graphite substrates. As indicated, the value of the initial contact angle for all systems was large, which ranged from 125° to 150°, and it mainly resulted from the resistance exerted by the aluminium oxide film against the movement of the triple line (Eustathopoulos et al. 1974). The variation of these curves almost showed the same trend, which the contact angle decreased at a relatively fast rate in the initial 60 minutes, and then remained unchanged nearly in the rest of the holding time in all cases. After holding 100 minutes, the equilibrium contact angle was stabilized at approximately 71°, 76°, 81° and 85° for the Al-10 mass%Cu/C, Al-20 mass%Cu/C, Al-30 mass%Cu/C and Al-40 mass%Cu/C systems, respectively, and the wetting decreased slightly with increasing the content of Cu. From the viewpoint of spreading kinetics, the kinetics of the rapid decrease of the contact angle in the initial 60 minutes was due to the occurrence of two phases: (1) the removal of the oxide film and (2) the interfacial reaction between the Al-Cu drop and the graphite substrate. At high temperature, the following reaction could occurred between the molten Al and the aluminium oxide film (Park et al. 1994):

$$\frac{4}{3}Al(l) + \frac{1}{3}Al_2O_3(s) = Al_2O(g) \quad \Delta G^o_{(1273K)} = +189kJ$$

(1)

According to this reaction, the equilibrium partial pressure of Al_2O at 1273 K was calculated to be approximately 1.7×10^{-3} Pa, which was only 0.1 percent of the total pressure in the furnace.

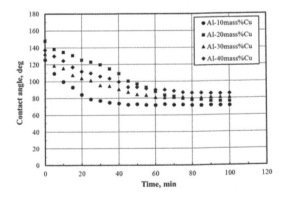

Figure 1.   Variation in the contact angle with time for the Al-Cu alloys on the graphite substrates at 1273 K.

Therefore, the oxide film was disrupted most probably and this reaction enabled the subsequent wetting to proceed. Figure 2 shows the backscattered electron composition (COMPO) images of the cross section of the interface between the alloy drops and the graphite substrate in all cases at 1273 K for 100 min. A new reaction layer was formed at the interface for all systems and its thickness varied with the samples, which decreased from about 55 μm to 10 μm with increasing the Cu content from 10 mass% to 40 mass%. Quantitative determinations performed by spot analysis of EPMA revealed that the element of Al and C existed at the reaction layer for all systems, and the atomic ratio of Al:C was approximately 4:3, suggesting a Al_4C_3 compound. Furthermore, combined with the result of XRD (Figure 3) after grinding the

solidified alloy drops away, the reaction layer was identified to be Al_4C_3 layer for all systems. Meanwhile, the intermetallic compound of $CuAl_2$ was also detected at the interface, which existed above the Al_4C_3 layer as white dendritic structures shown in Figure 2.

3.2 Surface tension of liquid Al-Cu alloys

The time dependence of the surface tension measured at 1273 K for liquid Al-Cu alloys is showed in Figure 4a. These curves described a similar change that the surface tension increased rapidly in the initial 15 minutes, then stabilized in the rest of the holding time, despite slight fluctuations. With respect to the lower surface tension values for all systems in the initial 5–15 minutes at 1273 K, previous investigators (Eustathopoulos et al. 1974, Anson et al. 1999) suggested that it resulted from a heavy oxide film formed on the surface of the drop and did not recommended as reliable values to adopt, due to this oxide film could significantly reduce the real surface tension. After 15 minutes, most of the oxide film was expected to be disrupted according to the reaction (1), and the stable values of the surface tension for the liquid Al-10 mass%Cu, Al-20 mass%Cu, Al-30 mass%Cu and

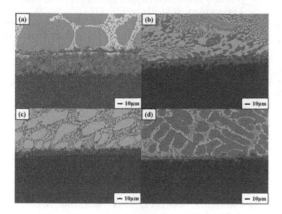

Figure 2. The COMPO images of the cross section of the interface between the alloy drops and the graphite substrate at 1273 K for 100 min: (a) Al-10mass%Cu/Graphite; (b) Al-20mass%Cu/Graphite; (c) Al-30mass%Cu/Graphite; (d) Al-40mass%Cu/Graphite.

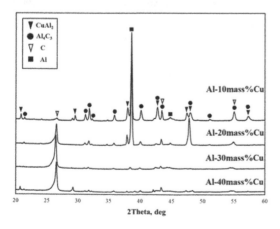

Figure 3. The XRD patterns of the interface between the alloy drops and the graphite substrate.

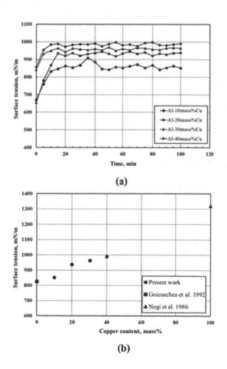

Figure 4. (a) Variation in the surface tension with time for liquid Al-Cu alloys at 1273 K; (b) the surface tension of liquid Al-Cu alloys with different copper contents.

23

Al-40 mass%Cu alloys reached to about 852, 938, 964 and 990 mN/m, respectively. These results highly agreed with the data measured by J. Schmitz (Schmitz et al. 2009). Figure 4b shows the surface tension of liquid Al-Cu alloys with different copper contents, and the data for liquid Al and Cu were taken from Ref. (Goicoechea et al. 1992, Nogi et al. 1986). Clearly, the surface tension of liquid Al-Cu alloys increased monotonously with increasing the Cu content. It should be note that under the present experimental condition, the oxide film was not possible to disappear completely, and even a perfect "unoxidized" surface might be covered with at least a single monolayer of oxide (Anson et al. 1999).

4 CONCLUSIONS

The wettability and surface tension of liquid aluminium-copper alloys on graphite substrates with Cu content of 10 mass%, 20 mass%, 30 mass% and 40 mass% were studied at 1273 K by sessile drop method under vacuum. The results showed that the wetting slightly decreased with increasing the content of Cu, which the equilibrium contact angles were 71°, 76°, 81° and 85° for the Al-10 mass%Cu/C, Al-20 mass%Cu/C, Al-30 mass%Cu/C and Al-40 mass%Cu/C systems, respectively. An Al_4C_3 layer was formed at the interface of Al-Cu alloys/graphite in all cases, and its thickness decreased with the increase of the content of Cu. The surface tension of liquid Al-10 mass%Cu, Al-20 mass%Cu, Al-30 mass%Cu and Al-40 mass%Cu were 852, 938, 964 and 990 mN/m, respectively, and it showed a monotonic increase with the increase of Cu content.

REFERENCES

Anson, J.P., Drew, R.A.L. & Gruzleski, J.E. 1999. The surface tension of molten aluminum and Al-Si-Mg alloy under vacuum and hydrogen atmospheres. *Metallurgical and Materials Transactions B*, 30(6): 1027–1032.

Eustathopoulos, N., Joud, J.C., Desre, P. & Hicter, J.M. 1974. The wetting of carbon by aluminium and aluminium alloys. *Journal of Materials Science*, 9(8): 1233–1242.

Goicoechea, J., Garcia-Cordovilla, C., Louis, E. & Pamies, A. 1992. Surface tension of binary and ternary aluminium alloys of the systems Al-Si-Mg and Al-Zn-Mg. *Journal of Materials Science*, 27(19): 5247–5252.

Keene, B.J. 1993. Review of data for the surface tension of pure metals. *International Materials Reviews*, 38(4): 157–192.

Lokker, J.P., Böttger, A.J., Sloof, W.G., Tichelaar, F.D., Janssen, G.C.A.M. & Radelaar, S. 2001. Phase transformations in Al-Cu thin films: precipitation and copper redistribution. *Acta Materialia*, 49 (8): 1339–1349.

Mukai, K. & Yuan, Z. 2000. Effects of boron and carbon on the surface tension of molten silicon under precisely controlled oxygen partial pressure. *Material Transactions, JIM*, 41(2): 331–337.

Nogi, K., Ogino, K., McLean, A. & Miller, W.A. 1986. The temperature coefficient of the surface tension of pure liquid metals. *Metallurgical and Materials Transactions B*, 17(1): 163–170.

Park, S.J., Fujii, H. & Nakae, H. 1994. Wetting of Boron Nitride by Molten Al-Si Alloys. *Journal of the Japan Institute of Metals and Materials*, 58(2): 208–214.

Plevachuk, Y., Sklyarchuk, V., Yakymovych, A., Eckert, S., Willers, B. & Eigenfeld, K. 2008. Density, viscosity, and electrical conductivity of hypoeutectic Al-Cu liquid alloys. *Metallurgical and Materials Transactions A*, 39(12): 3040–3045.

Poirier, D.R. & Speiser, R. 1987. Surface tension of aluminumrich Al-Cu liquid alloys. *Metallurgical and Materials Transactions A*, 18(13): 1156–1160.

Rotenberg, Y., Boruvka, L. & Neumann, A.W. 1983. Determination of surface tension and contact angle from the shapes of axisymmetric fluid interfaces. *Journal of Colloid and Interface Science*, 93(1): 169–183.

Schmitz, J., Brillo, J., Egry, I. & Schmid-Fetzer, R. 2009. Surface tension of liquid Al-Cu binary alloys. *International Journal of Materials Research*, 100(11): 1529–1535.

Zhang, M., Zhang, W.W., Zhao, H.D. & Li, Y. 2007. Effect of pressure on microstructures and mechanical properties of Al-Cu-based alloy prepared by squeeze casting. *Transactions of Nonferrous Metals Society of China*, 17(3): 496–501.

Advances in Civil, Architectural, Structural and Constructional Engineering – Kim, Jung & Seo (Eds)
© 2016 Taylor & Francis Group, London, ISBN 978-1-138-02849-4

A piezoelectric micro-vibration testbed and its calibration method

J.J. Dong, W. Cheng & Y.M. Li

School of Aeronautics and Engineering, Beijing University of Aeronautics and Astronautics, Beijing, China

ABSTRACT: In order to measure the micro-vibrations generated by large CMG (Control Moment Gyro) and SADA (Solar Array Drive Assembly), a testbed composed of eight piezoelectric sensors is invented. The working principle and dynamic calibration method of this testbed are specified. For the sake of assessing the performance, the natural frequency and test error of this testbed are obtained by experimental testing. The results show that the first order natural frequency of the testbed is 655 Hz and the maximum dynamic test error is ±9.4% in frequency domain (5–500 Hz), in addition, the resolutions of the three forces and three moments can reach 0.001 N and 0.0001 N·m respectively.

1 INTRODUCTION

The micro-vibrations caused by the motion parts of onboard spacecraft, for example, MWA (Momentum Wheel Assemblies), CMG, and SADA, can significantly degrade the pointing precision and image quality of spacecraft (Zhang, P.F. et al. 2011, Zhou, D.Q. et al. 2013). To achieve the goal of high point-ing precision, the micro-vibrations produced by the motion parts on spacecraft are required to be measured and suppressed. There are two major methods for micro-vibration ground testing of spacecraft structure now, separately by accelerometer and LSV (Laser Scanning Vibrometer) (Tan, T.L. et al. 2014). The target variables of the two methods are acceleration and displacement respectively, however, we can't get the disturbing force/moment of the motion parts through accelerometer or LSV directly. Generally, the disturbances generated by the typical motion parts of spacecraft are measured through the Kistler table which is made of piezoelectric sensors (Luo et al. 2013, Masterson et al. 2002, Zhou et al. 2012), whereas the inflexible installation interface of the table leads to difficulty in the testing of SADA especially with load in large flexible form (Chen et al. 2014). In addition, by the same token, the exorbitant structure height after installation of large CMG reduces the installed structure's stiffness, which leads to the limitation of test result in the frequency domain. In this paper, we invent a piezoelectric testbed for micro-vibration testing with a large and flexible installation interface and a hole inside the testbed for reducing structure height. The testbed is mainly composed of eight independent shear-type piezoelectric sensors and a load board, and each sensor can measure the shear forces in two orthogonal directions. By means of reasonable combination of the eight transducers, the testbed can measure forces (Fx, Fy and Fz) and moments (Mx, My and Mz) in three orthogonal directions of the center on the load board. Therefore, the measurement of micro-vibrations produced by large CMG or SADA in broader frequency band is realized.

Firstly, this article introduces the working principle and calibration method of the testbed. Secondly, the natural frequency and the test error of this testbed are obtained by experimental testing, and the results show that the testbed has high natural frequency, high resolution and small test error. The testbed in the paper provides the possibility for micro-vibration testing of large CMG and SADA in broader frequency range.

2 WORKING PRINCIPLE OF TESTBED

The operating principle of piezoelectric sensor is based on piezoelectric effect of some special dielectric. When the dielectric is under external load, there will be charge on the surface of dielectric. Therefore, the dielectric transforms force into charge which can be measured. As the piezoelectric component has the advantages of small size, low mass, simple structure, good reliability, high natural frequency, high sensitivity and well signal noise ratio, it has been applied to domains of acoustics, mechanics, medicine, aeronautics and astronautics (Pan et al. 2011). The testbed is mainly composed of eight piezoelectric sensors and a load board. Pictures of the testbed and its internal schematic are shown in Figure 1 and Figure 2 respectively.

The detailed structure of shear-type piezoelectric sensor adopted in this paper is shown in Figure 3. The left and right parts of the sensor are symmetrical with a steel plate in the middle, which

Figure 1. Picture of the testbed.

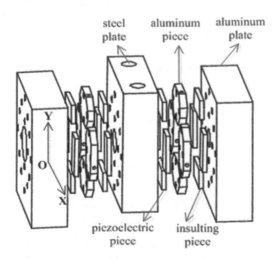

Figure 2. Internal schematic of the testbed.

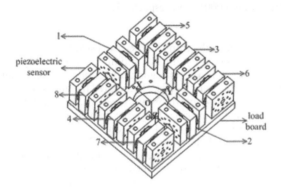

Figure 3. Schematic of the piezoelectric sensor.

is connected to the load board. The piezoelectric pieces are distributed on both sides of the steel plate. There are four pieces on each side, and half of them are polarized along the X axis (horizontal),

and the rest are polarized along the Y axis (vertical). By this way, the transducer can measure the shear forces in two orthogonal directions, along the X and Y axis separately. The lateral of piezoelectric pieces is the aluminum sheets which are used to transport charge generated by piezoelectric piece. Meanwhile, the insulting pieces outside aluminum sheets can prevent charge from transmitting to the aluminum plates which are used to assemble the sensor. Both the piezoelectric and insulting pieces are ceramic. The horizontal-polarized piezoelectric pieces of No.1 and No.2 sensors are applied to measure the force F_x, and the vertical-polarized piezoelectric pieces of them are used to measure the moment M_y. Similarly, the No.3 and No.4 sensors can measure force F_y, and moment M_x. In addition, the horizontal-polarized piezoelectric pieces of rest sensors are applied to measure the moment M_z, and the vertical-polarized piezoelectric pieces of them are used to measure force F_z. Therefore, we achieve the goal of dynamic measurement for three forces and three moments of the load board's center.

3 DYNAMIC CALIBRATION METHOD

The dynamic calibration of the testbed is carried out in frequency domain. Since the output of the testbed is voltage signal $U(\omega)$, we need a corresponding calibration matrix $W(\omega)$, to transform response signal $U(\omega)$ into exciting signal $F(\omega)$. The testbed can output six voltage signals corresponding to three forces and three moments. According to Xing, Y.F. et al. (2011), the relationship between the response and excitation of system is as follows:

$$X(\omega) = H(\omega)F(\omega) \qquad (1)$$

where, $X(\omega)$ is the response of system; $F(\omega)$ denotes the excitation of system; $H(\omega)$ is the FRF (frequency response function). So we can calculate the FRF after confirming the exciting and response signals of each force (moment) to form calibration matrix in theory. The process can be expressed as below:

$$U_{6\times1}(\omega) = diag(H_{6\times6}(\omega))F_{6\times1}(\omega) \qquad (2)$$

$$W_{6\times6}(\omega) = diag^{-1}(H_{6\times6}(\omega)) \qquad (3)$$

Equation (3) shows that the calibration matrix $W(\omega)$ is diagonal. It indicates that no coupling terms between the six voltage signals caused by the forces (moments).

However, we discover the phenomenon of coupling between the forces (moments) which can't be neglected. Hence, the decoupling calibration method isn't appropriate for testbed in this paper, and Chen et al. (2014) proposed a coupling calibration method. The calibration matrix of aforementioned method is a FRF matrix between the six response signals which are produced by the piezoelectric sensors and the loads on the equivalent center of these transducers. In this article, the equivalent center is the core of the load board's upper surface, O, nevertheless, it's difficult to apply moments on the center directly. So we install a calibration device with high stiffness after assuming the load board is a rigid body, and we choose sixteen points which are shown in Figure 4 for loading by exciting-hammer. According to the equilibrium theory of space force system, the sixteen exciting loads acting on the calibration device can be equivalent to three forces and three moments on the center of board. Figure 4 shows the schematic of the calibration device, where O is the center of the load board's upper surface and L_x, L_y, L_z are the arms of force of the loading points for O.

In the process of calibration, we applied sixteen exciting loads from F_1 to F_{16} on the calibration device, in the meanwhile, the Data Acquisition System (DAS) can collect FRFs between six voltage signals and each load to calculate the calibration matrix $W(\omega)$. According to Xing et al. (2011),

the response is the six voltage signals produced by the piezoelectric sensors and the excitation is the sixteen hammering forces, so Equation (1) can be written as:

$$U_{6\times16}(\omega) = H_{6\times16}(\omega)F_{16\times16}(\omega) \qquad (4)$$

It requires a transition matrix to transform the sixteen hammering forces to exciting loads on the center of the load board, O. The matrix is recorded as C_i, and its expression is shown below:

$$
\begin{aligned}
C_1 &= \begin{bmatrix} 0 & 1 & 0 & -L_z & 0 & -L_x \end{bmatrix}^T \\
C_2 &= \begin{bmatrix} 0 & 1 & 0 & -L_z & 0 & 0 \end{bmatrix}^T \\
C_3 &= \begin{bmatrix} 0 & 1 & 0 & -L_z & 0 & L_x \end{bmatrix}^T \\
C_4 &= \begin{bmatrix} -1 & 0 & 0 & 0 & -L_z & -L_y \end{bmatrix}^T \\
C_5 &= \begin{bmatrix} -1 & 0 & 0 & 0 & -L_z & 0 \end{bmatrix}^T \\
C_6 &= \begin{bmatrix} -1 & 0 & 0 & 0 & -L_z & L_y \end{bmatrix}^T \\
C_7 &= \begin{bmatrix} 0 & -1 & 0 & L_z & 0 & -L_x \end{bmatrix}^T \\
C_8 &= \begin{bmatrix} 0 & -1 & 0 & L_z & 0 & 0 \end{bmatrix}^T \\
C_9 &= \begin{bmatrix} 0 & -1 & 0 & L_z & 0 & L_x \end{bmatrix}^T \\
C_{10} &= \begin{bmatrix} 1 & 0 & 0 & 0 & L_z & -L_y \end{bmatrix}^T \\
C_{11} &= \begin{bmatrix} 1 & 0 & 0 & 0 & L_z & 0 \end{bmatrix}^T \\
C_{12} &= \begin{bmatrix} 1 & 0 & 0 & 0 & L_z & L_y \end{bmatrix}^T \\
C_{13} &= \begin{bmatrix} 0 & 0 & -1 & L_y & -L_x & 0 \end{bmatrix}^T \\
C_{14} &= \begin{bmatrix} 0 & 0 & -1 & L_y & L_x & 0 \end{bmatrix}^T \\
C_{15} &= \begin{bmatrix} 0 & 0 & -1 & -L_y & L_x & 0 \end{bmatrix}^T \\
C_{16} &- \begin{bmatrix} 0 & 0 & -1 & -L_y & -L_x & 0 \end{bmatrix}^T
\end{aligned}
\qquad (5)
$$

Therefore, the equivalent exciting forces, F', on O can be expressed as:

$$F'_{6\times16}(\omega) = C_{6\times16}F_{16\times16}(\omega) \qquad (6)$$

where, $C_{6\times16} = [C_1, C_2, \ldots, C_{16}]$.

Furthermore, the relationship between the six output voltages and the equivalent exciting forces can be written as:

$$W_{6\times6}(\omega)U_{6\times16}(\omega) = C_{6\times16}F_{16\times16}(\omega) \qquad (7)$$

We can get Equation (8) by substituting Equation (4) into Equation (7):

$$W_{6\times6}(\omega)H_{6\times16}(\omega) = C_{6\times16} \qquad (8)$$

Because the matrix $H_{6\times16}(\omega)$ is not square, we need to multiply both sides of Equation (8) by the generalized matrix of $H_{6\times16}(\omega)$, and the

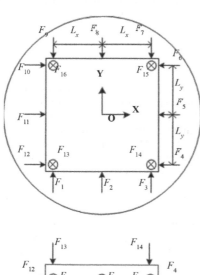

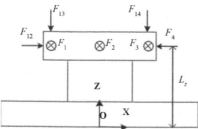

Figure 4. Schematic of the calibration device.

dynamic calibration matrix $W_{6\times6}(\omega)$ is expressed as follows:

$$W_{6\times6}(\omega) = C_{6\times16}H^T{}_{6\times16}(\omega)\left[H_{6\times16}(\omega)H^T{}_{6\times16}(\omega)\right]^{-1}$$

(9)

when, carrying out the micro-vibration measurement, we can obtain the equivalent loads on O by left multiplying six output voltage signals by the calibration matrix $W_{6\times6}(\omega)$, and the process is expressed as below:

$$F(\omega) = W(\omega)U(\omega)$$

(10)

4 TESTING OF NATURAL FREQUENCY

We can apply exciting force F_z on the center of calibration device by force hammer, which is depicted in Figure 4. As a result, the testbed will output a voltage signal mainly corresponding to excitation along axis, Z, and the DAS can collect the voltage signal to calculate FRF. We can get the natural frequency of testbed by analyzing the FRF which is shown as below:

Figure 5 shows the first order natural frequency of the testbed is 655 Hz. Thus, it is considered that the structure coupling of testbed won't influence the dynamic voltage signals within 500 Hz basically.

5 ERROR TESTING OF DYNAMIC MEASUREMENT

According to Section 3, we can obtain a dynamic calibration matrix. In the process of calibration, the sampling frequency is 2048 Hz, and the sampling

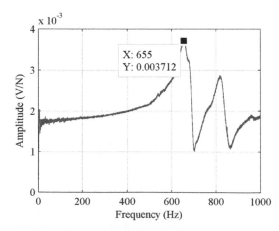

Figure 5. FRF of exciting force F_z.

time is 16 seconds. In order to test the error of the testbed's dynamic measurement, we apply extra hammering forces on the calibration device to conduct error testing after calibration. According to Equation (5), when hammering load acting point 3, the equivalent loads on O are F_y, M_x and M_z. And the equivalent loads corresponding to point 4 are F_x, M_y and M_z. At last, we need to hammer point 16 to add the remaining force, F_z. Therefore, we can get the error of dynamic measurement by comparing actual values and test results of the three forces and three moments when hammering points 3, 4 and 16 for another time.

We can get the hammering forces $F(\omega)$ and the output voltages $U(\omega)$ when hammering points 3, 4, 16 through DAS. Then the equivalent loads (actual values) $F'(\omega)$ can be obtained by right multiplying the hammering forces $F(\omega)$ by the corresponding transition matrix C_i for each point, and we can gain the test results $F'(\omega)$ through right multiplying the output voltages $U(\omega)$ by the calibration matrix $W(\omega)$. The process can be expressed as follows:

$$F'(\omega) = C_iF(\omega)$$

(11)

$$F''(\omega) = W(\omega)U(\omega)$$

(12)

So the error of dynamic measurement can be calculated by equation below:

$$\xi(\omega) = \frac{F''(\omega) - F'(\omega)}{F'(\omega)} \times 100\%$$

(13)

Because it has equivalent forces (moments) in common when hammering point 3, 4 and 16. So we choose the equivalent force F_y, moments M_x, M_z generated by hammering point 3, and force F_x, moment M_y through point 4, and force F_z through point 16 to test the error of the testbed. Figure 6 shows the comparison chart of the actual values and test results of the three forces and three moments from 0 Hz to 500 Hz. As the piezoelectric component has the problem of charge leakage in low-frequency band, the test results aren't reliable in low band. So we calculate the maximum test errors of the forces (moments) between 5 Hz and 500 Hz according to Equation (13), and the result is shown in Table 1.

Figure 6 shows the resolution of the testbed for force can reach 0.001 N and resolution for moment can reach 0.0001 N·m under 500 Hz. Table 1 shows the relative error of dynamic testing between 5 Hz and 500 Hz is within ±9.4%. According to the comparison chart of moment M_y in Figure 6, the maximum error of dynamic testing happens on 450 Hz which is 9-time power frequency.

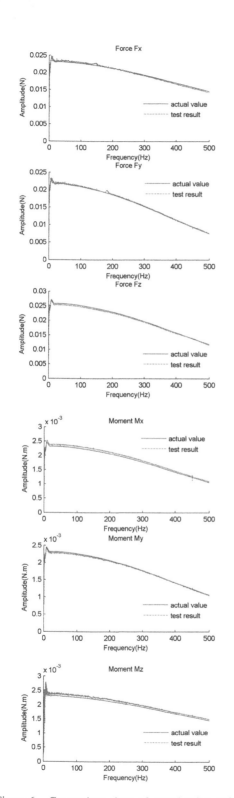

Figure 6. Comparison chart of actual value and test result.

Table 1. Maximum relative error of dynamic testing.

Force/ Moment	F_x	F_y	F_z	M_x	M_y	M_z
Max. Error (%)	3.8	5.6	−2.7	-9.4	3.9	6.5

6 CONCLUSIONS

We invent a piezoelectric testbed for micro-vibration testing in the paper, and the working principle and calibration method of testbed are specified. The testbed has features of high natural frequency, high resolution and low error in dynamic testing. The testbed will contribute to the micro-vibration research of large CMG and SADA. Moreover, the testbed in this article is coupled, and the coupling calibration method is complicated which isn't suitable for non-stationary signals. So we will research on the decoupling testbed in the future.

REFERENCES

Chen, J.P., Cheng, W. & Xia, M.Y. 2014. An Ultra-low Frequency Micro-vibration Testing Platform Based on the Strain-resistance Effect. *Journal of Vibration and Shock* 33(24): 79.

Luo, Q., Li, D.X. & Zhou, W.Y. 2013. Dynamic Modelling and Observation of Micro-vibrations Generated by a Single Gimbal Control Moment Gyro. *Journal of Sound and Vibration*, 332(19): 4505.

Masterson, R.A., Miller, D.W. & Grogan, R.L. 2002. Development and Validation of Reaction Wheel Disturbance Models: Empirical Model. *Journal of Sound and Vibration*, 249(3): 579.

Pan, X.T. & Wen, X.L. 2011 *Sensor Principle and Detection Technology*. Beijing: National Defense Industry Press.

Tan, T.L., Zhu, C.Y. & Zhu, D.F. 2014. Overview of Micro-Vibration Testing, Isolation and Suppression Technology for Spacecraft. *Aerospace Shanghai,* 31(6): 38–39.

Xing, Y.F. & Li, M. 2011 *Engineering Vibration Foundation*. Beijing: Beijing University of Aeronautics and Astronautics Press.

Zhang, P.F., Cheng, W. & Zhao, Y. 2011. Measure of Reaction Wheels Disturbance Considering Coupling Effect. *Journal of Beijing University of Aeronautics and Astronautics*, 37(8): 948–949.

Zhou, D.Q., Cao, R. & Zhao, Y. 2013. Micro-vibration Measurement and Analysis of a Series of Remote Sensing Satellites In-orbit. *Spacecraft Environment Engineering*, 30(6): 627.

Zhou, W.Y., Li, D.X. & Luo, Q. 2012. Analysis and Testing of Micro-vibrations Produced by Momentum Wheel Assemblies. *Chinese Journal of Aeronautics*, 25: 644.

Advances in Civil, Architectural, Structural and Constructional Engineering – Kim, Jung & Seo (Eds)
© 2016 Taylor & Francis Group, London, ISBN 978-1-138-02849-4

Effect of high volume fly ash on the chloride penetration resistance of Self-Compacting Concrete (SCC)

S.A. Kristiawan, S. As'ad & Wibowo
SMARTCRete Research Group, Department of Civil Engineering, Sebelas Maret University, Central Java, Indonesia

B.S. Gan
Computational Applied Mechanics, Department of Architecture, Nihon University, Tokyo, Japan

D.P. Sitompul
PT Wijaya Karya Tbk, Indonesia

ABSTRACT: Self-Compacting Concretes (SCC) have been produced incorporating high volume fly ash at various cement replacement levels. Their resistance against chloride ingress is investigated by carrying out salt ponding tests (AASHTO T259). To obtain chloride penetration profiles, concentration of chloride ions at various depth as specified by ASTM C1556 are determined using XRF method. The results show that a higher volume of fly ash content tends to decrease the resistance of SCC against chloride ingress at early age as confirmed by the linear increase of apparent chloride diffusion coefficient.

1 INTRODUCTION

Self-Compacting Concrete (SCC) is a special type of concrete with the following fresh properties criteria: flowability, fillingability and passingability under its own weight without a tendency of segregation. With these properties SCC could be placed and compacted without any or little vibrations. The key factors to obtain such properties are the use of high finer component, superplasticizer and sometime viscosity agent. If no other finer materials are incorporated in the mix except that of cement, the proportion of cement in SCC tends to be high. The reduction of this cement content is possible by partial replacement of cement with various mineral admixtures such as fly ash, Ground Granulated Blast furnace Slag (GGBS), silica fume, etc. (Dinakar 2012, Qiang et al. 2013, Wongkeo et al. 2014).

The advantages of using fly ash as cement replacement in SCC are due to its physical and chemical properties. The finer size of fly ash compared to cement is beneficial to promote this material as the filler that will occupy inter-particle voids. The spherical shape of fly ash is also useful to enhance workability of SCC. All these advantages offer opportunity to utilize fly ash as cement replacement at a high volume content in SCC (Bouzoubaâ & Lachemi 2001, Liu 2010, Sunarmasto & Kristiawan, 2014). Meanwhile, the pozzolanic reaction that takes place at hardened state will convert CaOH into CSH and causes

an increase in strength and durability at later age (Kosmatka et al. 2003, Douglass 2004).

The durability of concrete exemplifies the resistance of the concrete against degradation due to penetration of aggressive agents from the environment. The mechanisms by which aggressive agents may penetrate into concrete and cause degradation could be one or a combination of the following ionic transports: pressure-induced water flow, water absorption, water vapour diffusion, wick action, ion diffusion, gas diffusion and pressure-induced gas flow (Ferrera 2004). The resistance of concrete against penetration of the aggressive agents will be governed by pores characteristics/porosity of the concrete (Otieno et al. 2014, Simcic et al. 2015) which eventually influence the properties related to durability such as permeability, sorptivity, absorption and diffusion.

One of the aggressive agents that could induce durability problems is chloride ion. Penetration of chloride ion into reinforced concrete could initiate corrosion of the embedded reinforcement when the amount of this ion at reinforcement levels reaches a critical value. The time of corrosion initiation to occur is influenced by the rate of chloride ion penetration. Diffusion is the dominant mechanism by which chloride ion penetrates into concrete. The apparent diffusion of concrete has been the subject of many investigators and it is concluded that the individual components of concrete mixture influences the penetration of chloride ion (Du et al. 2014). SCC has different mixture proportion to that of normal concrete. Hence, it is expected that this type of concrete will offer dissimilarity in its resistance against penetration of chloride ion compared to normal concrete. Furthermore, incorporation of fly ash at high volume levels will modify the pores characteristic of SCC which consequently influences the chloride penetration resistance. Other factor which affects the chloride penetration resistance with respect to the use of fly ash is the improvement of chloride binding into CSH and AFm phase, which eventually reduces the degree of chloride penetration (Simcic et al. 2015). This paper investigates the penetration of chloride ion into SCC under steady state condition. The effect of fly ash content at high volume level is the main concern of this investigation.

2 MATERIALS AND METHOD

2.1 Mixtures proportion

SCC mixtures have been proportioned following the recommendation by Okamura and Ozawa (1995). The initial proportion was set as follows: Coarse Aggregate (CA) with a maximum aggregate size of 10 mm was determined at 50% of the total volume of solid ingredients; Sand (S) was limited to 40% of the total volume of mortar; volume ratio of Water/ Cement (W/C) was chosen in the range of 0.9–1; and finally the Superplasticizer (Sp) dosage was determined to meet the self-compacting concrete criteria. After few trials, the following proportion was obtained as per m³ of SCC: 738 kg of C, 579 and 703 kg of S and CA, respectively, 211 kg of W and 7.27 kg of Sp. The amount of cement was then reduced by Fly Ash (FA) substitution at 50–65% by weight of cement. The final mixtures proportion is given in Table 1.

2.2 Fresh properties of SCC

The flowability, passingability and fillingability of the SCC mixtures have been judged using a combination of the following test methods as described by Takada and Tangtermsirikul (2000): Flow table, J-Ring and V funnel test. The results of such tests are summarized in Table 2.

Table 1. Mixtures proportion of SCC.

ID	C (kg)	S (kg)	CA (kg)	W (kg)	Sp (kg)	FA (kg)
C50	369	579	703	211	7.37	369
C55	332	579	703	211	7.37	405
C60	295	579	703	211	7.37	442
C65	258	579	703	211	7.37	478

Table 2. Fresh properties of SCC mixtures.

Test method	Parameter	Results			
		C50	C55	C60	C65
Flow table	Diameter (mm)	680	655	780	730
J-ring	Diameter (mm)	545	505	660	625
V-funnel	Time (sec)	13.7	20.8	12.4	11.5

2.3 Chloride penetration test

Experimental investigation to determine chloride penetration into SCC was carried out based on the salt ponding test as specified in AASHTO T259. A specimen of concrete slab having a size of $200 \times 200 \times 100$ mm was cast in plastic-mould of $200 \times 200 \times 113$ mm for each SCC mixture. The specimens were cured for 14 days under saturated burlap and then left to dry for another 28 days. During the curing process, the specimens were kept in their plastic-moulds. The moulds were also mean to seal the three faces of the slabs preventing moisture evaporation from these faces the during salt ponding test. The only unsealed face of the slab was the upper surface where on top of it a 3% of NaCl solution was poured. The amount of 3% NaCl used for this ponding test was kept constant (steady) at a level of 13 mm above the top surface of slab (see Figure 1). To prevent a leakage of chloride solution through the perimeter of slab's surface, silicon rubber was used to seal this perimeter. The salt ponding test was carried out for 90 days. Laboratory environment during the test was recorded at 27–32° C and 65–85% RH.

2.4 Measurement of chloride profile

The chloride penetration profiles into SCC specimens were determined by the following procedure. First, a solution of 3% NaCl was discharged from the upper side of plastic-mould to terminate the 90 days of ponding test. The top of the specimens were then wiped using a clean and dry cloth to remove any remaining NaCl solution on the surface of specimens. After attaining dry surface condition the specimens were core-drilled to obtain cylinders of 20×100 mm, which were then cut into several slices representing, respectively, a depth of 0–1, 1–2, 2–3, 3–4, 4–6, 6–8, 8–10 and 10–12 mm from the top surface of slab. Each slice was crushed into powder for chemical analysis using XRF. The total concentration of Cl⁻ obtained from this chemical analysis at each depth was then plotted to demonstrate chloride penetration profiles into SCC specimen.

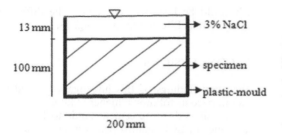

Figure 1. Illustration of salt ponding test.

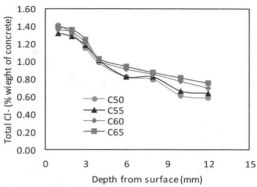

Figure 2. Chloride penetration profile.

3 RESULTS AND DISCUSSION

3.1 *Chloride penetration profile*

The chloride penetration profiles into SCC specimens are presented in Figure 2. The general trend is similar for all SCC specimens: at the near surface of the slab from which chloride ion is penetrated, the concentration of chloride ion is high. This concentration is decreased at a slightly diminishing rate as the depth from slab's surface increases. This trend is comparable to that findings in conventional concrete (Yu et al. 2015). It is also noted from Figure 2 that a higher fly ash content is likely to increase the total concentration of chloride ion even though the amount of increase is not significant. This may be related to the fact that the measurement of chloride ion penetration is started at the age of 42 days where at this age the pozzolanic reaction of fly ash has not effectively taken place yet. Hence, the expected reduction of porosity due to pozzolanic reaction is dismissed. Other researchers (Wongkeo et al. 2014, Sunarmasto & Kristiawan 2015) confirm that inclusion of high volume fly ash tends to increase porosity of SCC at early age. With a higher porosity, SCC containing higher fly ash will demonstrate lower resistance against chloride ingress. However, it should be noted that fly ash has significant impact on the chloride binding capacity. Even though the total chloride ingress into SCC with higher fly ash is greater, the amount of free chloride available in the pores system which directly affects reinforcement corrosion may be low.

The amount of chloride ion as presented in Figure 2 is a total amount of free chloride available in the pores solution and bound chloride. Wongkeo et al. (2014) show that replacement of cement in the mixtures of SCC with high volume fly ash tends to increase the resistance of SCC to chloride ingress due to improvement of binding capacity of this material.

This is related to an increase amount of C_3A when high volume fly ash is incorporated in the SCC mixture. Chloride ions can react with C_3A and C_4AF to form calcium chloroaluminates/ Friedel's salt and calcium chloroferrites which are stable forms and consequently reduce the free chloride available. If the amount of bound chloride is separated from that of Figure 2, the effect of higher volume fly ash to decrease the free chloride may be significant.

3.2 *Surface chloride concentration (Cs)*

At the beginning of ponding test, concentration of chloride ion at the surface of slab (Cs) is zero with the assumption none of the SCC ingredients containing chloride. After this surface is exposed to chloride solution, there is a buildup of chloride ion in this zone. The maximum amount of chloride concentration in the surface and the time taken to reach that maximum value depend on external chloride concentration, exposure condition and quality of concrete. In this investigation, only the quality of concrete i.e. owing to fly ash content is examined while other parameters are kept invariable.

The chloride concentrations at the surfaces (Cs) are projected from the curves fitting of data points of Figure 2. It is suggested in ASTM C1556 that the data point obtained from measurement of chloride at the exposure surface layer is omitted. By mean of non-linear regression analysis, the value of surface chloride concentration for all SCC mixtures are obtained and presented in Figure 3. It is clear that fly ash content has an influence on the buildup of surface chloride concentration. There is also a tendency to attain an optimum value of surface chloride concentration at around 55% fly ash content. It should be noted that a slower buildup of chloride concentration in the surface does not necessarily imply a lower ingress of chloride ion into the depth of SCC specimen.

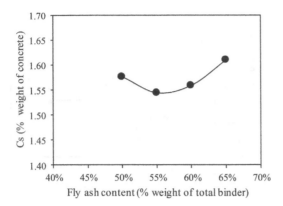

Figure 3. Surface chloride concentration.

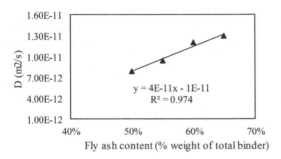

Figure 4. Apparent chloride diffusion coefficient.

3.3 *Apparent chloride diffusion coefficient (D)*

Apparent chloride diffusion coefficients (D) for all SCC have been determined by curve fitting of Equation 1 to the measured chloride penetration profile of Figure 2 by means of a non-linear regression analysis using least squares method.

$$C(x,t) = Cs\left(1 - erf\ \frac{x}{2\sqrt{Dt}}\right) \qquad (1)$$

where, C(x, t) = chloride concentration, measured at depth x and time t; and erf = error function. Figure 4 shows the relationship of apparent chloride diffusion coefficient of SCC with fly ash content. It clearly indicates that incorporating a higher fly ash to partially replace cement in SCC causes a linear increase of apparent chloride diffusion coefficient. The effect of high volume fly ash at early age (42 days) on apparent chloride diffusion coefficient should not be ignored as most service life predictions of chloride induced reinforcement corrosion (e.g. Life 365 model) assume that fly ash does not affect the diffusion coefficient at early age. However, at later age fly ash could be beneficial to reduce the diffusion coefficient as Life 365 model suggests.

4 CONCLUSIONS

The main findings of this investigation could be summarized as follows: a higher volume of fly ash to partially substitute cement tends to lower the resistance of SCC against penetration of chloride ion at early age. In term of surface Chloride concentration (Cs), it seems there is an optimum value of fly ash content resulting in a minimum value of Cs. However, a slower buildup of Cs does not necessarily imply a lower ingress of chloride ion into the depth of concrete. The overall resistance of SCC at early age against penetration of chloride ion is best shown by the apparent chloride diffusion coefficients (D). There is a linear increase of the value of D as fly ash content is greater.

ACKNOWLEDGEMENT

The authors would like to express sincerely gratitude to the Directorate General of Higher Education, Indonesia for providing the financial support through Hibah Unggulan Perguruan Tinggi scheme for the year of 2015 (Contract No. 339/UN.27.11/PL/2015).

REFERENCES

Bouzoubaa, N. & Lachemi, M. 2001. Self compacting concrete incorporating high volume of Class F fly ash: Preliminary results. *Cement and Concrete Research*, 31(3): 413–420.

Dinakar, P. 2012. Design of self-compacting concrete with fly ash. *Magazine Concrete Research*, 64(5): 401–409.

Douglass, R.P. 2004. *Properties of Self-Compacting Concrete Containing Class F Fly Ash*. PCA R&D Serial No. 2619, Portland Cement Association, Stokie, Illinois, USA.

Ferrera, R.M. 2004. *Probability-based durability analysis of concrete structures in marine environment*, PhD Thesis, University of Minho, Portugal.

Kosmatka, S.H., Kherkoff, B. & Panarese, W.C. 2003. *Design and control of concrete mixtures*. Engineering Bulletin 001, Portland Cement Association, Stokie, Illinois, USA.

Liu, M. 2010. Self-compacting concrete with different levels of pulverized fuel ash. *Construction and Building Materials*, 24: 1245–1252.

Okamura, H. & Ozawa, K. 1995. Mix design for self-compacting concrete. *Concrete Library of JSCE*, 25: 107–120.

Otieno, M. Beushausen, H. & Alexander, M. 2014. Effect of chemical composition of slag on chloride penetration re-sistance of concrete. *Cement & Concrete Composites*, 46: 56–64.

Simcic, T., Pejovnik, S., De Schutter, G. & Bosiljkov, V.B. 2015. Chloride ion penetration into fly ash modified concrete during wetting-drying cycles. *Construction and Building Materials.* http://dx.doi.org/10.1016/j.conbuildmat.2015.04.033

Sunarmasto & Kristiawan, S.A. 2015. Effect of fly ash on compressive strength and porosity of self-compacting concrete, *Applied Mechanics and Materials*, 754–755: 447–451.

Takada, K. & Tangtermsirikul, S. 2000. *Part IV: Testing of fresh concrete*, in A. Skarendahl, O. Petersson (Eds), Self-compacting concrete, State of the Art Report of RILEM Technical Committee, 25–39.

Qiang, W., Peiyu, Y. & Jingjing, F. 2013. Design of high volume fly ash concrete for a massive foundation slab. *Magazine Concrete Research*, 65(2): 71–81.

Wongkeo, W., Thongsanitgarn, P., Ngamjarurojana, A. & Chaipanich, A. 2014. Compressive strength and chloride resistance of self-compacting concrete containing high volume fly ash and silica fume. *Materials and Design*, 64: 261–269.

Yu, Z., Chen, Y., Liu, P. & Wang, A. 2015. Accelerated simulation of chloride ingress into concrete under drying-wetting alternation condition of chloride environment. *Construction and Building Materials*, 95: 205–213.

Advances in Civil, Architectural, Structural and Constructional Engineering – Kim, Jung & Seo (Eds)
© *2016 Taylor & Francis Group, London, ISBN 978-1-138-02849-4*

Characterization of watermelon rind as a biosorbent in removing zinc and lead

N. Othman, A. Abdul Kadir, R.M.S. Mohamed & M.F.H. Azizul-Rahman
Micropollutant Research Centre, Faculty of Civil and Environmental Engineering, University Tun Husseion Onn Malaysia, Malaysia

R. Hamdan
Faculty of Engineering Technology, University Tun Husseion Onn Malaysia, Malaysia

T.C. Chay
Faculty of Applied Science, Universiti Teknologi Mara (Arau), Malaysia

ABSTRACT: Rapid industrial development has discharged a huge volume of heavy metals to the environment. In some cases, heavy metals are present in high concentration and thus require treatment. An alternative option in treating heavy metals is using biosorption. Biosorption is a passive physical-chemical process where biomolecules of non-living biological materials, called as biosorbent, bind with heavy metal ions in aqueous solution. In this research, watermelon rind was chosen as a biosorbent. Characterization studies were conducted using SEM-EDX, XRF and FTIR to prove that the biosorption process occurred. The results from characterization studies confirmed the presence of heavy metals after biosorption due to the ion exchange process. These results supported the potential use of watermelon rind as a biosorbent for the removal of heavy metals in various wastewater.

1 INTRODUCTION

Development of industrial activities has resulted in accidental discharge of heavy metals to the receiving environment. The heavy metals discharged are non-biodegradable and considered to be toxic to the human and ecological system. Therefore, heavy metals should be removed from the polluted environment (Othman et al. 2013).

Various technologies have been explored for the removal of metals including chemical precipitation, lime coagulation, ion exchange, reverse osmosis and solvent extraction. However, these methods pose some drawback such as low selectivity, incomplete removal and producing large quantities of chemical waste (Othman et al. 2013, Ningchuan et al. 2011, Azizul Rahman et al. 2013, Schiewer et al. 2008).

One of the attempts is to use biosorption for metal removal. Biosorption is a biological treatment method for removing metals. This method is considered as environmental friendly due to no chemical usage (Ningchuan et al. 2011).

Several fruit wastes with high pectin content such as orange peel, banana peel and other vitamin C-based fruit have been explored as biosorbents (Schiewer et al. 2008, Feng 2011). Understanding the characteristic of the fruit waste is an important task for exploring the suitability of the waste as an adsorbent.

In this study on waste, watermelon rind (*Citrullus lanatus*) was used as an adsorbent material. This study presents the characteristic of watermelon rind as an adsorbent material.

2 MATERIALS AND METHODS

2.1 Biosorbent preparation

Waste watermelon rind was collected from fruit stalls around the UTHM campus and Parit Raja, Batu Pahat, Johor, Malaysia. The rinds were washed with distilled water for several times and soaked into 5% nitric acid (HNO_3) for about 24 hours to remove all the dirt particles and/or unwanted contaminants present on the surface of watermelon rind and further cleaned and soaked with deionized water few times until reaching the neutral condition. Then, the biosorbent was oven-dried at 60°C until constant weight. The dried rinds were ground using a laboratory ball mill to produce a particle size of 150 μm to pass through the sieve and preserved in airtight container for further analysis (Azizul Rahman et al. 2013).

2.2 Characterization study

Three tests were conducted for characterization study, namely X-Ray Fluorescence spectrometry (XRF), Scanning electron microscope-EDX (SEM-EDX) spectra and Fourier transfer infrared analysis (FTIR).

XRF was used to analyze oxide metals in the adsorbent. SEM-EDX was used to analyze the organic and inorganic contents in the watermelon. The presence of functional groups for metal removal was analyzed using FTIR.

2.3 Batch experiment at optimal conditions

A stock solution of zinc, lead and chromium was prepared in deionized water. The respective concentration needed was obtained through the serial dilution process. The pH value required for the test solutions was adjusted by adding a few drop of diluted 0.01M HCI or 0.01M NaOH using a pH meter. All samples were shaken at 125 rpm at room temperature. Then, the solutions were filtered using a 0.45 μm membrane cellulose filter. The concentrations of heavy metals contained in the wastewater were analyzed by using Atomic Absorption Spectrometry (AAS).

3 RESULTS

3.1 XRF analysis

XRF analysis was used to analyze the chemical composition in the watermelon rind before and after the chemical pretreatment (washing) with nitric acid (HNO_3) (Table 1) and also after biosorption (Table 2). The presence of silica oxide (SiO) and calcium oxide will help in the ion exchange

Table 1. Chemical composition and concentration of untreated watermelon rind.

Chemical	Untreated	Treated
Ori-g	7	7
Added-g	3	3
CO_2	0.1	0.1
K_2O	32.5	0.9
CaO	19.6	5.9
CI	6.9	0.8
P_2O_5	5.9	–
SO_3	5.2	8.3
SiO_2	26.0	83.7
MgO	2.5	–
MnO	0.6	–
MoO_3	0.6	0.3
ZnO	0.1	–

Table 2. Chemical composition and concentration of metal-loaded watermelon rind.

Chemical	Zn loaded	Pb loaded
Ori-g	7	7
Added-g	3	3
CO_2	0.1	0.1
CaO	0.2	0.2
SO_3	0.9	0.3
SiO_2	34.5	51.5
ZnO	64.2	–
Pb	–	47.47

process during biosorption. After treating with HNO_3, most of the oxide metals were removed after the pre-treatment process. A clean biosorbent is important to increase the efficiency of the biosorption process. Among various chemical modification methods, chemical pretreatment (washing) was selected due to its simplicity and efficiency. Acid washing can remove the mineral elements and improves the hydrophilic nature of the surface (Nadeem et al. 2008, Kratochvil & Volesky 1998). After biosorption, it clearly showed the presence of metals, namely zinc and lead. This indicated that some chemicals triggered the process of heavy metal removal.

3.2 SEM-EDX

Surface morphology, structure of the biosorbent and elemental analysis were determined by using Scanning electron microscopy (Liu et al. 2012, Iftikhar et al. 2009). Before biosorption, the SEM result indicated that the watermelon rind had a rough surface texture and high porosity surface structure could be distinctly noticed, making it possible for the various heavy metals to be adsorbed by different parts of the biosorbent (Figure 1). The SEM analysis showed the presence of two regions, namely white and dark regions, in the watermelon rind. The white area contained an inorganic element of silica, meanwhile the dark area was comprised of organic protein that was rich in carbon and oxygen (Aeslina et al. 2012). The EDX analysis proved the presence of Zn and Pb on the biosorbent surface (Figure 2a, b) that were not detected in the native biosorbent. This signifies the adsorption of heavy metal ions onto the watermelon rind.

3.3 FTIR analysis

The FTIR spectrum of the native biosorbent showed that carboxyl and hydroxyl groups were present in abundance. These groups contains

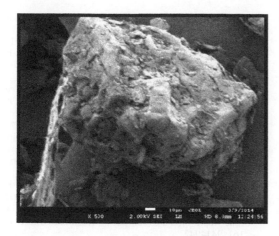

Figure 1. SEM micrograph of the acid-treated watermelon rind.

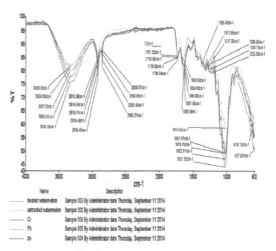

Name	Description
treated watermelon	Sample 003 By Administrator data Thursday, September 11 2014
untreated watermelon	Sample 002 By Administrator data Thursday, September 11 2014
Cr	Sample 006 By Administrator data Thursday, September 11 2014
Pb	Sample 005 By Administrator data Thursday, September 11 2014
zn	Sample 004 By Administrator data Thursday, September 11 2014

Figure 3. FTIR spectra.

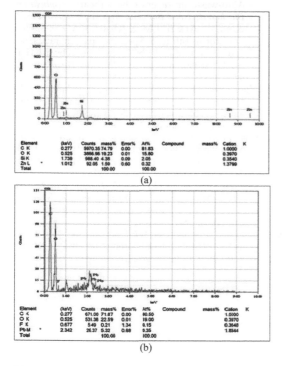

Figure 2. (a) & (b)- EDX analysis of Zn and Pb loaded.

biopolymers that may function as proton donors, hence deprotonated hydroxyl and carboxyl groups may be involved in sequestering the metal ions (Figueira et al. 1999). The changes in the functional group and surface properties of the biosorbent from the FTIR spectra of the metal-loaded biosorbent describe that the shift of functional

group bands is due to metal biosorption. These shifts may be attributed to the changes in counter ions associated with hydroxyl and carboxyl groups that are predominant contributors to the complexation of metal ions and ion exchange process (Gupta et al. 2010).

The FTIR spectra of the native biosorbent showed a peak at 3335.08 cm^{-1}. The changes observed in the band between 3500 and 3200 cm^{-1} may be assigned to the complexation of metal ions with the ionized O-H groups of free hydroxyl groups and the bonded O-H group in polymeric compounds such as alcohols, phenols and carboxylic acids present in pectin, cellulose and lignin on the watermelon rind. The changes in the band between 3000 and 2850 cm^{-1} and 2843 and 2863 cm^{-1} showed that the mechanism ion exchanges took place between protons and symmetric or asymmetric CH, CH^2 of aliphatic acids. Meanwhile the shifting of the band between 1750 and 1680 cm^{-1} and 1590 and 1750 cm^{-1} indicated the stretching vibration of the C = O bond of carboxyl groups. While the shifts demonstrated in the peaks between 1300 and 1000 cm^{-1} indicated the C-O stretching of—COOH (Figure 3). The same results were also observed in the adsorption of Pb(II) ions on cucumber peel, melon rind, orange peel and mango peel (Feng et al. 2011, Othman & Mohd Asharuddin 2013).

3.4 Removal of Zn(II), Pb(II) and Cr(II) under optimal conditions

The results of batch study performance under optimal values of all parameters for each metal are summarized in Table 3.

Table 3. Percentage of removal and uptake capacity of all parameters at optimal conditions.

Heavy metal	Removal (%)	Uptake capacity (mg/g)
Zinc	98	23
Lead	85	1.4

4 CONCLUSIONS

The watermelon rind is a potential biosorbent to remove metals with a percentage up to 85–95% for both lead and zinc. This is due to the characteristic of the rind with high silica content and porous microstructure to adsorb metals. The presence of functional groups such as alcohols and carboxylic acids is important for the ion exchange process between metals and these functional groups.

ACKNOWLEDGMENT

This research was supported by UniversitiTun Hussein Onn Malaysia (UTHM) and Ministry of Higher Education Malaysia through Fundamental Research Grant Scheme (FRGS) 2013 (Vot No. 1230).

REFERENCES

Aeslina, A.K., Othman, N. & Azimah, M.A.N. 2012. Potential of using Rosa Centifolia to remove iron and manganese in groundwater treatment, *International Journal of Sustainable Construction Engineering & Technology*, 3: 2180–2324.

Azizul Rahman, M.F.H., Mohd Suhaimi, A.A. & Othman, N. 2013. Biosorption of Pb(II) and Zn(II) in synthetic wastewater by watermelon rind (Citrullus lanatus), *Applied Mechanics and Material*, 455–466: 906–910.

Feng, N., Guo, X., Liang, S., Zhu, Y. & Liu. J. 2011. Biosorption of heavy metals from aques solutions by chemically modified orange peel. *Journal of Hazardous material*, 185(1): 49–54.

Figueira, M., Volesky, B. & Mathieu, H. 1999. Instrumental analysis study of iron species biosorption by Sargassum biomass. *Environmental Science and Technology*, 33(11): 1840–1846.

Gupta, V.K., Rastogi, A. & Nayak, A. 2010. Biosorption of nickel onto treated alga. Application of isotherm and kinetics model. *Journal of Colloidal and Interface Science*. 342(2): 533–539.

Iftikhar, A.R., Bhatti, H.N., Hanif, M.A. & Nadeem, R. 2009. Kinetic and thermodynamic aspects of Cu(II) and Cr(III) removal from aqueous solutions using rose waste biomass, *Journal of Hazardous Material*, 161: 941–947.

Kratochvil, D. & Volesky, B. 1998. *Advances in the biosorption oh heavy metals*, TIBTEECH.

Liu, C., Ngo, H.H. & Guo, W. 2012. Watermelon rind: Agro-waste or superior biosorbent?, *Application of Biochemistry and Biotechnology*, 167: 1699–1715.

Nadeem, R., Ansari, T.M. & Khalid, A.M. 2008. Fourier Transform Infrared Spectroscopic characterization and optimization of Pb(II) biosorption by fish (Labeo rohita) scales, *Journal of Materials Hazardous*, 156: 64–73.

Ningchuan, F., Xueyi, G., Sha, L., Yanshu, Z. & Jianping, L. 2011. Biosorption of heavy metals from aqueous solution by chemically modified orange peel, *Journal of Hazardous Materials*, 185: 49–54.

Othman, N. & Mohd Asharuddin, S. 2013. Cucumis Melo Rind as Biosorbent to Remove Fe(II) and Mn(II) From Synthetic Groundwater Solution. *Advanced Materials Research*, 795: 266–271.

Othman, N., Asharuddin, S.M. & Rahman, M.F.H.A. 2013. An Overview of Fruit waste as Sustainable Adsorbent for Heavy Metal, *Applied Mechanics and Materials*, 389: 29–53.

Schiewer, S. & Santosh, B.P. 2008. Pectin-rich fruit waste as biosorbents for heavy metal removal: Equilibrium and kinetics, *Bioresource Technology*, 99: 1896–1903.

Advances in Civil, Architectural, Structural and Constructional Engineering – Kim, Jung & Seo (Eds)
© 2016 Taylor & Francis Group, London, ISBN 978-1-138-02849-4

Moisture content assessment of heat-treated Malaysian hardwood timber: The case of Kapur (*Dryobalanops* spp.) and PauhKijang (*Ir-vingiamalayanaOliv* spp.)

N.I.F. Md Noh
Faculty of Civil Engineering, University Technology Mara (UITM), Selangor, Malaysia

Z. Ahmad
Institute of Infrastructure Engineering and Sustainable Management, University Technology Mara (UITM), Selangor, Malaysia

ABSTRACT: Heat treatment is one of the environmentally friendly ways to treat timber that will lead to the improvement of timber natural quality and equip the timber with new properties. It is an eco-friendly and alternative treatment method that will modify the properties of timber by using high temperature instead of using chemical preservatives as common practice. This paper presents the effect of heat treatment on physical properties such as moisture content for two types of Malaysia hardwood timber, namely Kapur and PauhKijang. Specially designed electronic furnace made up of ceramic fiber was used as an oven for the heat treatment process. The result indicates a reduction in moisture content for both species after heat treatment and the difference is significant. Low moisture content indicates a positive indicator, which theoretically leads to the increment of movement stability and maximizes the strength because mechanical properties increase when the moisture content is low.

1 INTRODUCTION

Malaysia is blessed with abundant of rainfall and its humidity is high. About two-thirds of Malaysia are covered by natural forest and tree plantations. Timber has an excellent record in stored carbon footprint released by other sources such as concrete. It is one of the oldest building materials that has been used to build human shelters (Al-nagadi 2012). However, there is a limit on timber usage based on its properties and characteristics. This limit has defined timbers as a building material that is not appropriate under all conditions (Patel 2011). The naturally durable timber is limited and considered expensive. Therefore, the use of non-natural durable timber is an alternative (Hall 2005). However, this timber needs to be modified or treated to help improve its durability properties. The most commonly used method of treatment for timber is by using chemical preservatives such as Copper Chromium Arsenic (CCA) (Hardin 2005). Due the usage of chemical preservatives that are toxic, many issues related to CCA are emerged. One of the key issues is related to humans where CCA can pose a threat to human health. EC Scientific Committee on Toxicity, Eco-toxicity, and the Environment (CSTEE) announced that CCA is both genotoxic and carcinogenic. CCA may cause the risk of cancer, especially in the lungs, bladder, kidneys and liver.

In light of this issue, an alternative treatment method needs to be explored and made available in Malaysia. One of the possible methods of treatment is heat treatment that uses heat rather than chemicals. This method of treatment will alter the substrate of a timber species by using high temperature, not by using chemical preservatives. Heat treatment is the most environmentally friendly way to treat a timber species that will be an alternative way to replace CCA.

When timber is treated with heat treatment (temperature range approximately 150°C–220°C) (Sundqvist 2004), the main purpose is to achieve new material properties rather than to dry the wood. The main aim of this treatment is to increase the biological durability and enhance dimensional stability. The temperature applied above 150°C in this treatment will lead to permanent changes in the timber properties and durability (Esteves & Pereira 2009). Johansson (2005) exposed spruce with heat treatment and reported that heat treatment can reduce the mechanical properties and may form inner cracks.

As stated in MS 544 Part 2 (2001), Kapur and PauhKijang are listed as timbers that require treatment to improve its properties and durability.

The objective of this paper is to assess the effect on one of the timber physical properties, namely the moisture content, of two Malaysian hardwood timbers, namely Kapur (*Dryobalanops* spp.) and PauhKijang (*Ir-vingiamalayanaOliv* spp.) treated by heat at temperatures of 150°C, 170°C, 190°C and 210°C within 1-hour duration.

2 EXPERIMENTAL WORK

This study utilized the solid hardwood timbers Kapur (*Dryobalanops* spp.) and PauhKijang (*Ir-vingiamalayanaOliv* spp.). Temperatures of 150°C, 170°C, 190°C and 210°C were determined for the heat treatment. A total of 15 samples of green (wet) timbers with Moisture Content (MC) more than 19% and a size of 50 × 90 × 1800 mm were prepared respectively for the above-mentioned heat treatment temperatures. All samples were weighted prior to the test.

These samples were stored in the conditioning room for 2 weeks prior to the heat treatment process. The conditioning room had 65% of relative humidity with a temperature level of 24°C. After the conditioning process, the reading of moisture content was taken by using a moisture meter. The readings were taken as the data for moisture content reading before heat treatment. Heat treatment was then executed on the samples after the conditioning process, which was conducted at Margin Heat Treatment and NDT Services Sdn Bhd, Shah Alam. A special electrical furnace acted as an oven was designed and prepared according to the size and number of the samples, as shown in Figure 1.

The 15 samples were put into the electrical furnace, as shown in Figure 2. The heat treatment was performed within 1-hour duration in an oven that was connected to the induction heating machine used to control the heat applied to the timber.

After the heat treatment, each sample from both species was cut from the three parts of mid-span

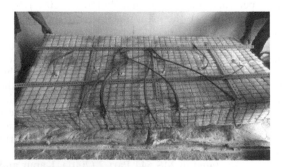

Figure 1. Specially designed electrical furnace made up from ceramic fiber.

Figure 2. Arrangement of the samples inside the electrical furnace.

Figure 3. Sample placement in the oven.

of the specimen for testing the moisture content. Moisture content of the timber is the amount of water inside the timber, expressed in percentage. The oven-drying method is the most universally accepted method for determining the moisture content of timber, but it is slow and requires cutting the wood. Specimens for determining the moisture content were prepared with a dimension of 25 mm thick, 50 mm wide and 90 mm long. The procedure for testing the moisture content (mc) is in accordance to the Malaysian Standard, MS 544:2001, "Method for Determination of Moisture Content of Timber" by using Equation (1):

$$mc = (m_i - m_{od})/m_{od} \times 100\% \qquad (1)$$

where m_i is the initial mass (in grams) of the test specimens and m_{od} is the mass (in grams) of the oven-dried test specimens.

The specimens were weighed as soon as it was cut by using a weight scale where all of them were cleaned by a brush to remove the dust or residue that may affect the weight of the specimen. The weight was then recorded before being placed in an oven for drying, as shown in Figure 3.

The oven temperature was maintained at $105 \pm 2°C$. The weight was then recorded again once all of the specimens were removed from the oven after 24 hours.

3 RESULTS AND DISCUSSION

Table 1 summarizes the result on the moisture content of hardwood timber Kapur before and after the heat treatment. The moisture content of the wet and dry timber was in the range provided by MS 544 as for the timber before the treatment. The moisture content was >19%, which showed that the timbers were in the green condition.

The moisture content seems to reduce once treated by heat, and the percentage of reduction does increase with an increase in temperature. The reduction in moisture content after treatment is caused by the reduction of the timber's hydroxyl where the heat reduces the timber water uptake that leads to the increment of movement stability of both swelling and shrinkage, which is a positive indication (Korkut et al. 2009).

To verify whether there is a significant difference in the moisture content before and after heat treatment, the t-test was performed with the hypothesis as follows:

H_0: No significant difference between before and after heat treatment

H_1: There is a significant difference between before and after heat treatment

The t-test analysis of Excel shows that the $t_{Stat} = 32.95 > t_{Critical}$ two-tail = 3.18. While the significance P = 0.00000614 is smaller than the P value of 0.05, which means the null hypothesis is rejected. Therefore, it can be ascertained that the moisture content before and after heat treatment differs significantly.

The same condition applies to the hardwood timber PauhKijang, which has a reduction in moisture content after heat treatment, as shown in Table 2. Based on the t-test, it can be concluded that the

Table 1. Moisture content value before and after heat treatment for hardwood timber Kapur.

Temperature of heat treatment (°c)	Average moisture content (%)		Percentage of reduction (%)
	Before heat treatment	After heat treatment	
150	19.72	10.12	48.68
170	19.96	9.97	50.05
190	19.77	9.03	54.32
210	19.87	8.94	55.01

Table 2. Moisture content value before and after heat treatment for the hardwood timber PauhKijang.

Temperature of heat treatment (°C)	Average moisture content (%)		Percentage of reduction (%)
	Before heat treatment	After heat treatment	
150	33.15	9.98	69.89
170	31.53	9.32	70.44
190	32.87	9.16	72.13
210	32.42	8.23	74.61

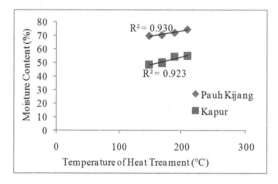

Figure 4. Moisture content (%) versus temperature of heat treatment (°C).

difference in moisture content before and after heat treatment of PauhKijang is significant. The t-test analysis of Excel revealed that the $t_{Stat} = 54.92$ is larger than $t_{Critical}$ two-tail = 3.18, where the significance P = 0.00000133 is smaller than the P value of 0.05, thus H_1 is accepted.

Reduction in moisture content indicates the positive impact on the timber. Other than improving its movement stability, the timber with low moisture content will lead to the reduction in fungus attack and decay inception (Muthike 2008). Both of these reasons are good for the timber to be a good material in the industry. In addition, the reduction in moisture content also will lead to increment in strength, as a dried timber with a moisture content lower than 19% will have improvement in its mechanical properties (Kohlen 2007).

The correlation between moisture content and the level of temperature of heat treatment is shown in Figure 4. A correlation coefficient nearing 1 indicates a strong positive relationship. A relatively strong relationship can be observed as the correlation between moisture content and temperature of heat treatment for Kapur and PauhKijang, where $R^2 = 0.9238$ and $R^2 = 0.9306$ respectively. It clearly shows that heat treatment causes a high reduction in moisture content that may lead to another improvement in the properties of both Kapur and PauhKijang.

4 CONCLUSION

Treatment of Malaysian hardwood timber is vital and heat treatment is another alternative way. Moisture content of Malaysian hardwood timber seems to decrease by heat treatment where the difference before and after heat treatment is significant. There is also a good correlation between moisture content and temperature of heat treatment. Reduction in moisture content indicates the increment of movement stability and strength. It is a good indicator that heat treatment is an alternative treatment that can help improve the properties of Malaysian hardwood timbers in an environmentally friendly way.

ACKNOWLEDGMENT

This research was financially supported by the Public Works Department Malaysia and Universiti Teknologi Mara (UiTM).

REFERENCES

Al-nagadi, E.M. 2012.*Concrete Construction Industry*, Cement Based Materials and Civil Infrastructure (CBM & CI), Saudi Arabia.

Esteves, B.M. & Pereira, H.M. 2009. Wood modification by heat treatment: A review. *BioResources*, 4(1): 370–404.

Hall, N.L. & Beder, S. 2005. *Treated Timber*, Ticking Time-bomb.

Hardin, R.A. & Beckermann, C. 2005. *Simulation of Heat Treatment Distortion*, in Proceedings of the 59th SFSA Technical and Operating Conference, 3(3), Steel Founders Society of America, Chicago, IL.

Johansson, D. 2005. Drying and Heat Treatment of Wood: Influences on Internal Checking. 3rd Nordic Drying Conference Karlstad, Sweden. June 15–17. 82-5942873-3

Joseph, G. & George, M. 2008. *Low Cost Timber Drying Method for Sawyers, Merchants and Other Users*, Kenya Forestry Research Institute (KEFRI) Forest Products Research Centre—Karura

Köhlen, J. 2007. *Reliability of timber structures*, Swiss Federal Institute of Technology, Institute of Structural Engineering, (May).

Korkut, S., Alma, M.H. & Elyildirim, Y.K. 2009. The effects of heat treatment on physical and technological properties and surface roughness of European Hophornbeam (OstryacarpinifoliaScop.), *Wood*, 8(20): 5316–5327.

MS 544: PART 2. 2001. *Code of Practice For Structural Used of Timber—Part 2: Permissible Stress Design of Solid Timber*, Department of Standard. Malaysia.

Patel, K.V. 2011.*Construction Materials Management On Project Sites*. In National Conference on Recent Trends in Engineering & Technology.

Sundqvist, B. 2004. *Colour Changes and Acid Formation in Wood During Heating. Luleå university of Technology*, Skellefteå Campus, Division of Wood Material Science, 2004:10 Doctoral thesis.

Advances in Civil, Architectural, Structural and Constructional Engineering – Kim, Jung & Seo (Eds)
© 2016 Taylor & Francis Group, London, ISBN 978-1-138-02849-4

Deformation behavior of the closed-cell aluminum foam based on a three-dimensional model

H.X. Wang & Z.Y. Wang
College of Materials Science, Yancheng Institute of Technology, Yancheng, Jiangsu, China

K. Yu
Analysis and Testing Center, Southeast University, Nanjing, Jiangsu, China

ABSTRACT: In order to assess and predict the deformation properties of the closed-cell aluminum foam metal, a simplified three-dimensional model for aluminum foam is established by finite element analysis software. The compression process can be seen from the simulated deformation behavior of aluminum foam. We can find that there are three stages of the compression process of the spherical pore aluminum foam, namely elastic, plastic collapse and densification stages. Microscopic deformation is highly irregular. It is a mixture of deformation mechanism of hole wall bending, folding, twisting and collapse. At the same time, stress-strain curves can be obtained. It is consistent with the existing experimental stress-strain curve.

1 INTRODUCTION

The ultra-light metal structure with various kinds of pores realizes the lightness and multifunction of structural material. A new type of spherical pore Al alloy foam with low porosity has become a new trend of research work. The mechanical properties of aluminum foam are mainly dependent on its density. But the size of the hole, the structure and distribution is also the important parameters of its mechanical properties. With the development of the finite element technique using the finite element simulation, a lot of experimental processes can be saved. In addition to the test, we can get the simulation experiment that is difficult to realize. It reduces the experiment cost and saves money. Therefore, this paper shows the compression performance of spherical aluminum foam by using the finite element software ANSYS/LS-DYNA.

2 THE DESIGN THOUGHTS OF FINITE ELEMENT SIMULATION OF ALUMINUM FOAM

By using the finite element software of foam metal computer simulation for quasi-static compression performance, the purpose of simulation is to assess and predict the mechanical properties and energy absorption performance of the foam metal under a certain condition of porosity. But concerning the structure of aluminum foam, it is a new type of porous structure function material with metal frame, large aperture and high porosity. It is extremely difficult to establish a complete physical model. So far, some quantitative analysis from the point of view of experimental research for the research of metal foam has only been reported. There is no ready-made theory of finite element simulation model for the aluminum foam.

There are a lot of characterization parameters of foam metal structure. For example, parameters are the pore diameter, porosity, density and hole degrees. The main reason for the simulation analysis conducted under a certain condition of porosity is as follows: porosity is one of the important parameters of foam metal performance. It is inversely proportional to the density of the foam metal. The mechanical properties of the foam metal are largely dependent on its density. In other words, the mechanical properties of the foam metal are largely dependent on its porosity.

3 THE STUDY OF THE SPHERICAL ALUMINUM FOAM WITH FINITE ELEMENT SIMULATION

3.1 *Establishing the finite element model*

Finite element analysis model is established based on the literature (Santosa & Wierzbicki 1998, Cheon & Megui 2004, Meguid et al. 2002), which put forward an improved unit cell model of the closed-cell aluminum foam. The closed-cell aluminum foam model is established, as shown in Figure 1(a). After its outspread on the xyz direction, the three-dimensional multi-layer model on the basis of the unit cell is obtained, as shown in Figure 1(b).

The specific size of the unit cell model is as follows:

t = 0.1 mm, d = 1 mm, w = 2 mm

Its relative density is given as follows:

$$\frac{\rho}{\rho_s} = 3\left(\frac{t}{w}\right) + \frac{\pi}{4}\left(\frac{t}{w}\right)\left(\frac{D}{w}\right)^2 = 0.16 \qquad (1)$$

where t is the wall thickness; D is the bore diameter; and w is the wing length connecting the adjacent holes.

Its compaction strain is given as follows:

$$\varepsilon_d = 1 - 0.6\frac{D}{w} = 0.7 \qquad (2)$$

The three-dimensional multi-layer model has a size of 8 mm × 8 mm × 8 mm.

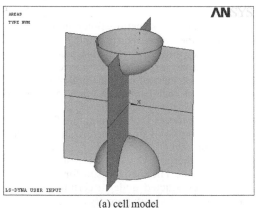

(a) cell model

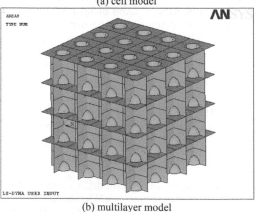

(b) multilayer model

Figure 1. Three-dimensional model of the aluminum foam.

3.2 Setting the material parameter

Industrial pure aluminum is selected as the matrix material. According to the stress-strain relationship curve, Power Law Plasticity Mode is used for substrate materials in the process of simulation. This model is mainly used for metal and plastic deformation analysis because this kind of material has an isotropic elasto-plastic deformation behavior. During the simulation process, it is assumed that the pure aluminum substrate material is not sensitive to the strain rate, so the coefficient of strain rate is zero. Power-reinforced plastic model parameter settings are presented in Table 1.

3.3 Meshing

Because the aluminum foam model belongs to the 3d model, the Belytschko-Tsay SHELL163 element of 3d unit type is chosen. The unit is a four-node quadrilateral element. Each node has six degrees of freedom. It is used for an explicit dynamic analysis. The mesh model is shown in Figure 2.

3.4 Defining the contact interface

Contact is a very common nonlinear behavior, and is a special and important subset in the nonlinear type. It has a highly nonlinear behavior. There are no contact elements in the ANSYS/

Table 1. Material parameters.

Density	2700 (kg/m³)
Elasticity modulus	68.9×10^9 (MPa)
Poisson's ratio	0.34
Strengthening factor	5.98×10^5
Hardening coefficient n	0.216

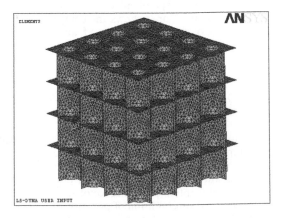

Figure 2. Meshing the model.

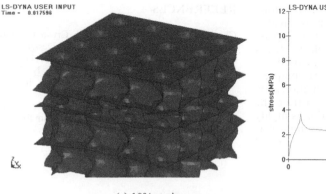

(a) 10% strain

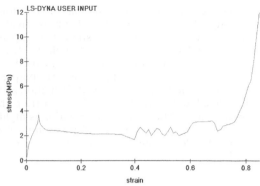

Figure 4. The simulation stress strain curve of the aluminum foam.

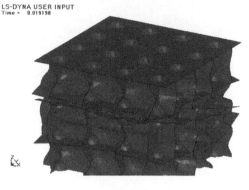

(b) 40% strain

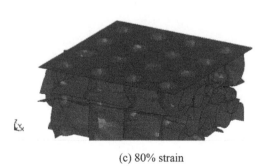

(c) 80% strain

Figure 3. The deformation distribution of the aluminum foam compression process.

LS-DYNA program. Possible surface contact, the type of contact and some parameters related to the contact are only needed to be defined. There are two kinds of common type of contact, namely surface-surface contact and point-surface contact. Because the single cell in the aluminum foam model of contact between wings is surface contact, surface-surface contact and the automatic contact type are chosen. Clamp with many on the surface of the cell unit is also face side of the contact.

Figure 4 shows the stress strain curve of foam aluminum obtained by simulating. This trend is consistent with a previous experimental result. There are three stages of the compression process of foam aluminum, as shown in the figure, namely the elastic zone, the plastic platform region and the densification region. Elastic deformation stage is mainly the mechanism of hole wall elastic bending. When the elastic strain increases to a certain value, the aluminum foam steps into the plastic deformation stage. The main characteristic of this stage is that stress almost remains the same with the increase in strain. Material damage first appears in the weakest region in the hole, and then extends in this layer. This layer hole wall shifts gradually from the elastic bending into plastic bending, folding and compaction. With the increase in the strain, the upper and lower layers of the compaction layer touch each other. Damage occurs in a new layer and the above process is repeated. The material is made by compaction step by step.

4 CONCLUSIONS

The research was conducted on the basis of the uniform distribution of holes. The three-dimensional model of the aluminum foam is constructed. Through the finite element method, the mechanical property of the aluminum foam is studied. Finally, the following conclusions can be drawn.

The compression deformation of the low-density closed-cell aluminum foam is divided into three phases, namely elastic collapse, plasticity and densification stages. The plastic deformation mechanism of the low-density closed-cell foam aluminum during the process of compression is as follows: concentrated deformation zone is formed and starts to collapse down from the middle layer. Microscopic deformation is highly irregular. It is a mixture of deformation mechanism of hole wall bending, folding, twisting and collapse.

REFERENCES

Cheon, S.S. & Meguid, S.A. 2004. Crush Behavior of Metallic Foams for Passenger Car Design. *International journal of automotive technology*, 5(1): 47–53.

Meguid, S.A. Cheon, S.S. & El-Abbasi, N. 2002. FE modelling of deformation localization in metallic foams, 38(7): 631–643.

Santosa, S. & Wierzbicki, T. 1998. On the modelling of crush behavior of a closed-cell aluminium foam structures, *Journal of the Mechanics and Physics of Solids*, 46: 645–669.

Mechanical and management engineering

Advances in Civil, Architectural, Structural and Constructional Engineering – Kim, Jung & Seo (Eds)
© 2016 Taylor & Francis Group, London, ISBN 978-1-138-02849-4

Research of the relationship between industrial structure and spatial layout in Valley economy—Cherry Valley in Shanhaiguan as an example

W. Wang
Urban Planning and Design, College of Urban and Rural Construction, Hebei Province, China

H.W. Li
Department of Architecture, College of Urban and Rural Construction, Hebei Agricultural University, Hebei Province, China

ABSTRACT: In recent years, the spatial organization relationship of element layout in mountainous areas with the valley as the carrier becomes the new development issue in mountainous areas. The booming valley economy shows tremendous vitality and achieves notable effect, which exhibits a wonderful prospect for the development in mountainous areas. However, meanwhile, in many areas, the industry layout is too decentralized and unreasonable, and has not yet formed the large-scale and systematization together with low spatial organization benefit and land resource utilization rate.

1 INTRODUCTION

With the development of the national economy, the industrial division of labor pattern is changing constantly. In this period, it is an effective method and an important way to effectively participate in market competition and to play a more effective way to improve the economic competitiveness of the region. (Lu 1991) The optimization and adjustment of industrial structure and the reasonable planning of spatial layout in the ditch area are very important to the economic development of the region, which influences the development of the economy in the future.

2 PLANNING CONCEPTS

Based on the domestic and foreign industrial structure, industrial layout, space layout theory, theoretical basis of research of industrial structure with ditch area economy, geographical domain specific structure with channel three industry, the land use planning and facilities standards must first phase should consider industry existence Valley economy is different from the overall planning of city planning Along with the economic system reform, agriculture is a variety of management in order to grow and process, accelerate and transform into intensive characteristic agricultural industrialization, in the industry, the formation of the economic and social structure is to promote the integration of workers and peasants, and

realize the economic comprehensive competitiveness of the region.

According to the characteristics of Cherry Valley, the existing resources and the concept of the planning, the integration and demonstration base of low carbon technology, the target oriented "low carbon comprehensive experimental base", and the carrier of low carbon industry cluster and characteristic industry chain, and the function of tourism complex are established.

2.1 *"Low-carbon development, structure adjustment, layout optimization and step-by-step implementation" concept*

The valley will be developed and the unified management of low carbon technology integration system, in different industries, different levels, low carbon technology choice is relatively reasonable, building a low-carbon industrial system in the whole valley range, so that the planning of Shanhaiguan "double" demonstration area has become a model for the development of low carbon technology the Chinese.

2.2 *Theories on efficient cyclic utilization of rural wastes and renewable energies*

Product processing cycle mode, waste material recycling mode, channel industry cycle mode, tourism agriculture cycle mode is very important. In real life, it is very important for the sustainable development of agriculture.

3 KEY PROJECT CONSTRUCTION

3.1 *"Happy Farm" manor*

3.1.1 *Project overview*
Located in the north of Waiyu Village, Shihe Town and near the Shanhaiguan Forest Farm, the ecological "Happy Farm" manor covers about 19 hectare, where the land is flat and neat, water conservancy facilities complete and traffic convenient.

3.1.2 *Construction objective*
According to the current land state and local climate conditions, to construct integrated service facilities such as agricultural product exhibition center, farming museum and cherry culture park and build an ecological theme manor with integrated functions such as vacation, entertainment, touring and sightseeing, in Shihe Town.

3.1.3 *Planning details*
The ecological "Happy Farm" manor is mainly divided into 6 functional areas, namely, management service area, agricultural product exhibition area, protected agricultural area, creative agritainment, sightseeing & picking area and happy farm.

Management service area: in this area, a 5000 m^2 farming museum and a 2000 m^2 cherry culture park are mainly constructed. They are both low-carbon, ecological, energy-efficient and environmental friendly modern buildings with leaf-oriented exterior. There into, the cheery culture park focuses on cherry culture exhibition and communication; on this basis, people's recognition and publicity on the cheery.

Agricultural product exhibition area: with 7000 m^2 building area, the indoor gives priority to agricultural product exhibition; while the outdoor is arranged with an open-air square and agricultural product sales booths so as to meet demands of visitors to purchase agricultural products.

Protected agricultural area: it is the only way leading to modern factory farming, environmental-safety agricultural production and non-toxic agriculture, and an efficient way for the agricultural products to break the seasonal characteristics of the traditional agriculture and realize out-of-season sales of agricultural products as well as further meet diversified consumer demands at different levels. Therefore, 3 intelligent greenhouses are specially constructed in this area to plant some high-end crops.

Creative agritainment: new technologies are applied to plant and cultivate new varieties of fruits and vegetables, and develop product production & transaction platform so as to bring new production projects and technical support for local residents.

Sightseeing & picking area: it is a new sightseeing & picking rest area, which provides a relaxing platform for busy people in cities to experience the farm life and understand agricultural knowledge. According to rich agricultural history & culture connotation, it is to provide a low-carbon, environmental-friendly and green organic agricultural product production and planting base for urban groups who are concerned with health and safety and build an ideal manor with healthy diet services.

Happy Farm: it is located in the south of the park. With popular online game as the design concept, it is divided into 4 theme farm areas (spring, summer, autumn and winter); spring, summer, autumn and winter festival squares are arranged, and 24-solar-term totem poles are arranged in the squares; activities are organized according to seasons. The plot mainly emphasizes the integration between the green space and human living; in addition, it utilizes elements such as green space, hard ground and sketches to meet production and activity demands of people at different levels.

3.1.4 *Capital source*
Guided by the government in Shanhaiguan District and self-raised by the project-undertaking enterprises.

3.1.5 *Benefit analysis*
Economic benefits: of the theme "Happy Farm" is utilized to attract visitors everywhere to appreciate the magic charm of the farming culture and local customs in the entertainment and experience, mobilize enthusiasm of the surrounding farmers and raise farmer income.

Social benefits: feature brands are built for the "Happy Farm" and ecological multi-function theme manor with integration of vacation, entertainment, touring and sightseeing in Shihe Town as well as create more jobs.

Ecological benefits: environmental protection is strengthened under the new agricultural transformation, pure green planting is adhered in every aspect, and crops are utilized to increase the surrounding greening rate to bring new elements for the ecological culture.

3.1.6 *Undertaking departments*
Guided by the government in Shanhaiguan District, coordinated by the government in Shihe Town and jointly undertaken by project-undertaking enterprises and farmers.

3.1.7 *Ecological "Happy Farm" manor planning*
On the basis of the original village, farmhouse, native integration of catering, shopping and other folk arts and crafts projects, the development of

folk culture food street The original scattered farmhouse combined into a certain scale of the industry, has more development potential, it is easy to attract more tourists to come here.

The deer Institute as the base, adding poultry Park, wild animal park and other parks, as young animal science tourism education base. The establishment of the museum to the fruit and vegetable teenagers to show high-tech agricultural achievements and the miracle of fruits and vegetables, and set the "happy farm" as the theme of the small amusement park in the museum, with fruit picking, for young people to play and learn many things in textbooks can not go to school, increase knowledge, know the hard labor.

3.2 Low carbon tourist center in Yansaihu Scenic spot

3.2.1 Project overview
The Low Carbon Tourist Center in Yansaihu Scenic Spot is located in Xinjian Village, Shihe Town, respectively backs on and toward the Shanhaiguan Forest Farm and Yansai Lake, with about 9 hectare planning area.

3.2.2 Construction objectives
To build comprehensive tourist center with integration of catering service, recreation, accommodation, meeting and business, create a recreateonal and enjoyable garden for visitors, thus achieve entertainment, vacation and leisure objectives.

3.2.3 Planning details
The Low Carbon Tourist Center in Yansaihu Scenic Spot is mainly divided into 4 functional areas below: integrated service area, waterfront cultural area, flower area and recreation area.

Management service area: wth about 10000 m² planning building area, Jiangnan architectural style is adopted for it with 3–4 floors dominated. Its integrated design of touring, shopping and entertainment, good image as well as high popularity and reputation attract visitors all around to experience the "homely and convenient" service. By reference to new technologies such as solar energy, rain collection and indoor aeration-cooling displayed in the Shanghai World Expo, it is aimed to build a real low-carbon economy service center.

Waterfront cultural area: a theme cultural square is arranged around the scenic Yansai Lake to be matched with the theme sculpture, which fully shows the magnificent Yansai Lake before visitors.

Flower area: various flowers are planted on the waterfront slope land. Viewed from a far, the sea of flowers is rippling. Walking toward the depth of the sea of flowers along the wooden trestle, people can't help feeling lost in the beauty of flowers.

Recreation area: by utilizing the existing forest, a recreation area with open and bright layout is provided for visitors adjoining the water surface. In the area, the water source is clear and stretching; between the stream and stones, the revetment is arranged in a way suitable to local conditions. In addition, stepping stones are interspersed in it, and they are crossed and wrapped with the surrounding pathways; all of them render a natural and fresh, interesting and charming sense.

3.2.4 Capital source
Guided by the government in Shihe Town and self-raised by the project investment enterprises.

3.2.5 Benefit analysis
Social benefits: with high popularity and reputation of the Yansai Lake, visitors all around are attracted; in combination with agritainments everywhere, the enthusiasm of the surrounding farmers is greatly mobilized.

Economic benefit: to build high-quality service core zone, attract visitors throughout the country and create a large number of jobs so as to drive the development of the leisure tourism industry in the whole ravine. (Song 2005).

Ecological effect: solar energy, rain collection and indoor aeration-cooling technology are utilized to introduce the design concept of low-carbon economy so as to reduce costs and promote optimization of the surrounding ecological environment and build a real low-carbon economy service center.

3.2.6 Undertaking departments
Guided by the government in Shihe Town and mainly undertaken by the project investment enterprises.

4 PUBLIC INFRASTRUCTURE CONSTRUCTION

Infrastructure construction shall highlight characteristics such as safety, ecology, economy, convenience and ornamental characteristics combine it with the objectives of the new rural construction and industry development with industry development service and low-carbon construction as its basic and core respectively.

4.1 Green road traffic system construction

Improvement of linking-up roads between the traffic arteries and nodes is accelerated and the accessibility of the scenic spot is improved so as to form rapid traffic network in the scenic spot. Infrastructures such as solar street lamps, solar insect killer and traffic signs are arranged; the greening design along the road, logo design, highway traffic signs and scenic spot signs are well completed.

4.2 Water supply and drainage system construction

I. Water shortage has been one of problems in Cherry Valley. Through pumping well construction, regional production water supply system is improved, water conservation engineering and rain collection engineering are carried out, and construction of drip irrigation facilities for arable land and orchards are strengthened. (Liu 2000).

II. Reservoir construction is carried out around the large cherry plantation; and 15 reservoirs with the standard storage capacity of 500 m³. The expected investment for this item is 7,500,000.

III. Water supply facilities shall be arranged for independent key tourist project construction sites so as to not only meet water consumption requirements of tourist management and service personnel as well as tourists, but also reserve necessary water supply for fire prevention etc. In areas with drainage facilities near towns and enterprises, the discharge systems for key projects shall be incorporated in the town drainage systems as much as possible; relatively independent, corresponding sewage treatment facilityies shall be arranged. The sewage may be discharged after it reaches corresponding standard through treatment. The planned investment for this item is 30 million.

IV. While rain drainage facility construction is strengthened, rain drainage collection means such as ditch, hardening roads, greenhouse and roof shall be fully utilized and rainwater collection system shall be established so as to effectively mitigate the drought. In villages and scenic spots, courtyard, pit-pond and sand-gravel pit may be used to extensively collect rainwater. The rainwater collection engineering shall be carried out at 20 places and that shall be carried out at 10 places by virtue of highway drainage system development. The planned investent for this item is 15 million.

V. Sewage treatment and cyclic utilization: scientific treatment and cyclic utilization shall be carried out for the waste water produced during the planting & breeding industry production and living, which not only saves precious water resources, but also reduce to emit harmful pollutants to the environment and realize the low-carbon economy objectives of low pollution and even zero pollution. (Zhang & Tan 2009).

4.3 Power & telecommunication facility construction

I. Solar streets lamps, scenery complementary lamps and landscape lamps are installed in constructed arterial roads and important scenic spot arteries. Altogether 300 sets are introduced with the planned investment for them of 3 million.

II. Corresponding substations are arranged in main tourist attractions (spots) and wire shall also be led to some relatively remote tourist attractions (spots).

III. Power grid facility construction. Pole line resource in the planning area is further integrated and two-pole system is carried out for power facilities (power supply, telecommunication and radio & television. For town centers and constructed scenic spots, underground line shall be provided so as to guarantee esthetic appearance and simplicity of the whole ravine. The planned investment for this item is 10 million.

IV. Corresponding telecommunication relays as well as iron signal transmitting towers of China Unicom, China Mobile and China Telecom are added, communication coverage ratio in the tourist area is increased and the communication quality is improved so as to guarantee smooth communication in the scenic spot. The planned investment for this item is 10 million yuan.

4.4 Service facility construction

Regional production and service facility construction is strengthened and levels of medical, sanitation, recreational, cultural and educational, scientific and technical facilities are enhanced so as to build an all-around service facility system.

4.4.1 Accommodation and catering facilities

Hotel-and-restaurant-dominated reception facilities are constructed; in line with the "Synchronization between the restaurant development and the ravine construction and visitor increment", the total quantity of restaurants in the Cherry Valley will be maintained at 10% growth rate in the future and the restaurant layout structure will be rationally adjusted. In main scenic spots, first-class restaurants and feature theme restaurants are arranged while in non-main scenic spots, economical restaurants and agritainments prevail with complete sanitary equipment and utensils as well as satisfactory sanitation conditions. The investment amount is subject to provisions on specific items.

4.4.2 *Shopping spot construction*

A batch of shopping spots for feature tourism commodities is constructed and their hardware facilities reach 3A standard. The investment estimation is subject to provisions on specific items. (Balamirzoev, M.A., Mirzoev, E.M.R. & Usmannov, R.Z. 2008).

4.4.3 *Sanitation facilities*

Public toilets are rationally arranged with satisfactory quantity and coordinated architectural style with the landscape environment. Regional wastes are recovered in classification so as to minimize the waste treatment volume, quantity of treatment equipment and treatment costs. The wastes are divided into recoverable waste, kitchen waste, harmful waste and other wastes. Environmental-friendly dustbins are arranged every 50 m in villages and scenic spots along the traffic lines. 2 waste treatment plants are constructed. The total investment for this item is 12 million.

4.5 *Safety facilities*

Some safety facilities shall be arranged in combination with the project development requirements and development degree by stages; unified management and one-by-one implementation shall be carried out for them so as to raise the regional capacities of comprehensive disaster mitigation and safety control.

I. Weather stations are established for coordination with meteorological departments. Special weather services are developed and implemented during the peak season of tourism, disasters such as typhoon, flood and rainstorm and they are timely reported; in addition, disaster prevention measures are timely taken. The investment amount is 1 million.

II. During the landscape design and ecological greening for the difficult road sections in mountainous areas, safety factors shall be specially considered and striking signs are set up at steep places. Guard rails and trestle bridges shall be arranged on outer side of the road. In rain and snow seasons, mountain road inspection and maintenance are carried out and importance is attached to the road safety in special sections (such as scenic spot access to mountains) and the scenic spot. The planned investment is 3 million. (Shafer, S., Bartlein, P. & Whitlock, C. 2005)

III. Police offices at corresponding levels are applied and arranged to be in charge of the public safety. Meanwhile, in sections with relatively poor conditions, emergency personnel and equipment are arranged. The investment amount is 1 million.

4.6 *River regulation*

Regional regulation for Dashi River is strengthened; through river regulation, beautification and greening, flood control and danger resistance capacities are enhanced and ornamental & recreational natures of the riverside are improved.

5 CONCLUSIONS

With the continuous development of social economy in China, especially with increasingly strengthened ecology status in the mountainous area which is regarded as an important geographyical unit, the development study on the mountainous area is related to the new pace of the modernization construction in China, so it is of great significance on the cause of socialism to attach importance to mountainous resource, ecology environment, land and industry development issues.

REFERENCES

Balamirzoev, M.A., Mirzoev, E.M.R. & Usmannov, R.Z. 2008. Concepts of soil-agriecological zoning of mountain regions using the example of Dagestan. *Eurasian Soil Science*, 41(6): 586–594.

Liu, L.L. 2000. Beijing mountainous area leading industry selection and layout research, *Regional research and development*, 2000(3): 61–65.

Lu, S. 1991. *China's mountainous area economic development model*, China Plans to Press.

Shafer, S., Bartlein, P. & Whitlock, C. 2005. *Understanding the Spatil Heterogenity of Global Environmental Change in Mountain Regions*. Holland: Springer Netherlands, 2005: 21–30.

Song, J.P. 2005. In the mountainous area of our country sustainable development model research, *Beijing normal university press*, 2005(6): 131–136.

Zhang, Y.F. & Tan, J. 2009. *Beijing Valley region economic development theory and practice*, Beijing: China meteorological press, 2009: 3–13.

Advances in Civil, Architectural, Structural and Constructional Engineering – Kim, Jung & Seo (Eds)
© 2016 Taylor & Francis Group, London, ISBN 978-1-138-02849-4

Implementation of European parliament and council directive 2012/27/EU into the Czech environment

E. Wernerova Berankova, F. Kuda & S. Endel
VŠB—TUO FCE, Ostrava, Czech Republic

ABSTRACT: This paper deals with problems related to the calculation of costs to heat housing units and areas where heating costs are allocated annually. This paper contains theoretical alternatives and attention is paid to newly established duty to install ratio meters for heating costs by which means it is possible to meet the principles of methodology and support the access of European Community to these problems. It describes the methods and instruments approved for application of methods and tools prescribed by European and Czech legislations.

1 INTRODUCTION

The immovable property owners in the Czech Republic have another worry. Before the beginning of a new calendar year, they had to cope up with compulsory installation of heat meters to register consumed heat on each heating element located in objects where the heat must be budgeted for the final consumer. These ratio-type heat meters serve for a more appropriate distribution of total costs for heating a house.

So it is not a matter of family house, weekend house or secondary residence owners, but this duty rises everywhere so that the object is used by more than one tenant. This duty was implemented in legal regulations with effect from January 1st, 2015. This paper focuses on the association of housing unit owners and on housing co-operatives (for more information, refer to Somorova 2014, Kuda & Wernerova 2014, Kuda & Berankova 2014, Trip & Burca & Leuca & Dudrik 2015).

2 METHODOLOGY

It is necessary to consider heat energy take-off, its consumption and subsequent allocation to final consumers so that three fundamental principles are maintained. These principles are fairness, professional background and verifiability. If any of the principles was infringed, it could cause displeasure to final consumers as well as supervisor organizations dealing with budgeting of heat energy costs. Consumers are concerned with equitable calculation of consumed heat costs and supervisor organizations are interested in the selection of the most suitable method for the determination of these costs, whereas both parties must have an oppor-

tunity for checking both input data for calculation of heat consumption costs and for checking the resultant values. Application of the three principles represents three connected circles forming together an inseparable part.

2.1 *Principle of justice*

This principle is considered as a general rule for application of law. In the area of heat energy management and primarily in its consumption and subsequent billing, the term "distributive justice" is used, i.e. the justice requiring that the distribution is applied to all parties according to the same measure. It does not mean that the heat consumption is equally divided but according to the same methodology and according to the same methods used that are fixed to the final consumer by the particular heat capacity consumed.

2.2 *Principle of professional background*

This principle respects technical conditions of methods used by which means we should meet the other principles of justice, lucidity and verifiability. This principle says that selected methods of application can be only the methods that are applicable and known by the expert public and recognized as acceptable.

2.3 *Principle of verifiability*

It holds true that inputs and outputs, featuring in the method according to which the costs are allocated, must be verifiable, justifiable and transparent. Each data occurring at the input or output must be retraceable or the data must be measurable and recorded. The use of the data that is inconsistent with the principle of verifiability and the values of which are not supported by any measurement or calculation in the methods are excluded.

Each determination of values by means of a partial calculation must be substantiated and the method of its calculation clarified.

3 HEAT ENERGY MEASUREMENT METHODS

It has been reported that two heat energy measurement methods apply in conditions of the Czech Republic most frequently (Skuhra 2011):

- Calorimetric method—relatively exact method but its disadvantage is the necessity to install two thermometers, one flow meter and one microprocessor on each radiator.
- Thermal measurement method of heat flow from radiator—less exact method.
- Heating Degree Days (HDD) method—relatively exact method based on the principle of measurement of temperature difference between the temperature maintained in the room (flat) and the external reference temperature.

The most frequent method on how heat energy is brought into the flat is by means of ascending pipes bringing heat from more than one point and from there to the heaters. This distorts the consumption calculation measurement and makes it difficult; so in this case, heat cost allocators for the determination of the consumption of room heating radiators (specified in more detail in technical standards ČSN EN 834 a 835) are installed directly on each radiator or heat indicators are installed on the draw-off pipe from the radiator or a meter with interior temperature sensor of a given heated room and with the sensor of outside temperature to the respective building with permanent continuous registration of temperature difference versus time interval, which is actually the application of the above-mentioned HDD method.

4 CURRENT LEGISLATIVE REQUIREMENTS

4.1 *Energy management*

In terms of legislation, these problems are covered by modification of the Act No. 318/2012 Coll. that modifies the Act No. 406/2000 Coll., on energy management. In §7, par. 4 of the modification states that:

"The developer, building owner or association of housing unit owners are bound to:

a. equip internal heat installations in buildings with devices for regulation and registration of

heat energy supply to end users to the extent specified by the implementing regulation; the end user is bound to facilitate the installation, maintenance and control of these devices."

For all respective subjects, this means that they are bound to install on each radiator not only thermoregulator heads but also ratio-type heat meters, which should ensure more appropriate allocation of heat consumption but they should also contribute to the reduction of heat consumption and lead to more economical heating. The end user is bound to facilitate installation, maintenance and control of the devices.

In most cases, the thermoregulator heads were installed on radiators before so it was necessary to complete the radiators on which ratio-type heat meters were not installed.

4.2 *Allocation of costs*

The problems related to allocation of costs are currently dealt with the Ministry of Regional Development Regulation No. 372/2001 Coll. by which rules for allocation of costs for heat energy and provision of domestic hot water among end users are fixed. It is fixed in this regulation, among others, that the costs of heat energy in the accounting unit per accounting period will be divided by the owner into basic and consumption components. The basic component is 40% to 50% and the rest of the costs are made up of the consumption component. The basic component is fixed according to the ratio of chargeable dwelling area or non-residential area to the total chargeable dwelling area and non-residential areas in the house. The consumption component is fixed according to values measured by the meters installed by attaching to radiators with the use of corrections and calculation methods that should reflect different intensities of heating in individual rooms given by their positions.

It implies from the physical essence of heat and its propagation that it will be more appropriate to set a higher portion of basic component in houses with thermal insulation where thermal losses are lower and factors such as sunlight and heat exchange with adjacent flats take effect. In houses with higher thermal losses, it will be better to fix a larger portion of consumption components. The association of owners and housing co-operatives should reflect the flats situated over basements or under roofs. These flats have the highest thermal losses.

4.3 *Heat cost allocators*

On the current market, there are many companies offering various types of meter. Meters of the same type from a different supplier are basically identical devices, differing in appearance and price. It is recommended to install measuring devices of

Figure 1. Example of heat cost allocator with radio data transmission.

Figure 2. Example of heat cost allocator without radio data transmission.

companies that will read the meters and perform allocation due to compatibility of meters and software for heat cost allocation.

Evaporative: owing to the fact that it is an obsolete and inaccurate measurement method, these meters are not currently demanded on the market despite lower purchase costs and a majority of companies carrying out installation and readout recommend electronic measurement.

Electronic:

– with radio data transmission (Figure 1),
– without radio data transmission (Figure 2).

5 CONCLUSIONS

In the Czech Republic, from January 1st, 2015, a fundamental and great change takes place in determination of heat consumption at SVJ, cooperative flats or other objects where heat costs must be allocated. The modification of Act No. 318/2012 Coll. imposes the duty to equip radiators with thermoregulating valves and heat cost allocators. The inducement for this fundamental step was undoubtedly due to the pursuit of introduction of fair insight into this sensitive subject matter. The goal of installation of ratio-type meters is to provide end heat consumers with more appropriate cost allocation. Currently, there are two types of consumers: those who save heat and try to minimize their consumption of heat energy and those who do not save heat energy or those who heat their neighbor's house helplessly. Up to now, the allocation of heat consumption has been carried out according to chargeable floor space where no heat cost allocators were installed on radiators. These mentioned cases are two extremes. The installation of ratio-type meters tries to find a way by which both consumers would pay only for what they have actually consumed. The newly suggested method is the subject of criticism. Opinions can be heard that the allocation according to ratio-type meters cannot be considered as a fair method so we can expect that the current state in setting the rules will be changed.

The other effect accompanying the ratio-type meters is the self-control of the occupant of the flat. The consumers, under the expectation of high additional payment at the end of the accounting period, think about heat consumption and begin to behave more economically and cut down heat energy waste through open windows.

ACKNOWLEDGMENT

This work was supported by funds for the conceptual development of science, research, and innovations for 2015 allocated to VŠB-TUO by the Ministry of Education, Youth and Sports of the Czech Republic.

REFERENCES

Kuda, F. & Berankova, E. 2014. Extending the life cycle of buildings using project and facility managements. *Applied Mechanics and Materials*, 584–586: 2291–2296.

Kuda, F. & Wernerova Berankova, E. 2014. EU approaches to unification of methodologies for determination of building object life cycle costing. *Advanced Materials Research.* 1044–1045: 1863–1867.

Skuhra, J. 2011. *Services provided in the management and operation of buildings*. Praha: Linde.

Somorova, V. 2014. Optimization of the Operation of Green Buildings applying the Facility Management. *Journal of civil engineering.* 9(1): 87–94.

Trip, N.D., Burca, A., Leuca, T. & Dudrik, J. 2015. Considerations on the analysis of an induction heating system. 2014 11th International Symposium on Electronics and Telecommunications, *ISETC 2014—Conference Proceedings.*

Advances in Civil, Architectural, Structural and Constructional Engineering – Kim, Jung & Seo (Eds)
© 2016 Taylor & Francis Group, London, ISBN 978-1-138-02849-4

Barrier-free use of structures in transport infrastructure within Czech standards environment

R. Zdařilová
VŠB-TU, Ostrava, Czech Republic

ABSTRACT: The fundamental prerequisite for active engagement of persons in community life is the accessibility of concourses and buildings, their utilization and the possibility of moving about freely. The barriers in traffic infrastructure are one of the most serious obstacles preventing persons with mobility limitations from moving about. The upcoming Czech technical standard Barrier-free use of transport structures—deals with problems of basic provision for usage of structures of transport infrastructure.

1 INTRODUCTION

The new upcoming Czech technical standard ČSN 73 6101—Barrier-free use of transport structures—deals with problems of basic provision for usage of structures of transport infrastructure stemming from the enactment of Regulation No. 398/2009 Coll. (2009) and from the existing Czech technical standards defining the conditions in the area of transport structures. This standard covers all all requirements for using transport structures by people with limited movement and orientation abilities that are defined in individual articles of related standards and put more precisely according to conditions of the existing legal environment and supplemented with illustrating figures.

This standard is designated for designers, local authorities, building offices and other organizations and it shall be in operation for preparation of project documents, for permitting procedure or reporting and performing structures, for issuing occupancy permits, for usage and during inspections of transport infrastructure structures in relation to securing the conditions of usage by persons with limited ability or orientation.

2 COMPARISON WITH INTERNATIONAL STANDARDS

The standard Barrier-free use of transport structures stems namely from the requirements of Regulation No. 398/2009 Coll. which defines also the fundamental requirements for barrier-free use of transport infrastructure structures. On this account, primarily the tactile adaptations for people with visual impairment are solved in a different way compared to international standards, e.g. DIN 18040-3:2014 Accessible buildings—Design principles—Part 3: Public transport

and free space (2014) or BS 8300: Design of buildings and their approaches to meet the needs of disabled people Code of Practice (2001).

The dissimilarity of solutions in the structures of transport infrastructure in the Czech environment namely consists in different requirements for independent movement and orientation in space for people with visual impairment, which stems from the principles of methods for training these people in the Czech Republic as indicated in more detail in Chapter 3 and which is illustrated in Figure 1.

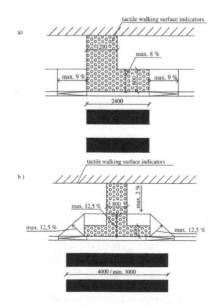

Figure 1. Example of comparison of different requirements for independent movement of blind people in terms of pedestrian crossing—solution a; example of standard solution according to BS 8300—solution b, the solution stemming from the Czech environment and ČSN 73 6101.

3 ACQUISITION PROCESS AND PROCESSING OF INFORMATION FROM BLIND PEOPLE IN TRANSPORT

Orientation of specific users, the people with visual limitations in the environment of transport structures is conditioned on acquisition process and processing of the information from the environment and for the purpose of planning and implementation of moving about the public space. Every movement is carried out by the blind purposefully and consciously with a great deal of concentration. The blind person is mobile when s/he is able to move about the space safely and surely when s/he makes use of movement techniques s/he has learned and information s/he has acquired. Mobility is a fundamental prerequisite of independent life of a visually impaired person and an important factor during his/her socialization and peace of mind. The excessive dependence of the visually impaired person on people who can see, resulting from failing to manage problems in the sphere of spatial orientation, complicates significantly and precludes essentially his/her independent life (2006).

The public space for the blind consists of the most diverse rememberable points, signs and lines. The movement of the visually impaired person through the public area takes place in a permanent contact with guiding lines and in keeping a constant gap. The contact with the guiding line is checked every 3–5 steps by extension of the stick swing towards the appropriate side to the guiding line. When crossing a road, the visually impaired person must find an important orientation point in the form of tactile paving treatment—a tactile ground surface indicator. Then the person changes the direction while walking and moves along this ground surface indicator at a width of 800 mm from one side or another, depending on the side the person is coming from. Here, it is very important to put the ground surface indicator on the centre line of the crossing so that safe area along both sides of the ground surface indicator can be ensured and the visually impaired person can enter the crossing area within the road and move along its centre line—see Figure 1. In the example of the solution according to BS 8300 conditions (2001), the blind person gets to the edge of the crossing when the guidance tactile paving is applied, or while crossing, the person is guided beyond the pedestrian crossing itself.

Tactile elements help the visually impaired people with independent movement and they must be designed and implemented according to a simple, logical and consistent plan stemming from the principles of the information acquisition.

4 PRESENT STATE OF REQUIREMENTS FOR BARRIER-FREE USE OF STRUCTURES IN TRANSPORT INFRASTRUCTURE

The requirements for barrier-free use of transport structures in the Czech environment are included namely in Regulation No. 398/2009 Coll. (2009). This regulation defines namely conditions for communication over land and public areas, both in the form of articles and namely in appendices 1 a 2 where general and technical requirements of the structures indicated are defined.

Within the scope of Czech technical standards, the problems of barrier-free use are split into individual standards that are not always in compliance with the Regulation No. 398/2009 Coll. and across these standards. These are the following standards:

ČSN 73 4959:2009	Platforms and roofs of platforms of state, regional and industrial railways
ČSN 73 6056:2011	Parking areas for road vehicles
ČSN 73 6058:2011	Small, multi-storey and mass garages
ČSN 73 6101:2004	Design of highways and motorways
ČSN 73 6110:2006	Design of urban roads, Change Z1 (2010)
ČSN 73 6201:2008	Standard specifications for bridges
ČSN 73 6380:2004	Railway level crossings and pedestrian crossings
ČSN 73 6425-1:2007	Bus, trolleybus and tramway lines halt—Part 1: Design of halts
ČSN 73 6425-2:2009	Bus, trolleybus and tramway lines halts—Part 2: Transfer junctions and stations

The main task of the newly prepared ČSN 73 6101 Barrier-free use of transport structures was namely to unify the requirements of the above-mentioned regulations based on user point of view stemming from fundamental principles of independent movement and orientation indicated in Chapter 3.

5 PROPOSAL OF REQUIREMENTS IN ČSN 73 6101 BARRIER-FREE USE OF TRANSPORT STRUCTURES

The standard applies to designing the barrier-free use of new structures of transport infrastructure, to changes in finished structures and to changes in structures before their completion. The standard applies to the structures of communication over land and public areas. The public area is namely referred to pavements, pavements in orchards and parks, squares, market places and similar pedestrian areas. The standard also applies to designing structures for railway and at the railway, cableway lines, and structures for air traffic and water transportation. The standard does not apply to tunnel building.

The requirements of this regulation are divided into two fundamental parts. The first part deals with fundamental elements of barrier-free solutions expressing elementary principles and system policies in relation to using the structures of transport infrastructures by the people with limited mobility or orientation. It is connected with problems of tactile, sound and visual elements of barrier-free treatments, representing the requirements of tactile elements, tactile surfaces, sound and visual elements applicable in transport structures.

Owing to differences in solutions for individual limitations, the requirements for individual types of disabilities are specified in this Czech standard, such as:

- solutions for people with limited mobility
- solutions for people with limited orientation abilities—people with visual impairment
- solutions for people with limited orientation abilities—people with hearing impairment
- The second part of the standard applies thoroughly the requirement for determination of basic technical parameters for individual types of limitation classified according to individual types of structures within transport infrastructure:
- Communications over land and public areas with requirements for pedestrians, foot-paths and cyclist traffic, pedestrian crossings, places for crossing, corridors for crossing car-track lane, public traffic service, parking areas and parts and equipment of local roads
- Railway structures and at the railway station with access to departure lounges, paved areas, access to platforms with low level and fly-over access, railway platforms
- Cable-railways with general principles for solution
- Air traffic structures with general principles for solution, access to departure lounges and their solution
- Water transportation structures with general principles for solution

The integral part of the application section was to define the requirements for orientation and information systems for the public namely in relation to terminals of transport structures, gangways, underpasses, platforms at the railway as well as signalization for operation of escalators, travelators and lifts.

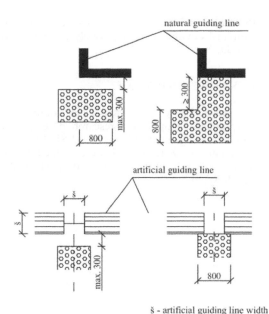

Figure 2. Example of solution how to connect a tactile element of the ground surface indicator to natural and artificial guiding lines.

The main objective of this regulation is the emphasis on the application of individual provisions with emphasis on graphic part with accompanying pictorial material—for example, see Figure 2. Another objective set by the standard is to define optimum solutions usable in technical practice, stemming from fundamental values and define simultaneously potential reductions in solutions in clearly defined crowded conditions with an accompanying illustration.

6 CONCLUSIONS

In the Czech Republic, blind people are trained in independent movement for a long time. Based on acquisition and processing of information from surroundings, it is necessary to accept these training methods having been applied for decades and these are well-developed within the framework of the European environment. The study of user requirements consequently brings about the conditions for a suitable tactile treatment of the exterior environment of structures within transport infrastructure allowing safe and independent movement of people with visual impairment. In this area, regulations and foreign examples cannot be adopted since the training practice in the CR is quite different, and on this account, it is necessary to solve legal and standard environments separately from foreign standards.

ACKNOWLEDGEMENT

The works were supported from sources for conceptual development of research, development and innovations for 2015 at the VŠB-Technical University of Ostrava which were granted by the Ministry of Education, Youths and Sports of the Czech Republic.

REFERENCES

BS 8300: 2001. *Design of buildings and their approaches to meet the needs of disabled people Code of Practice*, British Standards Institute.

DIN 18040-3: 2014. *Accessible buildings—Design principles—Part 3*: Public transport and free space.

Regulation No. 398/2009. *On general and technical requirements securing barrier-free use of structures*. Prague: Collection of Laws of the Czech Republic.

Wiener, P. 2006. *Spatial orientation of people with visual impairment*. Prague: Institute for Rehabilitation of the Visually Impaired Charles University Faculty of Humanities.

Advances in Civil, Architectural, Structural and Constructional Engineering – Kim, Jung & Seo (Eds)
© 2016 Taylor & Francis Group, London, ISBN 978-1-138-02849-4

Barrier-free use of transport interchanges

A. Bílková, B. Niemiec, D. Orsáková & R. Zdařilová
VŠB-TUO, Ostrava, Czech Republic

ABSTRACT: This article deals with the issues of barrier-free use of transport interchanges. It focuses on four basic criteria of a barrier-free environment, namely safety, accessibility, availability, and comprehensibility. These are presented using two examples of transport interchanges in the Moravian-Silesian Region (Ostrava—Main Station and Havířov—Bus Station). The objective of this research is to raise awareness of ample problems of barrier-free environment.

1 INTRODUCTION

Public transport is an integral part of the public areas of villages, towns and cities. The most essential and most important part of it is then transport interchanges, places where all types of transport and countless people (passengers) meet, moving constantly in the rhythm and intensity determined by the organization and connections of transport, as described by Lánská (2011). Due to the intensive use of these areas, it is necessary and important to assess these places not only from the viewpoint of transport, logistics, technology or economy, but also from the viewpoint of a barrier-free environment and the movement of a person at all stages of his or her life (child, adult, elderly person, person with limited or impaired mobility or orientation ability, and person with luggage—passenger). The areas and architecture of transport interchanges should therefore create a suitable and high-quality environment providing all of their users with all comfort and possibilities of movement (fast and safe transfer, easy to orient themselves, rest, services, etc.) required from public areas of towns and cities, as described by Jacura (2012). Due to the extensiveness of the problems, this paper will only deal with the criteria of barrier-free movement of people with mobility disabilities and visual impairments in specific areas of interchange junctions presented in two examples: Interchange junction Ostrava-Main Station and Havířov-Bus Station.

1.1 Transport interchanges

These are primarily points of contact of urban mass transport or bus and rail transport lines of regional importance. The connection of transport interchanges for urban mass transport ensures interconnection between functional components of a town or city such as housing, workplaces, and the other public utilities. Transport interchanges become an essential component of a public area, with very high demands on movement and orientation in an area due to a large number of different users. This fact implies the need to maximize the connection and accessibility of all communications (pedestrians, buses, trains, etc.) and boarding, exit and other areas to ensure that they meet all the criteria, as described by Ožanová (2008).

1.2 Criteria of a barrier-free transport interchange environment

Monitoring and using of four basic criteria defining the specific areas of transport interchange junctions in light of barrier-free environment became the objective of the research. These areas should guarantee all the people the following:

Safety: The smooth and intuitive movement of pedestrians is ensured, areas for crossing (pedestrian crossings and places for crossing) are created, and barriers against entry and quality of all surfaces (colour and tactile contrasts, anti-slip and functional surfaces, etc.) are ensured.

Accessibility: All areas are designed to allow for the movement of persons with limited or impaired mobility or orientation ability (person in a wheelchair, elderly person, passenger with luggage, etc.) and provided with the necessary elements (handrails, ramps, platforms, lifts, and tactile, acoustic and visual elements).

Availability: The main routes for the movements of pedestrians are arranged in such a way that distances to be covered are as short as possible.

Comprehensibility: All areas are equipped with comprehensible orientation signs (pictograms, directional signs, signs, etc.) and information systems (LED or LCD displays) are conveniently located from the acoustic and visual viewpoints.

2 RELATED LEGAL REGULATIONS

To evaluate transport interchanges, we must take account of the applicable legal regulations and Czech technical standards, especially:

Act No. 13/1997 Coll., on roads, as amended.

Act No. 183/2006 Coll., on town and country planning and the Building Code, as amended.

Regulation No. 398/2009 Coll., on general technical requirements ensuring barrier-free use of buildings, as amended, and also the Bezbariérové užívání staveb (Barrier-Free Use of Buildings) methodology (Renata Zdařilová, 2011) developed for this regulation.

ČSN 73 6425-2 Bus, trolleybus and tramway lines halts. Part 2: Transport interchanges and stations.

ČSN 73 6110 Design of urban roads.

ČSN 73 4130 Stairways and sliding ramps.

3 OSTRAVA—MAIN STATION TRANSPORT INTERCHANGE

The Main Station in Ostrava can be categorized as a transport interchange of a regional nature because it enables passengers not only to use urban mass transport, but also to go by rail. This makes it an important place within the public areas in Ostrava, with a high level of movement and concentration of people. In 2014, all outside areas related to urban mass transport undergone extensive reconstruction and adaptations to make them barrier-free. All boarding and exit platform edges are equipped with Kassel kerbs ensuring the optimal platform height and facilitating and accelerating the exit and boarding of per sons with impaired mobility as well as other users. All platform and transport interchange areas are designed in large format red tiles, completed with guide elements in the form of black guide lines and rubber guide strips with a width of approx. 0.8 m and with sufficient color and tactile contrast Figure 1. The ubber strips are used prima-rily to guide passengers to the main directions and communications in the transport interchange area (station hall and trolleybus and tramway stop). However, when they are used outside, shape deformations occur (the surfaces become corrugated) and prevent the smooth and safe movement of all passengers and become an obstacle.

The main directions of pedestrian traffic are wide and clear enough. The movement of passengers outside is preferentially designed on one level. In the areas of road crossings, the pavements are lowered to the road level to ensure that crossing does not cause problems for persons with limited mobility and orientation ability. The crossings and places for crossing are also equipped with features ensuring that these people cross safely (guide lines, warning strip, signalling strip, and crossing guide strip). Entering the station hall is ensured using the staircase and three ramps equipped with railing, directly connected with the main entrances and communication routes Figure 2, and the wide inclined surface between them, eliminating the height of the five staircase steps (0.8 m). For people with vision impairments, the entry to the station building is provided with an orientation voice beacon and the colour design of the siding round the entrance provides the visually handicapped with identification. Information boards are not provided with a voice informer and the blind are dependent on the information or assistance of passers-by. Acoustic orientation beacons generating appropriate warbles warning the blind of the given type of obstacle are installed in vicinity of staircases, lifts, and escalators. The outside area of the platforms and main routes is covered using a light 4.5 m high structure on columns. The presence of the columns is very problematic due to their subtle dimensions (easy to be overlooked) and irregular arrangement; therefore, the main routes are in unobstructed areas. A significant imperfection is the lack of contrast marking to draw attention to these obstacles. Some movement difficulties can also be caused by

Figure 1.   Material and color solution of surfaces.

Figure 2.   Main lines.

Figure 3. Columns intervening in the trolleybus stop area

the architecturally designed columns supporting the trolleybus stop roofing and limiting substantially the possibilities of the smooth movement of all passengers Figure 3 because the columns are dense. The stop platforms are designed in accordance with the applicable regulations, equipped with contrast strips and tactile gray signal strips (and also a black rubber guide strip near the tramway stop), stop signs and groups of benches.

The circular area among the main routes of movement, equipped with a sculpture, a water attraction (running water) and benches around the perimeter helps to make the area more pleasant and gives passengers the possibility of resting while waiting. A considerable imperfection of the outside area of this transport interchange is the lack of information and orientation systems for the public, including meeting the requirements for persons with limited mobility or orientation ability. In these areas passengers therefore have to rely on printed timetable information only, relating to urban mass transport only. Also moving inside the station building is problematic. The space of the building is divided into three height levels, accessible from staircases with a steep inclination. The communication with accesses to individual platforms (using long and steep staircases), which are not barrier-free is situated on the highest level. The independent movement of persons with limited mobility is therefore impossible there.

4 HAVÍŘOV—BUS STATION TRANSPORT INTERCHANGE

The Havířov Bus Station is an important arterial road of the Moravian-Silesian Region. Havířov, the second largest town in the region, interconnects Ostrava and the Těšín region. This arterial road has connections to Karviná, Orlová, and Frýdek-Místek. The bus station also connects regional bus service with local urban mass transport. The bus station is situated in the Podlesí town part, at the end of Havířov. The railway station is at the opposite end. The location of the bus station in relation to the Havířov Railway Station is not chosen properly and this was also a reason for much consideration given to changing its location. At last the management of at the existing location and to reconstruct it. The reconstruction took place in 2014 and included alterations to all outside places and the construction of the new station building. During the reconstruction, the emphasis was placed on alterations to make it barrier-free. Compared to the original building, the station has no waiting room inside any longer. That waiting room was replaced with a covered outside area with sixteen seats without backrests. The dominant feature of the entire terminal is the station building Figure 4. There is a wheelchair accessible toilet cubicle and a selling point of the carrier. The only area accessible to the public is the toilet area. The outside covered area, well accessible by two ramps with an adequate length and slope and equipped with railing is used as the waiting room.

All communication surfaces are covered with gray concrete 200 × 100 mm tiles; the tactile design of the signal and warning strips consists of red concrete tiles with the significantly shaped surface and sufficient colour and tactile contrast. Artificial guide lines Figure 5 are there to solve the organization of the movement of persons with visual impairments in the entire area of the

Figure 4. Bus terminal.

Figure 5. Artificial guide lines.

Figure 6. Columns with sufficient contrast marking.

transport terminal. The artificial guide lines consist of grey 200 × 200 mm SLP tiles. Due to this, they visually blend in with their surroundings and do not provide the required optical contrast.

The platform edges are made using Kassel kerbs ensuring the optimal platform height. The stop platforms are designed in accordance with the applicable regulations. The communication area as a whole is sufficiently wide and clear. The entire terminal is designed on one level. In the areas of road crossings, the pavements are lowered to the road level. The places for crossing are equipped with crossing guide strips for ensuring that persons with impaired orientation ability cross safely. The platform outside area is covered using a light steel column structure, roofed with corrugated sheets. This structure is problematic somewhat because it is too high and has a flat arrangement. In rainy and windy weather, it does not protect passengers from the weather. The columns are equipped with reflective targets Figure 6 to draw attention to this obstacle and thus they are much better identifiable for dim-sighted persons.

5 CONCLUSIONS

This paper is of an informative nature resulting primarily from the scope and comprehensiveness of this issue. Owing to the fact that this text will be continued with other papers (dealing with individual handicaps), it is focused primarily on the problems of people with mobility disabilities and visual impairments and on defining the criteria of barrier-free movement of these people. Monitoring and using of four basic criteria (safety, accessibility, availability, comprehensibility), defining specific areas of transport interchange junction in selected locations, became then the objective of the research. When designing or reconstructing transport interchanges, we must make sure that using them is intuitive, friendly, uncomplicated, and convenient for both regular and irregular

passengers with or without limited mobility or orientation ability. The need for the safe and smooth movement of passengers and transport is essential and this further implies the necessary space requirements for platform and communication areas that are specified in more detail and often specifically laid down by the relevant legal regulations. Considering the intensive use of these areas, it is necessary and important to assess these places thoroughly from the viewpoint of a barrier-free environment. It is apparent from the examples that it is not always easy to bring all operational, technical and architectural and aesthetic requirements resulting from the function of these public areas into accord. Thus, mainly the design of the roofing of the boarding and exit areas (insufficient protection against the weather, improper placement of columns, etc.), the use of surface colors (contrast), unsuitable materials, incorrect positioning of acoustic elements with comprehensible orientation signs and information systems, which are often reduced to basic pictograms, directional signs, and signs only, become problematic. From the viewpoint of the movement of persons with physical limitations, the design of both the reconstructed transport interchanges and newly created transport interchanges attempts to comply as much as possible with all requirements in accordance with the applicable regulations. These areas thus have all the necessary features for the movement of these people (guide lines, lowered or raised kerbs, crossing guide strips, warning and signal strips, contrast areas, etc.). Transport interchange areas are an integral and intensively used part of a public area and must be seen comprehensively and coherently; otherwise, they will not be able to provide their users (passengers) with sufficient comfort required for their proper and pleasant use.

ACKNOWLEDGEMENT

The work was supported by the Student Grant Competition VŠB-TUO. Project registration number is SP2015 / 187.

REFERENCES

Jacura, M. 2012. *Optimal Layout of Public Mass Transport Transfers Nodes*: 2–5. Prague: ČVUT in Prague, Fakulty of Transportation Science.

Lánská, M. & Čepa, M. 2011. A Change Junctions Innovation from the Point of View of Passenger Moving—Introduction. *Perner´s contacts,* (4): 205–231.

Ožanová, E. 2008. Integrated transport systems for mass pas-senger transport. *Proceedings of scientific papers Transactions of the VŠB—TU of Ostrava,* (1): 295–301.

Advances in Civil, Architectural, Structural and Constructional Engineering – Kim, Jung & Seo (Eds)
© 2016 Taylor & Francis Group, London, ISBN 978-1-138-02849-4

Qualitative parameters of a public place and their analysis using a case study of the complex of the former bituminous coal mine of Karolina in Ostrava

A. Bílková, B. Niemiec, D. Orsáková & R. Zdařilová
VŠB-TUO, Ostrava, Czech Republic

ABSTRACT: The paper deals with the monitoring of qualitative indicators and their direct influences on the use of public spaces. The monitored indicators are safety, accessibility, functionality, comprehensibility (easy orientation), comfort, and pleasure in staying. The purpose of the pa-per is to achieve a comprehensive view of the issues and to define basic qualitative indicators to ensure the functionality and usability of a public space, pointing out deficiencies found.

1 INTRODUCTION

The city of Ostrava is known for its industrial tradition. The end of mining is associated with the creation of industrial brownfields, which poses a potential threat to all cities with industrial history. Karolina and the Lower Area of Vítkovice belong among the most famous brownfields in Ostrava with great potential for development.

1.1 *Multipurpose centre of Nová Karolina*

Within the reclamation of the former bituminous coal mine area, the new multipurpose centre of Nová Karolina was built in a building area of 32 ha. It consists of retail and office premises, residential buildings, and large parking areas. All of this is supplemented with a leisure and sports facility and greenery. The reconstruction of the only two surviving buildings, called Trojhalí (Three Halls) today, has created new spaces for cultural, social, sporting and educational events accessible to the public. The buildings are the main point between the centre of Ostrava and the Lower Area of Vítkovice, evidencing the industrial history of this area.

2 QUALITATIVE PARAMETERS OF PUBLIC SPACES

In the Czech legal environment, a public place is defined by Act No. 128/2000 Coll. (2000), on municipalities (municipal establishment), as amended. Ac-cording to Sec. 34, public spaces shall mean all squares, streets, marketplaces, pavements, public green areas, parks and other places publicly accessible without restriction, i.e. places serving for public use regardless of the ownership of such places. Basically, a public place is a freely accessible environment where people of different sex, race, age group, social class, and religion, including people with physical limitations, can meet and spend their time. The issues of public places is currently more and more frequently in the spotlight all over the world due to the fact that their number and quality are decreasing. And the quality of public spaces is what is unconditionally connected with the stimulation of human activities, as described by Gehl (2010).

The main qualitative parameters of public spaces are safety, accessibility, functionality, comprehensibility (easy to orientate), comfort, and pleasure in staying there.

2.1 *Safety*

A feeling of safety is one of human's basic psychological needs. It gives him a feeling of certainty and protection against any accident or crime. The greatest potential hazard is traffic and crime. From the viewpoint of traffic, high-quality public places must, therefore, provide safety especially when crossing roads, as described by Gehl (2010).

According to ČSN P CEN/TR 14383-2 (2009), the security of public places can be increased, in relaxation to the crime rate, by improving their regime security without physical changes (patrols, shop-owners' cooperation with the police, improved maintenance, and the like), by improving the security system (private and public lighting, locks, alarm systems, and the like), by new interior equipment, and by improving details (the maintenance of trees, street furniture, public and private street lighting, and schedule of activities), or by the overall new design of the place (connected walkways, the location and schedule of activities, the shape and use of the area, marking to distinguish public areas from private are-as, and decelerating traffic flow, for example).

Also protection against unwanted sensory perceptions such as glare, noise, dust, air pollution, or odour is connected with a feeling of safety, as de-scribed by Gehl (2010).

2.2 *Accesibility*

The accessibility of the area is determined by two basic aspects—transport accessibility, including parking, and the possibility of being used by people with reduced mobility and ability of orientation. Accessibility to persons with reduced mobility (people in a wheelchair, the elderly, people walking on crutches, children, people with a pram, and the like) determines the design of so-called barrier-free routes, which must comply especially with the requirements concerning the maximum slope of communications and their height differences. The system of natural and artificial guide lines along with color and tactile contrasts is then crucial to people with impaired ability of orientation (blind and visually impaired).

2.3 *Comprehensibility*

The comprehensibility of an environment is connected with the ability of all users to orientate easily. Attention must be primarily turned to adaptation in order to get a comprehensible environment for the elderly, visually and hearing impaired people, and mentally disabled people. The orientation and information elements must always be clear and the directions and destinations of individual routes must always be legible.

2.4 *Comfort and pleasure*

Comfortable use of public places is determined by the relation between static and dynamic activities that take place there. Generally, the dynamic component is movement and the static component is staying. Comfort when moving is connected not only with the width and surface finish of individual roads and walkways, but also with to the possibility itself of carrying out any active recreational activities such as a game or sport. Static activities are ensured by having a passive rest in the form of sitting or standing while talking right on the street.

2.5 *Functionality*

A functional public place is characterized by being used as much as possible. The multipurpose nature

is one of the most important characteristics. Individual activities that take place in a public place, whether static or dynamic, must be connected with each other in such a way that they correspond and do not limit each other.

3 ASSESSMENT OF THE PUBLIC PLACE OF NOVÁ KAROLINA

The public place in the premises of Nová Karolina is very specific in relation to the various functional use of the buildings which shows Figure 1. The requirements concerning the adjacent outdoor areas are dependent on it. The following sections of this paper determine how criteria are fulfilled according to the above quantitative parameters using the selected rating scale.

3.1 *Safety*

In terms of traffic safety, the area concerned is solved very well. Road II/479 in the 28. Října (Street of 28 October) is one of the main arteries of Ostrava communications. This is a two-lane road with two-way traffic, with a tram line. All crossings leading from the urban mass transport stops in this street are equipped with light and acoustic signalling. To reduce the speed of cars, access to the under-ground parking is equipped with speed bumps and signing and marking regulating orientation and speed.

Another important element of safety is the pedestrian bridge starting at the bus station and

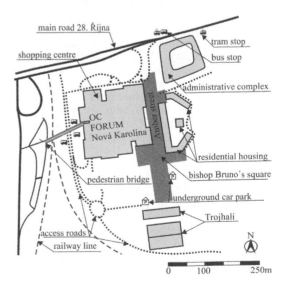

Figure 1. Scheme of the premises of Nová Karolina.

leading along the Místecká (Místecká Street) and its heavily frequented road. The pedestrian bridge runs to the last floor of the FORUM Nová Karolina building and enables pedestrians to enter safely the shopping centre. Amber street forming a central area among buildings, see Figure 2. It is solved as a pedestrian zone with exclusion of motor traffic, except supplying vehicles, the approach of them is time-limited. Which In light of securing the pedestrians, this solution is a great benefit. The movement of bikers is not defined here, however, the street width enables their mixed movement.

The public place of Nová Karolina is very good when assessed from the viewpoint of a feeling of certainty and protection against danger or criminality. The free open space between the buildings has convenient furniture, public lighting, ground shaping, and greenery. This space gives an impression that it is rather intended for relaxing and resting. It has no dark narrow streets or confined space that would cause the feeling of danger. The premises of Nová Karolina have commercial buildings that contain modern alarm systems directly connected to the city police. The city police increase the security of this public place by regular patrols.

3.2 Accesibility

The traffic accessibility of this locality was described in the Safety section. When we evaluate barrier-free accessibility for people with reduced mobility and ability of orientation, access to the premises of Nová Karolina from the urban mass transport stops in the 28. Října (Street of 28 October) is marked using natural and artificial guide lines in the form of warning and signalling strips, see Figure 3. However, both the absence of the contrast marking of the upper edge of the first and last steps and the inefficient visual contrast on the glass facade of the office building can however pose a potential threat to a visually impaired or blind person. People with reduced mobility are

Figure 2. View of the central area in the Jantarová (Amber street).

Figure 3. Pedestrian crossing from the urban mass transport stops.

Figure 4. Design of access to the buildings of Trojhalí (Three Halls).

provided with access along the sloped pedestrian way with a longitudinal slope of 1:12 max. Most entrances to individual buildings lead from the central area of Nová Karolina, along the Jantarová (Amber Street) to the buildings of Trojhalí.

Street furniture, greenery and game elements with a playground are located there.

Orientation is ensured by combining the systems of natural and artificial guide lines. The visually con-trasting design of the walking area, consisting in red paving, marks out the entrance to the shopping centre. From the viewpoint of the flatness of the ground, access to this environment is not limited for people with reduced mobility. However, the problem may be the design of the area that is immediately connected with the buildings of Trojhalí. As the entrances to these buildings are on a lower ground level, access to them is stepped using stairs and terraces to which ramps are added. Figure 4 shows that this stepped area may look like an undivided area with no stairs when visibility is reduced. This fact poses a great threat to people with impaired ability of orientation. For persons with reduced mobility, the barrier-free ramp on the left side is intended. The ramp meets the requirements of Regulation No. 398/2009 Coll.(2009).

3.3 *Comprehensibility*

Comfort is offered to pedestrians by the wide walkways with high-quality surfaces that provide enough room to meet. From the viewpoint of staying, furniture for static activities is placed there. For dynamic activities, there is a sports ground and a playground oriented towards Trojhalí.

3.4 *Functionality*

The classification of the place according to the nature of its operation and functional use has divided the public space into categories, see Figure 5. The residential category is located close to the residential complex, which is lucrative housing with small amenities in the parterre. These amenities are also in the parterre of the shopping center

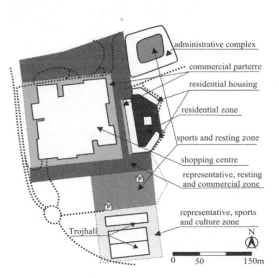

administrative complex

commercial parterre

residential housing

residential zone

sports and resting zone

shopping centre

representative, resting and commercial zone

representative, sports and culture zone

Trojhalí

N

0   50   150m

Figure 5.   Zoning of the public place.

✿✿✿ fully ✿✿ partially ✿ satisfying	commercial parterre	residential zone	sports and resting zone	representative, resting and commercial zone	representative, sports and culture zone
safety	✿✿	✿✿✿	✿✿✿	✿✿✿	✿✿✿
accesibility	✿✿✿	✿✿	✿✿✿	✿✿	✿
comprehensibility	✿✿✿	✿✿	✿✿	✿✿✿	✿✿✿
comfort and pleasure	✿✿	✿✿✿	✿	✿✿	✿✿
functionality	✿	✿✿	✿	✿✿✿	✿

Figure 6.   Evaluation of qualitative parameters of individual categories of the assessed public place.

and of the office complex. The representative, commercial and resting area goes through a large part to the sports and re-laxation facility. New elements for playing and furniture are added step by step in that place; unfortunately, they do not fit with the overall concept as regards their scale.

4   CONCLUSIONS

The mentioned article is based on partial results of the project Accessibility of Public Spaces of a Residential Environment being solved. The subject of the project is the issues of accessibility of these spaces and the overall examination and more detailed specification of selected urban public places. It is very important to realize that there are large numbers of requirements that apply to a public place in general and large numbers of qualitative parameters such as functionality, accessibility, understadibleness, convenience, and safety. This paper refers to their importance and indicates examples for improvement. For clearness, a case study was carried out in the New Karolina areas which may serve as a model for evaluation of similar public areas and lead to their upgrading. Within the framework of the given case study, this area is divided into individual categories according to function of buildings that are located in the area, see Figure 6. Pursuant to the survey based on observation and subsequent evaluation of the given criteria, we define that the public area of the New Karolina is safe, well accessible and understandable. Shortcomings were detected in the functional utilization of sports and relaxation zones and the representative and cultural neighbourhood of the Trojhalí buildings.

ACKNOWLEDGEMENT

The work was supported by the Student Grant Competition VŠB-TUO. Project registration number is SP2015/187.

REFERENCES

ČSN P CEN/TR 14383-2. 2009. *Prevention of crime—Urban planning and building design—Part 2: Urban planning*. Prague: Czech Office for Standarts.
Gehl, J. 2010. *Cities for people*. Washington: Island Press.
Regulation number 128/2000. *Act on Municipalities (municipal establishment)*. Prague: Collection of Laws of the Czech Republic.
Regulation number 398/2009. *Regulation on general technical requirements ensuring barrier-free use of buildings*. Prague: Collection of Laws of the Czech Republic.

Advances in Civil, Architectural, Structural and Constructional Engineering – Kim, Jung & Seo (Eds)
© 2016 Taylor & Francis Group, London, ISBN 978-1-138-02849-4

Analysis of optimal forms of shallow shells of revolution

L.U. Stupishin, A.G. Kolesnikov & T.A. Tolmacheva
Southwest State University (SWSU), Kursk, Russia

ABSTRACT: This paper contains an analysis of the most optimal forms of the middle surface of shallow shells of revolution. Consideration is given to the static action of evenly distributed vertical force for various support types. Critical force and stress in shells are found considering geometric nonlinearity. Critical force coefficient and stress of shells are found by Bubnov-Galerkin method. Calculations were done by means of "Maple" software.

1 INTRODUCTION

Middle surface of shallow shells of revolution can be described by the equation

$$F(\rho) = f_0 \rho^\xi, \qquad (1)$$

where f_0— is the rise of arch in the center of the shell,

$f_0 \leq \frac{a}{5}$.

a—radius of the shell base,

ρ—dimensionless radius of the shell, changing in the range of [0,1],

ξ—shape parameter of the shell middle surface, changing in the range (1,4) (Figure 1).

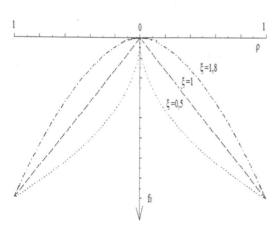

Figure 1. Dependence of the middle surface of shallow shell of rotation on the form parameter.

2 EQUATIONS FOR THE DETERMINATION OF CRITICAL FORCE COEFFICIENT AND STRESS FOR SHALLOW SHELLS OF REVOLUTION

Critical force coefficient and stress for geometric nonlinear shallow shells of revolution were found by Bubnov-Galerkin method (Stupishin & Kolesnikov 2014b; Stupishin & Nikitin 2014). Equations have been obtained for various types of boundary conditions. Beam functions are used for the approximation of boundary conditions.

Critical force coefficient can be described by the equation:

$$p = \frac{a^4}{Ef_0^4} p_{cr}, \qquad (2)$$

where,

$$p = \frac{1}{t^4} k_1 ((k_2 t^2 \psi^2_{(\xi)} - k_3 (k_0 + t^2 F_{(\xi)}))^{\frac{3}{2}} + k_4 t \psi_{(\xi)} (k_2 t^2 \psi^2_{(\xi)} - 1.5 k_3 (k_0 + t^2 F_{(\xi)}))); \qquad (3)$$

$$t = f_0/h, \qquad (4)$$

$$k_0 = \frac{\frac{4}{3} - 2\alpha}{12(1 - \nu^2)}, \qquad (5)$$

$$k_1 = \frac{2}{27} \left(\frac{S4^*}{S1} \right)^2, \qquad (6)$$

$$k_2 = \left(\frac{1}{S4^*} \right), \qquad (7)$$

$$k_3 = \frac{3S1}{(S4^{\bullet})^2} \qquad (8)$$

$$k_4 = \frac{1}{S4^{\bullet}}, \qquad (9)$$

$$\alpha = \frac{(3+\nu)c1+1}{(1+\nu)c1+1}, \qquad (10)$$

$$\begin{aligned}
S_1 = &\frac{1}{96}\left(\frac{1}{14} - \frac{\alpha}{2} + \frac{3}{2}\alpha^2 - 2\alpha^3 + \alpha^4\right. \\
&-\left(\frac{(c2-\nu)(1-4\alpha+6\alpha^2)}{(c2-\nu)+1} + \frac{1-20\alpha+18\alpha^2}{(c2-\nu)+1}\right) \\
&\left.\times\left(\frac{1}{8} - \frac{\alpha}{3} + \frac{\alpha^4}{4}\right)\right),
\end{aligned}$$

$$\qquad (11)$$

$$S_4^{\bullet} = \frac{\frac{4}{3} - 2\alpha}{16}, \qquad (12)$$

E—elastic module, ν—Poisson's ratio; c1, c2—characteristics depending on support type; $\psi(\xi)$, $F(\xi)$—functions depending on the size and shape of the shell and its support type, h—shell thickness.

Equivalent stresses in the shell are found by the fourth stress hypothesis. Stress coefficient can be described by the equation:

$$\overline{\sigma} = \frac{a^4}{Ef_0^4}\sigma, \qquad (13)$$

where,

$$\overline{\sigma} = \sqrt{\frac{1}{2}\left[\left(\overline{\sigma_r} + \overline{\sigma_v}\right)^2\right]}; \qquad (14)$$

$$\overline{\sigma_r} = \overline{\sigma_r^u} + \overline{\sigma_r^m}, \qquad (15)$$

$$\overline{\sigma_v} = \overline{\sigma_v^u} + \overline{\sigma_v^m}, \qquad (16)$$

$$\overline{\sigma_r^m} = N_r, \qquad (17)$$

$$\overline{\sigma_v^m} = N_v, \qquad (18)$$

$$\rho = \frac{r}{a}, \qquad (19)$$

$$\overline{\sigma_r^u} = \left(-\frac{1}{2(1-\nu^2)}\left(\frac{dw}{d\rho} + \nu\frac{w}{\rho}\right)\right), \qquad (20)$$

$$\overline{\sigma_v^u} = \left(-\frac{1}{2(1-\nu^2)}\left(\nu\frac{dw}{d\rho} + \frac{w}{\rho}\right)\right), \qquad (21)$$

r—distance from an arbitrary point on the middle surface of a shell to its rotation axis, changing in the range (0,a); N_r and N_v—dimensionless parameters of radial and circumferential forces, w—deflection of a median surface of shell.

3 ANALYSIS OF CRITICAL FORCE COEFFICIENT AND STRESS IN SHALLOW SHELLS OF REVOLUTION

The calculation method was based on the Maple software package. The changes of the parameters of shallow shells of revolution depending on the shell forms, support type and material were studied earlier (Stupishin & Pereverzev 2012).

Figure 2 shows how the critical force parameter changes under the action of various vertical forces: centered, circular and ring. Types of support: 1—fixed; 2—guided; 3—simple support.

The graphs demonstrate a similar dependence of the critical force parameter for different force forms. This suggests that in further calculations we can use evenly distributed vertical force.

Figure 3 shows the dependence of critical force on the middle surface form parameter.

It is obvious that when rigidity decreases in the radial direction, the most optimal shape of the shell approaches the cone. With increasing rigidity this shape approximates to sphere. Critical force increases when the rigidity of circumferential fixing decreases (Stupishin 1989).

Figure 4 shows design parameter spaces g, t and ξ for which we have built surfaces of limit stresses in the center of the shell (light area) and the stresses arising in the same section under a force that will cause a critical stability condition in the shell with the same shape (dark area) (Stupishin & Kolesnikov 2014a).

These graphs demonstrate that shallow shells should be analyzed for stability only when g, t and ξ parameters are such that the light area is located above the dark area. When parameters g, t and ξ correspond to the positioning of the dark area above the light one, the shell will be losing its strength faster, which means that such shells shall be analyzed for strength. If the parameters of a shell match the curve (track) at the intersection of these two areas, the shell shall be analyzed both for the strength and stability.

4 ANALYSIS OF OPTIMUM FORMS OF GEOMETRIC NONLINEAR SHALLOW SHELLS OF REVOLUTION

Optimization problems for shells of revolution can be written as:

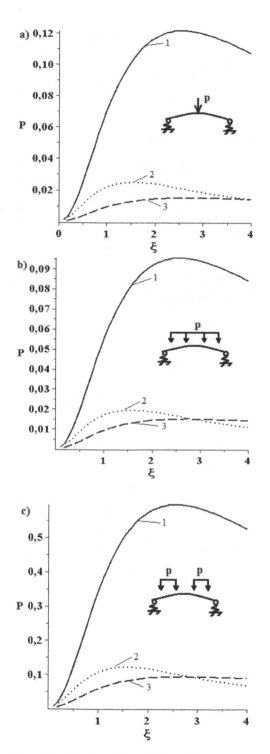

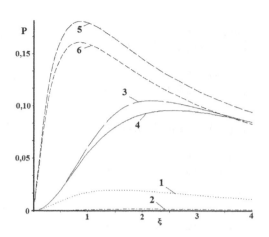

Figure 3. The dependence of the critical force and the shape parameter of the middle surface for metals (lines 1, 3 and 5) and ferro-cements (lines 2, 4 and 6): 1, 2 – fixed; 3, 4 – guided; 5, 6 – simple support.

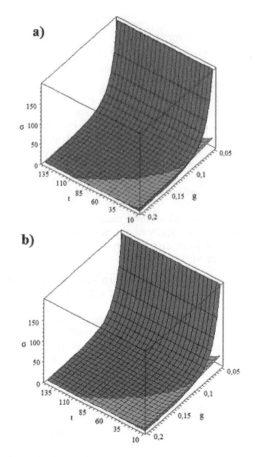

Figure 2. The dependence of the critical force on the shape parameter of the middle surface for fixed (line 1), guided (line 2) and simple support (line 3): a—centered force; b—circular force; c—ring force.

Figure 4. Dependence of the stresses on the following parameters of a shell: a) middle surface form parameter ξ and parameter $g = f_0/a$; b) relative thickness parameter $t = f_0/h$ and parameter $g = f_0/a$.

$$\begin{cases} p(\xi,t) \to p_{\max}; \\ V(\xi,t) - V_0 \leq 0, \xi \in G, t \in G. \end{cases} \quad (22)$$

$$\begin{cases} \sigma(\xi,t) \to \sigma_{\min}; \\ V(\xi,t) - V_0 \leq 0, \xi \in G, t \in G. \end{cases} \quad (23)$$

$$\begin{cases} V(\xi,t) \to V_{\min}; \\ p(\xi,t) - p_0 \geq 0, \xi \in G, t \in G. \end{cases} \quad (24)$$

$$\begin{cases} V(\xi,t) \to V_{\min}, \\ \sigma(\xi,t) - \sigma_0 \leq 0, \xi \in G, t \in G. \end{cases} \quad (25)$$

$$\begin{cases} V(\xi,t) \to V_{\min}; \\ p(\xi,t) - p_0 \geq 0, \xi \in G, t \in G; \\ \sigma(\xi,t) - \sigma_0 \leq 0, \xi \in G, t \in G. \end{cases} \quad (26)$$

$$G = \left\{ \xi : 0,5 < \xi_i < 2, i = \overline{1,n}; t : 0,001 < t_j < 0,1, \right.$$
$$\left. j = \overline{1,m} \right\}.$$

where,

$V(x_i)$—volume function,
$p(x_i)$—critical force function,
$\sigma(x_i)$—stress functions.

Optimization algorithms are realized in the "Maple" environment. The solutions that allow us to design rationally shaped shells depending on the changes in their form have been successfully obtained for both variants of the stated problems (Andreev 2012, Stupishin & Kolesnikov 2012).

In some cases just due to a rational shape of the envelope it is possible to save up to 35% of its material with a considerable weight reduction of the structure.

The offered method makes it possible to optimize geometrically nonlinear shallow shells due to optimal form distribution when there is a need to (Figure 5):

– optimize stress in the median surface of a shell for any kinds of support with a limitation on the volume of used material;
– optimize critical force for any kinds of support;
– optimize the stress in the middle surface of the shell for any kinds of support with a pre-set critical force value and a limitation on the volume of used material;
– optimize critical force at pre-set values of stresses in the middle surface of the shell and a limitation on the volume of used material;
– optimize the volume of a used material for any kinds of support at pre-set values of the stresses and critical force.

5 CONCLUSIONS

This methodology can be applied to determine critical force and stresses for geometrically non-linear shallow shells of revolution with variable form of the middle surface. The dependences demonstrate that it is possible to find an optimal relationship between the thickness and the form of geometrically nonlinear shallow shells of revolution by applying the criteria of maximum critical force and minimum stresses. This optimization algorithm can ensure considerable savings in the weight of building structures.

REFERENCES

Andreev, V.I. 2012. *Optimization of thick-walled shells based on solutions of inverse problems of the elastic theory for in-homogeneous bodies.* Computer Aided Optimum Design in Engineering XII (OPTI XII). WIT Press. 20012, 189–201.
Stupishin, L.U. 1989. Approximate method for determining op-timal form of geometrically nonlinear shallow shells of revolution under the conditions of stability, *News of Higher Educational Institutions. Construction.* 9: 28.
Stupishin, L.U. & Kolesnikov, A.G. 2012. Investigation of Op-timum Forms of Depressed Geometrically Nonlinear Shells of Variable Thickness, *Industrial and Civil Construction.* 4: 11–13.
Stupishin, L.U. & Kolesnikov, A.G. 2014a. Geometric Non-linear Orthotropic Shallow Shells Investigation, *Applied Mechanics and Materials.* 501–504: 766–769.
Stupishin, L.U. & Kolesnikov, A.G. 2014b. Geometric Nonlinear Shallow Shells for Variable Thickness Investigation, *Advanced Materials Research,* 919–921, 144–147
Stupishin, L.U. & Nikitin, K.E. 2014. Mixed finite element of geometrically nonlinear shallow shells of revolution, *Applied Mechanics and Materials,* 501–504: 514–517
Stupishin, L.U. & Pereverzev, M.U. 2012. A certain force at shallow shell of rotation optimal form based on pontryagin maximum principle, *Proceedings of the South-West State University.* 2–3: 187–189.

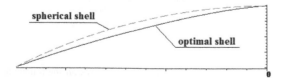

Figure 5. Middle surface of a spherical shell and the shape that is optimal to sustain critical force impact.

Advances in Civil, Architectural, Structural and Constructional Engineering – Kim, Jung & Seo (Eds)
© 2016 Taylor & Francis Group, London, ISBN 978-1-138-02849-4

Earth architecture: An eco-compatible solution for future green buildings

D. Benghida

Department of Architecture, Dong-A University, Busan, South Korea

ABSTRACT: Earth materials, including minerals, rocks, soil and water, are the oldest and have been the most widely used construction materials since more than 9000 years. This earthen architecture has stood the test of time and proved that it can stand for more than 2500 years like the Arg-e-Bam Citadel in Iran (the world largest sun-dried brick structure). Earth as a building material has almost disappeared during the 20th century due to the international development of the commercial concrete. This kind of architecture made from mud is until recently considered primitive, not suitable for our contemporary lifestyle, and usually described as the architecture for low-income societies. In this research, I will demonstrate that the unbaked mud is not only a durable construction material, but also the right construction material for all because it can respond to the sustainability challenges they face in terms of energy efficiency, comfort and eco-compatibility.

1 INTRODUCTION

Since the 1997 signed Kyoto Protocol, all industrialized nations and those with economic transition have been engaged to reduce the major anthropogenic greenhouse gas emissions. According to numerous studies, the fossil fuel energy is now coming to an end (Hirsh et al. 2005); we are close to its production peak and consequently close to the CO_2 emission peak (Alekkett 2007). In March 2015, the International Energy Agency announced that energy-related emissions for 2014 stopped at 32.3 billion metric tons, which is the same as in 2013 (IEA 2015). This is significant. Zhai & Previtali (2010) found that buildings are globally responsible for 45% direct or indirect energy consumption. Buildings and their users contribute by approximately one-third of energy-related global CO_2 emissions (Minke 2001). The massive urbanization, the excessive use of fossil fuels and the increasing degradation of the environment make it necessary to have a careful thought and new insights on the architecture and urban space design.

2 EARTHEN ARCHITECTURE

Unfired-mud construction is a widely present traditional architecture. It is quite easy to notice the presence of traditional mud architecture that is still standing and having a remarkable weathering resistance despite the time span. Obvious examples such as the ancient city Arg-e-Bam Citadel in Iran show the resistance and the potential to use this product of nature. Many adobe buildings are recognized as having a historical and architectonic value.

About one-third of the global population lives in the unbaked-mud buildings (Minke 2006). It is impressive to see how it could be an effective solution to the 21st century energy problem for local communities. Mud is one of the few building materials that can be recycled indefinitely; it can also be found in almost all parts of the world with low transportation, processing and pose cost. In addition, unlike the conglomerate concrete whose process of production alone is responsible for almost 13% of global CO_2 emissions (Benghida 2015), the process of mud brick making has near zero emissions. Brick is, indeed, the cheapest, most practical and convenient sustainable construction material. With the great value of bricks and their architectural technology, architects and other professionals should always keep bricks in the race.

3 CHARACTERISTICS AND PROPERTIES

The massive use of HVAC systems in our industrialized world changed the local traditional architectonic character to a McDonaldization of the architecture technologies and generated a high dependency of energy-use and CO_2 emissions (Benghida, 2014). The use of mud in architecture will generate climatic responsive buildings with a

Table 1. Comparison of different construction material properties (Fernandes et al. 2014).

Material	Thermal conductivity λ-value (W/m.°C)	Thermal storage capacity (Wh/kg°C)	Heat transfer time lag (250 mm thickness) hour	Global Warming Potential (kg CO_2/m$_3$)
Rammed earth/ adobe	1.00–1.20	0.23–0.30	10 / 9	38
Stone	2.30–3.50	0,22–0.24	5.5	26
Concrete	1,80	1.10	7	264
Hollow bricks	0.39–0.45	0,26	6	357

high relationship with the environment with a low embodied energy of 38 CO_2/m$_3$ (Fernandes et al. 2014) (Table 1).

The high mass properties, local availability and affordability of mud architecture make it one of the favorite construction materials in one-third of the world population (Minke, 2006); this is a climate responsive architecture.

The thermal mass property of mud brick makes it suitable for diurnal solar energy storage and a nocturnal re-radiation, due to its delayed heat-flow phenomenon, staying this way in the comfort zone and reducing the energy use.

Thermal conductivity, as shown in Table 1, the ability to conduct heat through the walls' thickness, reveals that stone gets the highest thermal conductivity rate with 2.30–3.50 W/mK, followed by concrete with 1.80 W/mK. This makes them the best suitable materials for heat storage. However, the short time Lag, respectively 5.5 and 7 hours, makes them release the thermal energy after it substantially warms up in a short period during the day. Rammed earth has a relative low thermal conductivity of 1.00–1.20 W/m.°C with a high time lag 10/9 h, making it the best construction material for hot climates because it minimizes the heat transmission inside buildings, although we can get more satisfying performance through increasing the walls' thickness (Fernandes et al. 2014).

The durability of an earthen structure will depend on its compressive strength, durability, thermal conductivity, and most of all its porosity (Clifton et al. 1979). It is known that mud brick buildings are subjected to deterioration by weathering, especially by wind and moisture.

Based on the study of Brown and Clifton (Clifton et al. 1978), mud brick is principally made of:

clay: 10–15%
silt: 10–15%,
sand: 70–80% and,
water: 25–30%

The excess or privation of one ingredient may cause the non-stability of the structure. The right mixture of the ingredients is extremely important. For example, the density of mud brick increases

Figure 1. The perch, a rammed earth architecture project, Sedona Arizona, 2008, by courtesy of The Construction Zone.

if there is an excess in sand, and the mud brick cracks if there is a higher percentage of clay. So the composition and microstructure of the mud brick is a case-sensitive issue. To make the mud brick stronger, it suffices to use a non-eco-compatible stabilizer, to apply a plaster, to dry it through baking, or to add natural fibers to its composition.

4 STRENGTH AND WEAKNESS OF MUD BRICKS

The great thickness of earthen walls acts as a buffer to the heat, holding it while the wall temperature rises. However, this process is only slowing down the heat transmission and deferring it later for several hours during the night on the opposite side after the sunset, after the thick wall warms up. The night-natural ventilation can also be an effective passive technology (courtyards) to lie in the thermal comfort zone during summer (Steele 2009).

Because of its low-cost and low aesthetics options, earthen architecture is perceived as a sign of poverty (Sheweka 2011). But this image of mud architecture does not do justice to it since it has the potential of creating contemporary houses (Figure 1).

Table 2. Strength and weakness of mud bricks.

Strength	Weakness
Suitable for hot dry climates	Not suitable for Humid climates (moisture absorption)
High compressive strength	Low tensile strength
High thermal resistance	Low response to weathering: shrinkage cracks + erosion
Global availability	Low acceptability: perceived as a sign of poverty
Zero waste	Low aesthetic value
Fireproof	Labor-intensive
Cost-effective	Low financial industry profit
Low sound transmission	Needs constant maintenance
Low embodied energy	Low governmental incentives

Brick is a popular and distinctive choice due to its advantages over other construction materials. Table 2 summarizes why it is appealing to homeowners and architects, and what disadvantages might be contributing to its dropping popularity (Benghida 2015, Revuelta-Acosta 2010).

5 AIR TEMPERATURE FLUCTUATION PERFORMANCE

During March 1964, Hassan Fathy undertook a building experiment at the Cairo Building Research Center (CBRC), using different construction materials (Figure 2)

It shows the air temperature fluctuation performance during 24 hours of two building models, with a usual shape and form: the first one was built with only mud bricks and a dome vaulted roof, with 50 cm thickness of the envelope; the second model was built with a precast concrete panels with a flat roof and 10 cm thickness of the envelope.

Knowing that the CBRC is located within the Giza area, and it has a hot desert climate (hot days and cold nights), the comfort zone varies between 18 and 27°C (Minke 2006).

The diagram shows that the indoor air temperature of the first model (mud bricks) varies from 21 to 23°C (during the total 24 hours), which did not exceed 2°C, keeping this way its stability in the comfort zone. On the other hand, the indoor air temperature of the concrete model reaches 36°C, which means greater than that of the mud brick model by 13°C, and greater than that of the outdoor air temperature by 9°C.

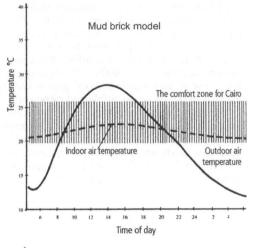

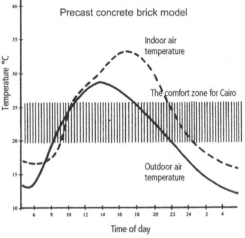

Figure 2. Comparison of indoor and outdoor air temperature fluctuation within 24 h period for the prefabricated concrete test model, for the mud-brick test model (Fathy 1986).

In fact, the concrete model enters within the comfort zone only a short time: one hour during the day (from 9 to 10) and three hours 40 minutes during the night (from 20:40 to 24:20). This is due the high thermal conductivity of concrete (0.9) in comparison with mud brick thermal conductivity (0.34) and, of course, with the thickness of the walls used in this experiment (five times thicker). Mud brick wall has a thermal resistance 13 times greater than the concrete panels (Fathy 1986).

6 CONCLUSIONS

Earth brick is an excellent example of eco-compatible construction material because the global availability and the *in situ* production make it cost-effective with low embodied energy. Its high thermal resistance makes it the favorite material, especially for hot dry climate populations, because it provides a high standard of indoor comfort. Brick has few weaknesses, which can be fixed through an appropriate use of sustainable technologies to keep this material 100% eco-compatible and responsive.

By educating architects about this traditional material, by implicating engineers and by improving the mechanical properties of the mud bricks against weathering and moisturizing, earth architecture will thrive in the near future, notably for its intrinsic advantages at reducing CO_2.

AKNOWLEDGMENT

This work was supported by the Dong-A University research fund.

REFERENCES

Alekkett, K. 2007. *Reserve Driven Forecasts for Oil, Gas & Coal and Limits in Carbon Dioxide Emissions: Peak Oil, Peak Gas, Peak Coal and Peak CO₂*, OECD/ITF Joint Transport Research Centre Discussion Papers, No. 2007/18, OECD Publishing, Paris. DOI: 10.1787/234435550466.

Benghida, D. 2014. *The Urban Identity Recovery in Seoul: The Case of the Outdoor Markets*. In Tostoes, A., Kimm, J.M., Kim, T.W. Expansion & Conflict. Presented at 13th Docomomo International Conference Seoul, 227–231. Korea: Docomomo Korea.

Benghida, D. 2015. *CO₂ Reduction from Cement Industry. Presented at International Conference on Civil, Architectural, Structural and Constructional Engineering*, Busan. CRC Press/Balkema: The Netherlands.

Clifton, J.R. Robbins, C.R. & Brown, P.W. 1978. Adobe. I: The Properties of Adobe, Studies in Conservation, *Maney Publishing*, 23(4): 139–146.

Clifton, J.R. Robbins, C.R. & Brown, P.W. 1979. Adobe. II: Factors Affecting the Durability of Adobe Structures, Studies in Conservation, *Maney Publishing*. 24(1): 23–39.

Fathy, H. 1986. *Natural Energy and Vernacular Architecture: Principles and Examples with Reference to Hot Arid Climates*. USA: The University of Chicago Press.

Fernandes, J. Dabaieh, M. Mateus, R. & Bragança, L. 2014. *The influence of the Mediterranean climate on vernacular architecture: a comparative analysis between the vernacular responsive architecture of southern Portugal and north of Egypt*, Presented at World SB14 Barcelona, pp. 1–7 [online] Available at: http://hdl.handle.net/1822/31403 [Accessed 7 Sep. 2015].

Hirsch, R.L. Bezdek, R. & Wendling, R. 2005. *Peaking of World Oil Production: Impacts, Mitigation, & Risk Management*. Science Applications International Corporation/U.S.Department of Energy, National Energy Technology Laboratory. USA 7 [online] Available at: http://www.netl.doe.gov/publications/others/pdf/Oil_Peaking_NETL.pdf [Accessed 7 Sep. 2015].

International Energy Agency. 2015. *Global energy-related emissions of carbon dioxide stalled in 2014*, March. [online] Available at: http://www.iea.org/newsroomandevents/news/2015/march/global-energy-related-emissions-of-carbon-dioxide-stalled-in-2014.html [Accessed 7 Sep. 2015].

Minke, G. 2001. *Manual de construcción para viviendas antisísmicas de tierra*. Research Laboratory for Experimental build, University of Kassel, Germany.

Minke, G. 2006. *Building with Earth: Design and Technology of a Sustainable Architecture*. Basel, Switzerland: Birkhäuser.

Murphy, D. 2012. Fossil fuels: Peak oil is affecting the economy already. *Nature*, 483(7391): 541–541. doi:10.1038/483541a.

Revuelta—Acosta, J.D. Garcia—Diaz, A. Soto—Zarazua, G.M. & Rico—Garcia, E. 2010. Adobe as a Sustainable Material: A Thermal Performance. *Journal of Applied Sciences*, 10: 2211–2216. DOI: 10.3923/jas.2010.2211.2216.

Sheweka, S. 2011. *Using Bricks as a Temporary Solution for Gaza Reconstruction*, Energy Procedia, Elsevier Ltd, 236–240.

Steele, J. 2009. *The Greenwood Encyclopedia of Homes through World*, Volume 1 History, From Ancient Times to the Late Middle Ages, 6000 BCE–1200(first edition). (Volume1). Connecticut: Greenwood Press.

Zhai, Z. & Previtali, J.M. 2010. Ancient vernacular architecture: Characteristics categorization and energy performance evaluation. *Energy and Buildings*, 42: 357–365.

Advances in Civil, Architectural, Structural and Constructional Engineering – Kim, Jung & Seo (Eds)
© 2016 Taylor & Francis Group, London, ISBN 978-1-138-02849-4

Introduction of thermal sharpening technique for improving classification accuracy in Landsat-8

S.B. Lee, M.H. Lee, Y.D. Eo & S.W. Kim
Department of Advanced Technology Fusion, Konkuk University, Seoul, South Korea

W.Y. Park
Agency for Defense Development, Daejeon, Korea

ABSTRACT: This paper shows the experimental results from thermal image sharpening for improving classification accuracy. Landast-8 OLI and TIR images are considered as a band combination for statistical image classification. Classification accuracy is accomplished in the two cases: one in which the TIR band is sharpened with panchromatic image, also applied to band combination, and the other in which the TIR band is re-sampled with 30 spatial resolutions. As results, TIR sharpening gives a positive effect to classification accuracy especially in urban regions. It also improves overall accuracy.

1 INTRODUCTION

A TIR (Thermal Infra-Red) band with a 8–12 μm wavelength can present the temperature of the ground surface (Figure 1). To make a more detailed temperature map, some research has investigated precise temperature extraction by using a TIR band (Zhang et al. 2006). The TIR band has not previously been used for image classification because the spatial resolution of the TIR is much lower than the multispectral bands. However, improvement of classification accuracy using the temperature information from the TIR band has been studied despite low spatial resolution (Warner & Nerry 2009). A few algorithms for improving the spatial resolution of the TIR image, such as image fusion, have been developed (Jeganathan et al. 2011, Jung & Park, 2014). This study performed an experiment on band combination including thermal images for landcover classification. The thermal sharpening effect on classification accuracy was analyzed.

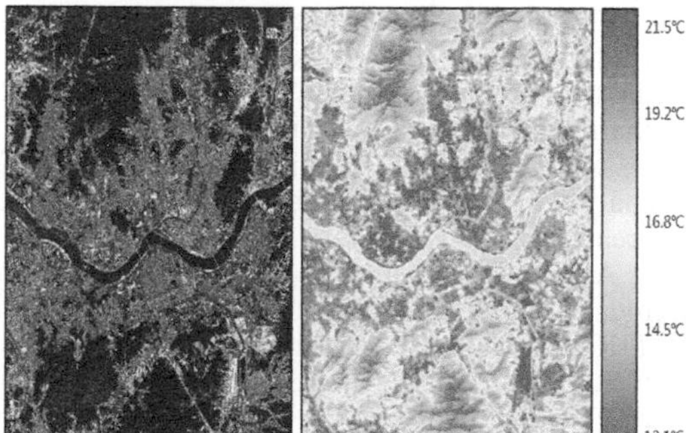

Figure 1. Study area (a) Landsat-8 Natural color image, (b) Temperature map from Landsat-8 TIR image.

2 DATA AND PREPROCESSING

2.1 *Study area and image data*

The area of study was the capital city, Seoul, in the Republic of Korea (Figure 1). The Landsat-8 (WRS-2, Path: 116, Row: 34) was used for image data. Landsat-8 is an Earth observation satellite launched in 2013 to acquire satellite images of 30 m spatial resolution with two sensors. The two sensors are composed of an OLI (Operation Land Image) and TIR (Thermal Infrared Sensor). These data are ac-quired from the USGS (http://earthexplorer.usgs. gov/).

2.2 *Atmospheric correction*

FLAASH (Fast Line-of-sight Atmospheric Analysis of Spectral Hypercubes) is used for multispectral bands. Atmospheric correction for the TIR band is used by using Equation (1) as developed by Coll et al. (2015).

$$\mathrm{Rad}_{ac} = \frac{Rad - L_U}{\epsilon \cdot \tau} - \frac{1-\varepsilon}{\epsilon} L_D \quad (1)$$

where, Rad_{ac} = corrected radiance; Rad = radiance; L_U = upwelling radiance; L_D = downwelling radiance; τ = transmittance; and ϵ = emittance.

Radiance, emittance, and transmittance are acquired by entering the geometric and atmospheric conditions into the NASA (National Aeronautics and Space Administration) website (http://atm-corr.gsfc. nasa.gov/). According to the NASA website, there are several problems in TIR Band 11 when calculating the above variables. Therefore, the TIR band 10 only has been used in this study.

3 THERMAL SHARPENING

The image sharpening method enhances the low spatial resolution in TIR or multispectral by merging the panchromatic image with high spatial resolution.

This is called TIR sharpening in the case of TIR images (Jeganathan et al. 2011). In this study,

(a)

(b)

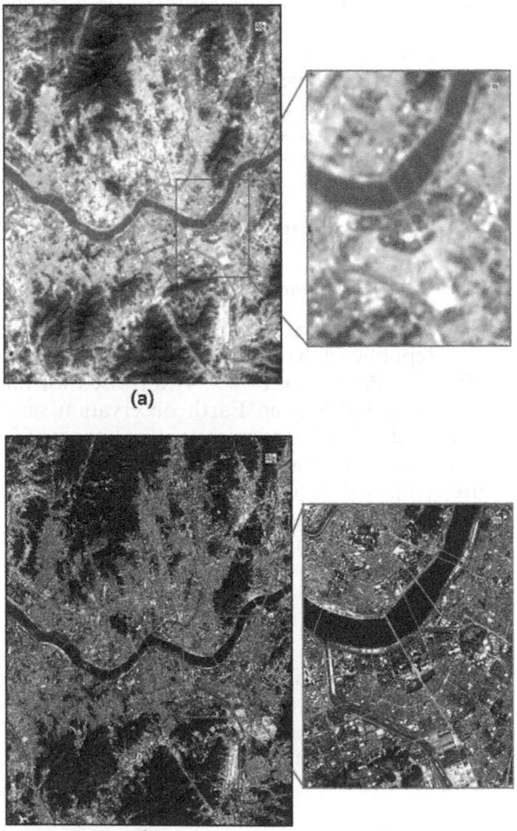

Figure 2.   TIR imagem, TIR sharpened image.

the band combination for landcover classification included the TIR image sharpened with better spatial resolution.

The GS (Gram-Schmidt) sharpening method was introduced for this study. GS can effectively preserve the DN (Digital Number) of the image and has shown the best results as compared to other sharpening methods (Choi & Kim 2010). GS was processed from ENVI software.

GS is a Component Substitution (CS)-based fusion technique, one of the most popular image fusion techniques, and it can be stated as shown in Equation (2) (Choi & Kim 2010). The GS fusion, technique known as the best among the CS-based algorithms, obtains coefficients using covariance and variance ratios and can be expressed as shown in Equation (3) (Laben & Brower 2000).

$$M_F(i) = (P_H - I_L) \cdot \omega(i) + M_L(i) \qquad (2)$$

$$\omega(i) = cov(I_L, M_L(i))/var(I_L) \qquad (3)$$

where, P_H = high-resolution panchromatic image; $M_L(i)$ = ith low-resolution multispectral image; $M_F(i)$ = high-resolution fusion image; I_L = virtual low-resolution satellite image; var = variance of pixels; and cov = covariance among the images.

4 EXPERIMENT AND ANALYSIS

MLC (Maximum Likelihood Classification) is used in image classification. MLC is a common method used to classify satellite images in remote sensing. MLC depends on the probability distribution of training data (Sugumaran et al. 2003).

Forest, water, field, paddy-field, grass, dry stream, and urban demarcations were chosen as landcover categories. Field and paddy-field were merged because of the limitation of the spatial and spectral resolution.

The dry stream area revealed the most significant improvement. However, since the location distribution in the study area was far more limited than other classes, the number of training pixels was low. Therefore, it cannot be said that the TIF sharpening effect is the greatest in the dry stream area.

As in Table 1 and 2, the classification result regarding urban, dry stream and grass classes with TIR sharpening shows better P.A. and U.A. than the results without TIR sharpening. The worst result is in the water area. When applying the resampling method in the TIR band, the classification accuracy in the water area was 94.9% in P.A. and 92.2% in U.A. But in the case of TIR sharpening, P.A. and U.A. decreased as 91.7% and 88.4%. This means the TIR sharpening does not have an effect

Table 1. Classification accuracy with TIR sharpening (%).

	Forest	Water	Crop	Grass	Dry stream	Urban	O.A
P.A	95.1	91.7	55.8	86.0	84.7	70.4	79.9
U.A	90.4	88.4	70.3	70.9	58.2	78.1	

Table 2. Classification accuracy without TIR sharpening (%).

	Forest	Water	Crop	Grass	Dry stream	Urban	O.A
P.A	94.9	95.5	50.5	86.3	80.2	66.4	78.3
U.A	92.2	88.6	61.2	65.6	46.5	76.8	

P.A: Producer's accuracy, U.A: User's accuracy, O.A: Overall accuracy.

on classification accuracy in the water area. The forest area also showed the same trend as the water area. With the exception of the forest and water areas, the other landcovers have a better accuracy than applying the resampling method.

When considering the aspect of results, the reason why forest and water areas have a low accuracy is the ground temperature.

There was almost no significant temperature difference between the surrounding pixels in the forest and water because the satellite images used in the study was acquired in October. For this reason, the forest and water areas have unclear boundaries, which are expected to cause TIR sharpening error.

5 CONCLUSIONS

In this study, whether the TIR band influenced in classification accuracy was tested through the image classification. According to an analysis of accuracy, overall accuracy increased in the case of applying TIR sharpening. The difference was found in each class. Especially, the forest area had insufficient effect on TIR sharpening. This trend was also shown in the water area. But the other landcovers such as crops, grass, dry stream, and urban areas increase obtained improved classification accuracy by applying TIR sharpening. The effect on TIR sharpening in the urban area is the best result in particular. Although there were poor results in forest and water areas, TIR sharpening facilitates image classification when considering overall classification accuracy.

In future studies, it is necessary to apply another sharpening method and find adequate methods for TIR sharpening. Moreover, considering the season, there should be further studies using diverse satellite images throughout the four seasons.

ACKNOWLEDGEMENT

This work is financially supported by Korea Ministry of Land, Infrastructure and Transport (MOLIT) as [U-City Master and Doctor Course Grant Program].

REFERENCES

Coll, C. Galve, J.M. Sanchez, J.M. & Caselles, V. 2010. Validation of landsat-7/ETM+ thermal-band calibration and atmospheric correction with ground-based measurements. *IEEE Transactions on Geoscience and Remote Sensing,* 48(1): 547–555.

Choi, J.W. & Kim Y.I. 2010. Pan-sharpening algorithm of high-spatial resolution satellite image by using spectral and spatial characteristics. *Journal of the Korean Society for Geospatial Information System,* 18(2): 79–86.

Jeganathan, C. Hamm, N.A.S. Mukherjee, S. Atkinson, P.M. Raju, P.L.N. & Dadhwal, V.K. 2011. Evaluating a thermal image sharpening model over a mixed agricultural landscape in India. *International Journal of Applied Earth Observation and Geoinformation,* 2(13): 178–191.

Jung, H.S. & Park, S.W. 2014. Multi-sensor fusion of land-sat 8 thermal infrared and panchromatic images. *Sensors.* 2(14): 24425–24440.

Laben, C.A. & Brower, B.V. 2000. *Process for enhancing the spatial resolution of multispectral imagery using pan-sharpening.* U.S. Patent 6011875, Tech. Rep. Eastman Kodak Company.

Lee, M.H., Lee, S.B., Kim, Y.M, Sa, J.W. & Eo, Y.D. 2015. Effectiveness of Using the TIR Band in Landsat 8 Image Classification. *Journal of the Korean Society of Survey, Geodesy, Photogrammetry, and Cartography.* in press.

Sugumaran, R. Pavuluri, M.K. & Zerr, D. 2003. The use of high-resolution imagery for identification of urban climax forest species using traditional and rule-based classification approach. *IEEE Transactions on Geoscience and Remote Sensing,* 9(41): 1933–1939.

Warner, T.A. & Nerry, F. 2009. Does single broadband or multispectral thermal data add information for classification of visible, near—and shortwave infrared imagery of urban areas? *International Journal of Remote Sensing,* 9(30): 2155–2171.

Advances in Civil, Architectural, Structural and Constructional Engineering – Kim, Jung & Seo (Eds)
© *2016 Taylor & Francis Group, London, ISBN 978-1-138-02849-4*

The control and simulation of MRAC based on Popov hyperstability theory in real-time substructure testing

L. Deng & W. Fan

School of Civil Engineering and Architecture, Southwest Petroleum University, Chengdu, China

ABSTRACT: In real-time substructure testing, if the loading actuator cannot achieve action command in the algorithm step timely and accurately or suffer disturbance in loading project, it will cause great distortion of test results. In order to prevent the occurrence of such a situation, an innovation of introducing Popov super stability theory into the load control in the test will be applied. A MRAC (Model Reference Adaptive Control) system is designed and carried out by MATLAB/Simulink to build its modeling and perform simulation. The results of simulation indicate that the design of the MRAC system based on the super stabilization theory has good tracking accuracy, response speed and online error correcting function, which meets the requirements of real-time substructure testing.

1 INTRODUCTION

Earthquake-resistance test of structures is not only an important way to reveal seismic characteristics of construction, but also the basis of seismic-resistance design. The traditional structure tests include pseudo static test, pseudo dynamic test and earthquake simulating shaking table test (Qiu & Chen 2000). In order to fit the development of structural seismic technology, Japanese scholars proposed the method of real-time substructure testing based on the pseudo dynamic test (Taknashi & Nakashima 1987). This new method of dynamic structure testing consists of two parts: a key part of the structure as the experimental substructure, which will perform physical loading in real time, and the remainder as the numerical substructure, which will be simulated by the mathematical model. The method of real-time sub-structure testing reduces the cost of large-scale seismic structural testing and can check the performance of the type of structures or components with speed and acceleration. It integrates the advantages of traditional anti-seismic tests, but also faces many crucial scientific issues. This study has attracted many domestic and foreign scholars' attention and obtained a fast development.

The question on stability, precision and response speed of loading control is one of the most important problems in real-time substructure testing. Model Reference Adaptive Control system (MRAC) is a classical control method, which is applied to the design of each control field in various industries. The electro-hydraulic servo system is mostly adopted in the structure experiment at present. With respect to the electro-hydraulic servo

system, a variety of mechanical, hydraulic and electronic nonlinear sets are met together in one of the nonlinear systems. If it uses routine PID control, it will be difficult to acquire satisfactory results. The MRAC system not only meets the need of the system's response speed accuracy and requirements of the premise but also improves the dynamic performance of the system. It has great adaptability to the change of system parameters, nonlinearities in some degree and can make online correction. The MRAC system based on the Popov hyperstability theory also has global stability and meets the test load's steady, accurate, fast targets, which improves the efficiency of the test.

The test uses LQR (linear quadratic regulator) control in the literature (Zhang 2008), in which the actuator can complete action instructions only by selecting certain reasonable gains in the integral algorithm step. If the integral step is relatively small or parameters of system are uncertain, this control method will be difficult to apply. Although the design in the literature (Deng 2007) is also the MRAC system, its foundation is Lyapunov theory. The limitation of this method is that the Lyapunov function is difficult to find, while the design of the MRAC system based on Popov theory (Popov 1973) has successfully overcome the above problems.

2 DESIGN OF THE MRAC SYSTEM

2.1 *System construction*

When designing the Model Reference Adaptive Control (MRAC) system by super stable theory, it

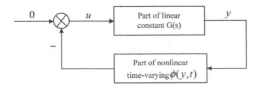

Figure 1. The nonlinear time-varying feedback system.

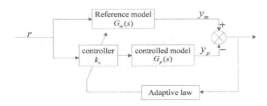

Figure 2. Diagram of the MRAC system.

is necessary to convert the controlled object into a nonlinear time-varying feedback system consisting of two squares, as shown in Figure 1. One box is the feedforward channel (transfer function $G(s)$) and the other one is the nonlinear feedback loop, using $\phi(y,t)$ to express.

The forward linear channel is made up of the reference model, and the nonlinear feedback loop consists of the controlled object with a nonlinear property and the proportional integral controller containing memory function, as shown in Figure 2.

2.2 Derivation of the adaptive law

2.2.1 Mathematical model

The research object adopts a second-order transfer function simplified by the power performance composed of experimental substructures, actuator and a controller, reported in the literature (Wang 2007), as given in Formula (1):

$$T_A(s) = \frac{\omega_A^2}{s^2 + 2\xi_A \omega_A s + \omega_A^2} \quad (1)$$

To simplify the derivation process, the study assumes that the reference model and the controlled object have the same poles, but the extracted rules also apply equal samples of different poles, which will be proved in the later simulation.

The mathematical model of the controlled object is as follows:

$$G_P(s) = \frac{y_p(s)}{u(s)} = \frac{b_p}{s^2 + a_1 s + a_2} \quad (2)$$

The reference model is as follows:

$$G_m(s) = \frac{y_m(s)}{R(s)} = \frac{b_m}{s^2 + a_1 s + a_2} \quad (3)$$

The controller is as follows:

$$G_e(s) = k_c \quad (4)$$

The generalized error is as follows:

$$e = y_m - y_p \quad (5)$$

The transfer function of the adjustable system, from Figure 2, is as follows:

$$G(s) = k_c G_m(s) = \frac{k_c b_p}{s^2 + a_1 s + a_2} \quad (6)$$

The error equation of equivalent is given in Equations (2), (3), (5):

$$(s^2 + a_1 s + a_2)e = (b_m - k_c b_p)r \quad (7)$$

2.2.2 Equivalent nonlinear feedback system

The nonlinear feedback system is obtained by transforming error Equation (7). In the system, the linear part is located in the anterior channel, and the nonlinear time variable part is located in the feedback channel. According to the super stable theory's requirements, the equivalent feedback system must meet two conditions: first, the anterior channel satisfies the positive realness; second, the feedback channel meets Popov integral inequality. In order to meet the anterior channel's positive realness, a compensator D (Landau 1969) is added. The design of the model is shown in Figure 3.

The linear block is as follows:

$$(s^2 + a_1 s + a_2)e = w_1, w_1 = -w \quad (8)$$

The feedback block is as follows:

$$w = (k_c b_p - b_m)r \quad (9)$$

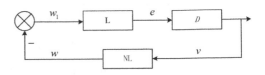

Figure 3. Equivalent nonlinear closed-loop system.

The function of the linear part (the anterior channel) is as follows:

$$L = \frac{e(s)}{w_1(s)} = \frac{1}{s^2 + a_1 s + a_2} \quad (10)$$

The compensator is D, in which e is input and v is output, $v = De$.

$$D = \sum_{i=1}^{q} d_i s^i \quad (11)$$

The general form of the adaptive control law:

$$k_c = \int_0^t \varphi_1(v, t, \tau) d\tau + \varphi_2(v, t) + k_c(0) \quad (12)$$

In order to make the output and input in the feedback block satisfy the Popov integral inequality,

$$\int_0^t vw \, dt \geq -r_0^2, \; r_0 \text{ is the initial value} \quad (13)$$

Introducing Equation (9) into Equation (13), we can obtain

$$\int_0^t v(k_c b_p - b_m) r \, dt \geq -r_0^2 \quad (14)$$

Introducing Equation (12) into Equation (14), we can obtain

$$\int_0^t vr\{b_p[\int_0^t \varphi_1 d\tau + \varphi_2 + k_c(0)] - b_m\} dt \geq -r_0^2 \quad (15)$$

Simplifying Equation (15), we can obtain

$$\int_0^t vr\{b_p[\int_0^t \varphi_1 d\tau + k_c(0)] - b_m\} dt \geq -r_1^2 \quad (16)$$

$$\int_0^t vr b_p \varphi_2 dt \geq -r_2^2 \quad (17)$$

2.2.3 *Determining D, φ_1, φ_2 to get the adaptive law*

2.2.3.1 Determining D
According to the Popov theorem, the forward channel meets the positive realness, so the forward channel (Li et al. 2010) of the equivalent system is as follows:

$$G(s) = h(s) = \frac{D(s)}{s^2 + a_1 s + a_2} = \frac{d_1 s + d_0}{s^2 + a_1 s + a_2} \quad (18)$$

where d_0, d_1 are constants.

Introducing $s = j\omega$ into Equation (18), we obtain

$$h(j\omega) = \frac{d_1(j\omega) + d_0}{(j\omega)^2 + a_1(j\omega) + a_2} = \frac{j\omega d_1 + d_0}{(a_2 - \omega^2) + j\omega a_1} \quad (19)$$

The real part of $h(j\omega)$ is

$$\text{Re}[h(j\omega)] = \frac{d_0 a_2 + (a_1 d_1 - d_0)\omega^2}{(a_2 - \omega^2)^2 + a_1^2 \omega^2} \quad (20)$$

$\text{Re}[h(j\omega)] > 0$, $d_0 > 0$, $d_1 > d_0/a_1$ will be satisfied. The compensator must meet the two conditions.

2.2.3.2 Determining φ_1, φ_2
From Equation (16), (17), φ_1, φ_2 can be solved.
Determine φ_2: Assuming that $\varphi_2 = k_2(t)vr$, $k_2(t) \geq 0$, Equation (17) will be satisfied. It is easy to take $k_2(t) = k_2$ (proportional coefficient), so $\varphi_2 = k_2 vr$.
Determine φ_1: The differential and integral calculus is given by

$$\int_0^t f'(t) k_1' f(t) dt = k_1' \int_0^t f(t) df(t)$$
$$= \frac{k_1'}{2}[f^2(t) - f^2(0)] \geq -\frac{k_1'}{2} f^2(0) \quad (21)$$

Suppose $f'(t) = vr$,

$$k_1' f(t) = b_p[\int_0^t \varphi_1 d\tau + k_c(0)] - b_m \quad (22)$$

and $f(t) = \frac{b_p}{k_1'}[\int_0^t \varphi_1 d\tau + k_c(0)] - \frac{b_m}{k_1'}$, we obtain

$$\varphi_1 = \frac{k_1'}{b_p} vr = k_1 vr, k_1 = \frac{k_1'}{b_p} \quad (23)$$

In conclusion, the adaptive law (Chen & Jiang 2009, Han 1986) is

$$k_c = \int_0^t k_1 vr \, dt + k_2 vr + k_c(0) \quad (24)$$

3 MODELING AND SIMULATION

3.1 *Step response of the dynamic model*

The study uses the simplified model of the second order system from Equation (1). According to the derived compensator D and controller k_c, a step command simulation system of the test is established in the Matlab/Simulink (Wang 2012, Shi 2014), as shown in Figure 4, in which the constants d_0, d_1 of the compensator must meet the conditions that make $\text{Re}[h(j\omega)] > 0$. Adjust the proportion coefficient to make sure that the requirements of φ_1, φ_2 are fulfilled.

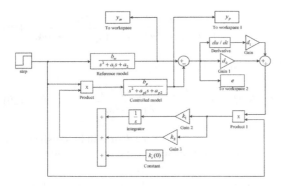

Figure 4.   Step control system diagram.

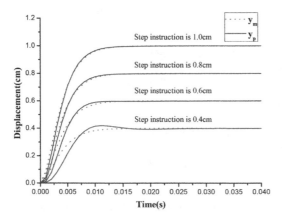

Figure 5.   Displacement response of several step instructions.

The parameters $\xi_A = 1.0$, $\omega_A = 500$ of the reference model are taken, and the transfer function of the reference model is

$$G_m(s) = \frac{250000}{s^2 + 1000s + 250000} \qquad (25)$$

The parameters of the controlled object are $b_p = 20000$, $a_{p1} = 200$, $a_{p2} = 20000$; the parameters of the compensator are $d_0 = 9.5$, $d_1 = 0.05$, $k_1 = 1500$, $k_2 = 5.5$, $k_c(0) = 0.05$. When the step orders are 0.4 cm, 0.6 cm, 0.8 cm and 1 cm. The displacement response curves of the system are shown in Figure 5.

The response of four kinds of step instruction, as shown in Figure 5, show that the designed MRAC system has good tracking accuracy and stability. From the analyses of data shown in the figure, the system can finish the action of a single-step instruction before 0.015 s, which is

much smaller compared with that reported in the literature (0.04 s) (Zhang 2008), and can greatly enhance condensed space of the integral algorithm step, effectively improving the accuracy of the test. From the smoothness of curves shown in Figure 5, the action of the MRAC system is very stable under the step instruction. Comparing several kinds of the response of the step instructions, if the adjustment coefficients k_1, k_2 are determined down, although step instructions change, the system is still able to achieve a specified displacement quickly and stably, which indicate that the MRAC control system based on the Popov hyperstable theory has a strong adaptability (Landau 1974).

In order to prove that the design of the MRAC system based on the Popov hyperstability theory in the load control of real-time substructure testing is a kind of good control methods, for the test substructure simplified the same controlled object, the step response of the different stiffness (K_E and $0.1 K_E$) (Deng 2007) under the control of the super stable MRAC system and the PID control system is calculated and their control effect is compared. The results are shown in Figure 6. It is obvious to find that the displacement track basically remains unchanged in the super stable MRAC system's control under the condition of the stiffness changing from K_E to $0.1 K_E$ shown in Figure 6, while the PID control system is unstable and cannot quickly perform the response displacement instruction under the same stiffness changing condition. So compared with the PID, the super stable MRAC system has stronger adaptability, stability and rapidity in response.

To further demonstrate the superiority of super stable MRAC, the controlled object simplified by the same test substructure is again taken into consid-

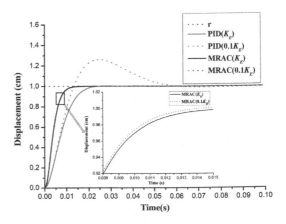

Figure 6.   The response of different rigidity under super stabilization MRAC and the PID control.

eration. Compared with the MRAC system (Deng 2007), the gain parameters are $\gamma_r = 4 \times 10^5$, $\gamma_{y0} = 4$, $\gamma_{y1} = 0.08$, based on Lyapunov in the same step command response (r = 1cm). The gain parameters in super stable MRAC are still $k_1 = 1500$, $k_2 = 5.5$. The step responses of the two control methods are shown in Figure 7. The response speed difference between the two kinds of MRAC control is not big, as shown in Figure 7, but MRAC based on Lyapunov shows a slight fluctuation on the stability when the control does not reach the target displacement, which is unfavorable for test, so super stable (Popov) MRAC in control is better than the MRAC based on Lyapunov on the stability.

In summary, in control load of the real-time substructure test, the step response of the super stable MRAC system can reach the desired displacement steadily, accurately and quickly. Its performance in the test is better than PID and the MRAC based on Lyapunov to some extent.

3.2 *Response under earthquake*

Because the study of the test is load control, the effect of numerical integration algorithm can be ignored. The standard center difference method

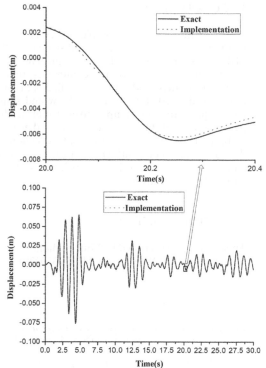

Figure 9. The comparison between the exact solution and the simulation results by super stable MRAC control.

(Wu et al. 2005, Wu et al. 2009) is adopted. The schematic diagram of the test is shown in Figure 8.

For the single degree of freedom system, $M_N = 5000$ kg; $K_N = 147.392$ kN/m, $C_N = 3.1432$ kNs/m are numerical substructure stiffness coefficient and damping coefficient. $K_E = 50{\sim}2000$ kN/m, $C_E = 0$, are experimental substructure stiffness coefficient and damping coefficient. The initial displacement, velocity and acceleration of the structure are all zero. El-Centro (NS, 1940) seismic wave is input load in the simulation, in which the peak acceleration is adjusted for 200 gals, and the integral step is set to $\Delta t = 0.04$ s. The simulation result using the MATLAB/Simulink is shown in Figure 9.

As can be seen from Figure 9, anastomosis degree of the two curves is very high, almost completely overlapping. The result proves that the MRAC system based on the Popov theory can be very good to ensure the test's precision and stability, which can improve the reliability of the test structure. It can reflect the dynamic characteristics of the structure, so as to provide the basis for anti-seismic design.

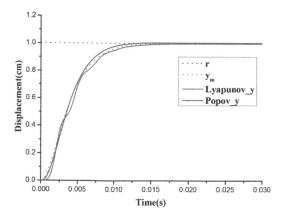

Figure 7. Step response under the two kinds of MRAC control.

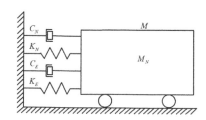

Figure 8. Schematic diagram of the structure with one degree.

4 CONCLUSIONS

Based on the Popov hyperstable theory in hybrid loading control of real-time substructure, the MRAC system is a relatively good control method, which can achieve response commands timely, stably, accurately and can be applied and promoted in the test. But the derived law and simulation application are comparatively simple in this paper, where still there are some places that need to be perfect, such as proportionate selection of the gains, the change influence on the test results and higher-order control law. Further research is required on these areas.

ACKNOWLEDGMENT

This paper was sponsored by Young scholars development fund of SWPU (GN: 201131010078).

REFERENCES

Chen, F. & Jiang, B. 2009. *Adaptive control and application.* Beijing: National Defence Industry Press, 111–121.

Deng, L. 2007. *Adaptive control method for real-time substructure testing.* Harbin Institute of Technology. China: Harbin.

Han, Y. 1986. The eigenvalues of positive definite integral operator. *Journal of Hebei University (NATURAL SCIENCE EDITION),* 01: 18–30.

Landau, I.D. 1969. A hyperstablility criterion for model reference adaptive control systems. *IEEE Trans. Autom. control,* 552–555.

Landau, I.D. 1974. A survey of model reference adaptive techniques—Theory and applications. *Automatica,* 10(4): 353–379.

Li, Y. Zhang, K. & Wang, H. 2010. *Adaptive control theory and appliction.* Xi An: Northwestern Polytechnical University Press, 81–83.

Popov, V.M. 1973. *Hyperstability of control systems.* Spring Verlag.

Qiu, F. & Chen, Z. 2000. *Structure test method of seismic.* Science Press. China: Beijing.

Shi, L.C. 2014. *MATLAB/Simulink system simulation super learning handbook.* People's Posts and Telecommunications Press. China: Beijing.

Taknashi, K. & Nakashima, M. 1987. Japanese Activities on On-line Testing, *Journal of Engineering Mechanics, ASCE,* 113(7): 1014–1032.

Zhang, T. 2008. *LQR control of electro hydraulic servo loading system and the application in real-time substructure testing,* Harbin Institute of Technology. China: Harbin.

Wang, Q. 2007. *Real time substructure testing method and its application.* Harbin Institute of Technology. China: Harbin.

Wang, Z. 2012. *MATLAB/Simulink and Simulation of the control system.* Electronic Industry Press. China: Beijing.

Wu, B. Bao, H. Ou, J. & Tian, S. 2005. Stability and Accuracy Analysis of Central Difference Method for Real-time Substructure Testing. *Earthquake Engineering and Structural Dynamics.* 34: 705–718.

Wu, B. Deng, L. & Yang, X. 2009. Stability of Central Difference Method for Dynamic Real-time Substructure Testing. *Earthquake Engineering and Structural Dynamics,* 38: 1649–1663.

Advances in Civil, Architectural, Structural and Constructional Engineering – Kim, Jung & Seo (Eds)
© *2016 Taylor & Francis Group, London, ISBN 978-1-138-02849-4*

3D design support and software compensation for rapid virtual prototyping of tractor rockshaft arm

Anil Kumar Matta & D. Ranga Raju
S.R.K.R. Engineering College, Bhimavaram, W.G, Andhra Pradesh, India

K.N.S. Suman
A.U.E.C, A.U, Visakhapatnam, Andhra Pradesh, India

ABSTRACT: The paper presents 3D layout configurator software tool to rapidly build up the 3D model of tractor rockshaft arm; The 3D model from the point cloud of scanned component is analyzed and developed. This paper also addresses accuracy improvement of rapid prototyping by software error compensation. This approach is inspired by the techniques developed over the years for the parametric evaluation of coordinate measuring machines and machine tool systems. A 3D model is developed and error compensation is applied to the files which drive the build model. The results show a significant improvement in dimensional accuracy of built parts.

1 INTRODUCTION

Raising requirements on component durability and reliability on one hand and the introduction of new products on the other hand, forces the automobile manufacturers to enhance their products continuously. In order to achieve these improvements, CAD systems, based on noncontact servos, have found their way into tractors and they contributed considerably to many new advantages in terms of durability and reliability such as tractor lift arm, steering system, seat cushion system, semi-active suspension control, tractor mounted bevameter and Roll over protective structure (ROPS). Adding to this most of the research projects are initiated with the aim of developing methodologies, tools and cooperative environments for rapid design and prototyping of automobile components (Tong et al. 2004). There are two approaches which can be used to improve the accuracy of a Rapid Prototyping process. The first approach seeks to eliminate the source of an error. The second approach strives to cancel the effect of an error and is known as error compensation (Nemeth et al. 2013). In the research conducted by A.K. Matta (Matta & Purushottam 2014), a prototypical model is developed which is capable of working under high temperatures and different loading conditions. From the review conducted by Matta et al. 2015 new technologies and concepts to Rapid Prototyping systems reduce the cycle and cost of product development.

This paper presents the results achieved in the development of a 3D model and error compensation in STL files for rapid prototyping of rockshaft arm.

2 COMPONENT SELECTION

A rock shaft is by definition a shaft that does not rotate completely around but in a particular arc. The rock shaft is the lift shaft of the power lift, it rocks about a half turn and that moves an attached bell crank lever to lift the cultivators or a three point hitch as shown in Figure 1. The rockshaft

Figure 1. Rear view of the tractor with arm in the box.

Figure 2.   Failure component.

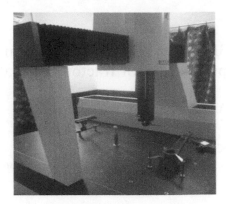

Figure 3a.   Coordinate Measuring Machine (CMM).

arm moves up and down which in turn provides motion to the agriculture equipment's. By providing this motion to the cultivator, the soil is plucked for the agriculture purpose. But the existing rock shaft model breaks and requires frequent replacement as shown in Figure 2.

2.1   *Creating 3D model from the point cloud*

Data processing of Rockshaft arm is investigated by Coordinate Measuring Machine (CMM) as shown in Figure 3a. The point cloud created by Laser scanning method contains all the data points which is processed for 3D visualization and design (Figure 3b). Existing software technologies are used to create CAD models from the point cloud. In the first method, the point cloud is imported into Pro-E as a reference. This cloud is used as a base to generate a CAD model. And the final data is exported as.prg.

In the second method, the Tangram software shown in Figure 3c is used which enables the recognition of features (plane, cylinder) from the point cloud. An output file is generated in the form of IGES file showing the cloud points of the part. The IGS file format is based on the IGES (Initial Graphics Exchange Specification) standard. This standard is used by a variety of CAD (computer aided drawing) applications. IGS files are saved in a text-based ASCII format. The IGS format has become the standard file format for transferring 3D model files between different CAD programs.

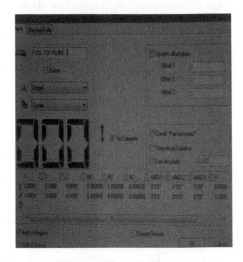

Figure 3b.   Point Cloud data from CMM.

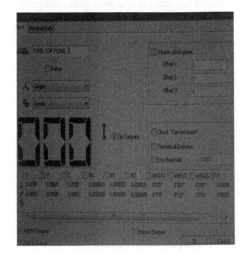

Figure 3c.   Tangram software.

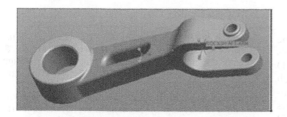

Figure 4. Modified design.

This allows a user who has created a model file in one application to send it to a user who does not have the same software program without worrying about whether or not the recipient will be able to open the file. The features are extracted from the cloud points and coordinates values x, y, z of each cloud point.

The contour surface is detected by the intersection of the recognized features. Then the closed surface is created using the solid modeling techniques.

2.2 3D analysis tool

ANSYS analysis tool is used to evaluate and improve the 3D design of the rockshaft arm. Loads of different magnitude is applied to accommodate a wide range of implements being attached. Static and fatigue analysis is carried out to detect the cause of fracture. Force is applied at fork side of the arm and fixed at another end. The analysis is carried out at loads of 3000 N, 4500 N and 6000 N. From the results, a modified design is created as shown in Figure 4 in which the material is added at locations where the stress concentration is high and sharp edges are modified by fillets.

The slot details are as follows:

Distance between centers: 60 mm
Radius: 10 mm
Radius of rounded edge: 4.75 mm
Extruded depth: 32.5 mm

The following figures display the meshing of the models with triangular elements.
No. of nodes: 31714
No. of elements: 17571

Volume before modification: 569100 mm³
Mass before modification: 4.4675 kg.
Volume after modification: 676650 mm³
Mass after modifying: 5.317 kg.

A comparison between the existing and proposed design is shown in Table 1 and Table 2 in terms of both structural and fatigue analysis. The equivalent stress induced due to static loading is reduced significantly by geometric modifications. Safety factor and no. of cycles before failure are also increased in fatigue analysis.

Table 1. Comparision of two designs for static structural analysis.

Load (N)	Existing Design Defor. (mm)	Existing Design Equ.Stress (MPa)	Proposed Design Defor. (mm)	Proposed Design Equ.Stress (MPa)
3000	0.007	115.04	0.00	65.07
4500	0.108	172.56	0.005	97.60
6000	0.145	230.98	0.007	130.14

Table 2. Comparision of two designs for fatigue analysis.

Load (N)	Existing Design Life	Existing Design Safety factor	Proposed Design Life	Proposed Design Safety factor
3000	1.94e5	2.173	1e6	3.84
4500	48096	1.448	1.488e5	2.56
6000	15854	1.082	1.23e5	1.92

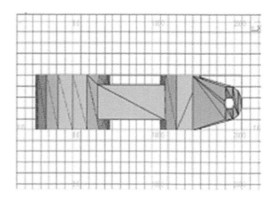

Figure 5. Compensation of STL file errors.

3 ERROR COMPENSATION

From a CAD design to a final prototype, the part is represented in several file formats: Pro/Engineer model, the STL file after triangulation, binary format or ASCII format, and the slice file generated by the slicing software. The error compensation needs to be addressed. The compensation software Desk Artes is used to improve the accuracy of the models, to demonstrate the capability of the compensation process and to improve profile accuracy and dimensional accuracy as shown in Figure 5. The overall size of the part and feature positions on the part show considerable improvement. The file errors are compensated with 512 triangles, length 256.2546 mm, width 93.1506 mm, height 50.8498 mm and surface 99.

4 CONCLUSIONS

The 3D modeling techniques presented in the paper offer much functionality. The analysis tools support the user to evaluate and improve the 3D designs; they help the selected component to match and fit into automobile systems. For the extension, the point cloud based modeling techniques may provide valuable support to rapidly create 3D models of existing objects.

Compensation is applied to the STL files. This allows control over all vertices on the contour of each layer and thus increasing the resolution of the model.

The presented 3D modeling techniques indicate the direction in which automobile design tools should advance in the future. In this way, the design tools may reach a similar functionality that current CAD systems provide for product design.

REFERENCES

Matta, A.K. & Purushottam, V. 2014. *Analysis of the Novel Brake Rotor using FEM*. 5th international & 26th All india Manufacturing Technology, Design and Research conference AIMTDR, IIT Guwahati, Assam, India, 12–14.

Nemeth, I. et al. 2013. *3D Design Support for Rapid Virtual Pro-totyping of Manufacturing Systems*. Forty Sixth CIRP Conference on Manufacturing Systems; Procedia CIRP, 7: 431–6.

Tong, K. et al. 2004. Software compensation of rapid prototyping machines. *Elsevier. Journal of Precision Engineering*, 28: 280–292.

Matta, A.K. et al. 2015. *The integration of CAD/CAM and Rapid Prototyping in Product Development: A review*. 3rd International Conference on Material Processing and Characterisation ICMPC, Elsevier, Journal of Materials today: Proceedings, 2: 3438–3445.

Advances in Civil, Architectural, Structural and Constructional Engineering – Kim, Jung & Seo (Eds)
© 2016 Taylor & Francis Group, London, ISBN 978-1-138-02849-4

Research on defining green road project management operating scope by the application of PDRI

A.P. Chang

Department of Environmental Technology and Management, Taoyuan Innovation Institute
of Technology, Taoyuan City, Taiwan, R.O.C.

ABSTRACT: This study focuses on the early project operation content of green road development, taking full account of the operation items of various stages of the life cycle, by using a special tool called the Project Definition Rating Index (PDRI), PDRI assessment framework, defining the need and scope of green road project development operations in the metropolitan area, and inquiring and setting various assessment indicators of the work to complete the building of the overall indicator assessment system. Based on a policy aspect, as well as data collection from instances of project case, our research program obtains the evaluation index operation item, and by semi-structured interviews, it obtains supplements insufficient evaluation index items and particular needs. The results of this study will be beneficial to the application in the metropolitan regional overall development project. Building the green road index evaluation system helps to avoid the environmental impact caused by land development, reduce damage to people's life safety as a consequence from needs for road maintenance, alleviate future maintenance and management expenses, and is in line with the expected benefit of green road project sustainable development.

1 INTRODUCTION

The term "green" is widely used in various industries and areas of expertise. A lot of studies on the scope of green roads such as the meaning, operations and performance evaluation have different definitions and final performance requirements. Taoyuan Zhonglu planning area is located in the center of the three urban planning areas of Nankan, Taoyuan City, and Neili, with a total development area of about 143 hectares. Major operations include the roads, five big pipeline projects, regional drainage and water conservancy projects, landscape engineering, flood detention and disaster prevention in Zhonglu Park. The whole project takes full advantage of the local environment and regional development needs, provides disaster prevention, disaster mitigation, emergency rescue and temporary shelter space, and brings into play the concept of sustainable environment.

In response to future regional development trends, regarding the operation scope of metropolitan green road development, including traffic control engineering, pavement and road engineering, common conduit, pedestrian environment and other projects, this study defines operating scopes at different stages of planning, design, construction, maintenance and management. The results of this study can be used by authorities as future reference for performance assessment in the segment expropriation project.

In response to future regional development trends, regarding the operation scope of metropolitan green road development, including traffic control engineering, pavement and road engineering, common conduit, pedestrian environment and other projects, this study defines operating scopes at different stages of planning, design, construction, maintenance and management. The results of this study can be used by authorities as future reference for performance assessment in the segment expropriation project.

2 LITERATURE

Many studies in the literature have divided green roads into three constructed forms of bridges, tunnels, and embankments, but the metropolitan road development carries more specific needs. This three infrastructures cannot meet the needs of the metropolitan road system (Chang et al. 2014a). Complete green roadwork must focus on the perfection of the operating range at various stages of the life cycle. According to "public construction work cost estimate provision manual", the life cycle of road is detailed, as shown in Figure 1 (Lin et al. 2011).

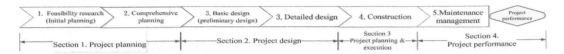

Figure 1. The name of the road project life cycle.

Full life cycle of public works, from the initial feasibility assessment, planning, design, construction to finally the maintenance management, at different stages has different management mechanisms (Lin et al. 2010a), and the quality management is mostly established in the construction phase. On the ground that quality management in the feasible research phase involves broader areas of expertise, with respect to each other, it is more difficult to set clear objectives of quality management and detailed job control content, and, likewise, the complexity is higher (Lin et al. 2010b).

Currently, Taoyuan City has consecutively completed Jingguo and Zhonglu two regional developments. Connotation species included in green road are numerous. This study is based on the metropolitan regional green road system as the main scope of the study, which also includes phases of work such as building road management information systems and public utility pipelines investigation and measurement (Chu et al. 2014a), the initial project planning and design, quality management control and test in the construction process, as well as road maintenance and management (Chu et al. 2014b).

At first, the Project Definition Rating Index (PDRI), was a research achievement made by the efforts of the construction project research team comprised by US Construction Industry Institute (CII) in 1994 (Chang et al. 2014b). PDRI is a checklist defined by a wide range of project work element scopes, to meet the needs of the project forward planning. PDRI is a simple and easy tool to provide project teams with a means to measure the objective assessment of the project during early project planning (Wang & Gibson 2010). PDRI was first used in 1999 in the construction industry project. After two revisions, it has a considered appropriate relationship between owners and contractors, and questioned the sustainable development of the project, so that the scoring of project forward planning assessment required to be automated (Chang et al. 2014b). The PDRI, in the early project planning process and development of strategy information, enables owners to obtain adequate assurance on risk and decision-making, especially the understanding and acceptance of the owners. Therefore, the objectives and guidelines for the project must be clearly defined in order to achieve the greatest success of the project (Chu et al. 2014c).

Project pre-planning process job elements are how good pre-planning of projects occurs. It will have a significant impact on cost, schedule and performance of equipment operation execution. So the better the initial planning of the project is, the more financial success the overall project achieves (Gibson and Hamilton 1994). The success in the initial stages of construction project's detail design is totally dependent on efforts made to define the management operation scope of the project during the period to define the scope of the project (Chang et al. 2013a).

3 BUILDING GREEN ROAD PROJECT MANAGEMENT ASSESSMENT MODEL

Application of the Project Definition Rating Index (PDRI) assessment model mainly divides metropolitan green road project management assessment model into four sections, of which job scope covers job items at various stages of all the life cycle. Building this evaluation mode will be conducive to the assessment basis when urban planning carries out projects related to sector expropriation. This article is limited to the length of the article. Various related milestone items and scopes of index work circumstantially are outlined in Table 1.

3.1 PDRI assessment model design

With regard to the PDRI assessment model structure hierarchy, its WBS job hierarchy grading layers take three levels in principle, and more detailed job decomposition items are included in the description for element. As shown in Figure 2, the first level is section, under which the second level is category. The third layer is element. Its operational item details are included in each element for description, in order to complete the construction of the whole project's PDRI assessment framework.

3.2 Set PDRI operation weights

After the establishment of PDRI evaluation mode is complete, sections, categories and elements will be placed in the PDRI definition rating scoring table as the basis for follow-up formulation of

Table 1. Metropolitan green road PDRI operating range.

SECTION I Feasibility Study

A   Project pre-planning and integrated planning

A 1   Master plan analysis
A 2   Correlation plan content analysis
A 3   Overall planning and basic design in resource area
A 4   Detailed design guideline and construction outline specification
A 5   Status survey and analysis for project area
A 7   Overall planning and basic design for rainwater drainage
A 8   Overall planning for running water and sewage project
A 9   Initially estimated project cost and time course planning

SECTION II Project Design

B   Planning and design concepts

B 1   Socio-economic environment
B 2   Natural environment
B 3   Landscape ecological environment

C   Issues Faced and countermeasures

C 1   Discussion topic formulation
C 2   Deliberation of discussion topics
C 3   Solution scheme

D   Design strategy and concept

D 1   Overall space framework deliberation
D 2   Zoning design ideas
D 3   Road perimeter planting design ideas
D 4   Disaster prevention system planning and design
D 5   Water conservancy project design ideas
D 6   Ecological design ideas
D 7   Road gallery landscape lighting design
D 8   Information, marking system design
D 9   Running water and sewage design
D 10   Spray irrigation system design
D 11   Common piping and conduit design

E   Project fund and financial plan

E 1   Project fund and budget details
E 2   Reasonable analysis on project budget
E 3   Analysis on difference of procurement contracting strategy composition
E 4   Bid price composition content research and analysis
E 5   The reasonableness of the bid price component

SECTION III Project Planning and Execution

F   Project program

F 1   Project supervision plan
F 2   Engineering overseeing program

Table 1. *(Continued)*

F 3   The overall construction plan
F 4   Human resources plan
F 5   Project administration management
F 7   Safety and health program
F 8   Project progress plan
F 9   Project quality control plan
F 10   Project procurement management
F 11   Project environmental management plan
F 12   Risk management program
F 13   Project finance plan
F 14   Project cost management

G   Project control

G 1   Safety assurance and control
G 2   Quality assurance and control
G 3   Risk control
G 4   Project purchasing control
G 5   Construction machinery
G 6   Construction workers
G 7   Project cost control
G 8   Project progress control
G 9   Environmental control

SECTION IV Project Performance

H   Project maintenance management

H 1   Maintenance management program
H 2   Operation maintenance execution cycle planning
H 3   Maintenance update plans
H 4   Operational maintenance expense plans

I   Project performance

I 1   Project planning accomplishment performance
I 2   Project design and process performance
I 3   Project efficiency and effectiveness
I 4   Project management executive performance

weight. This study adopts the "column" vector analysis method of the eigenvector weight. With an eye of setting mutual importance on weights between sections, categories and elements of the PDRI, Figure 3 shows a pairwise comparison of computing paradigm for the relative importance of category's element. It objectively defines the weights associated with sections, categories and elements and the importance priority among the projects of the job items.

A relative important pairwise comparison was made to measure the relative importance analysis between sections, categories, and elements, and the logic of the analysis results is required to pass

97

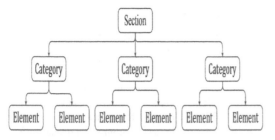

Figure 2. Schematic map of the PDRI work breakdown structure.

Pairwise comparison for the relative importance of category's element.

CATEGORY G.	G 1. List of equipment.	G 2. Equipment location map.	G 3. Equipment utility demand.	e-vectors.
G 1. List of equipment.	1.	1.	2.	0.411.
G 2. Equipment location map.	1.	1.	1.	0.328.
G 3. Equipment utility demand.	1/2.	1.	1.	0.261.
.	2.500.	3.000.	4.000.	1.000.
.	.		λ max = 3.054 C. I. = 0.027 <0.1.	

Figure 3. Eigenvector weight analysis paradigm.

Table 2. PDRI section and category weights.

Section weights		Category weights	
Section	Weights	Category	Weights
II Project Design	560	D. Design strategy and concept	205
III Project planning and execution	229	B. Plan design concept	157
I Project procurement risk	136	F. Project program	132
		E. Project funds and financial planning	117
IV Feasibility study	75	G. Project control	97
Total	1000	H. Project maintenance management	83
		C. Issues faced and countermeasures	81
		A. Project pre-planning and integrated planning	75
		J. Project performance	53
		Total	1000

consistency inspection (C. I. value). The maximum eigenvalue λ max and consistency test C. I. value are calculated, which increase the efficiency of the assessment model and decision-making. Based on the empirical rule, the integrated scale is proposed. The C. I. value of less than 0.1 is regarded as the matrix having validity, which denotes that the matrix is feasible and consistent (Chang et al. 2013b).

It is evident from Table 2 that when the mutual relative importance weights of each section and category are completed, the mutual importance order of the entire evaluation mode can be understood clearly. That is, from initiation to completion of the entire project, it is easy to identify the work items of which we especially have to take much account at the various stages of the entire life cycle. It also represents the proper proportion of effectiveness and performance in order that desired objectives of overall project performance are to be achieved.

4 CONCLUSIONS

This research, targeting the development of metropolitan green road projects, can precisely meet project requirements and expected development purposes. From the above analyses, the main conclusions and benefits can be drawn as follows:

1. Diminish pipeline manhole covers (hand hole covers), to shape a new urban appearance.
2. Pre-plan to set lead-in pipes for various utility pipelines on clients end.
3. Perfect bicycle road network system.

Regional zone expropriation project is one of the ways to thoroughly reinvent the old city, is also the process to evolve from the old thinking to new thinking, and is also the power to make the whole society improve the living environment and then become sophisticate and innovative. Promoting "green" and "smart" urban visions must be the future trend of urban development construction. To achieve this goal, it is necessary to have a sound overall public facility planning. Sustainable development of an advanced city affects dimensions even more extending to the social, economic, environmental and systematic aspects. We expect this study provides agencies with the assessment and reference for future public work of segment expropriation.

REFERENCES

Chang, A.P. Chou, C.C. Lin, J.D. & Hsu, C.Y. 2013a. Road Construction Project Environmental Impact Assessment Scope Definition Using Project Definition Rating Index (PDRI). *Advanced Materials Research,* 723: 885–892.

Chang, A.P. Lin, J.D. & Chou, C.C. 2013b. Analytic Network Process (ANP)-Selection of the Best Alternative in the Pro-motion of Participation in Infrastructure

Projects. *International Journal of Pavement Research and Technology*, 6(5): 612–619.

Chang, A.P. & Chu, T.J. 2014a. Road Service Quality Sophisticated Practice Road Information Survey Quality Control Operations, *Quality Monthly,* 50(9): 23–29.

Chang, A.P. Chou, C.C. & Lin, J.D. 2014b. To Enhance Quality Control by Using PDRI to Define Scope of Project Procurement, *Journal of Quality,* 21: 269–283.

Chu, T.J. Chang, A.P. & Chou, Y.S. 2014a. Road Service Quality Sophisticated Practice—Public Utility Pipeline Survey and Measurement. *Quality Monthly*, 50(5): 6–13.

Chu, T.J. Chang, A.P. & Lin, J.D. 2014b. Road Maintenance Quality Management Sophisticated Practice, *Quality Monthly*, 50(2): 19–24.

Chu, T.J. Chang, A.P. Hwang, C.L. & Lin, J.D. 2014c. Intelligent and Green Buildings Project Management Scope Definition Using Project Definition Rating Index (PDRI). *Advanced Materials Research*, 945–949: 3008–3011.

Gibson, G.E. & Hamilton, M.R. 1994. *Analysis of pre-project planning effort and success variables for capital facility projects SD-105.* Construction Industry Institute. Austin. Texas.

Lin, J.D. Chou, C.C. & Chang, A.P. 2010a. Explore the Quality Management Operations in the Road Project Engineering Feasibility Study Phase. *Pavement Engineering,* 8(4): 1–10.

Lin, J.D. Chang, A.P. Du, Y.C. & Gao, H.H. 2010b. Inquiry into Public Works Quality Management in the Whole Life Cycle Operation Maintenance Stage. *Pavement Engineering*, 50(5): 61–67.

Lin, J.D. & Chang, A.P. 2011. Explore the Quality Management Operations in the Project Engineering Feasibility Study Phase. *Journal of Chinese Institute of Engineers,* 84(1): 108–118.

Wang, Y.R. & Gibson, G.E. 2010. A study of pre-project planning and project success using ANNs and regression models. *Automation in Construction*, 19(3): 341–346.

Advances in Civil, Architectural, Structural and Constructional Engineering – Kim, Jung & Seo (Eds)
© 2016 Taylor & Francis Group, London, ISBN 978-1-138-02849-4

The conservation and revitalization of the historic district of Taizhou, Zhejiang Province

X. Chen & P. Zhang
The School of Architecture, Soochow University, Suzhou, Jiangsu, China
The College of Architecture and Urban Planning, Tongji University, Shanghai, China

ABSTRACT: North Xinjiao Street is the only traditional commercial street that remains from the Qing Dynasty in Jiao Jiang District in Taizhou, Zhejiang Province. Due to its special historic reasons, this historic street and its related vernacular community inherited the early form of opening port in the late coastal colonial city, boasting cherished valuable memories that are worthy to be preserved. The vernacular architecture conservation and regeneration group has conducted the conservation and regeneration strategy in the North Xinjiao Street, originally called "Haimen Old Street" (and popularly called "little Shanghai") in the urban renovation of Jiaojiang District since March 2000. This project preserved the key section of this 100-year-old street, which is 225 m long and covers 2.1 ha. In this project, the street is repaired, infrastructure facilities are improved, and the historic architecture are restored and renovated. Thanks to the conservation and regeneration design and practice, North JiaoXin Street has become an attraction of Tai Zhou, served for collective memory, nostalgic ceremony and scenery view with the integration of contemporary city life, which provided a reference example for the conservation and regeneration approach of vernacular blocks in the local city.

1 THE BACKGROUND OF THE HISTORIC DISTRICT

1.1 North xinjiao street in history

Jiao Jiang, a port city in the middle coast of Zhejiang, is located at the Taizhou Gulf. It is called "Haimen" (gate to the sea) and has a history of over 2,200 years. North Xinjiao Street was originally a path through the Qing Bo gate of Haimen. It was connected with port (traditionally called "Dao tou") by the Jiao River in the north.

During the Ming Guo age (the Republic of China era), North Xinjiao Street, as the major commercial street and traffic path in Jiao Jiang, was named as "Bei Da Street" (Big North Street), and its significance has been increasing. Later on, there were more and more branches along the section and more extensions to the length of the street. Inside the street, there are mostly residential houses, while along the street, there are mostly restaurants and inns. Because this area was always full of customers and reached the summit of the prosperous period at that time, people called it "little Shanghai". Its location of outgoing sea Port of Jiao Jiang River has determined that it could not be as competitive as the city Shanghai, which is at the outgoing sea port of the Yangtze River; thus, the street characteristics of earlier open-port city were well preserved. This little street had been flourishing until the new district was developed around the 1980s.

1.1.1 Social function

In history, North Xinjiao Street boasted diverse social functions. Except for commercial trade and city traffic, it played the role for social tradition activities. Specifically, around the Yang Fu Temple and Stage, there was one of the civic gathering spaces for the residents. When there came any festivals or temple fairs, it would be more hilarious, and, especially, the Haimen Gate Archway and Jie Guan Pavilion created the atmosphere of ceremony for the street (Figure 1). Moreover, North Xinjiao Street was also used for residential function as a high-density residential area with busy commercial functions as other traditional commercial community.

1.1.2 Space structure

The current North Xinjiao Street starts from Ren Min Rd from the south, which is constructed by filling City-guard River, ends in Xiang Yang Rd to the North, which lasts 225 m long and 4–6 m wide. The cornices of the architecture along the street are about 5–7 m high, and the scale of the street is appropriate in a natural form. The street is enlarged around Yang Fu Temple into a little plaza, which creates the highlight of the whole street space. The street is intricately close to the plaza to some extent and is overall in the direction of north-west to south-east. In addition, the street that is located south to the Yang Fu Temple is relatively narrow

Figure 1. The living situation in the 1990s.

Figure 2. The images of Chinese—and Western-style buildings.

and the architecture along the street mostly consists of timber structure, while the north section is wider and the architecture is made up of bricks with relatively more stories, which clearly displays the evolution of the street extending from old town to the riverside with the dam removed to the north. North Xinjiao Street is paved by the cordierite plate. Each of the central arrays of the plates is 160 cm × 80 cm, connected with each other horizontally. The plates at the two sides are relative small (120 cm × 60 cm), paved in vertical arrangement in several arrays. The pavement had been polished throughout the hundred years, and the plates are all complete except for some minor cracks.

1.1.3 *Architectural style (Figure 2)*

North Xinjiao Street boasts a diversity of architectural styles, which includes not only the traditional street shops and courtyard dwellings of Zhe Jiang Province, but also the Baroque style brick-stone gable that attached to the timber structure (locally called "shop in the wall"). We can even find the Western decorative architecture ornaments in some very traditional courtyard dwellings, which presents the Western tendency influence on the East after the opening of the city. It is interesting that the Chinese-style architectures are mostly located close to the old town, while the Western-style architectures are gathered in another side, which is close to the dam. The two kinds of style mixed with each other in the middle part of the street, displaying the

evolution of the street from old town to the dock, as well as the history of the opening port and prosperous business development of late Jiao Jiang. All these witnessed the value of North Xinjiao Street.

1.1.4 *Existing problems*

North Xinjiao Street is located in the old town, whether to demolish it or not was once a heated discussion. Although the government finally made the decision to preserve it, the renovation of other old towns still faced urgent challenges of the same kind.

The architectures are not repaired for years and are damaged to different extents. After the exposure to the sunshine and rainfall for a long time, the exterior facades are all in a poor condition, especially for the timber structure. Some delicate carvings are damaged due to the deterioration of the materials and some beams and columns are even tilted, thus causing security problems. The community lacks essential municipal pipes, and the life quality is poor. The water in some wells can be no longer used for drinking due to the serious environmental pollution, and there is a situation that several households share running water. Grey Water Drainage system is not supplied and traditional toilets are popularly used. Whenever it rains, the street will become water-logged because of non-smooth drainage. What's more, the electricity lines are mostly exposed and are not connected in a standard way, which exposes to severe potential risk of fire to the timber structure architecture along the street.

The typology of the blocks is relatively low and may easily get flooded. Jiao Jiang, as the port, is always impacted by floods and tornados. According to the historic record, Jiao Jiang suffered from tornados several times in the 1980s. Among these tornados, the No. 8923 in 1989 is a tide peak reaching 6.9 m (high level of Wu Song River), which has been the most destroying one since 1923. According to Xu Jinshan, a resident in Yang Fu Temple, his house was flooded to the bed level and the north part to the end of the suspension bridge were all flooded.

In terms of the social structure, most young people moved out to the new district, and this area is mostly occupied by mid-aged people and elders, with relatively low income and poor education. The social structure of this community is simple and lacks vitality or motivation for redevelopment. The traditional brands are disappearing, together with the decaying traditional public space, which causes the loss of traditional civil activities. With the fast expansion of the Chinese cities, this is also the common problem faced by all the urban traditional commercial streets. Therefore, the exploration in this project is of common value that could be used as a typical example.

2 SCOPE OF THE CONSERVATION PROJECT

2.1 *Determination of conservation scope*

In the conservation project of North Xinjiao Street, we conducted numerous on-site surveys and research before determining the final conservation scope. Influenced by the urban development planning, the comprehensive conservation in large scope is difficult to realize. As a result, we first determine the scope for strict conservation, which covers the main body of the old street, architectures as the street interface, as well as the traditional residential courtyard dwellings with a significant value around the street. Beyond this scope, we set up a control area, in which the architectures should comply with the original space structure, and coordinate with the historic districts regarding the scales, to perform as a transition to the surrounding urban space (Figure 3).

2.2 *Project procedure*

The total investment of this project is about RMB 80,000,000 ($US 11,420,000) and is divided into three phases: the pre-stage work, from the end of 2000 to the end of 2001, includes the onsite investigation and measured drawings, conservative planning standards making as well as specific restorative constructional drawings. From the beginning of 2002 to the beginning of 2003, the residents in the old street were relocated by the management department from the government,

Figure 3. Conservation tactics and boundary.

and the random constructions that affected the historic values with the instruction by the design teams were cleared up. From the first half year of 2003 to the first half year of 2004, the restoration work and associated facility construction of the historic community were carried out based on the restoration principles set by the design team. The whole project was completed by the end of 2004.

3 THE SIGNIFICANCE OF THE CONSERVATION WORK

3.1 *Preserving the historic heritage*

The architectures in North Xinjiao Street are of enormous historic value. With the application of the conservation project, the whole spatial structure and the historic buildings are preserved, inheriting the cherished historic heritage of the Jiao Jiang area.

3.2 *Preserving the city memory*

Each city goes through a circular process of born, develop, mature, decay, reconstruction, and then develop again. The historic events converted into the result of the architecture, and remain in the city, accumulating the hints throughout the history as the memory and context of the city (Chang 2011).

3.3 *Inheriting local traditional customs*

North Xinjiao Street is more than a significant commercial street of late Jiao Jiang. It is also a public space for local residents. In the past, whenever there were any festivals, the street would be hustle and bustle with all kinds of local custom and religious activities. Due to the fast expansion and development of contemporary Chinese cites, these traditions are disappearing with the decay of these ancient streets (Chang 2009). Thanks to this conservation and restoration project, not only the architectures are preserved, but the traditional public spaces are revitalized as well. As the design team, we are glad to see these spaces enriching the daily life of the local residents and regenerating the custom activities, which were previously in danger.

3.4 *Stimulating the local economic development*

The conservation of North Xinjiao Street not only preserves the tangible heritages, but also invigorates the traditional commercial business in the Jiao Jiang area, also attracting more tourism resource as well as stimulating the multi-oriented development of local economy. This project introduces more job opportunities and economic sources for the local residents, which improves the life quality of the old district.

4 CONCLUSIONS

The local vernacular culture could not be fully displayed by an individual architecture, but should be inherited by the architecture complex in the village that still remains the typical features. As stated in <Mexico Charter> that: the value of heritage is not only lies in the architecture, construction entity and physical form, but only in the utilizing approach and interpretation of the heritage, as well as the attached unphysical association (Chen 2003).

The architecture in North Xinjiao Street could not be regarded as distinguished individually, and some of them look even simple and crude. However, the uniqueness of this street lies in the fact that it represents the historic events, diverse architectural styles, as well as the consistency and integrity between space and time. The position of the architecture also displays the time when it was constructed, which forms the space of the street into a coordinate axis, displaying the architecture images and city features that have evolved since the opening of the city. Here, the earlier life styles still remains as well. At the end of the Qing

Dynasty and beginning of the Republic of China period when the whole nation faced the struggling revolution and complicated opportunities, this street, as both a splendid social stage and a significant culture character, promoted the development and collected the history of the whole city, which is regarded as an unforgettable memory of this city.

Due to the specific role in the Jiao Jiang history and the current situation, we define the future of North Xinjiao Street as a real and lively traditional commercial platform with full of local products. The historic street scenery should be strictly preserved while a very rigid preservation would be negative. In fact, as for this historic community that covers more than 10,000 m², it is impossible to preserve it like a specimen. Therefore, the active approach is to introduce the contemporary life that could extend the traditional customs to regenerate the old body of the community and to create a lively atmosphere. Our conservation proposal is as follows: First, to revitalize the local small-scale retails, as the basis of street regeneration later in a larger scale; then, reorganize and restore the physical environment, which is the technology approach to realize our goal.

Therefore, the conservation and reuse project of North Xinjiao Street not only preserves the historic physical value via restoration, what's more, it exerts an active influence on the residents' life, revitalizes the traditional civil custom activities, and regenerates the traditional commercial culture via the conservation of the decaying old urban community, which makes up for the gap between the new and old culture, and finally forms a continuous living force inside the city, thus creating a lively and active traditional commercial street in this contemporary city. In the background of urbanization in China, this project is no doubt a helpful typical case for other cities.

ACKNOWLEDGEMENTS

Supported by the National Natural Science Foundation of China (Grant No.51508361)

REFERENCES

Chang, Q. 2009. Reflective thinking of historic conservation. *Time Architecture*. 3: 118–121.

Chang, Q. 2011. *Reflection on architecture heritage's conservation and restoration*. Proceedings of the architectural heritage and the conservation of cultural heritage. Tianjin: Tianjin University Press.

Chen, Z.H. 2003. An International Charter for the conservation of building relics and historic sites. *World Architecture*, 3: 13–14.

Advances in Civil, Architectural, Structural and Constructional Engineering – Kim, Jung & Seo (Eds)
© 2016 Taylor & Francis Group, London, ISBN 978-1-138-02849-4

Industrial engineering methods to improve administrative processes

L. Šťastná
Škoda Auto a. s., Mladá Boleslav, Czech Republic

M. Šimon
University of West Bohemia in Pilsen, Univerzitní, Pilsen, Czech Republic

ABSTRACT: This article focuses on the improvement of administration processes using the method of industrial engineering. It highlights the main differences between manufacturing and administrative processes. Then, it shows the potential that could be found in the administrative processes. The main goal of this article is to introduce the work procedure using the administrative processes, in which the industrial Engineering methods are being used. The use of this procedure will make companies achieve better results in the efficiency of administrative processes.

1 INTRODUCTION

Nowadays, we encounter the use of Industrial Engineering methods, especially in the production spheres. Concurrently, a big potential use of these methods is just in administration processes. With the help of their application to administrative processes, companies can still increase the efficiency and competitiveness of the business and reduce their costs.

In most cases, the non-manufacturing processes do not add value to the final product, but significantly increase costs and delivery times. Most of these processes cannot be eliminated since they are necessary for further follow-up work. Precisely for this reason, the company should start working with these processes and make them more effective to achieve a maximum thinness of the company.

The aim of this article is to introduce industrial engineering methods to improve administrative processes and show a procedure how we can work with these processes.

2 MANUFACTURING COMPANIES AND ADMINISTRATIVE PROCESSES

2.1 Present state of administrative processes

Current status of administrative processes points that these processes significantly contribute the cost to the companies. It is obvious that it depends on the company. However, estimations normally show 60-80 percent to be assigned to administrative processes. Surveys made by API (Figure 1) (Mašín et al. 2007) show that:

– over 100 years, the labor productivity of work in production processes has been raised by 1000 percent. On the other hand, the productivity in administrative processes has been raised by 150 percent only.
– only 30 percent of complaints are caused by production processes and the rest "70p" are related to the development, services and administration.
– administrative processes render 50% of continuous time of order (Košturiak et al. 2006).

2.2 Comparison of manufacturing and administration processes

Based on the surveys, which were conducted in manufacturing companies, the following problems in administration process are defined:

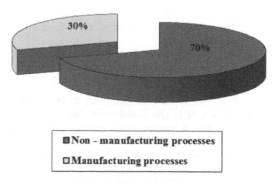

Figure 1.   Causes of complaints in non-manufacturing and manufacturing processes.

- Administrative processes are not dedicated as much attention as to manufacturing processes.
- Administrative processes generate 25–60% of costs on order, sometimes more.
- Many people working at administrative processes do not know concepts and methods of industrial engineering.
- Seldom in the administration process one can find a worker who is responsible for safety, quality, shipping and cost of this process.
- For workers performing administrative processes, changing their thought patterns can be difficult.
- People who work at administration are not accustomed to working at predetermined intervals.

In the above-mentioned lists, we can find a couple of examples of the problems that occur in administration processes.

This status shows in which critical conditions the administration processes are being found. And there is a great potential to increase the company's efficiency and reduce their costs. From this point of view, the company should start working with their administrative processes in the same way, as they are working with their production processes.

3 THE WORK PROCEDURE WITH THE ADMINISTRATIVE PROCESSES

3.1 Identification of objectives of methodology

Initially, during the creation of working methods using the administrative processes, it was necessary to define the objectives that will be achieved. These objectives were based on long-term analysis and requirements from daily practice. The analysis found that the biggest problem is that companies are not able to identify the problems in their administrative processes. That means if companies cannot identify their problem, then they can also not remove away this problem. This fact defined the first objective:

1. The problems identified in the administration processes.
   Based on the first aim, the second aim was set:
2. Find solutions to identify waste removal in these processes.

First, 81 types of waste were defined in these processes. It is the best-known and most occurring types of waste in administrative processes. These 81 types of waste were divided into 8 areas of waste, as it is in case of manufacturing processes. Individual types of waste were determined and adjusted based on a survey that was conducted by the company Fraunhofer IPA, and the book

written by Košturiak (Košturiak et al. 2006), and by identifying waste in the administrative processes according to Lareau (2012). All eight areas including the number and types of waste in each area are detailed in Table 1.

In the second step, it was necessary to find a tool to remove waste in these processes. Here, we decided to use the methods of industrial engineering as it is in production processes. One method of industrial engineering was assigned to each type of waste, which should eliminate the waste. In total, 20 methods were identified for the removal of waste from these processes in Table 2.

Table 1. Areas of waste in the administrative processes.

No.	Areas of waste	Total number
1	Excess of information	8
2	Stock in administrative processes	5
3	Flow of information	9
4	Inefficient administrative processes	16
5	Unnecessary activities in administration processes	11
6	Waiting	9
7	Mistakes	10
8	Unused human skills	13
		81

Table 2. 20 methods used in the administrative processes.

Pos.	Method
1	Jidoka
2	Balanced Scorecard
3	Kanban
4	5S
5	Standardization
6	SDCA
7	PDCA
8	Object Office Kaizen
9	Poka Yoke
10	Value stream mapping
11	Six Sigma
12	Teamwork
13	Theory of Constraints
14	Workshop
15	Visual management
16	Lean layout
17	Just in time
18	Total productive maintenance
19	Job rotation
20	Ergonomic

Methods were assigned on the following basis:

1. The method was assigned based on its utilization for the same problem in practice, i.e. its usage "regarding the given problem" is already verified in practice.
2. The method was assigned based on the assignment method for solving the problems according to Mašín et al. (2007).
3. Method was assigned on the basis of an assessment team of experts on methods of industrial engineering at the Department of Industrial Engineering.

3.2 The methodology for improving administrative processes, using industrial engineering methods

The whole process is divided into three phases. These phases are handled individually, but in sequence. To achieve the desired outcome, it is necessary to go through all the three phases of the methodology.

3.2.1 Preparatory phase

Preparation phase can be described as the actual state in which the company is located. The worker, who performs the implementation at this stage, will develop a complete picture of a company's administrative processes and get an idea of the condition of the processes. Also, he will know whether it will be necessary to implement any of the methods of industrial engineering.

The second step of this phase is to determine the region and deadline for creating the questionnaires. The area is defined as follows: the number of employees who will fill out the questionnaire. These workers must perform the same administrative processes. The questionnaire must be evaluated individually for each different position in the administrative processes. The distribution and filling out of the questionnaire are the last step of this phase.

3.2.2 Calculation phase

This phase is used to obtain calculated data, based on data input. Defined problematic processes are based on the evaluation questionnaires. The next step is the measurement of these problematic processes. It is the most important part of the first two phases, because the final result depends on the accuracy of data input. It is necessary to know the value of the problematic process before it is improved. Evaluation in the next step allows us to obtain the necessary feedback on the changes. For the evaluation of these problematic processes, it is recommended to use the following tools.

1. Method Value Stream Design in indirect Areas (VSDiA)
2. Method and tool for modeling ARIS

3. Method Activity-Based Costing
4. Motivational interviewing
5. Method Visual Office Kaizen

These tools will help us obtain the necessary input information about problematic processes.

3.2.3 Evaluation phases

At this stage, they are implemented methods of industrial engineering at problematic processes. As mentioned above, for each type of problem, one method is assigned, which will help remove the problem. That means after the identification of the problem, we assign a method for elimination based on the given table.

The next step is to present the results of the implementation of the employees. The evaluation step of the implementation is very important for the company. The management of the company will receive the information about what kinds of implementation methods are applied.

- How much time was reduced in this process (time).
- How many of funds were spared (finance).
- How to increase the efficiency of the process.
- How the number of workers has changed in these processes.
- The time spared for individual workers at this process (time).
- How to reduce the cost of office supplies (CZK).

Thanks to this information; in this way, we get the feedback on the real changes. Evaluation of the

Table 3. Individual steps of the methodology.

Phase	Step
Preparatory phase	
1	Determining the status of the company
2	Definition of the unit
3	Completing the questionnaire
Calculation phase	
4	Evaluation of the questionnaire
5	Finding of problem
6	Evaluation of present state of problematic process
Evaluation phase	
7	Implementation of methods on occurring problem
8	Checking and evaluation of implementation of the method
9	Continuous monitoring and improving all processes at company

process after the implementation of the methods must be done with the same tools as done for the problematic process at the second phase.

The last step of this phase is permanent control and improvement of all these processes at company. Phases and main steps are outlined in Table 3.

4 EXAMPLES OF USING THE METHODS OF INDUSTRIAL ENGEENEERING

In the below section, two simple examples are shown about the use of methods of industrial engineering in administrative processes.

4.1 Use of the method 5S in the office

At the car company, the method 5S is implemented to all secretariats (Figure 2). The same system was introduced to each of the secretariats because of fungibility. Furthermore, in the context of implementing this method, the system for print was set up. It was determined which documents should be printed and which documents should be saved on the shared disk.

There was a saving of up to 50 bales of paper per month. Of course, office supplies were reduced.

4.2 Use of the method kanban in office

Kanban in the office can be used, for example, for ordering of office supplies (Figure 3). The biggest advantage is that it can be used when the company has not saved funds for office supplies. Another advantage is that the staff have always the office supplies they need and it does not extend their work. Responsibility is too designated for these needs.

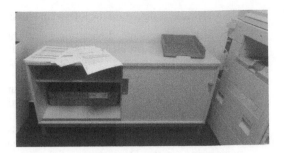

Figure 3. Area of the printer before implementation of the Kanban method.

Figure 4. Implementation of the Kanban system.

An example is given below for ordering of paper using the Kanban method. In order to save the employer's decision that due to the fact that color printing is not as frequently used, there will be only one printer for the entire department located in the corridor. Every office will have only one monochrome printer. After the implementation of the color printer, these two situations arise:

1. Around the printers, packages of papers were scattered, which were ordered by the secretary.
2. Paper for printers was not available

Neither of these situations is obviously not desirable. Therefore, the Kanban system was implemented (Figure 4). As part of the introduction of the Kanban system for office supplies, the value of office supplies was reduced by about CZK 150 000 at the department.

5 CONCLUSIONS

This article discussed the use of industrial engineering methods in administrative processes. The examples of waste occurring in administrative processes were mentioned. Then, a comparison between the administrative and manufacturing

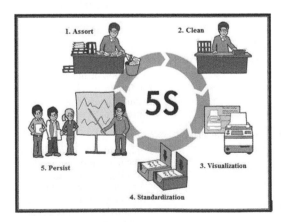

Figure 2. The individual steps of the method 5S.

processes was made. As it can be seen, these processes still have the potential to be improved. When companies begin to deal with all its processes, they can increase the efficiency of their processes, reduce their costs and increase their competitiveness completely. Therefore, they should start and improve their administrative processes as soon as possible, even using industrial engineering methods, as noted in the above examples.

This article gives only an outline of how companies can work with their administrative processes and how they can achieve their improvements. The possibilities and the use of industrial engineering methods in administrative processes require further research.

REFERENCES

Košturiak, J. a Frolík, Z. & collective. 2006. *Lean and innovative company*, ISBN 80-86851-38-9, EAN: 9788086851389, Location Unknown, Publisher: Alfa Publishing, 237.

Lareau, W. 2003. *Office Kaizen: Transforming, office Operations Into A Strategic Competitive Advantage*, ISBN 0-87389-556-8, I. title, Milwaukee: American Society for Quality, 174.

Mašín, I. Košturiak, J. & Debnár, P. 2007. *Improving* non-*production processes: Introductory program for service and process teams*. 1. edition. ISBN 80-903533-3-9, Location: Liberec, Institute of Technology and Management, 133.

Advances in Civil, Architectural, Structural and Constructional Engineering – Kim, Jung & Seo (Eds)
© 2016 Taylor & Francis Group, London, ISBN 978-1-138-02849-4

A comparative analysis of the industrialization of residential building performance standards—based on South Korea, Japan and Canada

S.J. Woo & E.K. Hwang
Korea Institute of Civil Engineering and Building Technology, South Korea

ABSTRACT: Recently, an attempt to construct rental housing using the modular construction method on unused site (upper area of public parking lot, etc.) in downtown area for housing vulnerable people is being made in Korea. Modular construction has an advantage to maximize site efficiency and minimize the construction period, minimizing the effect on urban traffic. The initial market was limited to school facilities and military bases, but due to recent expansion in modular construction market, it is being reviewed for various purposes such as apartment housing, hotels, and hospitals. According to the announcement of Korea Research Institute for Construction Policy in 2011, it is expected that the scale of domestic modular market will grow to 200 billion KRW in 2015 and approximately 940 billion KRW in 2020. Therefore, the purpose of this study is to review the current status of modular construction in Canada and Japan where can be considered as advanced countries of modular construction, compare and analyze relevant laws and regulations and use them as source material for the revitalization of domestic modular construction.

1 INTRODUCTION

1.1 Background

Recently, there has been rising interest in the use of industrialized buildings in Korea as a measure to resolve the housing shortage in downtown areas. Industrialized building is a housing construction method in which complex aspects of the construction process are carried out in a factory to minimize the work performed on the construction site. The Korean government has been promoting the performance improvement and revitalization of industrialized residential buildings by introducing the Recognition of Industrialized Housing in 1999. However, the outcomes from these efforts so far are very low, with only four systems recognized as industrialized housing.

1.2 Method of study

Thus, the performance standards of industrialized housing in Korea, Japan and Canada are compared and analyzed in this study with the aim of revitalizing the recognition of industrialized housing. The analysis is carried out on the recognition of industrialized housing in Korea, Industrialized Housing-Prefab Housing-in Japan, and CSA A277, CSA Z240 MH series, the standard of factory—built buildings in Canada. Pending problems in the domestic standards are identified after the comparative analysis to find the direction for the improvement of industrialized housing and modular architecture in Korea.

2 ANALYSIS OF INDUSTRIALIZED HOUSING CERTIFICATION SYSTEM RELATED STANDARDS IN SOUTH KOREA

2.1 General industrialized housing certification system

Industrialized housing in Korea was introduced in 1992 in order to improve the quality and revitalize industrialized housing, and it is defined as housing of which the main structure part is constructed in whole or part by using an industrialized method (e.g. panelized buildings) according to the performance standards and production standards provided by ordinance of the Ministry of Land, Infrastructure and Transport.

2.2 Performance evaluation items for industrialized housing certification system

The evaluation items for the industrialized housing certification system are divided into 7 items, including the safety performance of the structure, the fire resistance and fire prevention performance, evacuation safety performance and prevention of falling, thermal environment, sound environment and durability. Table 1 is Performance Evaluation items for industrialized housing certification system in Korea. And the details of each performance item are as following Table 1.

Table 1. Performance Evaluation items for industrialized housing certification system in Korea.

Performance Item		Evaluation method
Structural safety performance	Structure part Joint part	• Evaluate the safety of structure parts and connection parts for structural resistance as 'satisfactory' or 'unsatisfactory'.
Fire resistance and fire prevention performance resistance	Fire resistance	• Check the conformance of the building to evacuation rules through the design drawing or evaluate through the certificate.
	Fire prevention	• Evaluate with design drawing which specifies the material regulated in the evacuation rules for the building, or with a test report which confirms its performance
Evacuation safety performance and prevention of falling		• Check whether the evacuation safety performance and falling prevention performance meet the relevant standards
Ventilation and air tightness performance		• Evaluate with air tightness performance value according to the relevant design document and KS L ISO 9972 (insulation-measurement of air tightness of building).
Thermal environment performance	Insulation	• Check whether the heat transmission coefficient calculation for each part or the test report of heat transmission coefficient meets the standards of the heat transmission coefficient for the relevant construction area or not.
	Prevention of dew condensation	• Evaluate with the simulation result of dew condensation prevention performance using a proper program for ISO 10211
Acoustical environment	Boundary barrier between households	• Evaluate by checking the standard floor structure or with certificate recognized according to a certification standard of floor impact sound insulation structure for apartment houses
	Floor impact sound	• Evaluate by checking whether the boundary barrier structure meets the housing construction standards or with the certificate for noise reduction structure of wall according to the certification standard for noise reduction structure of wall
Durability performance	Rust prevention/ corrosion prevention	• Coating thickness of reinforcing bar: Check through design drawings or work specifications
		• Lumber: Check through the corrosion prevention and rust prevention treatment confirmation or the test report
	Water proofing/ drainage	• Evaluate as 'satisfactory' or 'unsatisfactory' based on required performance standards

Table 2. Current state of certified industrialized housing in South Korea.

Housing	Admission date	Properties	Construction Material
CHS industrialized housing	2010. 09	Inserted into the unit between the completed structure	Staco fireresistance panel, lightweight C-shape
MUTO Frame Module Type-1	2012. 12	Unit system (4 stories, 18 household)	Steel Concrete, plaster board
KMC Type-A House	2014. 06	Detached house	Steel-Concrete, plaster board
U'Vista House	2015.02	-	-

2.3 Current state of certified industrialized housing in South Korea

Table 2 is Current state of certified industrialized housing in South Korea but obtaining certification housing is only 4 type in korea and the detail is following Table 2.

3 ANALYSIS OF FACTORY—BUILT BUILDINGS RELATED STANDARDS IN CANADA

3.1 General certification standards for CSA factory—built buildings in Canada

In Canada, the standards for factory built buildings including industrialized housing and modular buildings have been regulated and observed. The standards for factory built buildings are provided

Table 3. Design and factory certification procedure for factory built buildings.

Definition and certification summary of factory built buildings	• A modular, manufactured, or panelized building that is built in a manufacturing plant before being transported to its point of installation.
	• Provides approved internal certification procedure inside factory; requires field inspection and tests applicable to the components of factory–built buildings

Table 4. Factory certification procedure for designed factory–built buildings (CSA A277).

Factory certification procedure for designed factory–built buildings (CSA A277)	1) Certification of the factory quality program
	2) Certification of the built product
	3) Auditing of the factory quality Program
	4) In-factory inspection of the built product

Table 5. Performance item (CSA Z240 MH Series).

Definition	Minimum requirements applied to the design of manufactured buildings, modular buildings completed in a factory, and panelized buildings.
Evaluation items	–Transporting requirements (vehicle requirements)
	–Structural requirements
	–Installation requirements of gas combustion equipment
	–Installation requirements of oil burning equipment
	–Cooling and heating loads and duct

in CSA A277 and CSA MH Z240. The definition and certification summary of factory built buildings are as following Table 3.

Factory–built buildings should be designed and constructed based on the following codes: National Building Code of Canada or the applicable local government building code, CSA Z240

MH Series, Canadian Electrical Code, Part I: Installation of electrical system, National Plumbing Code of Canada: Factory-installed pipes and structure, other applicable laws, standards and requirements.

The factory certification procedure for design factory–built buildings follows CSA A277. And following Table 4 is Factory certification procedure for designed factory–built buildings.

The design requirements for factory built buildings are provided in CSA MH Z240 and the detail performance item following Table 5.

4 ANALYSIS OF MODULAR BUILDING-RELATED STANDARDS IN JAPAN

4.1 General of modular building standards in Japan

In Japan, the term form housing rather than industrialized housing is used, and Law on the securing and promotion of quality of housing, etc. Article 33 Certification of suppliers for form housing is provided as the relevant standard. Table 6 is describing Certification of suppliers for form housing in Japan.

For performance evaluation in the form housing performance certification system, 10 items including structural safety, fire safety, deterioration mitigation (deterioration measure grade of building elements), maintenance and update (maintenance and update plan), thermal environment (energy saving grade), air environment (indoor air quality), light environment (simple opening ratio), sound environment (heavy and light floor impact sound blocking and transmission loss grade), considerations for senior citizens (senior citizen consideration measure grade) and crime prevention (intrusion prevention measure for openings) are evaluated.

Table 6. Certification of suppliers for form housing in Japan.

Definition	– Certification of suppliers for form housing (Law on the securing and promotion of quality of housing, etc. Article 33): certification that modular housing or housing manufacturer accordance with form housing performance. Certified form housing part is considered to be suitable for the type. Therefore, simplify the evaluation of individual housing performance.

5 COMPARISON OF THE INDUSTRIALIZED HOUSING PERFORMANCE STANDARDS OF KOREA, JAPAN, AND CANADA

5.1 Comparison of industrialized housing performance evaluation items of South Korea, Japan, and Canada

With regard to the evaluation items for the industrialized housing certification system, the following comparison of the performance standard items for industrialized housing reviewed previously are as following Figure 1.

In Canada, the transporting requirements are included in the performance standards, and the standards for interior material are sub-divided. In both Japan and Canada, the standards for maintenance have been prepared, but there are no standards for maintenance in Korea, so user-based standards are needed.

Also Figure 2 below shows Connectivity of industrialized housing performance evaluation items.

Performance item	South Korea	Japan	Canada
Structural safety performance	O	O	O
Fire resistance and fire prevention performance	O	O	O
Evacuation safety performance and prevention of falling	O	-	-
Ventilation and air tightness performance	O	O	O
Thermal environment performance	O	O	O
Sound environment	O	O	-
Durability	O	-	O
Light and visual environments	-	O	-
Consideration for senior citizen	-	O	-
Crime prevention	-	O	-
Deterioration mitigation (regarding building elements)	-	O	O
Maintenance and update	-	O	O

Figure 1. Comparison of industrialized housing performance evaluation items.

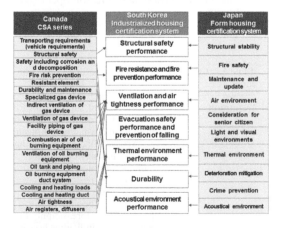

Figure 2. Connectivity of industrialized housing performance evaluation items.

6 CONCLUSIONS

The industrialized housing standards of Korea, Japan and Canada are compared and analyzed to find the direction for the improvement of the industrialized housing certification system in Korea and the detail overall analysis following Table 7.

The comparison of the laws in Canada, Japan and Korea in terms of the performance standards showed that the transporting requirements were included in the performance standards, and the subdivided standards and maintenance standards for interior materials were different from the standards in Korea. Therefore, the addition of user-based certification items was found to be a limitation, and it is necessary to examine the validity of the evacuation safety evaluation item, which is only evaluated in the industrialized housing certification system in Korea. Also, it is necessary to improve the relevant building laws-based performance standards. The performance items and evaluation method reflecting the

Table 7. Overall analysis.

Overall comparative analysis of performance standards	**Canada** – The transporting requirements are included in the performance standards. (Production-based in Korea) – Sub-divided standards on interior material – Standards on maintenance **Japan** – User based standards such as light environment, crime prevention, consideration for senior citizens and maintenance included
Seek direction for improvement of performance standards	– Limitation in the addition of user-based certification items – Necessary to examine the validity of evacuation safety (not included in the standards in other countries) – Performance standards based on the relevant laws (necessary to clarify the application time when revising the standards)

characteristics of industrialized housing and modular housing should be established again in Korea.

ACKNOWLEDGEMENT

This research was supported through funding from the 'Technical Development of Modular Construction in Mid-high Rise Building and Higher Productivity (2nd year KICT)' project by the Ministry of Land, Infrastructure and Transport in South Korea.

REFERENCES

KICT, 2015, *Technical Development of Modular Construction in Mid-high Rise Building and Higher Productivity*, Ministry of Land, Infrastructure and Transport.

Ministry of Land, *Infrastructure and Transport, 2012, Industrialized Housing Performance and Production Standards.*

Park, J.Y. & Lee, W.H. 2013, Research on the improvement of Performance Evaluation for Industrialized Housing, *Architectural Institute of Korea*, 33(2): 9–10.

Takeshi, T. 2010. Transition of the *Industrialized Housing-Prefab Housing-in Japan, its Industry and Engineering*, 231–313.

Hwang, E.K. & Lim, S.H. 2012. A Comparative Study on the Industrialized Housing Recognition System and the Housing Performance Recognition System, *Architectural Institute of Korea*, 32(2): 111–112.

Kim, H.S. & Ji, J.H. 2012. Research on the improvement and progressive measures of Modular Housing for Industrialized Housing Recognition System, *Korean Housing Association*, 24(2): 305–310.

Advances in Civil, Architectural, Structural and Constructional Engineering – Kim, Jung & Seo (Eds)
© 2016 Taylor & Francis Group, London, ISBN 978-1-138-02849-4

Analysis of overseas cases of determining construction supervision cost

E.K. Hwang & E.Y. Kim
Korea Institute of Civil Engineering and Building Technology, South Korea

ABSTRACT: This study aims to analyze cases of the U.S., the U.K. and Japan in order to establish clear criteria for appropriate compensation for construction supervision. Also it aims to identify improvements through analysis of local status and problem of criterion for supervision cost, and its comparison with overseas cases. There is no specific table of rate, which is calculated mostly based on lump sum contract and unit contract. Japan follows formula for calculation of cost plus fixed fee based on the Notification No. 15 from Ministry of Land, Infrastructure, Transport, and Tourism (MLIT). The U.K. has no specific rate system, but RIBA provides criteria information for calculation based on compensation. Appropriate cost system suitable for local status will have to be set up or introduced through such analysis of overseas cases.

1 INTRODUCTION

1.1 Background and purpose

For safety of a building, role of the supervisor, as well as the construction work itself, takes great significance. Also appropriate amount of supervision cost has to be provided based on the scope of work and responsibility, for sincere supervision by the supervisor. Elimination of supervision cost criteria system for private construction sector in 2008 sparked unfair contracts biased towards owner of the building. The problem is that supervision is considered as a supplementary work of construction design, leading to insufficient amount of supervision cost. This study aims to analyze status of criteria of supervision cost in the U.S. the U.K. and Japan in order to set up appropriate standards.

1.2 Method and scope

This study is based on analysis of local status and overseas cases, former of which is analyzed with survey on experts in the field and comparison of related laws, while the latter is studied through corresponding laws and literature of overseas.

2 THEOREATICAL BACKGROUND

2.1 Standards for construction supervision cost

Calculation methods of supervision cost in Korea are categorized into percentage of construction cost basis, cost plus fixed fee basis, and fixed amount-added method. Percentage of construction cost basis is comparatively easy and simple enough to enable easy integration into budgets but it involves risks of errors, making it difficult to consider characteristics of work such as difficulties or complexity. Cost plus fixed fee basis is adding up direct labor cost, direct expenses, indirect expenses, technical cost and value added tax according to the actual number of people in the work process. Fixed amount-added method is estimated by adding direct labor cost, direct expenses, overhead expense, technical cost, additional processing cost value added tax, and insurance (benevolent fee). Before Construction Technology Promotion Law was revised in Korea, fixed amount-added method had been used while costs for CM (Construction Management) were calculated according to percentage of construction cost basis.

2.2 Local status and issues

Survey on experts in the field was taken regarding the status of supervision cost. The experts were 75 members of the Korean Institute of Architects and were surveyed from November to December of 2014. To sum up the results, most respondents said current supervision cost is low compared to that of public construction projects by government. The reasons are that supervision cost must reflect the size of the building, difficulty of work and conditions of work, and even though public and private projects have differing work standards, respondents said that same standards have been applied, which is not appropriate. Also as the current supervision cost standards for small building does not even match those of cost plus fixed fee basis, so they require immediate improvement.

3 ANALYSIS OF OVERSEAS CASES

3.1 Standards for construction supervision cost in U.S

The U.S. has Uniform Building Code applied throughout the country and Brook Act regarding public projects. However, each state features different construction-related administration system. Construction supervision system in the U.S. differs from that of Korea in that the architect performs CA (Construction Administration) work in regards to the contract during the work. Roles of each principals involved in the project are clearly sectioned and operated—While construction design architect performs CA service, town or city government provide inspection, and the client (building owner) performs overall management through separate agreement with the CM provider or a consultant.

Supervision cost is calculated in the U.S. as following—unlike Korea, there is no specific rate system and market principle controls the fee overall. An appropriate supervisor is placed according to characteristics and difficulty of the work, and technical expertise and performance of the supervisor plays the biggest role in the calculation of the cost. Total supervision cost depends on the scope of the work provided by the supervision company, which is 3–10% of total construction cost. Payment is made differently such as lump-sum contract or unit price contract depending on characteristics of the work.

3.2 Standards for construction supervision cost in Japan

Construction supervision in Japan mostly is based on contract and is considered as extension of design, thus provided by design companies.

Standards for supervision cost in Japan are based on the Notification No. 15 from MLIT of 2009, which states 'the standard of compensation that can be invoiced for the work by the founder of an architect company'. The most noticeable aspect of the standards is that gross area determines the number of supervisors and their man-hours, which is to prevent reduction of supervisors derived from reduction of the whole construction cost. Japan currently applies two methods of cost plus fixed basis and simple calculation method, and the formulas are as following:

1. Cost plus fixed basis = Direct labor cost (P1) + Direct expenses (E1) + Indirect expenses (E2) + Special expenses (R) + Technical cost (F) + Consumption tax
2. Simple calculation method = [Direct labor cost (P1) X 2.0 (including other costs)] + Special

Table 1. Category of purposes building construction and work difficulty in Japan.

Purpose	Difficulty	
	Category 1 (General)	Category 2 (Complex)
Logistics	Garage, warehouse, multi-dimension parking lot, etc.	Multi-dimension warehouse/ logistics center, etc.
Office	Office, etc.	Bank, head office building, government office, etc.
Culture/ exchange/ public interest	Civil hall, assembly hall, community center, etc.	Movie theater, art museum, library, etc.
Other/Detached house	Detached house	–

expenses (R) + Technical cost (F) + Consumption tax
3. Notification No. 15 by MLIT categorizes purposes of a building as in following Table 1 and provides the number of designers and supervisors to place and man-hours by gross area.

Also Figure 1 below shows subdivided ratio range of cost by standard work of construction supervision set by the Notification by the MLIT and supervisors conduct whole or partial work according to contract with building's owner.

The ration ranges vary depending on the scale, but mostly standard construction supervision takes up 65~75%, while other standard work account for 25~35%, so cost standards can be applied to the contract with building's owner.

3.3 Standards for construction supervision cost in U.K

Basically, the U.K. government forces the constructor to conduct inspection and be responsible for the construction site, rather than establishing regulations or codes for construction management. Such code was developed by RIBA (Royal Institute of British Architects).

Standards in the U.K. for supervision cost serve more for negotiation and are considered as the issue of contract. There are no actual legal standards or recommendations. The standard is calculated based on complexity of the project, scope of work, method of ordering, construction cost and period. RIBA issues "Architects Fees" as the basic material to determine fair standards between architect and client. If estimated construction cost

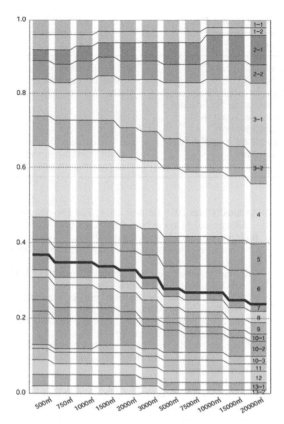

Figure 1. Subdivided ratio range of supervision cost by construction supervision.

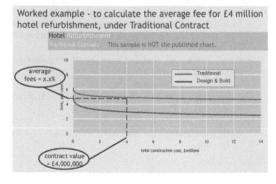

Figure 2. Example of calculation of supervision cost in UK.

according to the purpose, remodeling and ordering method is entered, then standard fee is calculated. Figure 2 shows an example of calculation.

Table 2 below shows standards provided by RIBA. Cost for from stages J to L1, which is similar to supervision work according to Korean Certified Architects Act is set as 25% of basic compensa-

Table 2. Cost calculation standard—RIBA Outline Plan of Work 2007.

Work stages		Cost calculation standard
A	Validation	Hourly labor cost +
B	Design Brief	basic cost
C	Design of work concept	35% of basic cost
D	Basic design	calculated based on percentage of construction cost
E	Working design	40% of basic cost
F	Order preparation F1	calculated based
	Order preparation F2	on percentage of
G	Bidding work	construction cost
H	Support for ordering	
J	Site preparation	25% of basic cost
K	Commencement/	calculated based
	completion	on percentage of
L	Completion support L1	construction cost
	L2	Hourly labor cost +
	L3	basic cost

tion. Besides the calculation method on service cost to construction cost, the U.K. and Europe use diverse methods depending on terms of contract between constructor and client. In the research of 25 member states in ACE (Architects Council of Europe), percentage of construction cost basis is 43%, lump sum method is 36%, time-based charge is 12% and others are 9%. Overall, these countries have the frame where both architect and client can benefit from the guides by developing the guide book for the client as well as presenting statistics for compensation for architect work.

3.4 Summary

Table 3 below shows analysis on cost calculation method of the U.S. the U.K. and Japan.

The U.S. has no specific rate system and market principles determine the cost. Supervision cost calculation is based the contract, as scope of work, so agreement of the parties determines the cost, which is because selection of supervision company is conducted in the form of CM—overall construction management—strictly based on technical expertise and contract.

Japan has clear calculation methods based on the regulations on Article 25 of Architect Law and Notification 15 of MLIT. Basically, cost plus fixed fee basis is utilized while other standards such as purpose of the building, difficulty of work and number of inspectors are applied.

Like the U.S., the U.K. also does not have a specific rate system and depends on market principles,

Table 3. Analysis on cost calculation method of the U.S the U.K. and Japan.

Category	U.S.	Japan	U.K.
Established standards	No specific rate system (Market principle)	Notification No. 15 by MLIT	No specific rate system (Market principle)
Reference for calculation	Characteristics and difficulty of the project/Technical expertise of supervisor	Regional architect company association "Guidelines for calculating work compensation"	RIBA, "Architects Fees"
Calculation method	Lump-sum Contract and unit price contract	Formula for cost plus fixed fee basis	Lump sum method, time charge method, percentage of construction cost

but refers to materials provided by RIBA annually for guidelines of calculation. Diverse methods such as lump sum method, cost reimbursable are applied.

4 CONCLUSIONS

This study has looked at how to calculate appropriate supervision cost through analysis of local and overseas construction supervision cost calculation methods as well as status of local cost calculation. Currently, Korean construction industry does not provide appropriate level of supervision cost matching the scope of work and responsibility. There is the need to introduce and establish supervision cost suitable for our current status, by summarizing the cases of the U.S., the U.K., and Japan.

ACKNOWLEDGEMENT

This research was supported through funding from the 'A Study of the Improvement of Safety Strengthening Comprehensive System for Buildings' project by the Ministry of Land, Infrastructure, and Transport in South Korea.

REFERENCES

A Client's Guide to Engaging an Architect, 2009, RIBA

Architects Fees 2015 Edition, 2014, *The Fees Bureau*, A Division of Mizra & Nacey Research Ltd.

Building Standard Law of Japan, 2013.

Construction Design and Management Regulation, 2007.

Criteria for service cost to architect firms, 2014, Tokyo Association of Architectural Firms.

Criteria of fee for construction project management (Notification by Ministry of Land, Infra-structure and Transport. No. 2014–298), Ministry of Land, Infrastructure and Transport.

Enactment and description of construction supervision, 2009, Association for distributing new architect system in Japan.

Korea Institute of Civil Engineering and Building Technology, 2014. *Research on construction project management supervision cost for application of cost plus fixed fee basis*, Ministry of Land, Infrastructure and Transport.

RIBA, 2013. Handbook of Practice Management Ninth Edition.

Scope of service and cost for architect regarding publicly bidding construction project, (Notification by Ministry of Land, Transport and Mari-time Affairs No. 2011–750), Ministry of Land, Infrastructure and Transport.

Scope of work and cost for founder of architect firm, 2009. *Notification by Ministry of Land, Infra-structure, Transport, and Tourism. 15.*

The American Institute of Architects, The Architect's Handbook of Professional Practice.

The Architectural Profession in Europe 2012 Chapter 3: Architecture-the Practice, 2012, Architects' Council of Europe.

Advances in Civil, Architectural, Structural and Constructional Engineering – Kim, Jung & Seo (Eds)
© 2016 Taylor & Francis Group, London, ISBN 978-1-138-02849-4

The effect of finger-joint profile and orientation on the strength properties of timber beam for selected Malaysian timber species

Z. Ahmad & L.W. Chen
Institute of Infrastructure Engineering and Sustainable Management, University Technology Mara (UITM), Shah Alam, Malaysia

M.A. Razlan & N.I.F. Md Noh
Faculty of Civil Engineering, University Technology Mara (UITM), Shah Alam, Malaysia

ABSTRACT: Finger-jointed lamella plays an important role in glued laminated timber (glulam). Information regarding finger-joint strength of temperate timber species has been well documented. However, little information is available on the characteristics and properties of finger-jointed lamella for the manufacturing of glulam using Malaysian tropical hardwoods, especially the effect of finger profile geometry for optimum strength. This study investigates the structural performance of finger-jointed lamella in the application for glulam by exploring the effect of finger-joint length on the bending strength of lamella for different densities of Malaysian timbers. A total of eight species were used in this research, namely Kapur, Merpauh, Resak, Bintangor, White Meranti, Jelutong, Kelempayan and Sesendok. Phenol-Resorcinol-Formaldehyde (PRF) was used as the adhesive for the joints. The dimension of the test specimens was $30 \times 100 \times 600$ mm. Finger lengths selected for horizontal finger-joint orientation were 15 mm and 25 mm. Bending test was conducted in accordance with the ASTM D-198. It was found that the Modulus of Rupture (MOR) of 25 mm and 15 mm vertical finger orientation was almost the same and the values were 20% higher when compared with 25 mm and 15 mm horizontal finger orientation. However, finger length and orientation did not have a significant effect on the Modulus of Elasticity (MOE). Among the Malaysian tropical timbers, finger joints from the low density Sesendok gave the lowest value of bending strength and MOE.

1 INTRODUCTION

Timbers are generally described as the normal sawn structural members that contain macroscopic defects such as knots and cracks of different shapes, sizes and orientations. From the very first housing, bridges and tools, timber has provided humans with a broad range of building products and materials for construction. There are a number of inherent characteristics that make timber an ideal construction material. Timber has more advantages compared with steel and concrete as it has a higher strength-to-weight ratio. It is also easy to work with as it can be easily fabricated and constructed using simple machinery. Besides, timber is also known for its superior insulator properties among all the structural building materials. Benefits from timber also come from its natural growth characteristics such as grain patterns, colors and its availability in many species, sizes and shapes, which make it a remarkably versatile and an aesthetically pleasing material.

In most timber structures, connections are one of the most important, but least understood, components. Connections provide continuity to the members and strength and stability to the system. Of the failures observed in timber structures, most are attributed to improper connection design, construction (fabrication) detail or serviceability. Timber connections are categorized into two types: mechanical and end joints. Few types of fastener used to connect a timber in mechanical joint are nails, screws, bolt and shear plate. As for end joints, three possible end joints may be considered in timber connection, namely butt joints, scarf joints and finger joints. Finger joints offer the best way of jointing wood, since they provide high strength, require less amount of wood and can be manufactured at high production rates (Nestic & Milner 1993; Sellers et al. 1998). Finger jointing also reduces variation in the final product, such as less tendency to warp because the grain is randomized over the length of the piece. Finger joint is a type of structural end joint used in glue laminated timber (glulam) to form long, continuous lamination out of individual pieces of timber. Finger-jointed connection has high potential to become an ideal method in timber connection, due to its clear economic advantage compared with scarf jointing (Madsen & Litterford 1964, Strickler et al. 1990).

To use timber more efficiently in the development of structural finger-jointed products, specific process parameters must be taken into account. The type of finger joints, the moisture content and the machining process must be controlled and optimized. Many timber-related factors are known to affect the strength of finger joints such as species, density and natural defects, while others are related to the gluing process. The types of adhesive, curing time and the pressure of application also have a great influence on the strength behavior of the assemblies. Finger-joint geometry has an effect on the performance of structural joints. Parameters such as tip width, pitch width, finger length and slope are interrelated and can influence the performance of the joint. Most information available on the finger joints is on end gluing of softwood in contrast to hardwood (Özçifçi & Yapici 2007, Gaspar 2004, Ayarkwa et al. 2000, Jokerst 1981).

Engineered timber product such as glulam is the solution to efficiently utilize timber for structural application. Even though in Malaysia there are glulam factories, studies on the effect of glue types, finger-joint geometries and machining processes on glulam are not available. Currently, the manufacturers are manufacturing glulam using the finger-joint machine for producing non-structural components. Due to the lack of test data, the performance of glulam produced has not been optimized. Hence there is a need to investigate the performance of finger-jointed timber manufactured using Malaysian tropical timber.

Therefore, the purpose of this study is to determine the effect of finger length on the bending properties of Malaysian tropical timber from different Strength Grouping (SG). The timber species used are Merpauh, Kapur, Kelempayan, Sesendok, Resak, Bintangor, White Meranti and Jelutong. The data collected for bending performance include the maximum load data, MOR and MOE.

2 EXPERIMENTAL WORK

Nine species from different strength grouping (SG) were identified as a raw material for this study and were representative of timber in strength grouping of SG4 to SG7 and also because of the availability of the timbers (MS 544 Part 2, 2001). These nine species identified were Merpauh, Kapur, Resak, White Meranti, Bintangor, Jelutong, Sesendok and Kelempayan.

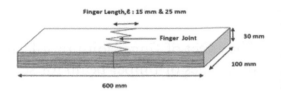

Figure 1. The dimension of the vertical finger-jointed specimens for bending test.

The timbers used for this study were sourced from reserved forests in UiTM Jengka. The processing of the samples was performed at the timber laboratory in UiTM Jengka. Prior to sample manufacturing, timber samples were visually graded in accordance with MS 1714 (2003) and the manufacturing of finger-jointed beams was in accordance with MS 758 (2001) respectively. The timber beams were finger-jointed using Phenol-Resorcinol-Formaldehyde (PRF) adhesive supplied by Dynea Nz Ltd in vertical lay-up with two different finger lengths (15 mm and 25 mm) with the dimension of 30 mm × 100 mm × 600 mm, as shown in Figure 1.

The controlled specimens also were prepared from timber samples with a size of 30 mm × 100 mm × 600 mm without any finger joint. The resin was thoroughly mixed by hand on a bed of ice in order to prevent the resin from hardening.

Then, the prepared timber samples with finger profile were dipped into the resin and further enhanced with brush in order to ensure evenly distributed resin on the surface of the fingers and covered over a length of at least ¾ of the finger length. The resin on the samples was air-dried for 60 seconds before mating (jointing the two ends together). Then, the pressure was applied to the jointed beam in order to form good bonding. The finger-jointed specimens were cured under a temperature of 30°C for over 48 hours before further processing. During curing, no pressure was applied to the specimens.

Four-point bending tests were performed on the finger-jointed beam using a 100-ton capacity Universal Testing Machine (UTM) UTS 348 in accordance with ASTM D 198 at a cross-head speed of 1mm/min, such that failure occurred within 10 to 15 minutes of the test period.

3 RESULTS AND DISCUSSION

The bending strength properties of beams with and without finger joints in terms of MOR and MOE are shown in Figures 2 and 3 respectively.

As shown in Figure 2, the MOR values of finger-jointed beams are lower than the those of solid timber beams. The differences in the MOR values are in the range of 50–120%. In general, the MOR values

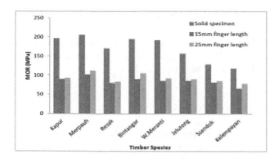

Figure 2. Modulus of Rupture (MOR) of jointed timber beams with different finger lengths.

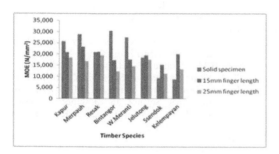

Figure 3. Modulus of Elasticity (MOE) of jointed timber beam with different finger lengths.

for beams with 15 mm finger length are lower than those for all timber species. The largest difference in MOR is 115% for the species White Meranti and the least difference is 55% for the species Kelempayan. The highest MOR value for solid timber beam is for Merpauh as well as for the finger-jointed beam. Merpauh has a fine, tight grain and large medullar rays with a small tracheal structure. This characteristic may allow for better glue penetration, thereby enhancing the bonding of Merpauh.

However, it did not seem to differ much in mean MOR for these two finger lengths. The 25 mm finger length gives the highest mean MOR that might be related to its comparatively highest effective glueline area or shear area.

Selbo (1963) and Rakness (1982) indicated that to obtain high joint strength, finger must be sufficiently long and the slope sufficiently low so that a large enough effective glue joint area is provided to withstand a shear load approaching the tensile strength of an uncut net effective section of wood. This observation contradicts with the work carried out by Wolforel (2012) on finger-jointed Radiata pine, which showed that shorter finger lengths are slightly stronger than longer ones, but require greater precision in manufacture. It also found that joint strength increases with timber stiffness or density.

The timbers used in this study were from the strength group of SG4 to SG7. Analysis on MOR based on strength grouping shows that the value decreases in the same order as the strength grade of the timber. This relates to the density of timber, as shown in Table 1. Density is an important physical characteristic of timber that affects its strength properties. Density can be a good indicator of timber mechanical properties, provided that the section is straight grained and free from knots and defects.

To determine the difference in the MOR values between the control and jointed beams, the ratios between the MOR of the different joint length specimens (FJ) and the MOR of solid timber (S), defined as joint efficiency (FJ/S), are expressed in percent, as tabulated in Table 2. The results obtained seem to suggest the existence on the effect of density on the bending strength. These values suggest that joint efficiency increases with density.

Lower density timber has higher bond strength compared with higher density timber due to the penetration of the adhesive into timber (Muhammad Azlan et al. 2012). Ayarkwa et al. (2000) studied the effect of densities on finger-jointed timber beam and reported that gluing of hardwoods with density in excess of 700 kg/m[3] produced uncertain results, while hardwoods lower than this density appeared to be predictable in performance.

Figure 3 shows the MOE values for the tested beams, which includes the solid and finger-jointed specimens. There is no proper trend in the MOE values for solid and finger-jointed beams. For timbers in SG4 until SG5, the MOE values of control specimens are higher than those of jointed beams. However, for the lower strength groupings SG6 and SG7, the MOE values of solid beam are lower than those of finger-jointed beams. The MOE values of all the jointed specimens ranged from 8720 to 23,278.98 N/mm[2]. The difference in the MOE values between the solid and finger-jointed beams is smaller than that in the MOR values. At the same time, the

Table 1. Joint efficiency values with respect to density of timbers.

Timber species	Density (kg/m[3])	FJ/S (%)	
		15 mm	25 mm
Kapur	899.7	45.8	47.1
Merpauh	882.0	49.6	54.5
Resak	810.9	47.1	49.1
Bintagor	804.4	46.2	53.8
White Meranti	783.4	44.6	47.9
Jelutong	610.7	55.2	57.4
Sesendok	540.2	63.2	66.3
Kelempayan	500.0	55.5	65.8

MOE values for beams with 15 mm finger length are higher than that for beams with 25 mm finger lengths. This result contradicts the values of MORs where beams of 25 mm finger length attain higher MOR values than those of 15 mm finger length.

The highest values of MOE for solid timber and finger-jointed beam were attained by the species Bintangor and Kelempayan respectively. The largest difference in MOE is 43% for Bintagor and the least difference is 5.5% for Jelutong. In this case, the difference in MOE values between the two types of finger length can be considered small, which indicates that MOE values were not significantly influenced by finger length profiles. Dutko et al. (1982) stated that finger-joint profile does not influence the MOE of the jointed specimen as it does MOR. This may be due to the fact that the strength is a local phenomenon while stiffness is more global and, therefore, less sensitive to joint properties. This led to the conclusion that finger profile geometry has less significant influence on the stiffness of the finger joints than the MOR values.

4 CONCLUSIONS

This study was conducted to determine the effect of finger length and timber species on the joint strength of finger-jointed beams. Based on the tests conducted, several conclusions can be drawn as follows:

- Control specimen (solid specimen) attained higher bending strength compared with finger-jointed beam
- When the length of the finger joint increases from 15 mm to 25 mm, the bending strength also increases.
- As for finger-jointed beams, higher MOR values were achieved using timber of higher strength group and density. This finding is very much contributed by the strength of the timbers. However, in terms of efficiency of jointing, lower density timber is considered more efficient.
- The beams with finger length of 25 mm exhibited marginally higher MOR than 15 mm-length finger-jointed beams.
- The difference in MOE values between the two types of finger length can be considered small, which indicates that MOE values were not significantly influenced by finger length profiles.
- The MOR and MOE values for beams of 25 mm finger length are marginally higher than those for beams with 15 mm finger length (irrespective of density). Based on this finding, it can be concluded that 25 or 15 mm finger length can be used in the production of Malaysian hardwood glulam.

ACKNOWLEDGMENT

This research was financially supported by the Public Works Department Malaysia and Universiti Teknologi Mara (UiTM).

REFERENCES

Ayarkwa, J., Hirashima, Y. & Sasaki, Y. 2000. Effect of finger geometry and end pressure on the flexural properties of finger jointed tropical African hardwoods. *Forest Products Journal*, 50(11/12): 53–63.

Madsen, B. & Litterford, T.W. 1962. Finger joints for structural usage. *Forest Products Journal*, 12(2): 68–73.

Raknes, E. 1963. Gluing of wood pressure-treated with water borne preservative and flame retardant, Inst Wood Sci, 22–44.

Gaspar, F. 2004. *Scientific Report. Cost Action number: E34.* Short term scientific mission at NTI-Cost E34.

Azlan, H.M., Ahmad, Z., Hassan, R. & Jamadin, A. 2012. *Shear Strength of Steel and CFRP to Timber Interface.* Proceedings of the 2012 IEEE Symposium on Humanities, Science and Engineering research (SHUSHER 2012), 24–27 July 2012, Kuala Lumpur, Malaysia.

Jokerst, R.W. 1981. *Finger-jointed wood products.* Forest Products Laboratory, Res. Pap. FLP 382. USDA Forest Serv., Prod. Lab. Madison, WI. USA.

Malaysia Standard, 2001. *Code of practice for structural use of timber: Part 2: Permissible Stress Design of Solid Timber*, Malaysia, MS544:2.

Malaysia Standard, 2001. *Glued laminated timber: performance requirements and minimum production requirements*, First Revision, Malaysia, MS758, 2001.

Malaysia Standard, 2003. *Specification for visual strength grading of tropical hardwood timber*, Malaysia, MS 1714: 2003.

Strickler, M.D. 1980. Finger-jointed dimension lumber—past, present and future, *Forest Products Journal,* 30(9): 51–56.

Selbo, M.L. 1963. Effect of joint geometry on tensile strength of finger joints, *Forest Products Journal.* 13(9): 390–400.

Özçifçi, A. & Yapici, F. 2008. Structural performance of the finger-jointed strength of some wood species with different configurations. *Construction and Building Materials*, 22(7): 1543–1550.

Dutko, P., Stellar, S. & Kozelouh, B. 1982. *Research into and experience of the use of finger-joints in timber structures in Czecho-slovakia*, In: Proc. of Seminar on Production, Marketing, and Use of finger-jointed sawnwood. C.F.L. Prins,.ed. Timber Committee of the United Nations Economic Commission for Europe. Martinus Nijhof/Dr. W. Junk Publishers, The Hague, Boston, London. 36–39.

Nestic, R. & Milner, H.R. 1993. The use of laminating technology to overcome shortages of large sections of solid timber, *Journal of the Institute of Wood Science,* 13(2): 380–386.

Sellers, T. Jr., Mcsween, J.R. & Nearn, W.T. 1988. Gluing of Eastern Hardwoods: A Review, USDA Forest Service. Southern Forest Experiment Station. GTRSO-71.

Advances in Civil, Architectural, Structural and Constructional Engineering – Kim, Jung & Seo (Eds)
© 2016 Taylor & Francis Group, London, ISBN 978-1-138-02849-4

Integrity assessment of railway components using ultrasonic testing technology

J.G. Kim
Korea Railroad Research Institute, Uiwang, South Korea

ABSTRACT: Recently, several different types of NonDestructive Evaluation (NDE) techniques are employed to characterize and/or control defects in railway industries. Among them, ultrasonic testing is one of most reliable NDE techniques. Especially, ultrasonic testing is one of powerful NonDestructive Evaluation (NDE) techniques, and it is well applied to the fields of railway industries. In this investigation, the ultrasonic testing technique for railway field was summarized and its applications were introduced. Before introducing the ultrasonic testing in detail, several different types of NDE techniques in the railway engineering, such as magnetic particle testing, infrared thermography, fiber optic sensing technology, radiography were also briefly introduced. The actual application and current research on ultrasonic in railway fields include railway axle inspection, railway bogie defect evaluation, and rail defect inspection. The rail defect inspection is accomplished by the rail inspection car with the speed of about 20 km/h. More advanced techniques have been employed with laser-air coupled probe, guided wave technique, etc. In this paper, the current employed ultrasonic testing methods were introduced, and actual applications were provided. Moreover, the current research and development status of ultrasonic equipment, the suggestion for the future, and research efforts in the field of railway industries will be introduced and provided.

1 INTRODUCTION

As increase in the speed of train, the running safety of the railway rolling stocks has become one of important issues. Also, in order to guarantee the safety during running, the effective maintenance issues have made the progress on applications of nondestructive evaluation techniques in railway area. In the areas of railway rolling stocks, there has been every effort to detect flaws and/or abnormality in railway components, which were generated from in-service or manufacturing process.

In order to guarantee the safety of railway systems, it is important to control defects or flaws using appropriate NonDestructive Evaluation (NDE) techniques. Among several different types of NDE techniques, including Acoustic Emission (AE), Magnetic particle Testing (MT), radiography (RT), etc., Ultrasonic Testing (UT) is one of powerful NonDestructive Evaluation (NDE) techniques, and it is well applied to the fields of railway industries. In this investigation, the ultrasonic testing technique for railway field will be provided, and its applications will be introduced. Before introducing the ultrasonic testing in detail, several different types of NDE techniques in the railway engineering, such as magnetic particle testing,

infrared thermography, fiber optic sensing technology, radiography, will also be briefly introduced.

2 NDE TECHNIQUES FOR RAILWAY APPLICATIONS

Several different types of NDE techniques are available for defect evaluation of railway components and/or systems. Figure 1 presents the examples on the several applications of NDE techniques such as magnetic particle testing, Infrared (IR) thermography, ultrasonic testing, and radiography. Besides those techniques, recently, the sensing techniques including fiber optic sensor and Surface Acoustic Wave (SAW) sensor, are newly used for the integrity evaluation of railway systems and/or components.

3 ULTRASONIC INSPECTION IN RAILWAY APPLICATIONS

3.1 Ultrasonic testing of railway wheelsete

The railway wheelset is one of important running units in railway rolling stock. The wheelset is composed of wheel and axle, and two wheels are

(a) Magnetic particle testing of railway bogie

(b) Infrared thermographic image of bolster

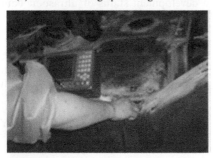

(c) Ultrasonic testing of bogie

(d) Radiography of bogie

Figure 1. NDE techniques for railway applications.

pressed at both ends of an axle. The periodical maintenance is critical to avoid unexpected accident by initiation and propagation of fatigue flaws during operation of vehicle. When the flaws are

Figure 2. The ultrasonic inspection of railway wheelset with immersion tank (ROTECO 2006).

exist on the surface or inside axle, the control of flaws is critical to avoid premature failure of axle. The ultrasonic testing is well applied to inspect the wheelset as shown in Figure 2. Generally, the wheels are inspected after every 600,000 km running without separation from axle by the pulse-echo technique. The ultrasonic C-scan is conducted for axle bearings, which are separated from the axle for the test, in immersion tank. Since the axles are hollow axle, the ultrasonic probe is inserted into the hollow axle, and the inspection is performed with rotating the probe.

3.2 Ultrasonic testing of weldments on railway carbody

Since the railway carbody is fabricated with arc welding technique between each parts to reinforce the carbody structure as shown in Figure 3, the ultrasonic inspection of weldments is performed using ultrasonic C-scan with pulse and transmitter transducers. In Figure 3, an example view of ultrasonic C-scan results is shown with amplitude gate control.

3.3 Automated ultrasonic wheelset inspection

For the effective flaw inspection of railway wheelset, an automated ultrasonic testing methods are applied. As shown in Figure 4(a), a combined automated testing system with ultrasonic probe, phased array probe, eddy current coil, etc. is used for the inspection of wheel, axle, or wheel-disc. In DB as

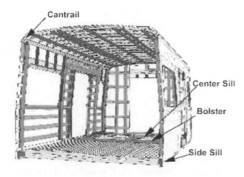

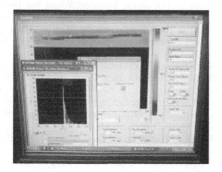

Figure 3. The structure of railway carbody (left) and ultrasonic C-scan results. (right) (ROTECO 2008).

shown in Figure 4(b), an automated ultrasonic testing system with phased array technique is used for the inspection of railway wheelset. Through the multiple application of 8 probes and 64-channel, the system provides effective ultrasonic testing with reduced testing time.

3.4 Ultrasonic testing equipment for rail

For railway rails, defect control on rail surface and regular maintenance of rail system including sleeper, points, ballast, etc. are critical to avoid railway accidents and/or derailment. Therefore, the regular inspection on rail is performed using rail inspection cars on a regular basis. A portable ultrasonic equipment are also available as shown in Figure 5. The equipment provides an ultrasonic inspection with A-scan and B-scan with dual rail inspection.

3.5 Guided wave ultrasonic inspection for rails

Recently, the guided wave ultrasonic inspection is used for various industrial applications. Since the guided wave ultrasonic makes possible defect inspection for long distances, this technique is very useful for the long distance rail inspection as shown in Figure 6.

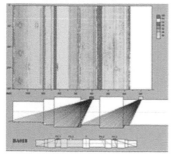

(a) Automated ultrasonic testing system (Automated Ultrasonic Railroad Wheel Set Inspection).

(b) Phased array ultrasonic testing system in DB (Wolfgang Hansen)

Figure 4. Railway wheelset inspection with automated ultrasonic testing system.

Figure 5. Portable ultrasonic equipment for rail inspection (Advances in Portable Ultrasonic Test Equipment).

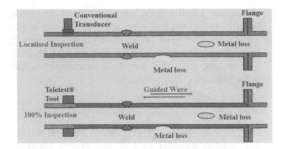

Figure 6. Guided wave ultrasonic inspection as compared with conventional ultrasonic testing (LRUT).

Besides the ultrasonic testing methods as described above, more advanced techniques have been employed with laser-air coupled probe, ultrasonic broken rail detector, ultrasonic rail inspection based on piezoelectric probe, etc.

4 CONCLUSIONS

In this investigation, currently employed ultrasonic testing methods were introduced, and actual applications were provided. Moreover, the current research and development status of ultrasonic equipment, the suggestion for the future, and research efforts in the field of railway industries will be introduced and provided.

ACKNOWLEDGEMENT

This work was supported by a grant from the R&D Program of the Korea Railroad Research Institute, Republic of Korea.

REFERENCES

Advances in Portable Ultrasonic Test Equipment, Markus Nottelmann, Sperry Rail Service, https://www.arema.org /files/library/2011_Conference_Proceedings/Advances_in_Portable_Ultrasonic_Test_Equipment.pdf.

Automated Ultrasonic Railroad Wheel Set Inspection, NDT Systems & Services GmbH & Co, Germany, http://www.ndt-systems.de/fileadmin/ndt-global/downloads/2_brochures/K.057e.v02_Flyer_AURA_lowRes.pdf.

Railway Rolling Stock Technology, ROTECO, 125, 2006.

Railway Rolling Stock Technology, ROTECO, 133, 2008.

The application of Long Range Ultrasonic Testing (LRUT) to inspect railway tracks, NDT Technology Group, TWI, http://www.monitorail.eu/publication/files/The%20application%20of%20 Long%20 Range%20Ultrasonic%20Testing% 20(LRUT)%20 to%20 inspect%20railway%20tracks.pdf.

Wolfgang Hansen, Ultrasonic testing of railway axles with phased array technique, http://www.ndt.net/article/wcndt2004/pdf/railroad_inspection/266_hansen.pdf.

Advances in Civil, Architectural, Structural and Constructional Engineering – Kim, Jung & Seo (Eds)
© 2016 Taylor & Francis Group, London, ISBN 978-1-138-02849-4

Analysis of kitchen hood noise characteristics in apartment houses

H.K. Shin, K.W. Kim, J.O. Yeon & K.S. Yang
*Building and Urban Research Institute, Korea Institute of Civil Engineering and Building Technology,
Republic of Korea*

ABSTRACT: In this study, we analyzed the characteristics of noises generated by the use of kitchen hoods in apartments and the noise distribution in the living room. We studied three types of hoods—built-in, chimney, and independent—according to their structure and analyzed the characteristics of noises. The noise level distributions in the kitchen and living room were analyzed by simulations, and the ratio of area that exceeded the recommended level of 45 dB(A) in the living room was calculated. In the front of the hood, the noise level ranged between 53.5 and 59.5 dB(A) in the high frequency range of 500 to 1,000 Hz. Of the entire living room area, the area at which noise levels exceeded 45 dB(A) was 78 to 100% for the chimney, 88 to 97% for the built-in type, and 89 to 97% for the independent type.

1 INTRODUCTION

Recently apartments have become more air-sealed to save energy. The importance of ventilation has increased for the purpose of discharging all sorts of toxic substances (Kabir & Kim 2011). In particular, the kitchen hood is recognized as an equipment component essential for removing toxic substances and smells produced by cooking, and for reducing humidity.

As the standard of living of residents in apartments has been increasing, there has been a demand for noise reduction in equipment used to create a comfortable room environment. In Europe, kitchen hood products that meet NC-35 have been commercialized (Hong et al. 2005), but the kitchen hoods in Korea have often been found to produce noise that is too loud for apartment equipment.

Accordingly, there have been studies on the design of kitchen hoods that will create a quiet residential environment. Choi et al. (2012) measured pressure losses due to the types and sizes of kitchen exhaust systems such as the hood, filter, and duct, so that the measurement results could be used them as the base line data when designing kitchen hoods. Song et al. (2005) evaluated the performance of eight kitchen hood models by measuring their operation noise and capture efficiency.

In addition, in their studies to reduce kitchen hood noise levels, Hong et al. (2005) and Lee et al. (2003) attached a perforated plate system on the hood's side and compared noise reductions with respect to the plates' porosity and cavity depth. Kim (2004) analyzed noise reduction performance by attaching a damping panel on the front of the hood.

This study analyzed the noise characteristics in three kitchen hood models with respect to their type and performance control. Additionally, we examined the noise distribution levels in the kitchen and living room when a hood was being operated in an apartment, representing them quantitatively, and we calculated the area of ratio at which they exceeded the recommended level.

2 KITHCEN HOOD NOISE MEASUREMENT AND PREDICTION SIMULATION

2.1 Hood noise measurement outline

In order to analyze the operating noise of a kitchen hood heard by a resident, a hood was installed and its noise was measured in the same apartment with 84 m² area. We selected the three types of hood categorized on the basis of their structure, as presented in Figure 1—chimney, independent, and built-in. The chimney and built-in types are made of the same material, so they can be divided according to their shade. The built-in hood is advantageous in that its upper part, excluding the lower portion that collects indoor air, can be designed with the same finishing material as the kitchen's storage closets.

The hood performance intensity could be controlled. The chimney-type hood is controlled by levels 1, 2, and 3, while the independent and built-in types are controlled by levels 1 and 2.

The hood noise was measured at a point 1.0 m in front of the hood, and at a height of 1.2 m in the center of the kitchen by using the 1/2 UC-59

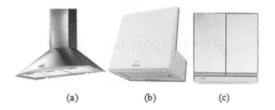

Figure 1. Kitchen hoods (a) Chimney type (b) Independent type (c) Built-in type.

Table 1. Noise level measurements and predictions at the center of living room.

Type—Intensity	Measured values dB(A)	Predicted values dB(A)
Chimney		
Level 1	47.5	47.6
Level 2	51.8	51.9
Level 3	53.1	53.3
Independent		
Level 1	49.5	48.8
Level 2	49.9	49.8
Built-in		
Level 1	47.9	48.2
Level 2	49.1	49.8

microphone, manufactured by Rion, and the Harmonie frequency analyzer of 01 Db. The equivalent noise level was measured for minute of the hood's operation.

2.2 Hood noise prediction simulation outline

The hoods' noise levels were measured in the living room, but not in the other rooms in which the noise levels were relatively low owing to the sound insulation of their doors. The kitchen with the hood is open toward the living room, whose area is 20 m² (4.5 m × 4.5 m). The noise-level-prediction simulation program Raynoise 3.1 was used, by applying acoustic absorptivity and transmission loss of sound in the ceiling, floor, and door materials. Predictions were made by setting up gridlines 0.5 m apart for a height of 1.2 m in the living room. The number of rays was set to 2,000 and the number of reflections was set to 10.

The level of the sound source of the operation was measured at 1.0 m in front of the hood. Table 1 presents noise level values that were actually measured at the center of the living room, and values predicted by simulation. The predicted and measured values had differences within 1 dB(A). The WHO defines the noise level at which residents can understand 100% of the conversations among one another below 45 dB(A) (World Health Organization 1980). We, therefore, verified whether this condition was met when the hood was being operated.

3 ANALYSIS ON THE CHARACTERISTICS AND DISTRIBUTION OF KITCHEN HOOD NOISE

3.1 Noise characteristics with respect to the type and intensity of the hood

Characteristics of the noise measured at 1.0 m in front of the hood during its operation, for the chimney, built-in, and independent types are as shown in Figures 2, 3, and 4, respectively. The chimney-type hood generated noise levels of 53.5 dB(A) at level 1, 58.1 dB(A) at level 2, and 59.5 dB(A) at level 3. At levels 2 and 3, there was a difference of within 3 dB(A) over the entire frequency range. Regardless of its operating intensity, noise levels were high in the range of 315 to 1,000 Hz and started to decrease as the frequency increased beyond 1,250 Hz. In addition, the noise level was relatively high in the region of 125 Hz.

The built-in hood showed noise levels of 54.5 dB(A) at level 1, 56.3 dB(A) at level 2, and a similar level across the entire frequency range. Unlike the chimney-type and independent hoods, it decreases at lower frequencies.

The independent hood generated noise levels of 55.0 dB(A) at level 1 and 56.1 dB(A) at level 2, showing little difference across the entire frequency range. In addition, it has a relatively high level at 125 Hz and from 500 to 1,000 Hz.

3.2 Noise level distributions in the kitchen and living room

The noise level distribution in the Kitchen and living room was visualized in order to understand the hood

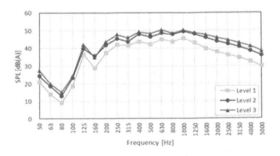

Figure 2. Noise characteristics of the chimney-type hood.

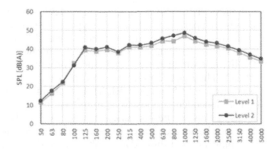

Figure 3. Noise characteristics of the built-in hood.

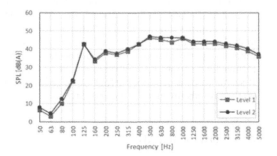

Figure 4. Noise characteristics of the independent hood.

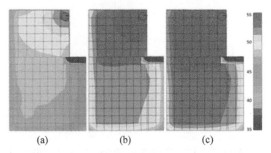

Figure 5. Noise level distribution for the chimney-type hood. (a) Level 1 (b) Level 2 (c) Level 3.

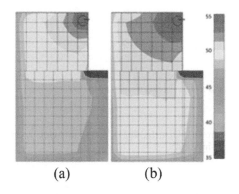

Figure 6. Noise level distribution for the built-in hood. (a) Level 1 (b) Level 2.

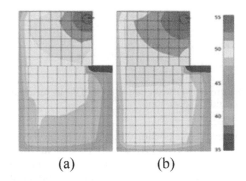

Figure 7. Noise level distribution for the independent hood. (a) Level 1 (b) Level 2.

noise a resident hears. The noise level distribution during the operation of the chimney-type, built-in, and independent hoods are as shown in Figures 5, 6, and 7, respectively. The circles in Figures 5–7 are the locations of the hood. Regardless of the hood type, the upper right region of the living room shows a low noise level, of around 35 dB(A), which is believed to be caused by a sort of shadow zone because it is not directly connected to the kitchen.

In the case of the chimney-type hood, there was a large difference in noise level between levels 1 and 3. While at level 1, the noise level exceeded 50 dB(A) only at locations in the kitchen adjacent to the hood, at level 3, major portions of the kitchen and living room showed noise levels of over 50 dB(A). On the other hand, the independent hood generated a higher noise level than built in hood at level 1, and a similar level at level 2.

3.3 Evaluation of hood noise in the living room

To evaluate the level of noise made by the hood in the living room, the percentages of the living room area were calculated according to noise level, as in Figures 8–10.

As seen in Figure 8, the region where the chimney-type hood generated noise levels of 47 dB(A) were 45.8% area at level 1. At level 2, 39% of the area experienced 51 dB(A), and at level 3, 46.6% of the area experienced 53 dB(A), which was the highest portion. Areas where the noise level exceeded 45 dB(A) were 78.0%, 99.2%, and 100% of the living room area, for levels 1, 2, and 3, respectively.

The noise level distribution of the built-in hood in the living room is shown in Figure 9. At level 1, 37.3% of the entire area experienced 48 dB(A), and at level 2, 42.4% of the area experienced 49 dB(A).

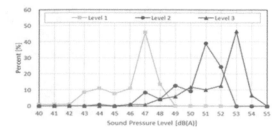

Figure 8. Distribution of noise in living room from chimney-type hood.

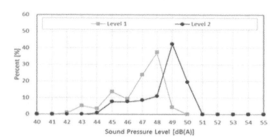

Figure 9. Distribution of noise in living room from built-in hood.

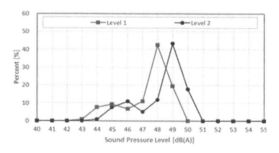

Figure 10. Distribution of noise in the living room from the independent hood.

At level 1, the portion where the noise level exceeded 45 dB(A) was 88.1%, and at level 2, it was 96.6%.

Figure 10 shows the noise distribution in the living room caused by the independent hood, indicating that the noise levels are different between levels 1 and 2, but the distributions are similar. At level 1, 37.3% of the area experienced 48 dB(A), and at level 2, 42.4% experienced 49 dB(A). Also, the portion that exceeded 45 dB(A) was 88.1% and 96.6%, for levels 1 and 2, respectively.

4 CONCLUSIONS

In this study, we aimed to analyze the characteristics of noise generated by the operation of the chimney-type, built-in, and independent kitchen hoods in apartments. For each type of hood, we evaluated the noise characteristics, noise distribution across the living room, and the proportion of the living room area that experienced more than 45 dB(A). The noise level at 1 m in front of the hood was between 53.5 and 59.5 dB(A). The chimney-type and independent hoods were found to make high levels of noise in the frequency range of 500 to 1,000 Hz, and the noise characteristics were similar as well. The built-in hood showed a distribution in which the noise level decreased as the frequency decreased, under 125 Hz. Also, the quantification of noise level distributions in the living room showed that the level exceeded 45 dB(A) at 78–100% of the living room area. It appears that the level of kitchen-hood noise in the living room was higher in the tested apartment than in apartments of other shapes because the former living room had an open layout to kitchen. It seems to be a need for further noise reduction in kitchen hoods through further research.

ACKNOWLEDGEMENT

This research was supported by a grant (15RERP-B082204-02) from Residential Environment Research Program funded by Ministry of Land, Infrastructure and Transport of the Korean government.

REFERENCES

Choi, S.H. & Lee, G.T. 2012. An Experimental Study on the Kitchen Ventilation System Effectiveness by Character of Static Pressure Loss of Each System Component in Apartment Building, *Journal of the Architectural Institute of Korea*, 269–276.

Hong, B.K. Song, H.Y. Lee, D.H. Lee, C.K. & Kim, D.Y. 2005. *A Research on the Noise Reduction of Ramge Hood for Household*, Proceeding of the KSNVE Autumn Conference, 449–452.

Kabir, E. & Kim, K.H. 2011. An investigation on hazardous and odorous pollutant emission during cooking activities, *Journal of Hazardous Materials*, 188(1–3): 443–454.

Kim, C.H. & Choi, Y.S. 2004. Noise Reduction of Range Hood in Kitchen Ventilation System, *Proceeding of the KSNVE Spring Conference*, 848–851.

Lee, D.H. & Kwon, Y.P. 2003. Estimation of the absorption performance of multiple layer perforated panel systems by transfer matrix method, *Journal of Sound and Vibration*, 278(4–5): 847–860.

Song, J.W. Kang, I.K. Kim, T.H. Lee, K.R. & Shin, Y.S. 2005. Method to Select the Optimal Gas Range Hood, *Journal of Korean Society Living Environment System*, 12(1): 2–18.

World Health Organization, 1980. Environmental Health Criteria, No. 12: Noise.

Advances in Civil, Architectural, Structural and Constructional Engineering – Kim, Jung & Seo (Eds)
© 2016 Taylor & Francis Group, London, ISBN 978-1-138-02849-4

Analysis on the correlation between bang machine and rubber ball

K.W. Kim, J.O. Yeon, H.K. Shin & K.S. Yang
Building and Urban Research Institute, Korea Institute of Civil Engineering and Building Technology,
Republic of Korea

ABSTRACT: In apartment housing, floor impact sound is generated by actions of children, walking of adults and so on. As diverse technologies and methods are applied to reduce floor impact sound, an accurate performance evaluation is required. As heavyweight impact sound among the floor impact sound includes low frequency domains, a bang machine and a rubber ball are used as the standard impact source. The rubber ball has been developed to have the characteristics more similar to actual impact sources than those of the bang machine. In this study, the correlation between the bang machine, a heavyweight impact source, and the rubber ball was analyzed by evaluation index. As the coefficient of determination of the bang machine and the rubber ball is low in the reverse-A single number quantity, the correlation between the two heavyweight impact sources is thought to be not high. The coefficients of determination of the two impact sources were analyzed to be high at 25 Hz and 63 Hz, respectively, showing values of 0.89 and 0.92. The bang machine and the rubber ball showed high coefficients with the reverse-A values at 63 Hz and 125 Hz, respectively. The deviation from dB(A) of the floor structures of which the equal performance can be exhibited by the bang machine is shown to be 10 dB. It is thought that, even though the floor structures are evaluated to have the same performance, the magnitude of noise the residents feel will be different.

1 INTRODUCTION

The vibration generated inside a building is transferred to the adjacent households as noise through the floor plate and the walls. Houses are constructed using the materials such as concrete and wood. The structure borne sound generated by the vibration transferred through structures causes inconvenience to the residents in a way of noise and vibration. The floor impact sound transferred through the floor structure is the most uncomfortable noise among the noise generated inside apartment housings.

Installation of a floating floor structure for which a resilient material is used is the general method used to reduce floor impact sound (Cremer et al. 1988). The materials used as resilient materials include glass wool and rock fiber that are fiber materials, and EPS (Expanded Polystyrene), EVA (Ethylene Vinyl Acetate), and EPP (Expanded Polypropylene) that are produced by expanding a material (Bettarollo et al. 2010). A double-floor structure built by installing two floor layers is also used. In order to evaluate the floor impact sound insulation performance of such various methods, standard impact sources are used. The standard impact sources are divided into lightweight impact sources and heavyweight impact sources. As a lightweight impact source, the equipment called a tapping machine is used and is used to evaluate medium and high frequency domains. As heavyweight impact sources, a bang machine which is a tire and a rubber ball are used. The heavyweight impact source is the impact source applied only to Japan and Korea. The heavyweight impact source is the equipment produced simulating the sound of children jumping and walking of adults, and was produced in Japan (Tachibana 1998).

Among the heavyweight impact sources, the rubber ball was produced in place of the bang machine, and its impact force is smaller than the bang machine. There is a research result that the rubber ball can better simulate the impact source which actually takes place than the bang machine (Yoo & Jeon 2014, Park et al. 2010). Though the two heavyweight impact sources are for evaluation of low frequency domains, it is pointed out that there are differences from the actual impact source. Rubber ball is accepted by the ISO (International Standard). The necessity for evaluation of heavyweight impact sound has been recognized in Europe and the related characteristics have been also studied (Andrea et al. 2015).

In this study, the characteristics of the bang machine and the rubber ball that are the heavyweight impact sources have been compared with each other, and their correlation has been analyzed based on the values measured in an apartment housing of concrete wall structure.

2 METHODS

The bang machine and the rubber ball used as heavyweight impact sources are prescribed in Korean standard in Korea. The minimum performance standard for the floor structure of apartment measured using the bang machine is prescribed by law. As to the evaluation standard, a method using reverse-A curve is applied. Table 1 shows the legal standard for floor impact sound. As for the heavyweight impact sound, 50 dB measured by the bang machine is set as the minimum grade standard. For lightweight impact sound, 58 dB is the minimum grade standard.

The bang machine and the rubber ball show differences in the impact force and the exposure level, and it has been pointed out that the impact force of the bang machine is higher than the impact force actually generated in apartment housings showing a value of 4200 N. Therefore, the rubber ball, of which the 1500 N impact force is similar to the actual impact force is developed. Evaluations of the two impact sources are shown to be different from each other. The noise level of the bang machine is measured to be higher at 63 Hz and the rubber ball is characterized by high noise level at 125 Hz.

The correlation between the evaluation method [reverse-A or dB(A)] and the frequency was analyzed on the basis of the result of measurement made for the two impact source, in order to review the similarity between the two impact sources. To this end, the measurement results of 42 floor structures were analyzed. The floor structure of the measurement object was measured using the bang machine and the rubber ball. The floor structure where the measurement was made was an apartment of a concrete wall structure, and the measurement was made using 1/3 Octave. The measurement result was synthesized using 1/1 octave and the result of each frequency was synthesized into dB(A). Also, the reverse-A single number quantity was calculated from the result of 1/1 Octave.

Figure 1 shows the cross section of the floor structure measured. The differences between the 42 structures include the difference between the resilient materials used and the structures where

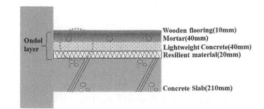

Figure 1. Section of typical floor structure.

lightweight concrete is excluded. A floor structure in the apartment was made of 210 mm of concrete slab, 20 mm of resilient material, 40 mm of lightweight aerated concrete, 40 mm of mortar, and 10 mm of wooden flooring. Green circular part is called Ondol layer in South Korea. The measured houses were structured with a box-frame structure mode where walls and floors were shared with each other.

The correlation between each factor and the impact source was analyzed based on the measurement result.

3 RESULTS AND DISCUSSION

3.1 Correlation by evaluation index

Performance of floor impact sound insulation is evaluated using the reverse-A single number quantity. Figure 2 shows the result of analyzing the correlation between the reverse-A single numbers of the bang machine and the rubber ball. As shown in the figure, the result of the bang machine was analyzed to be higher than that of the rubber ball. As the impact force of the bang machine is higher than that of the rubber ball, it showed a higher value also in the result of the performance evaluation. The coefficient of determination between the bang machine and the rubber ball was not high showing a value of 0.32, and the deviation from the levels of the two impact sources was shown to be big even when the performances were the same. The difference was shown to be 5 dB or bigger even when the performances were the same. As to the performance of the 42 floor structures, that of the bang machine showed performance distribution between 44 and 56 dB and that of the rubber ball between 37 and 48 dB.

Figure 3 shows the result of analyzing the correlation between the two impact sources and the dB(A) evaluation index. dB(A) was calculated by synthesizing the results of 1/1 Octave and 1/3 Octave. dB(A) is the result of correcting the value of the level when humans hear the sound with ears and was calculated to be higher than the reverse-A single number quantity. As they are obtained by

Table 1. Legal standard for floor impact sound.

Grade	Lightweight impact sound($L'_{n,AW}$)	Heavyweight impact sound($L_{i,Fmax,AW}$)
1	$L'_{n,AW} \leq 43$ dB	$L_i F_{max,AW} \leq 40$ dB
2	43 dB $< L'_{n,AW} \leq 48$ dB	40 dB $< L_{i,Fmax,AW} \leq 43$ dB
3	48 dB $< L'_{n,AW} \leq 53$ dB	43 dB $< L_{i,Fmax,AW} \leq 47$ dB
4	53 dB $< L'_{n,AW} \leq 58$ dB	47 dB $< L_{i,Fmax,AW} \leq 50$ dB

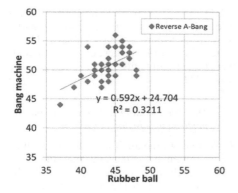

Figure 2. Correlation between impact sources and reverse-A single number quantity.

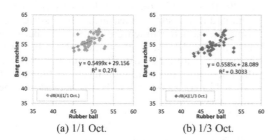

(a) 1/1 Oct. (b) 1/3 Oct.

Figure 3. Correlation between impact sources and dB(A).

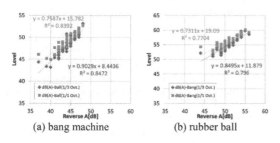

(a) bang machine (b) rubber ball

Figure 4. Correlation between Reverse-A single number and dB(A).

simply synthesizing the result by frequency, 1/1 Octave and 1/3 Octave show almost similar patterns. It can be seen that the distribution of the levels of 1/3 Octave result is wider than that of 1/1 Octave. Also in the dB(A) evaluation index, the coefficient of determination between the two impact sources was analyzed to be not high.

Figure 4 shows the result of comparing the reverse-A single number and the dB(A) result using the same impact source. The result of the bang machine was shown to have more inclined linear regression line than that of the rubber ball. The 1/1 Octave and 1/3 Octave had almost similar patterns showing only a little difference in the level. The

coefficient of determination between the reverse-A single number and the dB(A) when the bang machine was used was lower than when the rubber ball was used. Though the coefficient of determination between the two impact sources was very low showing a value of less than 0.3, it can be seen that the coefficient of determination between the two evaluation indexes of the same impact sources is high.

3.2 Correlation by frequency

Figure 5 shows the correlation between the impact sources by frequency. The analysis was carried out for the range between 25 Hz and 125 Hz using 1/3 Octave. At low frequencies, the variation in the level showed a wide performance distribution of about 35 dB. However, at 50 Hz, the variation in the level was only 10 dB. The coefficient of determination by frequency was analyzed to have a higher value than the coefficient of determination by evaluation indexes. The coefficients of determination at 25 Hz and 63 Hz were 0.89 and 0.92, respectively. It can be seen that the coefficient of determination at 63 Hz was the highest among those of other

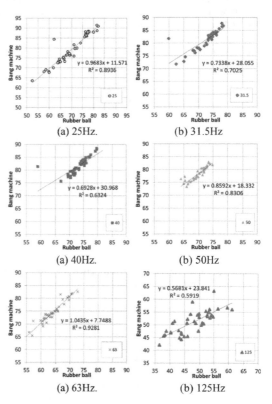

(a) 25Hz. (b) 31.5Hz

(a) 40Hz. (b) 50Hz

(a) 63Hz. (b) 125Hz

Figure 5. Correlation between impact sources by frequency.

frequencies in particular. The bang machine and the rubber ball showed the highest correlation at 63 Hz. If the level at 63 Hz is known, the level of the two impact sources can be inferred.

Table 2 shows the result of analyzing the correlation between the result of measurement made by frequency using 1/3 Octave and the evaluation index. The correlation analysis was carried out for the measurement result using Excel program, and the values inside the table represent the correlation coefficients. The red letters in the table represent the correlation coefficients of 0.7 or higher. The correlation coefficient between the bang machine and the rubber ball was shown to be high at the same frequency. The correlation coefficient between the bang machine at 63 Hz and the dB(A) of the bang machine was shown to be 0.745, and that between the bang machine at 63 Hz and the reverse A-bang was shown to be 0.763.

The indexes which showed a high correlation coefficient with the reverse A-Bang were shown to be 63 Hz-bang, 63 Hz-ball, and dB(A)-bang. The items of which the correlation coefficient was the highest among the evaluation indexes were reverse A-Ball and dB(A)-ball, and the value was analyzed to be 0.920. It can be seen that the correlation between the Reverse A evaluation method and the dB(A) evaluation method is the highest.

Table 3 shows the result of analyzing the correlation between the results measured using 1/1 Octave. It is almost similar to the result in Table 2, and the correlation coefficient values were higher than those of Table 2. The indexes which showed the highest correlation coefficient were 63 Hz-bang and dB(A)-Bang, and the value was 0.922.

3.3 Deviation of floor structure with the same performance by impact source

Table 4 shows the analysis result of the maximum value, minimum value, and the standard deviation of each index carried out for only the floor structures of which the reverse-A single number quantity of bang machine is 49 dB among the 42 structures measured. The number of the structures of which the performance was analyzed to be 49 dB was 7 in total. The levels by frequency obtained as a result of measuring the 7 structures using the bang machine were shown to be between 4.0 and 9.9 dB and that of the rubber ball was between 3.1 and 22.9 dB. The difference between the maximum and minimum levels of the rubber ball at 125 Hz was shown to be the biggest and the value was 22.9 dB.

Such deviation also appeared in the result of synthesizing the value into dB(A), and the rubber ball showed a deviation of 10 dB(A). The dB(A) level also showed a big difference of 10 dB(A) in the floor structures which were classified as having the same performance using the bang machine. It means that, though such difference is classified as having the same performance in the reverse-A evaluation, there may be a difference in the degree of the noise felt by the residents. Such deviation in performance is because the determining frequency of the reverse-A evaluation method is 63 Hz in most cases. The rubber ball showed a deviation between 40 and 48 dB.

Table 5 shows the result of analyzing the floor structures of which the performances were found to be the same using the rubber ball. The rubber ball was 44 dB, at which time the result of the bang machine was in the range between 49 and 54 dB. Though the result is similar to that of Table 4, the dB(A) value deviation was between 4 to 8.2 dB, which was lower than that of the bang machine. It can be seen that the deviation from the rubber ball is smaller than that of the bang machine in the floor structures of which the performances were analyzed to be equal.

However, this result is based on the limited number of measurements made for 42 structures, and the result may not be the same in diverse situations. It is thought that additional analyses are required for more diverse structures.

Table 2. Correlation coefficient between Frequency and evaluation Index (1/3 Oct.).

	31.5-Bang	63-Bang	125-Bang	250-Bang	31.5-Ball	63-Ball	125-Ball	250-Ball	dB(A)-Bang	dB(A)-Ball	Reverse A-Bang	Reverse A-Ball
31.5-Bang	1											
63-Bang	0.145	1										
125-Bang	-0.448	0.449	1									
250-Bang	-0.144	-0.113	0.447	1								
31.5-Ball	0.838	0.127	-0.510	-0.326	1							
63-Ball	0.183	0.963	0.328	-0.185	0.235	1						
125-Ball	-0.480	0.347	0.769	0.120	-0.481	0.248	1					
250-Ball	-0.366	0.006	0.512	0.557	-0.473	-0.092	0.605	1				
dB(A)-Bang	0.233	0.745	0.452	0.327	0.114	0.700	0.158	0.212	1			
dB(A)-Ball	-0.129	0.463	0.597	0.307	-0.157	0.437	0.741	0.723	0.551	1		
Reverse A-Bang	-0.006	0.763	0.644	0.313	-0.095	0.713	0.334	0.250	0.892	0.560	1	
Reverse A-Ball	-0.224	0.459	0.605	0.233	-0.253	0.444	0.789	0.620	0.436	0.920	0.567	1

Table 3. Correlation coefficient between Frequency and evaluation Index (1/1 Oct.).

	31.5-Bang	63-Bang	125-Bang	250-Bang	31.5-Ball	63-Ball	125-Ball	250-Ball	dB(A)-Bang	dB(A)-Ball	Reverse A-Bang	Reverse A-Ball
31.5-Bang	1											
63-Bang	0.143	1										
125-Bang	-0.387	0.130	1									
250-Bang	0.012	0.366	0.250	1								
31.5-Ball	0.804	0.167	-0.506	-0.161	1							
63-Ball	0.233	0.876	0.215	0.410		1						
125-Ball	-0.302	0.028	0.846	0.044	-0.368	-0.079	1					
250-Ball	-0.199	0.104	0.642	0.408	-0.335	-0.011	0.754	1				
dB(A)-Bang	0.308	0.922	0.229	0.514	0.212	0.753	0.076	0.203	1			
dB(A)-Ball	0.056	0.463	0.469	0.417	0.033	0.485	0.653	0.800	0.523	1		
Reverse A-Bang	-0.073	0.890	0.503	0.436	-0.120	0.663	0.330	0.341	0.878	0.565	1	
Reverse A-Ball	-0.161	0.379	0.672	0.250	-0.188	0.358	0.844	0.778	0.382	0.916	0.567	1

Table 4. Statistical analysis between the floor structures of 49 performance using bang machine.

	31.5-bang	63-bang	125-bang	31.5-ball	63-ball	125-ball	dB(A)-bang (1/3)	dB(A)-ball (1/3)	Reverse A-bang	Reverse A-ball
Max	83.9	73.1	56.0	75.8	63.8	61.2	53.6	53.1	49	48
Min	77	69.1	46.1	68.5	60.7	38.3	51.5	43.2	49	40
Stdev	2.47	1.37	3.78	2.40	1.09	7.38	0.78	2.97	0	2.48

Table 5. Statistical analysis between the floor structures of 44 performance using rubber ball.

	31.5-bang	63-bang	125-bang	31.5-ball	63-ball	125-ball	dB(A)-bang (1/3)	dB(A)-ball (1/3)	Reverse A-bang	Reverse A-ball
Max	84.6	80.7	55.1	75.9	69.7	54.4	59.7	50.8	54	44
Min	76.3	65.5	47.1	68.8	57.0	40.3	51.5	46.8	49	44
Stdev	2.86	4.18	2.43	2.34	3.72	4.37	2.24	1.10	1.58	0

4 CONCLUSIONS

The floor structure of the apartment also generates noise due to vibration. The floor impact sound insulation performance of floor structure is evaluated using lightweight impact sound and heavyweight impact sound, and a bang machine and a rubber ball are used as the heavyweight impact sources in Korea.

The correlation between various evaluation indexes were analyzed using the bang machine and the rubber ball used as the heavyweight impact sources. The result of analyzing the correlation is as follows:

1. In the reverse-A single number rating index, the coefficient of determination between the bang machine and the rubber ball was analyzed to be low showing a value of 0.32. The correlation between the two heavyweight impact sources is thought to be not high.
2. The correlation between the dB(A) and the reverse-A single number rating was found to be higher than that of between the two impact sources.
3. The coefficients of the two impact sources were analyzed to be high at 25 Hz and 63 Hz showing values of 0.89 and 0.92, respectively. The correlation coefficient of the bang machine from the reverse A value of 63 Hz was high, and that of the rubber ball was high at 125 Hz.
4. The deviation between the maximum and the minimum levels when the floor structures were of the same performance was shown to be high for both impact sources, and the dB(A) deviation of the floor structures of which the performances were found to be the same using the bang machine was shown to be 10 dB. It is thought that, even though the floor structures are evaluated to have the same performance, the residents may feel a difference in their performances.

ACKNOWLEDGEMENT

This research was supported by a grant (15RERP-B082204-02) from Residential Environment Research Program funded by Ministry of Land, Infrastructure and Transport of Korean government.

REFERENCES

Andrea, P. Anna, R. & Alessandro, S. 2015. Transmission of impact noise at low frequency: A model approach for impact sound insulation measurement, Proceedings of 22th International Congress on Sound on Vibration.
Bettarello, F. Caniato, M. Di Monte, R. Kaspar, J. & Sbaizero, O. 2010. Preliminary acoustic tests on resilient materials: comparison between common layers and nanostructured layers, Proceedings of 20th International Congress on Acoustics.
Cremer, L. Heckel, M. & Ungar, E.E. 1988. Structure-Borne Sound. 2nd, Ed, Sprinter-Verlag, Berlin.
Park, B. Jeon, J.Y. & Park, J.H. 2010. Force generation characteristics of standard heavyweight impact sources used in the sound generation of building floors, Journal of the Acoustical Society of America, 128(6): 3507–3512.
Tachibana, H. 1998. Development of new heavy and soft impact source for the assessment of floor impact sound of building, Internoise 98, Christ-church, New Zealand.
Yoo, S.Y. & Jeon, J.Y. 2014. Investigation of the effects of different types of interlayers on floor impact sound insulation in box-frame reinforced concrete structures, Building and Environment, 76: 105–112.

Advances in Civil, Architectural, Structural and Constructional Engineering – Kim, Jung & Seo (Eds)
© 2016 Taylor & Francis Group, London, ISBN 978-1-138-02849-4

Economic feasibility of fan-assisted hybrid ventilation system in apartment housings in Korea

Y.M. Kim, J.E. Lee & G.S. Choi
Korea Institute of Civil Engineering and Building Technology, Gyeonggi, South Korea

ABSTRACT: Heat exchanging ventilation systems are widely distributed in apartment housings in Korea; however, most occupants rarely use the systems because of the electricity cost, maintenance cost and noise by operation of the ventilation systems. The hybrid ventilation system can be considered as an alternative to counter such issues because it does not require any additional energy or maintenance from introduction of the external air. The target hybrid ventilation system for this study is an auxiliary fan-type system that utilizes both the auxiliary fan and the natural ventilation port, of which the former is to be used if the natural ventilation port does not provide enough ventilation. The energy simulation tool was conducted to evaluate heating and electrical energy consumption from operation of a ventilation system in winter, and economic feasibility was calculated based on the results of the energy simulation.

1 INTRODUCTION

1.1 *Ventilation system of households in Korea*

Even though the waste heat exchanging ventilation system is widely in use, modern shared residential buildings provide the advantage of reduced load and capacity of heat source facility, thanks to exchanged heat. It also suffers the issues such as consumption of electricity to retrieve waste heat, high maintenance cost to replace heat exchanger, need for a ventilation system for kitchen and bathroom and excessive noise, all of which leads to few cases of actual application of the system by shared residential buildings.

The hybrid ventilation system can be considered as an alternative to counter such issues because it does not require any additional energy or maintenance from introduction of the external air. Such a system will have minimal influence from outdoor air conditions as the air will be introduced to the air duct that maintains acoustic pressure. It also connects the air duct with exhaust ports of the kitchen and bathroom to secure a certain amount of ventilated air of the kitchen and bathroom and to prevent backdraft. Also, the system is expected to provide great anti-condensation performance as absolute moisture amount in daily life is controlled by ventilation for each room. Especially, the system has the benefits of overcoming natural ventilation opening's technical difficulty of uneven ventilation efficiency, cost reduction from initial investment, thanks to the overall simple structure and high applicability to old shared residential buildings and thanks to the utilization of the existing air duct.

This study includes energy consumption analysis and economic feasibility review through simulation with the aim of identifying actual applicability of the hybrid ventilation system.

1.2 *Fan-assisted hybrid ventilation system*

The target hybrid ventilation system for this study is an auxiliary fan type system that utilizes both the auxiliary fan and the natural ventilation port, of which the former is to be used if the natural ventilation port does not provide enough ventilation. For controlling the amount of ventilated air, a differential pressure controller and motor inside the power fan installed at the top of the shared air duct maintains the differential pressures of the central air duct and outside on the real-time basis so that the amount of ventilated air is maintained at 0.5 times/hour.

The system fulfills target indoor ventilation amount by controlling the ventilated air from the room by the motor-operated constant air damper installed at each household. If the damper is closed, the wind pressure and temperature differences of indoor and outdoor are utilized as driving power so that outdoor air is introduced through the natural ventilation port installed at the bottom of the window frame. Figure 1 shows the power fan that is installed on the rooftop for the hybrid ventilation system and the diagram with the operation mechanism of the system.

Figure 1. Power fan on the rooftop and the diagram of the fan-assisted hybrid ventilation system.

2 ENERGY CONSUMPTION ANALYSIS

The commercial simulation tool, TRNSYS, was employed for the analysis of heating and electrical energy consumption from the operation of a ventilation system in winter, from November to February for standard apartment buildings. Figure 2 shows the floor plan of the target apartment, and Table 1 presents the conditions of target households and operational features of ventilation system for simulation of energy consumption.

The "Construction standard for health-friendly house" by the Korean government required that waste heat recovery-based ventilation systems have a bypass and preheater installed. Depending on the manufacturer, a bypass may have different operation conditions by the temperature of the outdoor air to prevent condensation or freezing in and outside of the product and electric heating elements; thus, two conditions (HE 1 and HE 2) were established. It allowed the heat transfer element to pass through by activating the preheater if the outdoor temperature is lower than 5°C after raising it above 5°C (HE 3). The hybrid threefold ventilation system was set to be operating regardless of the outdoor temperature. Table 2 presents the operation conditions of each ventilation system for energy consumption simulation. HE 1 and HE 2 were assumed to bypass the heat exchanger if the outdoor temperature goes below −10°C and −15°C respectively so that 0.5 times of ventilating per hour is constantly maintained.

Table 3 presents the energy and ventilation performance analysis of each system per household. In case of applying HE 1 and HE 2 systems, ventilation load can be reduced through waste heat recovery, and winter heating energy could be saved by 6.2% and 7.2% respectively compared with the hybrid system. Regarding the amount of ventilated air through the system, the 3-type hybrid ventilation system could be available regardless of the temperature of the outdoor air so that it could deal with 100% of necessary amount of ventilation

Figure 2. Floor plan of the target apartment.

Table 1. Condition of the target apartment.

Contents	Condition	Remark
Gross area	80 m²	–
Volume	192 m³	–
Occupants	3	2 adults, 1 child
Schedule of occupants	Adults	weekday 19:00–08:00 weekend 17:00–11:00
	Child	weekday 17:00–08:00 weekend 17:00–11:00
Heating temperature	23°C	
Ventilation rate	0.5 ACH	According to the schedule of occupants
Infiltration	0.2 ACH	constant
Efficiency of heat exchanger	0.739 × ambient temperature + 65.929	
Heat source	District heating	–
Efficiency of heating equipment	90%	–

while the user was in the room. HE 3 used the preheater to maintain 100% operation rate. Both HE 1 and HE 2 systems conducted total heat exchange depending on the outdoor air temperature. Therefore, the amount of ventilation through bypassing or ventilation only without total heat exchange was relatively high. This resulted in only 66.0% and 79.8% of the required amount of ventilation that was supplied to indoor through heat exchanging in January when the outdoor air temperature was low. Operation conditions, given in Table 2, vary depending on the indoor temperature and absolute humidity; thus, it can be determined that the operation rate of actual total heat exchanger may be lower.

Table 2. Operation condition of the ventilation systems.

	Ambient condition	Supply fan	Exhaust fan
HB	Constant	–	ON
HE I	T > 5°C	ON	
	0°C < T ≦ 5°C	50 min ON, 10 min OFF	ON
	–5°C < T ≦ 0°C	45 min ON, 15 min OFF	
	–10°C < T ≦ –5°C	40 min ON, 20 min OFF	
	T ≦ –10°C		bypass
HE II	T > 0°C	ON	
	–5°C < T ≦ 0°C	80 min ON, 10 min OFF	ON
	–10°C < T ≦ –5°C	50 min ON, 10 min OFF	
	–15°C < T ≦ –10°C	OFF	
	T ≦ –15°C		bypass
HE III	Preheater is on when T < 5°C	ON	ON

Table 3. Energy and ventilation performance.

Ventilation system	Heating energy [KWH/ year]	Ventilation through the system (A) [m³]	Ventilation through the bypass (B) [m³]	A/(A + B) [%]
HB	9,148	276,768	0	100.0
HE1	8,576	227,960	48,808	82.4
HE2	8,492	255,376	21,392	92.3
HE3	8,135	276,768	0	100.0

Table 4. Energy cost.

Energy	Cost
District heating	85.93 KRW/Mcal
Electricity	up to 300 kWh: 187.9 KRW/kWh
	up to 400 kWh: 280.6 KRW/kWh

Table 5. Electricity consumption.

System	Ventilation system [W/CMH]	Exhaust fan only (including bypass) [W/CMH]
Hybrid	0.14	–
HE 1.2.3	0.31	0.15

Table 6. Operation cost for each system.

Ventilation system	Heating cost [KRW]	Electricity cost for ventilation system [KRW]	Electricity cost for bypass and preheating [KRW]	Total [KRW]
HB	750,978	9,060	0	760,038
HE1	704,092	15,993	1,959	722,044
HE2	697,139	18,119	896	716,154
HE3	667,861	43,681	113,013	824,555

3 ECONOMIC FEASIBILITY ANALYSIS

Table 4 presents the unit cost of both heating energy and electrical bills resulting from utilizing the district heating system. Heating cost was based on the winter household heating cost announced by the Korea District Heating Corporation, while electricity cost was based on the billing system for low pressure household electricity of Korea Electric Power Corporation. As the progressive rate was applied to electricity, the monthly average usage from 2010 Energy Consumption Survey was also used as the standards, which results in the application of the cost for usage lower than 300 kWh for March, October and November while cost for usage lower than 400 kWh for January, February and December.

Table 5 present electricity consumption by each system. As multiple households use one fan, the hybrid system results in rather lower amount of usage, while HE 1, 2 and 3 systems have a bigger amount of usage as ventilation fans are simultaneously operated due to the characteristics of heat exchanger. Also under the assumption that if the total heat exchanger only uses the exhaust function or bypass, the exhaust fan of the total heat exchanger is only used, so that only 50% of the electricity for total heat exchanging was used. Power for the preheater of HE 3 was applied with the same cost as that of the total heat exchanger.

Table 6 presents the operation cost per household by each system through converting the energy consumption data given in Table 3 into the cost through Tables 4 and 5. Therefore, the energy cost for heating and ventilation in winter by household turned out to be KRW 754,575 (hybrid system), KRW 743,475 (HE 1 system), KRW 738,854 (HE 2 system) and KRW 824,555 (HE 3 system). So, it indicated the reduction in operation cost by KRW 11,100 (HE 1 system) and 15,721 (HE 2 system) compared with that of the hybrid system, but the H2 system showed an additional cost of KRW 69,980 compared with that of the hybrid system.

Initial investment for the hybrid system was KRW 749,250, while both HE 1 and HE 2 systems required KRW 1,015,940. With these costs, economic feasibility of the systems, as shown in Figure 3, are obtained by analyzing calculated winter heating and ventilation energy consumption.

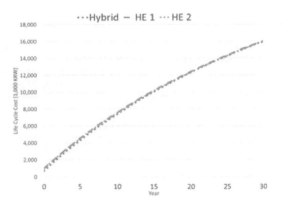

Figure 3. Economic feasibility.

The inflation rate applied herein is 2.93%, the average of 7 years from 2007 to 2013. Thus, it took more than 30 years and 24 years for the HE 1and HE 2 systems respectively to recover the invested amount into the hybrid system. Feasibility analysis for HE 3 has not been conducted as it required the biggest initial investment and operation cost among the three systems.

4 CONCLUSIONS

This research analyzed winter season and annual heating and electricity consumption through the two cases of waste heat recovery ventilation system and three types of hybrid ventilation system for a standard apartment building through TRNSYS energy simulation. Applicability of the three types of hybrid ventilation system to a shared residential building was analyzed through economic feasibility review.

Approximately 6~8% of heating energy reduction could be expected from using the total heat exchanger compared with the hybrid ventilation

system. Given the additional cost of electricity that the total heat exchanger uses, annual cost reduction was about KRW 11,100~15,721 compared with the case of using the hybrid ventilation system. It was analyzed to take about more than 24 years in order to recover the additional initial investment of KRW 250,000. Also, it was shown that using a preheater resulted in the biggest electricity bill due to the power it consumes.

ACKNOWLEDGMENT

This research was supported by a grant (15RERP-B082204–02) from the Residential Environment Research Program funded by the Ministry of Land, Infrastructure and Transport of Korean government.

REFERENCES

Chang, H. 2006. *Long Term Energy Simulation of Ventilation System for Apartment Houses*, Proceeding of Winter Annual Conference of the SAREK, 114–119.

Hwa, T. 2000. Ventilation Performance Evaluation in the Work Place according to the Diffuser arrangement Methods, *Magazine of the SAREK*, 29: 17–23.

Kim, J. et al. 2009. A Study on Current Problems of Heat Recovery Ventilator Operated by Occupants in High-rise Apartment Houses, *Journal of the Architectural Institute of Korea*, 23: 249–256.

Kim, H. & Park, J. 2009. *Effects of Ventilation System Operation on Annual Energy Consumption in Apartments*, Proceeding of Autumn Annual Conference of the Architectural Institute of Korea, Planning and Design.

Kwon, K. et al. 2006. Study on Ventilation Performance of Multi-room in Apartment, *Journal of the SAREK*, 399–404.

Lee, J. et al. 2015. The Ventilation Performance Analysis of a Fan-assisted Hybrid Ventilation System, *Journal of KSLES*, 22(3): 454–459.

Advances in Civil, Architectural, Structural and Constructional Engineering – Kim, Jung & Seo (Eds)
© 2016 Taylor & Francis Group, London, ISBN 978-1-138-02849-4

Numerical analysis of the fire accident at Happy Song KTV in Taipei City

C.S. Lin, J.P. Hsu & S.C. Hsu
Department of Mechanical Engineering, Yuan Ze University, Taoyuan, Taiwan

ABSTRACT: In Taiwan, due to the limited land area and dense population, high-rise buildings are the most common form of buildings and are used for diverse purposes, such as entertainment venues. Disaster prevention in high-rise buildings is difficult to achieve, and rescue operation of such fire accidents is also challenging. This study adopted Fire Dynamics Simulator (FDS), the dynamic simulation software for fire accident developed by the National Institute of Standards and Technology (NIST) in the U.S., to analyze the fire scene of a fire accident at Happy Song KTV in Taipei City. The hazard analysis included the computer numerical simulation on the air flow phenomenon driven by the buoyancy of the fire accident, as well as the distribution of the temperature, heat radiation, dense smoke on the fire scene, in order to reconstruct the fire scene. The results can serve as a reference for the fire prevention and fire safety improvement of high-rise buildings.

1 INTRODUCTION

A number of major fire accidents had occurred in Taiwan over the past two decades, such as the fire accidents in Hua Chi Hotel in Kaohsiung, Lun Ching Café and Carlton Salon in Taipei, Wellcome Restaurant in Taichung, Happy Song KTV in Taipei City, and ALA Pub in Taichung City. The fire accidents had claimed huge property losses, as well as caused high casualties. It is clear that building safety in urban areas is essential to the public's lives and properties. Both governmental and private organizations should attach more importance to the safety problems. The research of fire accident is a complicated field, and the use of computer simulation is conducive to hazard analysis. This study used the Fire Dynamics Simulator (FDS) to reconstruct the fire scene at the Happy Song KTV in Taipei City. The simulation results showed that different situations represent different physical phenomena. Different ventilation conditions and partition materials imply the flash burning trigger time at a low-intensity fire. These factors can all affect the flash burning (Wang et al. 2008). Although there are various definitions about flash burning, they can be classified into three categories, containing the mathematical, visual and some dependent terms respectively (Francis & Chen 2012). The flammable materials in buildings determine the intensity of fire where the key influencing parameters include the thermal properties of the room partition materials and the sizes of the room ceiling openings on the 1st floor (Lin et al. 2008). The factors causing personnel casualties include the temperature and harmful gases. The fire accident

growth and smoke movement are the basic and necessary components. In the fire accident risk analysis, a numerical model has been proposed to predict the fire accident growth and smoke movement. These models can be grouped into two basic types, namely the field model and regional model. However, the mixed model can be used as well. Compared with the regional model, field model or CFD (Computational Fluid Dynamics) model, the FDS is relatively new and more complex (Zhang & Hadjisophocleous 2012). Smardz (Xin et al. 2005) evaluated the FDS expansion on the smoke spreading from a small room into a nearby large space (hotel building) and then took out the powered-ventilation system adopted. In order to simplify the restrictions of simulation, the NIST and BFRL developed the FDS (Fire Dynamics Simulator), the dynamic simulation software for fire accident, by using the relevant previous studies conducted on fire accident situations (McGrattan et al. 2010). It is hoped to have a better knowledge of the fire accidents of such category happening in public places, thus achieve the goal of prevention.

2 CASE DESCRIPTION

This case is a fire happened at 2:08 a.m. on April 17, 1995, in Happy Song KTV, located at Sec. 2, Hankou Street, Taipei City. The fire scene is a 5-floor building that was still in business by the time of fire occurrence. The first group of rescue personnel arrived at the scene to found that the fire was burning vigorously at the entrance on the 1st floor and the outer wall of 2nd, 3rd and 4th floors on the west side of

Table 1. Summary table of fire accident case information.

Time of Occurrence	2:08 a.m., April 17, 1995
Location of Fire Accident	Sec. 2, Hankou Street, Taipei City
Record of Attendance	45 fire engines, 13 ambulances (3 teams of medical personnel from the Department of Health, Taipei City Government) and 262 people arriving at the scene for rescue
Number of Casualty	11 fatalities, 3 people in danger and 8 people injured
Property Damage	Over 3 million NTD
Cause of Fire	Arson
Origin of Fire	Stairs on 1st floor
Meteorological Data	22–29°C, fine, breeze, wind force less than F3

the building for the origin of the fire was on the 1st floor and the fire spread quickly. A large amount of dense smoke came out from each floor rising up at a speed of 3 to 5 meters per sec. The whole building was soon full of smoke. The fire spreading was very quick as the most materials on the scene were flammable. The rescue personnel carried out rescue at the risks of dense smoke and high temperature and controlled the fire spreading at 2:49 a.m. and put out the fire at 3:05 a.m. It caused a total of 11 fatalities, including 1 occurring on the scene and 10 after being sent to the hospital for emergency treatment, and 8 people injured. The origin of fire generated fire and dense smoke instantly before the elevator on the 1st floor and the fire spread upward along the stairs. The information of this fire accident case is summarized as shown in Table 1.

3 RESEARCH METHOD

FDS, developed by the U.S. NIST, was used in this work. FDS can construct the geometric boundary conditions of a fire to fully simulate the combustion reactions of flammable objects and dynamic changes in thermal boundary conditions and smoke layers within a site. Smokeview can be employed to display the 3D representation of the changes in smoke, heat, and other gases over time in an event of a fire. FDS can provide substantial amounts of information for the fire engineering research. This information includes flame diffusion, smoke movement, fuel or wall surface heat transfer, and the effects of sprinkler activation, which facilitates the development of a fire safety design for smoke control and sprinkler/sensor systems. FDS relies on Navier-Stokes equations to calculate the flow field and heat transfer changes that fires cause in buildings.

4 MODEL ESTABLISHMENT

The case under this research is a 24 m (length) × 10 m (width) × 12.5 m (height) building, based on which a 3D numerical model of FDS is established. However, 2 m is reserved in the front, back, left, right and height of the building in order to observe the fire burning outside the building and the flow of smoke. Therefore, the region of simulation computation is designed as 28 m (X) × 14 m (Y) × 16 m (Z). The lattice point obtained in the lattice point independence test in this case is 108 (X) × 54 (Y) × 64 (Z) and used for the simulation of this base case to verify the conformance with the situation of the original fire scene. Happy Song KTV is in an old-fashioned 5-floor complex where the 3rd–5th floors are occupied by Happy Song KTV as shown in Figure 1, having a total of 43 rooms of various sizes and a floor height of 2.5 m.

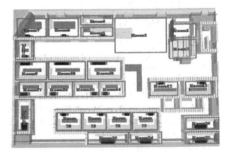

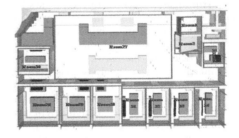

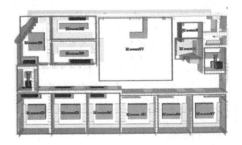

Figure 1. Layout of 3rd, 4th and 5th floors.

144

5 SIMULATION RESULTS

5.1 *The fire scene verification*

5.1.1 *Verification of flash burning on fire scene*

There was flame seen come out from the KTV doorway not long after a man came downstairs. Then the fire spread out suddenly and up to the floors above very soon, giving out very dense smoke. The fire and black smoke sprang out of Happy Song KTV shortly after the suspect set the fire to it. The simulation results are as shown in Figure 2: The Simulated Transient State Diagram of Temperature and Dense Smoke on The Fire Scene at the 90th sec based on the above description of the time point. The fire and dense smoke spreading can be seen at 3rd-floor Lobby at around the 90th sec. Since a large quantity of flammable gases is accumulated on the scene, the entire fire scene would be engulfed by the fire within 1 to 2 sec after all flammable materials are ignited automatically by high temperature when the temperature continues to rise to more than 500°C under the condition of fire burning in a sealed space, turning the scene

into a sea of fire and possibly causing serious casualties. It can be seen from the following temperature curve diagram that the flash burning in Happy Song KTV happened at around the 90th sec, bring the temperature up to 1000°C for a time.

5.1.2 *Verification of fire exposure of concrete on fire scene*

When the fire temperature continues to rise up to 580°C, the concrete in a high-temperature event leads to complicated result. Because each structure has a different thermal expansion coefficient, the internal structure of concrete would be seriously damaged at about 650°C (1200°F), thereby causing the overall concrete damage (SherifYehia & GhanimKashwani 2013). The modulus of elasticity would be reduced by 50% at around 427°C (800°F). Figure 3 is the temperature diagram of simulation results. The temperature of the 3rd-floor outer wall has reached above 600°C when the concrete structure is seriously damaged at around the 90th sec and the temperature is maintained as long as over 100 sec. In summarization of the

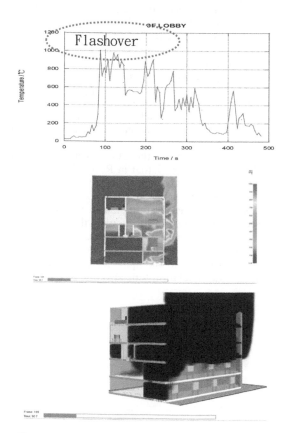

Figure 2. Simulated transient state diagram of temperature and dense smoke on fire scene at the 90th sec.

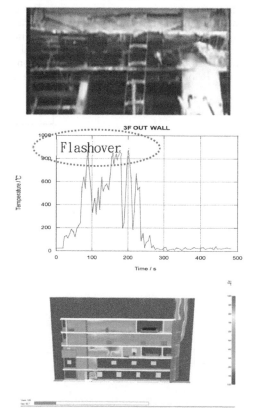

Figure 3. The simulated transient state diagram of 3rd-floor outer wall and temperature.

above data, the cement flaking phenomenon may occur to the concrete through the high-temperature burning after such a long time.

5.2 Discussion of fire dynamics

Figure 4 is the temperature curve diagram of the outside of the elevator (Room 1), 3rd-floor lobby (Room 3), KTV Room 207 (Room 9) and KTV Room 210 (Room 12). The outside of elevator is the fire-setting point of this arson case. Because the floor outside the elevator was laid with carpet, an intense fire was started immediately after the arsonist poured gasoline over the carpet and lighted the carpet adsorbing the poured gasoline based on the dangerous nature of gasoline, which went wild and caused a fatality in the elevator (Room 1). Consequently, the temperature has risen to over 250°C within about 6 sec. The fire spread upwards in the staircase instantly along the carpet laid on the floor. Moreover, the vertical staircase caused the dense smoke on the fire scene to increase and accumulate at the ceiling. The smoke layer became thicker and started to accumulate downwards.

At around the 70th sec of the simulation time, the temperature in the staircase is over 500°C and the fast burning results in a sea of fire. It can also be see from the temperature curve diagram in Figure 5 that the temperatures of the 4th-floor lobby (Room 27), KTV Room 312 (Room 28), KTV Room 317 (Room 33) and 5th-floor lobby

(Room 37) have been climbing after the 200th sec of the simulation time, with the highest even up to 1000°C or so.

The 5th-floor lobby (Room 37) is an open plane area out of the staircase, containing sofas, tea tables and a lot of combustible materials helping the burning. The temperature has already exceeded 200°C before it reached the 100th sec of the simulation time and even 500°C at around the 200th sec of the simulation time. After the surrounding glass breakage due to the high temperature, a lot of outside air flew in, intensifying the internal flaming fire and causing the temperature to continue to rise to about 1000°C or so. The people rescued out of the scene were seriously injured with about dozens of them in critical situations. It could be seen from this curve diagram that this temperature has already reached the fatality level.

As shown in Figure 6, the outside of the elevator (Room 1) is the first scene, no CO concentration is generated before the 250th sec of the simulation time due to the intense fire. However, after a certain time of burning, the temperature of the accompanying dense smoke rose sharply up to over 500°C and then dropped down gradually with the floor carpet burned up. In the 3rd-floor lobby (Room 3) and KTV Room 210 (Room 12), the CO generation started before the 50th sec of the simulation time due to the large amount of dense smoke generated by the origin of fire on the 1st floor and the concentration has risen to 500 ppm~600 ppm at the 150th sec of the simulation time. The generated concentration has reached 500 ppm, the level able to cause poisoning, and maintained at this level till the simulation was over.

It is known from the CO concentration curve diagram in Figure 7 that in Room 27, 28, 33 and 37, due to the large amount of dense smoke generated by burning at the origin of fire on the 1st floor, the CO generation started before the 50th sec of the simulation time with the time for the 4th-floor lobby (Room 27) being earlier due to the through staircase. Therefore, the concentration rose sharply up to 400 ppm~500 ppm at the 100th sec of the simulation time also due to the burning

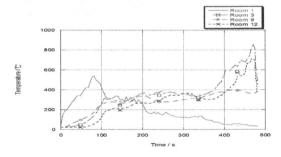

Figure 4. Temperature curve diagram of room 1, 3, 9 and 12.

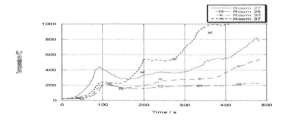

Figure 5. Temperature Curve Diagram of Room 27, 28, 33 and 37.

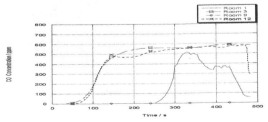

Figure 6. CO Concentration Curve Diagram of Room 1, 3, 9 and 12.

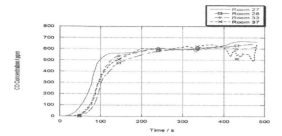

Figure 7. CO Concentration Curve Diagram of Room 27, 28, 33 and 37.

accelerated by the flammable materials. With the continuous burning at the origin of fire and due to the extremely flammable materials everywhere in Happy Song KTV, the CO concentration remained high and continued to rise to about 700 ppm until the end of the simulation time.

6 CONCLUSIONS

This study collected information from the fire brigade on the Happy Song KTV fire accident and used FDS for dynamic simulation software of the fire scene.

1. The simulation revealed that the fire spreads very quickly and may cause injury and fatality if it is not handled carefully. Therefore, the consumers and employees in entertainment venues should keep alerted and alarmed for arson and the timely disposal can prevent the occurrence of disasters.
2. The quantity of flammable materials affects the intensity of the fire. Use fire resistant materials as possible; do not allow the scooter parking in the arcades; do not install iron gratings on the window to make the escape difficult.
3. The simulation indicated that the temperature and gas distribution could reach the very dangerous level within a short time of 90 sec. The time is critical. One should first find out the emergency exits and evacuation routes in the area when going to a strange environment.

REFERENCES

Francis, J. & Chen, A.P. 2012. Observable characteristics of flashover, *Fire Safety Journal*, 51: 42–52.
Lin, C.S., Wang, S.C. & Chou, K.D. 2008. *Fire Damage Investigation Using FDS in a Basement Building*, The 3rd International Symposium on Energetic Materials and their Applications, 04.171.
McGrattan, K.B., Forney, G.P., Floyd, J.E. & Hostikka, S. 2010. *Fire Dynamics Simulator (Version 5), User's Guide*, NIST Special Publication 1019–5, NIST.
Sherif, Y. & Ghanim, K. 2013. The performance of Structures Exposed to Extreme High Temperature—An Overview, *Open Journal of Civil Engineering*, 3(3): 154–161.
Wang, S.C., Lin, C.S. & Yu, C.C. 2008. *Effect of Wall Material and Ventilation on Flashover Phenomenon in a Compartment Fire*, The 3rd International Symposium on Energetic Materials and their Applications, 64, 2008.04 172.
Xin, Y., Gore, J.P., McGrattan, K.B., Rehm, R.G. & Baum, H.R. 2005. Fire dynamics simulation of a turbulent buoyant flame using a mixture-fraction-based combustion model, *Combustion, and Flame Journal*. 141(4): 329–335.
Zhang, X. & Hadjisophocleous, G. 2012. An improved two-layer zone model applicable to bothpre—and post-flashover fires, *Fire Safety Journal*, 53: 63–71.

Advances in Civil, Architectural, Structural and Constructional Engineering – Kim, Jung & Seo (Eds)

The comprehensive utilization of building material waste based on the green building system

H.Y. Zhao & W.M. Guo
School of Design, Jiangnan University, Wuxi, China

ABSTRACT: Building material waste output in China is huge while the recycled rate is very low. How to reduce the generation of material waste has become an urgent problem. In this paper, different methods of reusing the material waste have been compared. According to the present situation of the construction industry, the principle of direct reuse should be put forward, and the whole process optimization of the building construction, deconstruction and demolition should also be enhanced. Some cases of the material mixed design were analyzed. Finally, this article proposed to develop design concepts with the rules of 3R, and to promote institutional environment construction. Only in this way can the maximum effect of waste utilization be realized.

1 INTRODUCTION

As environment resources are rapidly reducing, maintaining sustainable development of construction industry and improving the utilization level of resources are becoming a global concern. In the process of urbanization in China, building waste has become a public nuisance. Old city reconstruction and municipal engineering have caused a large number of demolition of old houses and generated a lot of construction wastes, but a lot of material waste can still be used. As the related research data displayed, the Chinese building waste resource utilization rate is less than 5%, which is far lower than the average level of developed countries. The annual construction waste emission is about 15.5 million tons to 24 million tons, which is about 40% of the total amount of urban waste. The new technology about how to reduce and utilize of construction waste is imminent.

2 METHOD CLASSIFICATION OF THE BUILDING MATERIAL WASTE REUSE

2.1 Different levels of classification

The reuse methods of the material waste can be divided into direct and indirect reuse. Direct reuse refers to retaining the prototype of material and simple processing such as material cutting, decontamination and other methods for processing. Then, the material waste can be reused, but regeneration utilization energy consumption is still large. The methods for material waste changing back into raw materials need more sophisticated processing technology such as crushing and melting. It saves only the energy consumption of resource exploitation while the former holds all the curing energy, and the emission reduction effect is better.

Methods of reusing materials can be used with respect to three levels: system level, product level and material level.

1. System level refers to the repeated use of the product in accordance with the original function. Adaptation to meet the needs of the people by the continuous transformation;
2. Product level is the product being sent back to the factory to break it down into basic components, where the usable parts can be used to manufacture new products;
3. Material level refers to the collection of products and crushed into basic materials. It would be taken as the replacement of the raw materials used in the production of new products.

After comparing the energy cost of collecting, transporting and processing products, resource energy consumption shows that reuse is better than remanufacturing and remanufacturing is better than recycling.

2.2 Technology based on material separation

The most important component of construction waste is the damaged concrete and the wall material (sintered brick, block, mortar). These materials cannot be directly used. It must go through a certain separation and process methodology, which always increase energy consumption. European scientists studied successfully a heating method to remove the mortar from sintered brick in the tunnel kiln. After the mortar separation, the brick products still meet the technical standard. A brick enterprise was founded in 2004 in Handan city, China where waste clay brick can be crushed and pressed into the brick again. The raw materials mainly come from the broken bricks, which first require screening, classification and impurity removal. After two levels of fragmentation, it is mixed with fly ash and cement, and then pressed for different types of blocks. There is no need to heat the coal in the production process, so the sewage and gas emissions are very small to meet the requirements of cleaner production. This technology processing more than 40 million tons of construction waste and annual output reached 150 million standard bricks.

Under normal circumstances, the waste concrete can be formed as recycled aggregate by a certain proportion and gradation after recycling, crushing, screening and grading. At the same time, the concrete broken into small stones can also be used for broken stone hardcore or flooring material. It can be added water reducer 4.5 kg/ m^3 or additive 10%~20% if the intensity is good. All the waste also can be pulverized as fine aggregate in mortar for the strength can reach above 5 MPa, and all these processes can replace the partial output of cement.

Making full use of construction waste to develop artificial aggregate becomes a new trend. The scrap of concrete waste, brick, tile and other wastes can replace the natural sand stone. Taking the green concrete for example, which contains glass, the demolition of concrete and other recycled materials, and their special ratio, ensures the larger porosity between the particles that have the potential for rainwater infiltration and plant growth.

3 OPTIMIZATION OF THE WHOLE PROCESS

3.1 Sorting and separation in removal process

Building waste generally consists of broken concrete, clay brick, sand, dust, wood, plastic and scrap metal. Preliminary classification of the different materials is necessary in the process of the buildings demolition, and a small part of the material is needed to be excluded in advance (e.g. plastic, paper, glass and metal). It is very important to adopt reasonable technology in the process of removing and demolition. Bricks can be very complete if the detailed dismantling methods were taken, with the utilization rate of more than 70%; if bulldozers are used to tear down, the brick recycling rate is less than 50%. Due to the immature of the large-scale recycling and utilization environment in China, most of the demolition allows the

Table 1. Main restricting factors of construction waste sorting.

	Restricting factors	Coping strategies
a	Waste transportation and emission costs increase	Reasonable organization arrangement
b	Equipment investment	Good management coordination
c	Construction period extension	Clean up in time; reduce waste mixture
d	Site restriction	Reasonable arrangement of processing frequency and position
e	Effect on normal construction activity	Rational organization arrangement
f	Lack of recycling market	Corresponding regulations Should be formulated to encourage construction waste recycling
g	Environmental constraints in the project area	To do the protective measures according to the regulations

structural material to fall from the sky. Most bricks are damaged seriously with the reutilization rate of less than 10%.

Although the classification of waste is very troublesome, their uncontrolled disposal results in long-term pollution costs, resource overuse and wasted energy. Table 1 analyzes the restriction factors and the corresponding strategies of waste classification, such as the increase in transportation and emission costs by the implementation of classification and placement. If the construction waste is unprofitable or has lower profits after sorting, it will increase the cost of the project. Second, because the construction waste classification and sorting rely on the ground, the more narrow the site is, the more reluctant the implementation of the classification the developer will be. In China, the site environment of different regions is very different. If the requirements of the construction pace are fast, the problem is more prominent. In China, most of the construction waste recycling is carried out by private enterprises, which have just simple equipment and low technology content. Thus, a lot of wastes that can be recycled into the recycling market are abandoned.

In this paper, 51 valid questionnaires were collected through the questionnaire survey conducted in Jiangsu Province, the effective rate of which was 68%. The different control level of each factor is obtained for the present situation of the classification of the field.

On the basis of the comparison of these schemes, we can arrive at a conclusion that the bottleneck is not at the technical level, but in the market cultivation and related government policy support for recycling of material waste.

3.2 Structure design based on separation and recovery

When a building is demolished, most of the materials are discarded, and along with them, the embodied energy is lost. If buildings were designed for disassembly, rather than demolition, greater proportions of building materials could be salvaged for reuse. The usage on the amount of too many composite materials has a bad influence on material separation. One of the hallmarks of the green building system is the convenience of the separation and recovery of all materials. The structural form of welding or other type is too hard to be disconnected; thus, it should be avoided. For example, the construction of the bolt type, socket type and jig type is convenient for the direct use of the old structural components. At the same time, using prefabricated technology and material modularization can effectively reduce the waste generation. Most building wastes are caused by the material cutting process, while it is easy to be handled in the factory, so promoting modular design of the building components helps reduce waste, and the universal property of the component can also improve the efficiency of reuse.

4 INNOVATION OF THE GREEN BUILDING DESIGN SYSTEM

4.1 Mixed use concept of old and new

Construction waste minimization at the design stage is a key strategy in effective waste reduction. It is necessary to focus on the rules of 3R again. Design concept should fully comply with the rules of 3R: reducing, reusing and recycling. The reduction is the process of reducing waste generation and pollution. It is the most basic way to prevent and reduce pollution. Reuse means the use of items as many times as possible to prevent the material becoming garbage too fast. Recycling is to put the raw materials into new production. The importance of the principles in the circular economy is not an equal relationship.

According to Germany's "circular economy and waste management act", which came into force in 1996, the priority order in the treatment of the waste problem is to avoid the generation (i.e. the reduction), repeated use (i.e. reuse) and the ultimate disposal (i.e. recycling). The reduction is one of the most effective and efficient methods of waste management, so it is unwise to define it as a waste too early, which is very obvious in practice.

According to the actual situation of China's construction, the consumption of fast-building renewal is very large. The rapid change in lifestyle has caused the growth in consumption of building materials. The original construction products that have not reached the service life will be eliminated by new products too soon; the solution to the problem is to change the current situation of overconsumption and to promote a more efficient way of life and to reduce the generation of construction waste from the source. Retaining original buildings and continuing to use and promoting the old-new mixed use of architecture should be an important method. It can be divided into two levels in design. One is to retain the old architecture (building frame, components, external wall skin) and promote the use of organic mixture, which makes some of the old building materials extend its service life. The other is the old building materials, i.e. components to be used elsewhere.

Therefore, it should be advocated to change the concept that only the new material is beautiful. To establish an aesthetic model based on the ecological and natural life style, Figure 1 shows the construction of a traditional garden. Brick and stone wastes after morphological classification, ingenious design, and careful arrangement can also be constructed as a beautiful gravel pavement in the garden. It is called the "Colorful street" pavement in the Chinese vernacular dwellings. It not only is beneficial to rainwater infiltration and plant growth but also becomes a beautiful decoration. The traditional way of building, which coincides with climate characteristics and environmental conditions, shows that China's traditional residence has a fine tradition of green design.

In fact, stone leftover materials and old brick can be mostly used for surface layer pavement, after the water mill slightly down, the polychromatic stone aggregate surface layer can also become a very beautiful design effect, but most people think it can only be used for building non-essential parts and outdoor paving part. Some of the characteristics of the old materials are far better than some

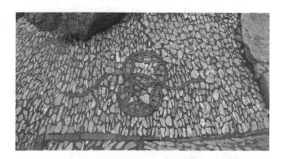

Figure 1. Traditional stone paving in a garden.

151

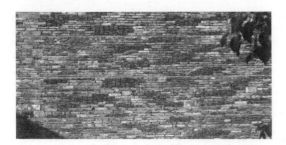

Figure 2. Exterior wall design in Xiangshan Campus, China academy of art.

new materials, such as the texture of wood and metal corrosion. Which have been widely used by designers?

Take the architecture design in China academy of fine art campus for example. The reuse of the old brick for the appearance of the building is particularly significant. For building the exterior wall, tens of thousands of pieces of old brick are used. The use of the old brick buildings in modern architecture can evoke the memories of traditional culture. Figure 2 shows that a designer has a strong originality concept, the simple and old wall in contrast with the newly refined steel showing that the beauty of the history of waste materials.

4.2 Function conversion design

The combined effect of multi-design strategies is more significant than the simple sum of the effects of all single-design strategies, and thus demands a full study of the relevant and potential design strategies that lead to a combined effect. A variety of reuse methods can be created through the functional transformation of the building materials. For example, brick slag is a very good mineral material, and according to the methods developed specifically, sintering bricks can be processed into decorative particles as an inorganic material to cover exposure area of planting pit. It can also be used for flower cultivation. These methods are more flexible and conducive to environmental protection.

In addition, different building wastes can be used flexibly in different environments. Outdoor garden, interior decoration, environmental art and other industries can greatly expand the use of construction waste. Clever design can make it to be a new product. Classification of waste materials is done by size, color, texture, light and shade, and reflective rate. The use of contrast, staggered, interspersed, repeat and other techniques can form a very rich and sophisticated outdoor landscape space. Construction waste can also be built in landscape pavilion, flower racks, tables and chairs, landscape wall, booth and sculpture.

5 CONCLUSIONS

All buildings have a specific lifetime, which can be divided into construction, operation and demolition phases. In summary, this paper puts forward the comprehensive utilization of building materials, which means to change the current straight-type ideas (production, construction, waste, reusing waste) to the comprehensive method (green production, extended using, recycling, multiple reusing). The core idea is to turn material resources into good account and extend the service life of materials. It saves the curing energy and reduces carbon emissions. (Embodied energy is the total energy consumption of building products in the process of raw materials mining, production, transportation and other products consumption. The common unit of measurement for energy density is mega joules per kilogram.)

The protection and reuse of building material resources is a comprehensive system. It involves multidiscipline crossing and integration. Construction industry concept and design ideas also should be changed. Setting out from the design source and extending building's service life from multiple aspects is a simple solution for complicated problems.

One of the major constraints is that people's living standards is improving and will be more reluctant to use the construction material waste. On the other hand, although there are many advantages, waste recycling is receiving inadequate recognition and support of the government interiority. Public buildings are subject to state regulations, quality and safety inspection departments are not allowed to use construction waste as load-bearing components in a new building, and scrap materials was excessively limited. It is also urgent to develop new policies to encourage the reuse of waste; only in this way can the maximum effect of waste utilization be realized.

REFERENCES

Abdol, R.C. & Bruening, S.F. 2003. Deconstruction and materials reuse in the United States, *The future of sustainable construction*. 11(5): 56–62.

Jan, O.F. 2006. Cut and paste. *Architecture*. 6: 74–79.

Lu, W.S. & Yuan, H.P. 2011. A framework for understanding waste management studies in construction. *Waste Management*. 7(31): 1252–1260.

Tan, X.N. 2011. Research toward the behavior of construction wastes minimization. *Xi'An architectural technology*. 6: 25–26.

Wang, J., Li, Z. & Tam, V. 2014. Critical factors in effective construction waste minimization at the design stage: a Shenzhen case study, China. *Resources, Conservation and Recycling*. 82: 1–7.

Advances in Civil, Architectural, Structural and Constructional Engineering – Kim, Jung & Seo (Eds)
© 2016 Taylor & Francis Group, London, ISBN 978-1-138-02849-4

The influence of nutrient medium and growth regulators on in vitro morphogenesis of *Thuja occidentalis* L.

D. Zontikov & S. Zontikova
Kostroma State University named after N.A. Nekrasov, Kostroma, Russia

R. Sergeev & A. Shurgin
Volga State University of Technology, Yoshkor-Ola, Russia

ABSTRACT: The method of clonal micropropagation of *Thuja occidentalis* L. is described in the present study. The activity of morphogenesis depending on the season of the in vitro culture introduction of donor explants isolated in spring, summer and autumn was assessed. The influence of MS, Q&L and Anderson media with the addition of growth regulators 2-ip at 1–3 mg l⁻¹ and NAA at 0, 1–0, 3 mg l⁻¹ on the effectiveness under in vitro culture introduction was also investigated. The dependence of the speed of shoot growth at the stage of micrografting on the kind and concentration of cytokinin, BA at 0, 5–1, 5 mg l⁻¹; 2-ip at 1, 0–3, 0 mg l⁻¹ and Kinetin at 1, 0 mg l⁻¹ was shown.

1 INTRODUCTION

Thuja occidentalis L. is widely used for landscaping, building and in the pharmaceutical industry as the source of valuable biologically active substances. Nowadays there are a lot of various ornamental types of this species that differ in form, color, height. They are used for landscaping not only of squares, parks and streets but also of living quarters and workrooms. It proves the high importance of this species.

After the analysis of literature it is evident that investigators paid much attention to the peculiarities of clonal micropropagation of certain species of genus Thuja as well as to the problems of low net reproduction, difficulty in rooting and relatively low growth rate in in vitro culture (Murashige T.A. (1962), Thorp T.A. & Nour K.A. (1993), William, R. (1997)).

Thus, the identification of factors influencing the activity of in vitro morphogenesis of T. occidentalis L. is an important and rather an urgent task.

2 MATERIALS AND METHODS

Explants of T. occidentalis L. and its two ornamental species Globosa and Smaragd were used for in a vitro introduction. They were isolated from the plants in June, August, and October.

At first the isolated shoots were washed in a water solution of potassium permanganate and sterile distilled water. Sterilization of plant material under conditions of the laminar box was carried out in 70% ethanol and NaCIO solution at 3% (Zontikov D (2014)), followed by three-time rinsing in sterile water. Explants that are the segments of a young shoot 2,0 to 5,0 cm long were placed on the nutrient medium (Figure 1). 20 donor explants were held over for each variant of the experiment.

For the activation of the morphogenesis MS (Quoirin, M. & Lepoivre, P.(1977)), with the addition of glycine at 2 mg l⁻¹; thiamin at 0,5 mg l⁻¹; pyridoxin at 0,5 mg l⁻¹; Quoirin and Lepoivre Thorpe T.A. (1977) with the addition of thiamin at 0,4 mg l⁻¹; Anderson (1980) with the addition of adenine at 80 mg l⁻¹; theamine at 0,4 mg l⁻¹ were used. All used media were added with growth regulators

Figure 1. The appearance and the relative size of shoots used for introduction in the culture.

2-ip (2-isopentyladenine) at 1,0–3,0 mg l⁻¹; NAA (α-naphthylacetic acid) at 0,1–0,3 mg l⁻¹ and also meso inositol at 100 mg l⁻¹, sucrose at 20 g l⁻¹; agar at 5,0 g l⁻¹. The pH was adjusted to 5,5–5,8.

In the experiment on the investigation of the influence of growth regulators at clonal micropropagation various types of *T. occidentalis* species 10 metameres for each experimental variant were held over. For the clonal micropropagation, basic nutrient medium Quoirin, and Lepoivre was used. BA at 0,5; 1,0 and 1,5 mg l⁻¹; 2-ip at 1,0; 2,0 and 3,0 mg l⁻¹; Kinetin at 1,0 mg l⁻¹ were used as growth regulators while the concentration of sucrose was 20 g l⁻¹, agar – 5,0 g l⁻¹; the pH was 5,6–5,8. Shoots 3 to 5 mm long were used in the experiment. They had been cultivated for 30 days on the nutrient medium without growth regulators.

In the experiment on the investigation of the influence of growth substances on the processes of root formation such substances were used: NAA at 0,1; 0,2; 0,3 mg l⁻¹; and IAA at 0,5; 1,0; 1,5 mg l⁻¹, while the concentration of sucrose was 20 g l⁻¹; agar –5,0 g l⁻¹; the pH – 5,6–5,8; and also there was placement on the nutrient medium without growth regulators. Before the experiment, all the shoots had been cultivated for 30 days on Quoirin and Lepoivre medium with 2-ip at 1,0 mg l⁻¹.

Clonal micropropagation was held every 50–60 days, by that time the shoots were 3 to 7 cm high.

3 RESULTS AND DISCUSSION

Clonal micropropagation of T. occidentalis L. includes the number of successive stages. At each stage the composition of the nutrient medium and growth regulators and the time of cultivation change. The first stage is the induction of bud formation by means of the usage of various nutrient media and growth regulators. In case the nutrient medium is chosen correctly, numerous buds appear at the ends of the shoots. The second stage is the growth of the new shoots (Figure 2)

out of the newly formed buds and their lengthening. The third stage is the formation of roots of the microshoots.

The formation of new buds occurs in the angles of cataphylls on the ends of the shoots. In the experiment on the investigation of the influence of the nutrient medium and growth regulators on the process of bud formation most of buds were formed on Q&L medium with 2-ip and NAA at 2,0 and 0,2 mg l⁻¹ respectively. Their average number was 4 ± 0,6 per explants. The fastest bud formation was marked on MS medium with 2-ip at 2,0 0,2 mg l⁻¹ and NAA at 0,2 mg l⁻¹. The formation of the first buds was marked on the 45th day of cultivation (Table 1).

For in vitro introduction the planting material was selected at different seasons, the different capability to morphogenesis was noticed at that. The effectiveness of selection of explants in autumn,

Table 1. The influence of the kind of the nutrient medium and growth regulators on the formation of the new buds on the explants of T. occidentalis L.

Nutrient medium	Growth regulators, mg l	Bud formation, day		The average number of buds, pieces	
		$\bar{x} \pm S_{\bar{x}}$	C.V.,%	$\bar{x} \pm S_{\bar{x}}$	C.V.,%
MS	2-ip 1,0	52 ± 0,7	4,1	2 ± 0,6	4,3
	2-ip 2,0	51 ± 0,5	4,4	2 ± 0,5	5,4
	2-ip 3,0	55 ± 0,4	3,2	3 ± 0,3	4,7
	2-ip 1,0 + 0,1 NAA	53 ± 0,5	5,7	2 ± 0,4	5,7
	2-ip 2,0 + 0,2 NAA	44 ± 0,4	6,1	3 ± 0,3	4,6
	2-ip 3,0 + 0,3 NAA	49 ± 0,4	5,9	3 ± 0,5	4,2
Anderson	2-ip 1,0	50 ± 0,5	4,6	2 ± 0,3	5,6
	2-ip 2,0	57 ± 0,6	5,3	2 ± 0,4	6,2
	2-ip 3,0	55 ± 0,4	4,5	2 ± 0,5	5,4
	2-ip 1,0 + 0,1 NAA	56 ± 0,7	4,4	2 ± 0,3	5,6
	2-ip 2,0 + 0,2 NAA	54 ± 0,4	4,6	3 ± 0,5	5,3
	2-ip 3,0 + 0,3 NAA	56 ± 0,5	4,3	2 ± 0,4	5,4
Q&L	2-ip 1,0	51 ± 0,5	5,1	2 ± 0,4	6,7
	2-ip 2,0	52 ± 0,7	5,3	2 ± 0,4	6,5
	2-ip 3,0	52 ± 0,6	4,5	2 ± 0,6	5,1
	2-ip 1,0 + 0,1 NAA	50 ± 0,7	4,6	2 ± 0,4	6,8
	2-ip 2,0 + 0,2 NAA	49 ± 0,8	4,7	4 ± 0,6	6,9
	2-ip 3,0 + 0,3 NAA	51 ± 0,7	5,0	3 ± 0,5	5,7

Figure 2. A) The formation of new buds in angles of cataphylls of T. occidentalis—day 45; B) the development of shoots in 40 days after the first replanting.

Table 2. The influence of the terms of in vitro introduction on the activity of morphogenesis of Thuja occidentalis plants.

Month of isolation of the donor material	Forms (cultivars)	Number of formed buds per explants $\bar{x} \pm S_{\bar{x}}$	C.V., %
June	T. occidentalis L.	2 ± 0,3	5,1
	T. occidentalis Globosa	1 ± 0,4	4,2
	T. occidentalis Smaragd	1 ± 0,4	5,4
August	T. occidentalis L.	4 ± 0,4	5,5
	T. occidentalis Globosa	3 ± 0,5	6,0
	T. occidentalis Smaragd	2 ± 0,4	5,2
October	T. occidentalis L.	8 ± 0,5	4,3
	T. occidentalis Globosa	6 ± 0,5	4,2
	T. occidentalis Smaragd	5 ± 0,4	4,7

Table 3. The influence of growth regulators on the growth and formation of shoots of various types of T. occidentalis at the clonal micropropagation (Quoirin and Lepoivre medium).

Forms (cultivars)	Growth regulators, mg l	Beginning of shoot growth, day $\bar{x} \pm S_{\bar{x}}$	C.V.,%
T. occidentalis L.	2-ip 1,0	8 ± 0,2	3,6
	2-ip 2,0	9 ± 0,2	6,2
	2-ip 3,0	14 ± 0,1	3,1
	6 BA 0,5	10 ± 0,1	4,5
	6 BA 1,0	16 ± 0,1	6,4
	6 BA 1,5	21 ± 0,2	4,6
	Kinetin 1,0	23 ± 0,1	5,6
T. occidentalis Globosa	2-ip 1,0	10 ± 0,1	6,1
	2-ip 2,0	14 ± 0,2	4,8
	2-ip 3,0	17 ± 0,1	4,1

Forms (cultivars)	The average height of shoots in 55 days, cm $\bar{x} \pm S_{\bar{x}}$	C.V.,%	The average number of shoots formed in 55 days, pieces. $\bar{x} \pm S_{\bar{x}}$	C.V.,%
T. occidentalis L.	5,0 ± 0,2	4,3	3,2 ± 0,2	22,6
	4,3 ± 0,2	5,4	4,1 ± 0,1	21,2
	4,1 ± 0,1	3,2	4,4 ± 0,1	19,3
	3,9 ± 0,1	4,1	2,8 ± 0,2	21,4
	3,2 ± 0,1	4,4	3,3 ± 0,2	12,3
	3,0 ± 0,2	5,2	3,4 ± 0,2	23,2
	3,3 ± 0,1	5,6	2,9 ± 0,1	14,8
T. occidentalis Globosa	4,6 ± 0,1	5,0	2,7 ± 0,2	18,2
	4,2 ± 0,2	5,7	3,9 ± 0,1	17,9
	4,0 ± 0,1	4,1	3,8 ± 0,1	19,1
	4,0 ± 0,2	4,3	3,2 ± 0,2	21,5
	3,2 ± 0,2	8,6	3,8 ± 0,3	9,9
	3,1 ± 0,1	10,2	3,3 ± 0,1	15,7
	2,9 ± 0,2	13,9	2,3 ± 0,2	16,4
T. occidentalis Smaragd	4,7 ± 0,2	4,6	3,0 ± 0,1	9,6
	4,4 ± 0,2	6,2	3,5 ± 0,1	7,3
	3,1 ± 0,1	3,1	3,5 ± 0,2	12,1
	3,7 ± 0,1	4,5	2,5 ± 0,2	10,8
	3,2 ± 0,1	3,4	3,1 ± 0,2	15,2
	3,1 ± 0,2	4,2	3,0 ± 0,2	13,7
	3,4 ± 0,1	5,6	2,6 ± 0,1	7,9

namely in October was experimentally established, as the biggest number of newly formed buds per explant was stated at this time. Depending on the form (cultivar) 5 to 8 buds were forming on the explants whereas 1 to 2 buds were forming in July (Table 2).

In 40–50 days, there was the first replanting of the formed buds on the fresh nutrient medium. It was stated that without replanting the newly formed buds considerably increased in size without shoot development and in some time they died. After the replanting on the new nutrient medium shoots developed from buds, and new buds were forming on the place of a slice. During the second transit cataphylls placed crosswise subopposite were observed in all the variants of all the forms of T. occidentalis. During the second transit shoots were not more than 0,5 cm long.

The usage of various growth regulators during the clonal micropropagation showed that the application of 2-ip as a growth regulator let get better rates concerning the average height of shoots (5,0 ± 0,2 cm), the beginning of their growth (8 ± 0,2 day) and the average number of shoots (4,4 ± 0,1 pieces) (Table 3). During the third transit (after the introduction into the culture) shoots didn't ramify, only some of them had one, less often two new shoots on the tylosis formed on the place of slice. And during the fourth transit branching of shoots was observed and 2 to 5 new shoots parted from the tylosis.

With the addition of BA to the nutrient medium more compact shoots were observed. It is much more difficult to micrograft them and at the output we get fewer microshoots. Apart from that at the concentration 1,5 mg l^{-1} some shoots were vitrified. On the nutrient medium with Kinetin such shoots were observed: they grew slower as in other variants, were strongly crooked, often grew inside the medium (Figure 3).

Unlike many other species of conifers (Pinophyta) T. occidentalis relatively easy roots in vitro, however, the speed with which it happens is defined with the type of growth regulators and their

Figure 3. The appearance of shoots of *T. occidentalis* L. under the influence of growth regulators A) 6 BA at 1,0 mg l; B) 2-ip at 2,0 mg l; C) Kinetin at 1,0 mg l.

Table 4. The influence of growth regulators on the growth and formation of roots at the clonal micropropagation of various types of *T. occidentalis* (Quoirin and Lepoivre medium).

Forms (cultivars)	The average length of roots in 55 days, cm		The average number of roots formed in 55 days, pieces	
	$\bar{x} \pm S_{\bar{x}}$	C.V.,%	$\bar{x} \pm S_{\bar{x}}$	C.V.,%
T. occidentalis L.	4,1 ± 0,1	5,6	1,8 ± 0,2	30,2
	3,3 ± 0,2	4,3	2,1 ± 0,2	27,1
	3,1 ± 0,1	6,1	2,5 ± 0,1	29,5
	3,3 ± 0,1	5,2	1,2 ± 0,2	22,5
	3,2 ± 0,2	4,5	1,5 ± 0,3	19,7
	3,0 ± 0,2	4,3	2,2 ± 0,3	23,3
	2,8 ± 0,3	9,6	2,9 ± 0,1	12,9
T. occidentalis Globosa	3,7 ± 0,1	6,4	2,1 ± 0,1	12,1
	3,3 ± 0,1	6,8	1,8 ± 0,1	11,9
	3,0 ± 0,1	7,1	2,0 ± 0,1	13,1
	3,1 ± 0,2	5,2	2,2 ± 0,2	20,2
	3,0 ± 0,2	8,4	2,4 ± 0,3	18,8
	3,0 ± 0,1	7,8	2,3 ± 0,3	19,3
	1,9 ± 0,3	22,5	2,3 ± 0,2	13,2
T. occidentalis Smaragd	4,2 ± 0,1	5,3	1,9 ± 0,1	10,2
	3,4 ± 0,1	7,2	2,0 ± 0,1	8,3
	3,1 ± 0,1	4,3	2,3 ± 0,1	9,9
	3,7 ± 0,2	6,1	2,3 ± 0,2	10,5
	3,4 ± 0,2	4,4	2,1 ± 0,2	11,4
	3,1 ± 0,2	3,9	2,0 ± 0,1	10,6
	2,6 ± 0,3	17,7	2,6 ± 0,3	25,1

concentration. In this study, the fastest growth of roots was marked on the nutrient medium with NAA at 0,1 mg l^{-1}. On the explants of T. occidental L. roots began to form on the 14th day of cultivation and in 55 days the length of the roots was 4,1 cm. By T. occidentalis Globosa and Smaragd, the growth of roots was observed on the 18th and 17th day of cultivation respectively (Table 4).

Figure 4. A) The beginning of root growing of *T. occidentalis*, B) roots before rooting into the peat tablets, C) rooted plants.

After the rooting the shoots were put out of the culture bottle, washed off the nutrient medium from the roots with the running water and placed in peat tablets Jiffy-7 and transferred to the microframe for adaptation and following completion of growing (Figure 4).

Thus, it is experimentally stated that at the selection of first explants for in vitro introduction before the dormancy the greatest number of new buds forms on each of explants. It is stated that for the induction of morphogenesis and the clonal micropropagation the usage of Quoirin and Lepoivre medium with 2-ip at 2,0 mg l^{-1} and NAA at 0,2 mg l^{-1} let get the greater net reproduction than on MS and Anderson media for the plants of T. occidentalis of all the investigated species. At the same time on the rooting stage, it is effective to use Quoirin and Lepoivre medium with 2-ip at 1,0 mg l^{-1} and NAA at 0,1 mg l^{-1}.

REFERENCES

Anderson 1980. Tissue culture propagation of red and black raspberries Rubus idaeus and R. occidentalis. *Acta Horticulturae.* 112: 13–20.

Kabir, M.H., Roy, P.K. & Golam, A. 2006. In vitro Propagation of Thuja occidentalis Through Apical Shoot Culture, *Plant Tissue Culture, and Biotechnology.* 16(1): 5–9.

Murashige, T.A. 1962. Revised Medium for Rapid Growth and Bio Assays with Tobacco Tissue, *Physiologia Plantarum.* 15: 473–497.

Quoirin, M. & Lepoivre, P. 1977. Etude de milieux adaptes aux cultures in vitro de Prunus.—*Acta Hortic.* 78: 437–442.

Thorp, T.A. & Nour, K.A. 1993. In vitro shoot multiplication of eastern whitecedar (Thujaoccidentalis). In vitro Cell. *Developmental Biology*, 29: 65–71.

Thorpe, T.A. & Coleman, W.K. 1977. In vitro culture of western redcedar (Thuja plicata). *Botanical Gazette*, 138(3): 298–304.

William, R. 1997. *Thomas Growing conifers*, Washington: Brooklin Botanic Garden PRESS, 112.

Zontikov, D., Zontikova, S., Sergeev, R. & Shurgin, A. 2014. Micropropagation of highly productive forms of diploid and triploid aspen. *Advanced Materials Research*, 962–965: 681–690.

Advances in Civil, Architectural, Structural and Constructional Engineering – Kim, Jung & Seo (Eds)
© 2016 Taylor & Francis Group, London, ISBN 978-1-138-02849-4

Exploring the suitability of a survey instrument to measure the walkability of streets in Doha

K. Shaaban & D. Muley
Qatar University, Doha, Qatar

ABSTRACT: This paper proposes a survey instrument to measure walkability indicators of streets using a rating scale. The survey instrument included sections related to location details, sidewalk characteristics, pedestrian crossing characteristics, sidewalk facilities, and driver behavior. This tool was tested on one two crowded streets in two different neighborhoods with dissimilar characteristics in Doha, Qatar. The results showed that the survey instrument was effective in measuring the various walkability indicators, which can be further useful in developing a walkability score to rate the walkability of streets.

1 INTRODUCTION

Even though sustainable transportation is getting importance for commuter trips as well as non-commuter trips; making people walk is a challenge for transport planners. Also, walking is a widely accepted physical activity to keep people healthy and reduce risks of chronic diseases (Edwards & Tsouros 2006, Coffee et al. 2013). Walkability of an area plays a vital role and is greatly affected by built and natural environment features such as pleasurability, comfort, safety, accessibility, and feasibility along with pedestrian perception of the area (Lindelöw et al. 2014, Livi & Clifton 2004, Shigematsu et al. 2009).

In the past, some studies have measured walkability at street level using different methods; Stevens (2005) used mobile GIS, Parks & Schofer (2006) used secondary data, Park & Kang (2011) used direct street measurement, and Gori et al. (2014) used secondary data and GIS.

The purpose of this paper is to design a survey instrument and apply it to check its accuracy and suit-ability for street walkability measurements. The remaining paper is arranged in three sections; description of the survey instrument, site description and street selection, walkability indicator measurements, results, and conclusions.

2 SURVEY INSTRUMENT

The survey instrument was designed to measure walkability indicators along selected streets of neighborhoods of Doha using qualitative and quantitative measurements. Around 30 walkability variables were chosen for walkability measurement using a thirteen-page survey instrument.

The survey form was filled manually and street details were recorded using a mobile application. The survey instrument was divided into five parts; location details, sidewalk characteristics, pedestrian crossing area characteristics, sidewalk facilities, and driver behavior. The first section collected details related to the location of the street section on a sketch related to latitude and longitude of the start point and end point, street width, street length and median width (if available). Remaining sections contained different measures and were rated on a scale of three or four.

In the second section, the sidewalk characteristics for each side were measured at every 200 ft or 70 steps distance with the help of nine variables. These variables included, number of obstructions, missing sections, broken or cracked sidewalks, blocked sidewalks, cleanliness of sidewalk, width of crosswalk, wheelchair accessibility, presence of trees, and the distance from the pavement.

The third section measured the pedestrian crossing area characteristics using five variables; availability of safe and clearly visible crossing points, presence of pedestrian signals, availability of curb ramps, and the level difference between footpath and the pavement.

The fourth section measured the sidewalk facilities using six variables; presence of seats/benches, shelter from rain or sun, land use along the corridor, presence of security personnel, availability of public transport stop, and street lighting.

The fifth section measured the driver behavior by measuring eight characteristics, from observations taken for 10 cars, right of way given, obeying of stop/yield sign, traffic lights, speed of driving, stopping location of on the street, availability of mid-block crossing, volume of traffic, and presence of traffic calming measures.

3 SITE DESCRIPTION

3.1 *Case study sites*

Two neighborhoods, Old Airport and The Pearl, in Doha, Qatar were selected for testing the designed survey instrument. These two areas have distinct characteristics. The Old Airport is an older residential neighborhood area with a large number of businesses and retailers and minimal green space. It is located on the southeast of Doha and covers an area of 1304 acres. Old Airport has many standalone villas, some compounds, and a school. The streets are well connected by a grid pattern. Generally, middle-class people, mostly single men, tend to live in this neighborhood.

On the other hand, The Pearl is a newly developing artificial island with a coastline where high-income people live. This new suburb is located on the northeast of the main Doha city center and covers an area of 985 acres. The first residential use of the neighborhood started in 2012. It has high-end apartment buildings with luxury shops and restaurants mostly suitable for couples and singles. The access to The Pearl is made easy by a double lane road. Figure 1 shows an aerial view of Doha showing locations of both the neighborhoods.

1.1 *Selection of streets*

In the first phase, the walking behavior of people in the two neighborhoods was observed for two days, a weekday and a weekend, for three time periods, namely morning, afternoon, and evening using video recording. The times for survey started from 7 am, 12 pm and 6 pm till recording was finished for morning, afternoon, and evening respectively. It should be noted that the video recording covered on all the streets in the neighborhood following a predefined path, only once in each time period. The number of pedestrians observed for each period for both neighborhoods is shown in Table 1. Around 1,910 pedestrians were observed in the Old Airport neighborhood and 709 pedestrians were observed in The Pearl neighborhood during the survey days. Approximately 10% more pedestrians were observed on the weekend than on the weekday in both the neighborhoods. The time of day variation for both neighborhoods indicated that the pedestrian volume was higher during the evening time in all cases except for the weekday case for the Old Airport neighborhood where number of pedestrians during morning and evening were almost equal.

The location of each pedestrian was traced on the map for all time periods. From this data, the pattern of spatial distribution was observed. From the observations, one street was chosen where most of the pedestrians were seen walking during all time periods. Particularly, Al Matar Al, Qadeem Street,

Table 1. Number of pedestrians for different time periods.

Conditions		Old Airport	The Pearl
Weekday	Morning	343	50
	Afternoon	216	76
	Evening	347	209
Weekend	Morning	200	56
	Afternoon	188	84
	Evening	616	234

Old Airport

The Pearl

Figure 2. Spatial distribution and street selection of study sites.

Figure 1. Location of the two neighborhoods.

spanning 1.7 km, in the Old Airport neighborhood and Pearl Boulevard, spanning 1.4 km, at The Pearl neighborhood were seen as the busiest streets in the selected neighborhoods. An example for one time period for both the neighborhoods is shown in Figure 2. In the second phase, the designed survey instrument was tested for these two streets. The details are presented in the following section.

4 WALKABILITY INDICATORS OF STREETS

4.1 Data collection

A recorder visited the survey site with the survey instrument and walked along the street from the start point to the endpoint to note down various listed walkability variables. The recorder also took photographs of each segment and made notes of all the observations. This process was carried out for both sides of the streets separately. Around 180 and 90 minutes were required for Al Matar Al Qadeem Street, Old Airport and Pearl Boulevard, The Pearl respectively. The characteristics of two streets were measured in 32 and 24 segments for each side of Al Matar Al Qadeem Street and Pearl Boulevard respectively.

4.2 Results from walkability street measurements

The sidewalk characteristics were measured in different segments. Table 2 shows the average score for sidewalk characteristics measured to indicate the quality of sidewalks on the selected streets. All the indicators were marked on a scale of three or four with the highest value showing the best conditions. From the results, it can be seen that Pearl Boulevard has wider and well-designed sidewalks which were pleasant for pedestrian use, only two obstacles were seen. For some segments, extensive use of landscape features was observed. For Al Matar Al Qadeem Street, sidewalks were mostly obstructed by merchandise from the adjacent shops or entrance ramps or steps leaving a narrow section for pedestrians use. In addition, some cracks were observed in the sidewalks on Al Matar Al Qadeem Street.

The results for pedestrian area crossing, sidewalk facilities, and driver behavior were measured for each side separately and shown in Table 3. The pedestrian crossing characteristics indicated that although pedestrian paths are separated from vehicle path, both the streets have poor treatments at crossing points with partially blocked view of traffic, absence of pedestrian signal, and poor ramp facilities. It should be noted that Al Matar Al Qadeem Street has mainly T intersections

Table 2. Average score of walkability indicators of sidewalk characteristics.

Characteristic	Old Airport			The Pearl		
	Side 1	Side 2	Over-all	Side 1	Side 2	Over-all
Obstructions in Sidewalk^	3.78	3.97	3.88	4.00	4.00	4.00
Missing sections in Sidewalk^	3.63	3.69	3.66	4.00	4.00	4.00
Cracks/breaks in sidewalks^	3.38	3.56	3.47	4.00	4.00	4.00
Blocked sidewalk#	2.28	2.66	2.47	2.71	2.83	2.77
Cleanliness of sidewalk^	3.94	3.91	3.92	4.00	3.92	3.96
Width of Sidewalk #	0.88	1.00	0.94	1.83	2.38	2.10
Wheelchair accessibility^	3.53	3.56	3.55	4.00	4.00	4.00
Trees along sidewalk#	1.09	1.03	1.06	1.79	1.96	1.88
Shoulder width#	1.84	1.00	1.42	2.46	2.92	2.69

\# measured on a three point scale.
^ measured on a four-point scale.

while The Pearl Boulevard has T intersections and roundabouts.

The land use on Al Matar Al Qadeem Street is mainly commercial use while The Pearl Boulevard has mainly residential land use along its length. Both streets did not have any rest areas or shelter to make the pedestrians comfortable while walking, but they had some patrolling available, which reflects a sense of safety. No bus stops were observed on the Pearl Boulevard, but more than one bus stop were observed on Al Matar Al Qadeem Street indicating more opportunity for public transport users to walk on the street. Further, The Pearl Boulevard was well lit by street lights but Al Matar Al Qadeem Street was illuminated using few street lights and businesses on the street.

The observed driver behavior indicated that drivers on the Al Matar Al Qadeem Street were driving moderately, obeying traffic rules, and creating a better environment for pedestrians compared to the Pearl Boulevard drivers. This might be because of the shared use, by cars and some heavy vehicles, of the street and presence of traffic calming measures like speed humps on Al Matar Al Qadeem Street.

Overall, the Pearl Boulevard has better sidewalk characteristics but more aggressive driver behavior compared to the Al Matar Al Qadeem Street. When the pedestrian crossing characteristics and

Table 3. Score of walkability indicators.

	Characteristic	Old Airport			The Pearl		
		S1	S2	Over-all	S1	S2	Over-all
Pedestrian crossing	Clear view of traffic at crossing points	2	2	2	2	3	2.5
	Safe crossing points	1	3	2	1	2	1.5
	Pedestrian signal	1	1	1	1	1	1
	Quality of curb ramps	2	2	2	1	1	1
	Elevated sidewalk	Y	Y	Y	Y	Y	Y
Sidewalk facilities	Seats for rest areas	1	1	1	1	1	1
	Shelter from rain/sun	1	1	1	1	1	1
	Land use	C*	C*	C*	R*	R*	R*
	Patrolling along street	2	2	2	2	2	2
	Public transport stops	3	3	3	1	1	1
	Street lighting	2	2	2	3	3	3
Driver behavior	Right of way to pedestrians	2	2	2	1	1	1
	Obey traffic signs	2	2	2	1	1	1
	Obey traffic signals	1	2	1.5	1	1	1
	Driving Speed	1	3	2	2	1	1.5
	Location of stop	2	1	1.5	3	1	2
	Midblock crossing	N	N	N	N	N	N
	Traffic condition	3	3	3	3	4	3.5
	Traffic calming measures	3	3	3	1	1	1

*C denotes commercial land use, R denotes residential land use, S1 means Side 1, S2 means Side 2.

sidewalk facilities are considered, the score was almost the same for both neighborhoods with Old Airport slightly outscoring The Pearl for public transport stops. From this comparison, it can be concluded that improved sidewalks are provided for new neighborhoods but pedestrian area crossing characteristics and sidewalk facilities need to be improved. However, these finding should be used with caution as only two streets are considered in this study. These findings can be further validated by measuring walkability for more neighborhoods across Doha.

5 CONCLUSIONS

Looking at the walkability scores, this survey instrument was successful in measuring an indication of walkability on the studied streets. A significant advantage of this survey instrument is that it is capable of measuring the micro level (street level) walkability indicators efficiently. The outcomes are beneficial in developing a walkability score and a scale for measuring the walkability of the streets using empirical measurements on foot. The main limitation of this methodology is that the instrument is resource intensive and requires skilled personnel to assess the walkability features accurately.

ACKNOWLEDGEMENT

This research was made possible by UREP grant #15–085–3-020 from the Qatar National Research Fund (a member of Qatar Foundation). The statements made herein are solely the responsibility of the authors.

REFERENCES

Coffee, N.T., Howard, N., Paquet, C., Hugo, G. & Daniel, M. 2013. Is walkability associated with a lower cardiometabolic risk? *Health & Place*, 21: 163–169.

Edwards, P. & Tsouros, A. 2006. The solid facts: Promoting physical activity and active living in urban environments the role of local governments. *World Health Organization.* ISBN 92-890-2181-0.

Gori, S., Nigro, M. & Petrelli, M. 2014. *Walkability indicators for pedestrian friendly design.* TRB 2014 Annual Meeting.

Lindelöw, D., Svensson, Å., Sternudd, C. & Johansson, M. 2014. What limits the pedestrian? Exploring perceptions of walking in the built environment and in the context of every-day life. *Journal of Transport & Health*, 1: 223–231.

Livi, A.D. & Clifton, K.J. 2004. *Issues and methods in capturing pedestrian behaviors, attitudes and perceptions: Experiences with a community-based walkability survey.* TRB Annual Meeting, 2004.

Park, S. & Kang, J. 2011. *Operationalizing Walkability: Pilot Study for a New Composite Walkability Index Based on Walker Perception.* TRB 2011 Annual Meeting.

Parks, J.R. & Schofer, J.L. 2006. Characterizing neighbor-hood pedestrian environments with secondary data. *Transportation Research Part D*, 11: 250–263.

Shigematsu, R., Sallis, J.F., Conway, T.L., Saelens, B.E., Frank, L.D., Cain, K.L., Chapman, J.E. & King, A.C. 2009. Age Differences in the Relation of Perceived Neighborhood Environment to Walking. *Medicine & Science In Sports & Exercise*, 41(2): 314–321.

Stevens, R.D. 2005. *Walkability around neighborhood parks: an assessment of four parks in Springfield, Oregon.* Terminal Project, Department of Planning, Public Policy and Management and the Graduate School, University of Oregon.

Advances in Civil, Architectural, Structural and Constructional Engineering – Kim, Jung & Seo (Eds)
© 2016 Taylor & Francis Group, London, ISBN 978-1-138-02849-4

New method of examination room desks region segmentation based on template matching

Y. Ding, Y.H. Li & B. Li
Kunming University of Science and Technology, Kunming, China

ABSTRACT: The desks region of the examination room plays an important role in locating students' position, and it is also the basic work of analyzing students' behavior. Based on the characteristics of the examination room desks region segmentation, this paper proposed a new method of examination room desks region segmentation based on template matching, which used perspective transformation to unify the size of the matched desk region, and finally used the k-means algorithm to correct the template matching results. The experimental results indicate that in the application of examination room desks region segmentation, the new method has a better noise immunity, accuracy and applicability.

1 INTRODUCTION

In computer vision theory, image segmentation, feature extraction and object recognition are very important. Image segmentation is the basic work of feature extraction and object recognition. The results of image segmentation will directly impact on feature extraction and target recognition (Xu et al. 2010). Thousands of image segmentation algorithms have been proposed, such as Levner I., and others have proposed the classification driven watershed algorithm (Levner & Zhang 2007). Comaniciu D. put forward a robust approach toward feature space analysis of mean-shif (Comaniciu & Meer 2002), which is based on statistics originated in the work of Geman and Geman (1984). These methods are the more typical image segmentation method. But they have their own characteristics, such as the specific environment and application areas. Currently, we do not have a method to suit a variety of image segmentation (Yang 2013).

In this paper, the segmentation of the examination room desks region is the premise and foundation work of intelligent monitoring of the examination room, which is also a key step to locate the positions of candidates. The quality of segmentation and the accuracy of position will directly impact on the analysis of candidates' behaviors. Therefore, this paper proposed a new image segmentation method.

2 NEW METHOD OF EXAMINATION ROOM DESKS REGION SEGMENTATION

Examination room desks region segmentation has some characteristics, such as a large proportion of

image brightness, a plurality of static and similar objectives, and a variety of image background. The traditional image segmentation methods are only applicable to the specific situations, such as thresholding method, morphological segmentation method and the spatial clustering algorithm based on feature. These methods cannot solve the problems of the examination room. Thus, this paper proposes a new method of examination room desks region segmentation based on template matching. It not only uses perspective transformation to pretreat the matching image, but also applies the k-means algorithm to correct the results of template matching. These treatments have improved the accuracy and applicability of segmentation. The process is detailed below.

2.1 Perspective transformation

Projection transformation is a transformation of linear or planar from general position to the parallel or perpendicular with the projection plane position. Perspective transformation is the part of the projection transformation, which maps the three-dimensional spatial information into two-dimensional space, so that the image can get the effect of orthographic angle. Equation (1) represents the process of perspective transformation (Xie et al. 2013):

$$\begin{bmatrix} x' \\ y' \\ 1 \end{bmatrix} = M \times \begin{bmatrix} x \\ y \\ 1 \end{bmatrix} \qquad (1)$$

Figure 1. Original image.

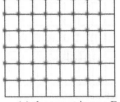

(a) the target image F

b) the template T

Sub-image F

(c) the matching process

Figure 2. Template matching.

where M is the perspective matrix; h_{ij} is the element of matrix; i is the row and j is the column. The elements of M are given in Equation (2):

$$M = \begin{bmatrix} h_{11} & h_{12} & h_{13} \\ h_{21} & h_{22} & h_{23} \\ h_{31} & h_{32} & 1 \end{bmatrix} \qquad (2)$$

Eight pairs of precise coordinate points are selected. Four pairs are experience coordinate points, such as $d_1[310, 220]$, $d_2[505, 218]$, $d_3[285, 495]$, $d_4[505, 500]$. The other four pairs are the artificially marked vertices of maximum external quadrilateral; this Quadrilateral contains all desks, as shown in Figure 1. Then, the L-M damped least squares method is used to solve the values of h_{11}, h_{12}, h_{13}, h_{21}, h_{22}, h_{23}, h_{31} and h_{32} (Zhou et al. 2005).

2.2 Template matching

In this paper, we can look for desk areas by template matching. These areas have the same size as the template, so we can achieve the purpose of segmenting the examination room desks regions. The principle of matching is as follows: (1) F is the target image and T is template. They are shown in Figure 2. In target image F, the vertex of the upper left corner of the every little rectangle represents a pixel (marked by a black speck in Figure a). (2) First, T is superimposed on F, then T is starts from the first pixel to move on F, and calculates similarity of sub-image F, sub-image F covered with T. When this similarity is greater than the matching threshold, it shows that matching is successful. The matching process is shown in Figure c. (3) Sub-image F represents the matching success-

Figure 3. Perspective image.

Figure 4. The result of template matching.

ful desk area, represented by dark colored areas in Figure c.

In the actual matching process, because the variable light seriously affects the result of matching,

162

this paper uses the normalized matching method. It is given in Equation (3):

$$R(x,y) = \frac{\sum_{x'y'}\left[T(x',y') - F(x+x', y+y')\right]^2}{\sqrt{\sum_{x'y'}T(x',y')^2 \bullet \sum_{x'y'}F(x+x', y+y')^2}} \quad (3)$$

R (x,y) represents the matching similarity. (x,y) is pixel coordinate of the target image F. (x',y') represents the pixel coordinate of the template T.

2.3 *K-means algorithm*

K-means algorithm is a clustering method, and also the main image segmentation methods. This algorithm has the biggest advantages of briefness and celerity. As shown in Figure 4, it has many results of template matching, so that it is not accurate positioning desk area of the image. In this paper, using the k-means algorithm to cluster all results of meeting template matching threshold in a certain range, this range belongs to the same desk area (the range is marked by the round mark in Figure 4). Finally, the smallest similarity R(x,y) is selected as the final desk area from the respective cluster.

Based on the above description, using perspective transformation to process Figure 1, the result is shown in Figure 3. Then, according to the empirical value to set the matching threshold value, it is 0.3. Using template matching to process Figure 3, the result is shown in Figure 4. Finally, using the k-means algorithm to process Figure 4, the result is shown in Figure 5.

Figure 3 shows that through perspective transformation, it has a better ability to solve the problem of different sizes of desk areas in examination room monitoring image, and makes an essential preprocessing to the next step.

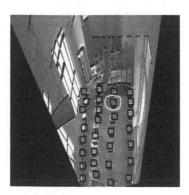

Figure 5. Result of the k-means algorithm.

Figure 5 shows that using the k-means algorithm can more accurately locate the position of the desk area.

3 EXPERIMENTAL ANALYSIS

The experimental image is taken from the surveillance images of the National Education Examination Standardized Exam, which has some characteristics, such as a large proportion of image brightness, a plurality of static and similar objectives, and a variety of image background. The relevant simulated experiments were conducted under the environment of operating system with windows 7 and microcomputer with 1G memory, using the visual studio 2012 tool. In order to validate the effect of this method, we use the thresholding method, morphological segmentation method, fuzzy c-means method and the method to segment desk area of the same examination room. The result of relevant simulated experiment is shown in Figure 6.

Figure 6 shows that this method obtains better results and is more accurate than the other three methods. The results of examination room desks region segmentation, the advantages and disadvantages of each method are described in detail in Table 1.

(a) the binary image after thresholding method

(b) the binary image after morphological method

(c) the binary image after fuzzy c- means method

(d) the binary image after by this method

Figure 6. Comparative results.

163

Table 1. The results of different methods.

Methods	Advantages	Disadvantages	Accuracy
Thresholding method	A small amount of calculation, stable performance and to fit image segmentation of difference between target and background	Ignoring the spatial structure of the image distribution, cannot distinguish similar ground and desktop, it is difficult to set threshold (as shown in Figure 6a)	20%
Morphological segmentation method	Considering the spatial structure of the image distribution, has a good response to the faint edge and more robust	Gradient operator is susceptible to the impact of noise and quantify, easily producing a local minimum, easily leading to oversegmentation (as shown in Figure 6b)	66.7%
Methods	Advantages	Disadvantages	Accuracy
Fuzzy c-means method	Is the unsupervised statistical algorithms, has robustness to noise and variable light	Depending on the cluster center value, ignoring the spatial structure of the image distribution, and easily leading to oversegmentation (as shown in Figure 6c)	70%
This method	Considering the spatial structure of the image distribution, more robust, suitable for segmentation with complex background and uneven brightness	Artificially setting perspective transformation coordinates, requiring the sample of template, and determining the match threshold by *a priori* knowledge (as shown in Figure 6d)	83.3%

4 CONCLUSIONS

This paper proposed a new method of examination room desks region segmentation based on template matching. The results indicate that in the examination room desks region segmentation applications, this new method has more anti-disturbing, more accuracy and more application.

In the examination room desks region segmentation applications, this method has achieved good results. But image brightness and template are still important factors of affecting the segmentation result. Therefore, in order to improve the accuracy of segmentation, we will make further adjustments and research on this method.

REFERENCES

Comaniciu, D. & Meer, P. 2002. Mean shif: a robust approach toward feature space analysis. *IEEE Transaction on Pattern Analysis and Machine Intelligence*, 24(5): 603–619.

Geman, S. & Geman, D. 1984. Stochastic relaxation, Gibbs distribution and the Bayesian restoration of images. *IEEE Transaction on Pattern Analysis and Machine Intelligence*, 6(6): 721–741.

Levner, I. & Zhang, H. 2007. Classification-driven watershed segmentation. *IEEE Transaction on Image Processing*, 16(5): 1437–1445.

Xie, Y. Cai, X. & Zhang, X. 2013. Target handoff algorithm based on perspective projection transformation and dynamic candidate strategy. *Journal of Computer Applications*, S1: 170–173.

Xu, X.Z. Ding, S.F. & Shi, Z.Z. et al. 2010. New Theories and Methods of Image Segmentation. *Acta Electronica Sinica*, 38(2): 76–82.

Yang, Y. 2013. *Research on Unsupervised Segmentation Method of Color-Texture Image Based on Multi-Scale Structure Tensor*. Huazhong University of Science and Technology, 6–12.

Zhou X.S. Gupta, A. & Comaniciu, D. 2005. An information fusion framework for robust shape tracking. *IEEE Transaction on Pattern Analysis and Machine Intelligence*, 27(1): 115–129.

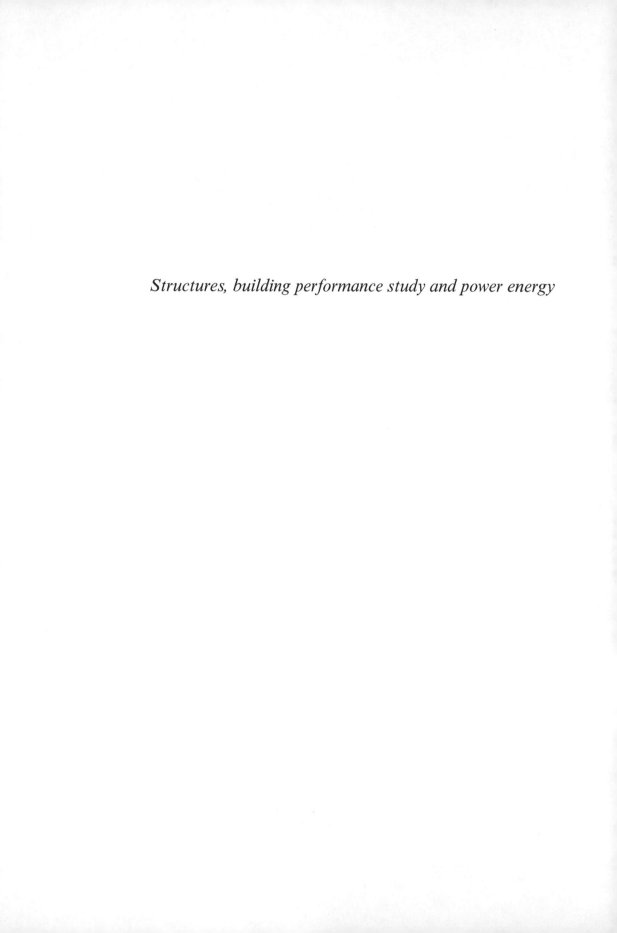

Structures, building performance study and power energy

Advances in Civil, Architectural, Structural and Constructional Engineering – Kim, Jung & Seo (Eds)
© 2016 Taylor & Francis Group, London, ISBN 978-1-138-02849-4

Study on the elastic dynamic buckling behavior of the plane frame under marine environment

Y.Y. Liu & Y.F. Fan
Dalian Maritime University, Dalian, China

Q.Y. Wang
Sendin Engineering Co. Ltd., Taiyuan, China

Z.J. Han
Taiyuan University of Technology, Taiyuan, China

ABSTRACT: Based on Hamilton principle and the effect of stress wave, the dynamic buckling governing equation of the portal frame is built. The expression of critical buckling conditions and the bucking mode equations of the beam-column are obtained including lateral sway and no-sway. Taking the corrosion of steel caused by marine environment into consideration, the critical buckling conditions were analyzed by a program written in MATLAB. The buckling parameter, the effective length factor (μ) and the regulation of critical buckling loads in various seawater areas are obtained. Theoretical analysis results indicate that the effective length factor (μ) decreases with increase of the linear stiffness ratios of beam-column (K). Compared with no-sway portal frame, the lateral sway portal frame has higher effective length factor under the same linear stiffness ratio of beam-column. The critical buckling loads of steel frame decrease 12% in immersion zone, but decrease 16% in mean-tide zone where steel is seriously corroded. Although they have the same buckling length, the dynamic buckling loads of no-sway portal frame is 1.4–1.6 times of lateral sway in various seawater conditions. 6 times of lateral sway.

1 INTRODUCTION

Along with the development and utilization of Marine resources, a large number of steel is used in ocean engineering, such as ships, offshore drilling platform, bridge and port engineering. However in marine environment the durability of steel structure is easily affected by the corrosion in marine environment, it widely received extensive attention of scholars. Chen (2012) gave corrosion laws in two materials (typical carbon steel and low alloy steel) in three zones (immersion zone, mean-tide zone and splash zone) by an experiment in Xiamen. Zhang (2006) focused on the mechanical behavior of corroded steel bar in various environments, and constructed the mathematical models of street-strain for corroded steel bars. Wu (2008) used artificial climate acceleration corrosion method to simulate the mechanical properties of non-uniform corroded steels under natural climate condition, and the constitutive relation model of corroded steel with corrosion rate was obtained. Shi (2012) carried out indoor salt spray accelerated corrosion tests to analyze the mechanical property of Q235 steel with and without coating respectively, the results showed that the plate thickness was an important factor influencing the behavior of H-section beams subjected to corrosion.

So far, many researches are carried out the influence of marine environment on the durability and mechanical properties of corroded steel, but less studies focus on dynamic buckling of corroded steel structure. Jing (2008) presented a non-linear analysis of buckling carrying capacity of corroded steel bars based on ANSYS, and proposed a corresponding calculating model. Mahmoud (2013) made a research and showed that under seismic loading the long term uniform corrosion has considerable influence on the dynamic buckling of steel tanks with different height to diameter ratios under seismic loading. Kubiak (2007) dealt with the local, global and interactive dynamic buckling of thin-walled structures subjected to compressive rectangular pulse load, and proposed a new criterion for the loss of stability leading by critical amplitude of pulse load. Sakar (2012) devoted to the analysis of static and dynamic buckling of multi-span. And the effects of beam to column length ratio, beam to column cross-section ratio and beam to column moment of area ratio were investigated by using the finite element method.

Based on the current researches, the dynamic buckling of no-sway and sway steel frame are investigated by theoretical analysis, and the mode equations and the critical buckling conditions of the plane steel frame were obtained. Considering the effects of steel corrosion caused by marine environment, calculating the critical buckling conditions were calculated by programs, the scope of the effective length factors μ, buckling parameters, the regulation of corrosion and the dynamic buckling loads of the steel frame were also obtained.

2 CRITICAL DYNAMIC BUCKLING CONDITIONS OF STEEL FRAME

2.1 Critical dynamic buckling conditions of no-sway steel frame

The steel frame schematic diagram under step load is shown in Figure 1. Assuming that the beam-column axial inertial and small elastic deformation is regardless according literature of Yang (1990) literature of Han (2010), the buckling equation of dynamic buckling column can be written as Equation (1).

$$EIw'''' + Nw'' + \rho A\ddot{w} = 0 \tag{1}$$

w is the lateral displacement of steel column (beam), k_1 and k_2 are as follows:

$$k_1^2 = N/EI, k_2 = \rho A/EI \quad w'''' + k_1^2 w'' + k_2\ddot{w} = 0 \tag{2}$$

Solve the equations above, the solution of Equation (1) is shown in Equation (3) below.

$$w(x,t) = A_0(e^{\sqrt{-\lambda/k_2}t} - e^{-\sqrt{-\lambda/k_2}t}) \times$$
$$(A\sin r_1 x + B\cos r_1 x + C\sin r_2 x + D\cos r_2 x) \tag{3}$$

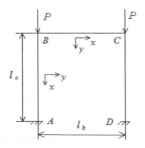

P P
B C
l_c
A l_b D

Figure 1. Portal frame.

Equation (4) should satisfy the following boundary conditions.

$$\begin{cases} w(0,t_{cr}) = Y(0)T(t_{cr}) = 0 \\ w'(0,t_{cr}) = Y'(0)T(t_{cr}) = \theta_B \end{cases}$$
$$\begin{cases} w(0,t_{cr}) = Y(0)T(t_{cr}) = 0 \\ w'(0,t_{cr}) = Y'(0)T(t_{cr}) = \theta_B \end{cases} \tag{4}$$

t_{cr} is the critical buckling time, l_{cr} is the critical buckling length, c is the velocity of stress wave.

Putting Equation (4) into Equation (3), the dynamic buckling mode Equation of no-sway steel frame is acquired, written as Equation (5).

$$Y(x) = \frac{\theta_B}{G}\left\{\frac{r_2 - r_1\sin r_1 l_{cr}\sin r_2 l_{cr} - r_2\cos r_1 l_{cr}\cos r_2 l_{cr}}{a}\sin r_1 x\right.$$
$$+ \frac{r_2\sin r_1 l_{cr}\cos r_2 l_{cr} - r_1\cos r_1 l_{cr}\sin r_2 l_{cr}}{a}\cos r_1 x$$
$$+ \frac{r_1 - r_2\sin r_1 l_{cr}\sin r_2 l_{cr} - r_1\cos r_1 l_{cr}\cos r_2 l_{cr}}{a}\sin r_2 x$$
$$\left. - \frac{r_2\sin r_1 l_{cr}\cos r_2 l_{cr} - r_1\cos r_1 l_{cr}\sin r_2 l_{cr}}{a}\cos r_2 x\right\} \tag{5}$$

$$a = 2r_1 r_2 - (r_1^2 + r_2^2)\sin r_1 l_{cr}\sin r^2 l_{cr}$$
$$- 2r_1 r_2\cos r_1 l_{cr}\cos r_2 l_{cr}$$
$$G = A_0(e^{\sqrt{-\lambda/k_2}t_{cr}} - e^{\sqrt{-\lambda/k_2}t_{cr}}) \tag{6}$$

Axial force of beam will be neglected. For making a static analysis to the beam of no-away frame, the control equation of beam can be written as Equation (7). For solving Equation (7), the mode equation of beam should be obtained, as Equation (8).

$$w_b(0) = 0, w_b'(0) = \theta_B, w_b(l_b) = 0, w_b'(l_b) = -\theta_B \tag{7}$$

$$w_b(0) = 0, w_b'(0) = \theta_B, w_b(l_b) = 0, w_b'(l_b) = -\theta_B \tag{8}$$

Equation (8) should satisfy the following boundary conditions:

$$w_b(0) = 0, w_b'(0) = \theta_B, w_b(l_b) = 0, w_b'(l_b) = -\theta_B \tag{9}$$

And the mode equation of beam can be written as Equation (10).

$$w_b = \theta_B(x - \frac{x^2}{l_b}) \tag{10}$$

Putting $w_c''(0,t_{cr})$, $w_b''(0)$ into the moment equilibrium equation of node B, the critical dynamic buckling conditions of no-sway frame can be obtained. And in the following formulas $K = I_b l_c/I_c l_b$.

$$\frac{l_c(r_1^2 - r_2^2)(r_2 \sin r_1 l_{cr} \cos r_2 l_{cr} - r_1 \cos r_1 l_{cr} \sin r_2 l_{cr})}{2 r_1 r_2 - (r_1^2 + r_2^2) \sin r_1 l_{cr} \sin r_2 l_{cr} - 2 r_1 r_2 \cos r_1 l_{cr} \cos r_2 l_{cr}} = \frac{2 I_b l_c}{I_c l_b}$$

(11)

2.2 Critical dynamic buckling conditions of sway steel frame

The dynamic buckling equation of sway steel frame is Equations (1). Its solution is Equations (5). And Equations (5) should satisfy the following boundary conditions for sway steel frame.

$$\begin{cases} w(0,t_{cr}) = Y(0)T(t_{cr}) = \Delta_0 \\ w'(0,t_{cr}) = Y'(0)T(t_{cr}) = \theta_B \\ w(l_{cr},t_{cr}) = Y(l_{cr})T(t_{cr}) = 0 \\ w'(l_{cr},t_{cr}) = Y'(l_{cr})T(t_{cr}) = 0 \end{cases}$$

(12)

$$\Delta_0 = \frac{1 - \cos k_1 l_{cr}}{k_1 \sin k_1 l_{cr}} \theta_B$$

Putting Equations (12) into Equations (3), the dynamic buckling mode equation of sway steel frame is acquired.

Axial force of beam will be neglected. And the static analysis is still suitable for the beam of sway frame. The control equation and the mode equation of beam are still Equations (7) and Equations (8).

$$y_c = G\{[\Delta r_2(r_2 \cos r_1 l_{cr} \sin r_2 l_{cr} - r_1 \sin r_1 l_{cr} \cos r_2 l_{cr})$$
$$+ \theta_B(r_2 - r_1 \sin r_1 l_{cr} \sin r_2 l_{cr} - r_2 \cos r_1 l_{cr} \cos r_2 l_{cr})] \sin r_1 x$$
$$+ [\Delta(r_1 r_2 - r_2^2 \sin r_1 l_{cr} \sin r_2 l_{cr} - r_1 r_2 \cos r_1 l_{cr} \cos r_2 l_{cr})$$
$$+ \theta_B(r_2 \sin r_1 l_{cr} \cos r_2 l_{cr} - r_1 \cos r_1 l_{cr} \sin r_2 l_{cr})] \cos r_1 x$$
$$+ [\Delta r_1(r_1 \sin r_1 l_{cr} \cos r_2 l_{cr} - r_2 \cos r_1 l_{cr} \sin r_2 l_{cr})$$
$$+ \theta_B(r_1 - r_2 \sin r_1 l_{cr} \sin r_2 l_{cr} - r_1 \cos r_1 l_{cr} \cos r_2 l_{cr})] \sin r_2 x$$
$$+ [\Delta(r_1 r_2 - r_1^2 \sin r_1 l_{cr} \sin r_2 l_{cr} - r_1 r_2 \cos r_1 l_{cr} \cos r_2 l_{cr})$$
$$+ \theta_B(r_1 \cos r_1 l_{cr} \sin r_2 l_{cr} - r_2 \sin r_1 l_{cr} \cos r_2 l_{cr})] \cos r_2 x\}$$
$$/ [2 r_1 r_2 - (r_1^2 + r_2^2) \sin r_1 l_{cr} \sin r_2 l_{cr}$$
$$- 2 r_1 r_2 \cos r_1 l_{cr} \cos r_2 l_{cr}]$$

(13)

According to the boundary conditions:

$$w_b(0) = 0, w_b'(0) = \theta_B, w_b(l_b) = 0, w_b'(l_b) = \theta_B$$

(14)

The detail mode equation of beam can be obtained that:

$$w_b = \theta_B x - \frac{3\theta_B}{l_b} x^2 + \frac{2\theta_B}{l_b^2} x^3$$

(15)

Putting $w_c''(0,t_{cr}), w_b''(0)$, into the moment equilibrium equation of node B, the critical dynamic

buckling conditions of sway frame is acquired. According to the boundary conditions:

$$\begin{cases} (l_c/k_1)[(r_1^3 r_2 + r_1 r_2^3)(1 - \cos r_1 l_{cr} \cos r_2 l_{cr}) \\ -2 r_1^2 r_2^2 \sin r_1 l_{cr} \sin r_2 l_{cr}] \tan(k_1 l_{cr}/2) \\ + l_c(r_1^2 - r_2^2)(r_2 \sin r_1 l_{cr} \cos r_2 l_{cr} - r_1 \sin r_2 l_{cr} \cos r_1 l_{cr}) \end{cases}$$
$$/[2 r_1 r_2(1 - \cos r_1 l_{cr} \cos r_2 l_{cr}) - (r_1^2 + r_2^2) \sin r_1 l_{cr} \sin r_2 l_{cr}]$$
$$= 2 I_b l_c / I_c l_b$$

(14)

2.3 Dynamic buckling parameters and effective length factors μ of steel frame

The Equations (11) and Equations (14) are calculated using program of MATLAB, the buckling parameters k_1 can be received. Because of $k_1^2 = N/EI_c$, the critical buckling loads of steel frame can be written as $P_{cr} = k_1^2 EI_c$. Comparing with Euler formula $P_{cr} = \pi^2 EI_c/(\mu l_c)^2$, the effective length factors μ of column can be written as $\mu = \pi/k_1 l_c$. It shows that as long as the regulation between the effective length factors μ, section moment of inertia and time is clear, the dynamic buckling loads P_{cr} can be gotten.

In the following case, the length of beam and column is 6 m. Some parameter values are as follows: the ratio of beam rigidity $K = 1$, dynamic factor $\lambda = n_1^2 n_2^2 \pi^4/l_c^{4}$[12], $n_1 = 1$, $n_2 = 2$. The value of k_1 and the effective length factors μ of sway and no-sway steel frame with different buckling length is listed in Table 1.

The Table 1 indicates that the effective length factors decrease with the increase of bending length of columns. In addition, under the same ratio of beam rigidity, the factor of sway steel frame is larger than no-sway steel frame, it also

Table 1. The buckling parameters k_1.

l_{cr}	Sway		No-sway	
	k_1	μ	k_1	μ
0.5	22.3122	0.2816	17.5938	0.3571
1.0	11.1364	0.2821	9.4248	0.3333
1.5	7.4110	0.2826	6.2832	0.3333
2.0	5.5483	0.2831	4.7124	0.3333
2.5	4.4307	0.2836	3.7699	0.3333
3.0	3.6856	0.2841	3.1416	0.3333
3.5	3.1534	0.2846	2.6928	0.3333
4.0	2.7542	0.2852	2.3562	0.3333
4.5	2.4437	0.2857	2.0944	0.3333
5.0	2.1954	0.2862	1.8850	0.3333
5.5	1.9921	0.2867	1.7136	0.3333
6.0	1.8228	0.2872	1.5708	0.3333

169

means that the critical buckling loads of no-sway steel frame are bigger than sway steel frame. Therefore, it makes the structure tends to be stable by adding lateral resistances to improve the stiffness in practical engineering.

3 CORROSION EFFECT OF SEAWATER TO THE SECTION PROPERTIES OF STEEL FRAME

Considering the seawater corrosion regulations of immersion zone, mean-tide zone and splash zone in Xiamen given by literature (Chen et al. 2012), an analysis of steel frame about the corrosion effect can be made. In the following case, the length of steel column and beam are 6 m, the size of H-section are 500/200/6/10 mm, and 921steel is adopted. The development of corrosion depth is shown in Figure 2, the evolution law of corrosion section moment of inertia is shown in Figure 3, the evolution laws of ratio of beam rigidity with time in

Table 2. Relation between average corrosion rate and exposure time for 921steel (Chen et al. 2012).

Materials	Sway	No-sway	
921steel	Corrosion rate	Confidence	Prediction
Immersion	$A = 0.1375t$ -0.366	0.903	0.05
mean-tide	$A = 0.2146t$ -0.635	0.986	0.04
Splash	$A = 0.0661t$ -0.358	0.867	0.02

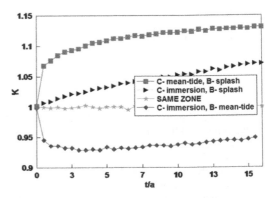

Figure 3. Relations between the ratio of beam rigidity and exposure time.

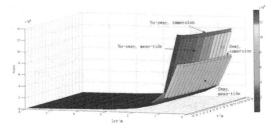

Figure 4. Relations between dynamic buckling loads and exposure time.

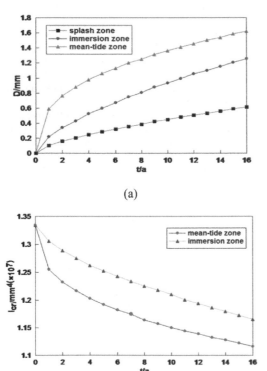

(a)

(b)

Figure 2. (a) Relation between corrosion depth and exposuretime of H-section, (b) Relation between the effective length and exposure time of H-section.

different zones are shown in Figure 4. It is thought that the elastic modulus remains the same in the process of corrosion (Han et al. 2010).

According to the analysis, the corrosion depth of steel frame increase gradually with time. After 16 years, the corrosion length in mean-tide zone is expected to reach 1.6 mm, in immersion zone 1.2 mm, in splash zone 0.6 mm. Because of the effect of corrosion, the section moment of inertia in immersion zone decreased 12.6%, in mean-tide zone decreased 16.2%. It is shown in Figure 2.

When the beam and column is in various zones separately, some different changes among the range of 1 ± 10% appear to the ratio of beam rigidity with the increase of time, and tend to be gentle. As the beam is in splash zone and column is in mean-tide zone, the ratio of beam rigidity is larger. It is shown in Figure 3.

It can be known that the variability (10%) about the ratio of beam rigidity will not have a big impact on the analysis. So in the following analysis the corrosion of the column will be emphasized. For the convenient calculation the ratio of beam rigidity.

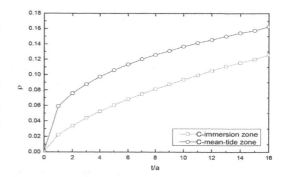

Figure 6. Decrease rates of dynamic buckling loads with exposure time.

4 CRITICAL ELASTIC DYNAMIC LOADS OF CORRODED STEEL FRAME

The regulations between dynamic loads and corrosion time are gotten in Figure 4. The dynamic buckling loads of no-sway steel frame with bending length of 0.5 m, 1 m and 3 m are detailed in Figure 5 when the steel columns are in mean-tide zone, immersion zone and splash zone separately. And the declining rates of buckling loads are shown in Figure 6.

After calculating the dynamic buckling loads decrease with the increase of corrosion time. For no-sway steel frame with bending length of 0.5 m, the dynamic buckling loads decrease from 1.4×10^6 kN to 1.2×10^6 kN in immersion zone, decrease from 1.4×10^6 kN to 1.1×10^6 kN in mean-tide zone. For the sway steel frame, the dynamic buckling loads decrease from 8.5×10^5 kN to 7.4×10^5 kN in immersion zone, decrease from 8.5×10^5 kN to 7.1×10^5 kN in mean-tide zone.

According to the information of Figure 4 and Figure 5, it can be shown that the dynamic buckling loads of no-sway steel frame are 1.4–1.6 times of sway steel frame with the same bending length.

The regulations between the effective length factors and time can be ignored under given conditions. The declining rates of dynamic buckling loads are relevant to the corrosion rates and irrel-evant to the bending length whether the frame has lateral movement. It shows that the dynamic buckling loads decrease 16% in mean-tide zone, but decrease 12% in immersion zone. It demonstrates that the corrosion of steel in mean-tide zone has greater influence on the durability of steel frame.

5 CONCLUSIONS

The work was devoted on the influence of marine corrosion to the durability of steel frame, and a series of theoretical analysis were made. The following conclusions are draw from this work.

1. The expressions of critical buckling conditions and the bucking mode equations of the beam and column are obtained including lateral sway and no-sway. Solving the critical conditions, the dynamic buckling loads and the effective length factors are got.
2. It indicates from Table 1 that the effective length factors decrease with the increase of ratio of beam rigidity. The factors of sway steel frame are larger than no-sway steel frame under the same ratio of beam rigidity.
3. Along with the development of corrosion, the moment of inertia decrease gradually, in splash zone the corrosion is minimal, in immersion zone is worse and in mean-tide zone is worst. The ratios of beam rigidity K change from 1.0 to 0.9 and 1.1 separately.

The dynamic buckling loads decrease with the increase of corrosion time, decrease 16% in mean-tide zone, and decrease 12% in immersion zone. And the dynamic buckling loads of no-sway steel frame are 1.4–1.6 times of sway steel frame with the same bending length. Adding lateral resistances to improve the stiffness in practical engineering, the structure will tend to be stable.

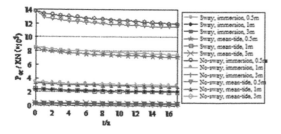

Figure 5. Details about the relationships between dynamic buckling loads and exposure time.

ACKNOWLEDGEMENT

The studies in this paper are financially supported by a National Natural Science Foundation of PR China (Grant No. 51178069), Program for New Century Excellent Talents in University (Grant No. NCET-11–0860).

REFERENCES

Chen, X.F. et al. 2012. Corrosion of Carbon Steel and Low-alloy Steel Exposure in Xiamen Natural Seawater. *Equipment Environmental Engineering,* 9(6): 21–24.

Han, Z.J. et al. 2010. The static buckling of elastic portal frame with hinged-hinged support based on the Hamilton principle. *Journal of Xi'an University of Architecture & Technology,* 42(4): 480–485.

Jiang, F.C. et al. 2008. Non-linear Analysis of Buckling Carrying Capacity for Corroded Reinforcing Bars Based on ANSYS. *Journal of Tongji university (natural science),* 36(8): 1045–1066.

Kounadis, A.N. 2002. Dynamic buckling of simple two-bar frames using catastrophe theory. *International Journal of Non-Linear Mechanics,* 37(7): 1249–1259.

Kubiak, T. 2007. Criteria of dynamic buckling estimation of thin-walled structures. *Thin-Walled Structures,* 45(10–11): 888–892.

Maheri, M.R. & Abdollahi, A. 2013. The effects of long term uniform corrosion on the buckling of ground based steel tanks under seismic loading. *Thin-Walled Structures,* 62: 1–9.

Sakar, G. et al. 2012. Dynamic stability of multi-span frames subjected to periodic loading. *Constructional Steel Research,* 70: 65–70.

Shi, W.Z. et al. 2012. Experimental study on influence of corrosion on behavior of steel material and steel beams. *Journal of Building Structures,* 33(7): 53–60.

Wu, Q. & Yuan Y.S. 2008. Experimental study on the deterioration of mechanical properties of corroded steel bars. *China Civil Engineering Journal,* 41(12): 42–47.

Yang, G.T. 1990. An experimental study of dynamic behavior of structures under impact loads. *Chinese Journal of Theoretical and Applied Mechanics,* 22(3): 374–378.

Zhang, W.P. et al. 2006. Stress-Strain Relationship of Corroded Steel Bars. *Journal of Tongji University (Natural Science),* 34(5): 586–592.

Advances in Civil, Architectural, Structural and Constructional Engineering – Kim, Jung & Seo (Eds)
© 2016 Taylor & Francis Group, London, ISBN 978-1-138-02849-4

Analysis of residential characteristics and energy saving

Yangluxi Li

School of Architecture, Syracuse University, NY, USA

ABSTRACT: Zhenyuan is an ancient city with a multicultural blend of immigrants. The culture, the local natural geography and climate environment formed the unique society. This article describes the characteristics of this area and makes suggestions to protect the culture and encourage sustainable energy use.

1 THE GENERAL SITUATION OF THE ANCIENT CITY OF ZHENYUAN

Zhenyuan is an ancient Miao city with a long history. It is located in Miao and Dong autonomous region of southeast, Guizhou Province (Figure 1). It is said that two Miao ancestors from Chiyou tribe had settled in the central plains about 5 thousand years ago. It began to set up county in Qin Zhaowang 30 years (227BC). This area celebrates 2238 years of history. In 1986, the State Council approved it as one of the first national historical and cultural cities (Figure 2).

From Qin and Han Dynasties to late Ming Dynasty and the early Qing Dynasty, Zhenyuan was the gate to go west to Dianqian, or as far as Burma, India and other Southeast Asian countries. All routes in this direction required passage through this city. It is also known as China's "southern silk road" (Figure 3). Due to the convenience of water transportation on the Wuyang River, it also can be referred to as "the important town of water silk road".

Because of its central geographical position, Zhenyuan became a strategic point of politics, military and trade. This city is often called "the key of Dianchu, the door and window of east Qian". It is a frontier fortress, and commercial city, which takes advantage of the military to create business prosperity. The cultures of the central plains, Xiang Chu, Wu Yue and some foreign countries came together with the iron heel and merchant ships many years ago. The locale diversity mixed with the local ethnic culture, made Zhenyuan multicultural, and known as the "Immigrant's City".

The stirring Wuyang River flows through Zhenyuan from the west to the east. It makes an "S" shape that looks like a taiji figure. This is a landscape town with a long history and rich culture. Beyond defining building types, it bears the weight of a residential culture with local history. It is the continuation of an ancient city's historical context.

Figure 1. The location of Zhenyuan in Guizhou Province.

Figure 2. A photo of Zhenyuan taken by a French missionary at the end of 19th century. (Information from Zhenyuan exhibition).

Figure 3. The model of Zhenyuan ancient county.

2 CITIZEN OF ZHENYUAN

2.1 *Alleys*

As a historical and cultural city, Zhenyuan is famous for its unique ancient residents. If each ancient residence is an old story, the crossed, long and narrow deep alleys that link them would be the veins and textures of Zhenyuan.

The width of the ancient alley is about 5 or 6 feet, or 2 meters. It is located on an upward sloping hillside, and linked by horizontal alleys. The single banked wall houses are arranged on the far-reaching and complex "maze" structure (Figure 4). Viewing from the inside, the criss-crossed, winding, bare and deteriorating steps ascend the deep ancient alleys. Blue bricks, black tiles, high-banked wall houses, and a flowing ancient well can be seen down the walkways. Around every corner, the alleys reveal the unknown beauty of the city's residences (Figure 5).

The families of Zhenyuan never set up doors parallel or perpendicular to the alleys or the hall opposite of the alley. The door's direction is intentionally turned angular to the streets. This is known as "slanting door evil (oblique) way". This approach is an ancient FengShui tradition. "South for the first" is not the only common phrase, but also "opportunity makes the thief".

2.2 *The facade and plane of the buildings*

The unique characteristic of Zhenyuan's ancient residences houses is blue bricks, black tiles, and high fire seal.

These residential buildings are mostly shaped as quadrangles. Some of the structures are built hillside and enclosed by high walls. This construction is good for seclusion and privacy (Yang et al. 2014a). Close to the outside and opened to the inside and the mountain terrain, the old courtyards are scattered high and low. That gives the space to rich taste and deposited the time's process. It makes the houses that were built in different periods blend together well.

The roof level of Zhenyuan ancient residences never protrudes outward, and brick walls surround them. The windows, which face the alleys, are always small and high. The magnificent horse head wall's height is several meters, up to half the height of the wall (Feng & Mou, 2007). Additionally the mountain flowers fall and are well proportioned. They are typical Huizhou-style horse head walls (Figure 6). It not only reveals awareness of fire prevention, but also is protective similar to a castle's walls, allowing businessmen to protect their property and personal safety (Yang, 2010).

2.3 *The buildings' style*

The ancient residences of Zhenyuan are the product of multiple cultural integrations. It has both Huizhou architecture style and the layout of mountain buildings (Li & Wu, 2010). The local Miao stilts style can be seen here as well.

The impact of tourism has resulted in the development of imitation Ming and Qing Dynasty style along the Wuyang River (Figure 7). The layout of the buildings is adjacent to the river, along the street, in front of the river and in back of the street. In order to add more space in the back of the buildings which are adjacent to the river (Qian, 2014), the cantilever style of stilts is used to not only expand the space, but also to enhance the space effect.

Figure 4.   The residence inn.

Figure 6.   The typical Huizhou-style horse head wall.

Figure 7.   The Ming and Qing Dynasties styles' buildings which are adjacent to the river.

Figure 5.   The far-reaching alleys.

The mountain-Huizhou architecture style residences are along the mountains, and the main structure consists of a post and panel. The Huizhou style's masonry is used on the exterior of the structure and a wooden stilts style is used on the interior (Lin, 2007). The combination of the two styles used in the structure reflects the heroic spirit of merchants with the peaceful elegance of the Jiangnan courtyard, and the contrast of the Miao stilts (Figure 8). This is the perfect fusion of wooden and stone structures that can accommodate the various uses of the building (Mochida & Lun, 2008).

2.4 *The detail of buildings*

The residential outer architectural decorations of Zhenyuan are simple but the interior decorations are marvelous and fascinating (Yang, 2013a). The difference between interior and exterior decorations are obvious. Having the more ornate décor on the inside was influenced by the cultures belief in hiding rich inside the walls of "slanting door evil way".

In Zhenyuan residences, woodcarving is a very important symbol of Huizhou architectural features. These perfect woodcarvings contain characteristics of humanism and of an aesthetic nature (Zhang et al. 2004). It is the perfect combination of ancient culture in China and art decoration. In residences' woodcarvings, the elements used most are the dragon, phoenix, unicorn, fish, insects, magpie, flowers, trees, books, people, fret and other kinds of auspicious patterns (Kubota et al. 2008). These woodcarving elements generally do not separate, always matching with one another, and can be combined to form a complete picture (Figure 9). The most amazing thing about this decoration is that optimistic Confucian characteristics of ethical reference hide on the back of every woodcarving. This culture's expression of woodcarving meaning comes from the classical aesthetic ideas, which are represented by Confucianism (Yang et al. 2014b). These exquisite woodcarvings, which decorate the buildings or furniture, are not only used as a kind of adornment on the shelf, but they also have an influence in people's lives and act as a cultural expression. The people living among the carvings can feel it anytime, and anywhere (Yang et al. 2014b).

Figure 8. The timber structure compound of the Yuan's family.

Figure 9. The door's detail of wood carving on the compound of the Fu's family.

Figure 10. The wood carving of Qinglong gully. <Yang warriors>.

At Zhenyuan, not only in residences, but also on the streets and alleys, you can see the fascinating and elegant woodcarvings anywhere (Figure 10). People usually use the line carving, circle carving, relief, openwork and some other techniques of expressions according to components of the buildings' necessity (Huang, 2005). Woodcarving uses just the color and texture of the wood itself to show the carving with no paint. The simple and elegant woodcarving makes the residences have a more antique flavor.

3 RECONSTRUCTION OF RESIDENCES— THE RESIDENCE INN OF ZHENYUAN

3.1 *The general pattern of the inn*

The residence inn of Zhenyuan is a successful case of protection and transformation of residences (Wu & Huang, 2002).

The inn consists of quadrangle dwellings. A wooden structure style courtyard is the main focal point. The quadrangle dwellings contain three courtyards, and are two-layer buildings. Walking into the gate, there's a central room in the first courtyard. It has now been transformed into the reception hall for the inn.

Through the gate and to the right you can find the second courtyard, which is the most important residential part of the residence (Chen & Qin, 2008). Now, it has been transformed into be the inn's guestrooms. The rooms are a standard snap well building, which has three bays. One is Mingtang, and the other two are private. The Mingtang is the room that is a set corridor around the middle hall, and connects the two sides and the back of the guest room (Li, 2015). There's a patio at the top of the middle of the Mingtang. The rain from the slope roof drain ditch is commonly known as the "four water to tank". There is a stone pan to accommodate the rain below, that relates to the saying, "rich water not flow outer". The room has windows on three sides to make use of the patio for lighting and ventilation. In order to avoid dampness indoors when it rains, the patio is designed very small (Qiu et al. 2007). The room's layout has axial symmetry, which made the room center on the patio, forming basic space units.

Entering from the gate, the second courtyard is in front of the wing-room. The main function of this room is for the daily life of the internal staff. It is a three-section compound and two-layer building composed of a wooden structure (Figure 11). This room fully retained the original old buildings' structure and features (Yang, 2013b).

The quadrangle dwellings, which consist of three courtyards, rely on the mountain and takes advantage of its power. It is well proportioned with the terrain's ups and downs. On the side towards yard of the second layer, there is a viewing corridor. All the separating wall's pillars and floors are made of wood because the wood is not only very light, but also antique and full of expressive design (Han & Guo, 2008).

It is a positive and dynamic way to protect and transform the old residences of Zhenyuan to an inn. The transformation updated and changed its using function. It injected new vitality and life to the traditional residences through giving them new function. This vitalization helps maintain the historic culture of Zhenyuan's residences. Rich economic value can be acquired from the restoration, while the old residences can express its deep, vivid historical and cultural connection (Simiu & Scanlan, 1996).

3.2 *The modern transformation of buildings*

Under the influence of traditional geomancy, the residences' inn of Zhenyuan has adopted the simple energy-saving design. This general layout made full use of the terrain to create rich and delightful spaces. The architectural design uses the patio to ventilate, produce lighting, and avoid interior dampness. The residences are cohesive to local conditions, and use local materials. However, due to modern day conveniences there is an increasing demand for a better quality of life (Jeong & Bienkiewicz 1997). The traditional residences still need improvement to meet these modern needs.

3.2.1 *The temperature preservation and sound insulation of the walls*

The rooms are constructed with a wooden structure. The external walls of the rooms are made of natural wood (Figure 12). The inner walls are a modern and simple style for the interior design and decoration. The uses of characteristics from modern life are reflected in the traditional atmosphere.

By not destroying the original wooden structure, the combination of the wood's firm structure and thermal inertia with the added heat and sound insulation interlayers, the goal of thermal and sound insulation can be achieved (Figure 13) (Qian, 2015).

3.2.2 *The heat preservation and thermal insulation of roofing*

Roofing is an area that dissipates a lot of heat. The heat preservation of the roofing is very important. The traditional wooden structure's roofing material of Zhenyuan residences is black tile. It neither insulates the heat in summer nor preserves the heat in winter. Due to the room's interior design and decoration, a ceiling is also installed. Between the ceiling and roof, there is added thermal inertia and heat preservation interlayers that better the indoor temperature condition and comfort of living.

Figure 12.   The wooden walls of the room.

Figure 13.   The wooden floors of the room.

Figure 11.   The wing-room of Sanhe countyard.

Without destroying the original roofing, the new materials and new technology to transform and improve the tiles of roofing can be used, thus, the performance of the roofing's control of the indoor temperature and thermal inertia performance could be improved (Figure 14) (Utley & Cappelen, 1978).

3.2.3 *The windows*

The windows of Zhenyuan residences inn are very transparent. Analyzing the enclosed structure, they are the weakness of heat preservation. The windows that open inward have no sealing measures because they use the patio for light and to ventilate air. Since the windows open inward, the activities in the corridors outside the rooms will disturb the guests and their privacy (Yang et al. 2014c). In the summer, the air from the patio will flow smoothly in the rooms. However, in the winter, because of the poor seal of the windows, the cold wind will come into the rooms, and cause higher energy use from the air heating system.

To start the improvement of the Zhenyuan residences traditional building, the first step is improving the air tightness of the windows to reduce the loss of heat transfer and improve the heat preservation performance in the winter. Since the residences are wooden structures, expanding the ratio of windows and walls appropriately would help as well. It would become more convenient to improve the lighting condition and airflow in summer. The shutters can add to the décor and increase the privacy, so that guests' requirements are met (Xin & Zhao, 2004). It would also help to use the transparent glass tiles on the roofing and form a small skylight to make up for the insufficient light due to the inward opening windows in the Zhenyuan residences (Figure 15).

3.2.4 *Equipment*

In the inn's rooms of antique flavor, there are many modern, basic equipments and systems, such as water heater, water supply and drainage system, air-conditioning heating system, telephone, television and other household appliances (Figure 16).

Since there was less knowledge of building energy saving, Zhenyuan residences ignored the cost of energy and resource consumption. That

Figure 14.   The tiles that the residences used.

Figure 15.   The transparent windows of the room.

Figure 16.   The modern sanitary equipment of the room.

causes the resident buildings' high-energy consumption, and increased the economic burden of heating and cooling today (Li, 2012). For example, since the residence buildings have no outward windows, and the local climate is damp and humid when the air is unventilated in the rooms. Although there's a patio to assist with the air freshening and sunlight conditions in the room, it is not enough. In order to improve the comfort of the buildings, air conditioning is always used through an exhaust system and heating system in the rooms, which consumes quite a lot of energy.

Zhenyuan residences' local and traditional architectural features are gradually formed according to its local climate and natural resources (Su & Zhang, 2010). The affection of multiple cultures, the local, traditional architectural features influenced its formation. With the development of the modern economy and fast paced lifestyle, along with the great changes of the interconnected information network, people's idea of living has changed. The pursuits of their quality of life have become increasingly important. These changes have impacted the traditional ways of living. To protect and transform the ancient residences, we should not only repair but also protect the building's appearance (Maier et al. 2009). Therefore, "change in form but not in content" is not advisable. We should increase the modern living facilities, improve the basic equipments, try to use new technology and new materials, reduce the energy consumption, make the traditional residences' cultural significance continue and developed to the ecological and energy saving living conditions (Zhang, 2000).

4 CONCLUSIONS

Zhenyuan residences developed under the specific conditions of history, cultures, natural geography, climate, and the environment. Zhenyan's remote geographic location and the block traffic decided its locked and independent cultural and geographical elements. Making it incredibly unique, the local customs and culture also merged together with other immigrant cultural. Gradually, the residence buildings that reflect the multiple historical contexts were formed. The residences reflect the society, politics, economics, aesthetics, values, ethics, and religion of that time. With the development of tourism, the value of Zhenyuan residences' historical, cultural, and social values are being recognized.

ACKNOWLEDGEMENT

This work was financially supported by the patents of 201320138568.0 and 201320138582.0.

REFERENCES

Chen, H. & Qin, F.Z. 2008. *Analysis on the structure and regional architectural characteristic of the ancient city Zhenyuan*, Science and Technology innovation Heraid.

Feng, M. & Mou, J. 2007. *Analysis local traditional way of living protection about the historical Street in ancitent city-a case of the ancient city Zhenyuan of Guizhou Province*. Architecture of central China.

Gottfried, D.A. 2003. Blueprint for green building economics. *Industry and Environment*, 26(2): 20–21.

Han, S. & Guo, B. 2008. Building natural ventilation the temperature effect. *Building Energy*, 18–22.

Huang, J.J. 2005. Ancient city planning and architecture conforming to nature-a discussion on the human and ecological environment of zhenyuan, *Journal of Guizhou University*, 3(19).

Jeong, S.H. & Bienkiewicz, B. 1997. Application of autoregressive modeling in proper orthogonal decomposition of building wind pressure. *Journal of wind engineering and industrial aerodynamics*, 69: 685–695.

Kubota, T. Miura, M. & Tominaga, Y. 2008. Wind tunnel tests on the relationship between building density and pedestrian-level wind velocity: Development of guidelines for realizing acceptable wind environment in residential neighborhoods. *Building and Environment*, 43(10): 1699–1708.

Li, X.G. & Wu, Y.Q. 2010. *Discussion on energy-saving mesures for fly-ash cement particleboard poduction line*, Mechanic Automation and Control Engineering (MACE), 2010 International Conference on, 5140–5142: 26–28.

Li, Y. 2012. Analysis of Planning of Neighborhood Communication Space in the Livable Community, *Applied mechanics and Materials*, 174–177: 3018–3022.

Li, Y. 2015. Interpreting solar house—using International Solar Decathlon works as example, *Advanced Materials Research*, 1092–1093: 567–572.

Lin, X.D. 2007. *Green Building*. Beijing: China Building Industry Press.

Maier, T. Krzaczek, M. & Tejchman, J. 2009. Comparison of physical performances of the ventilation systems in low-energy residential houses, *Energy and Buildings*, 41(3): 337–353.

Mochida, A. & Lun, I.Y.F. 2008. Prediction of wind environment and thermal comfort at pedestrian level in urban area. *Journal of Wind Engineering and Industrial Aerodynamics*, 96(10): 1498–1527.

Qian, F. 2014. Insulation and Energy-saving Technology for the External Wall of Residential Building, *Advanced Materials Research*, 1073–1076(2): 1263–1270.

Qian, F. 2015. Analysis of Energy Saving Design of Solar Building-Take Tongji University solar decathlon works for example, *Applied Mechanics and Materials*, 737: 139–144.

Qiu, N.I. Yang, Q. & Yan, C. 2007. *Research of the Ecological and Livable Residential Estates in Dong Guan*. Guangdong Landscape.

Simiu, E. & Scanlan, R.H. 1996. *Wind effects on structures: fundamentals and applications to design*. John Wiley.

Su, X. & Zhang, X. 2010. Environmental performance optimization of window–wall ratio for different window type in hot summer and cold winter zone in China based on life cycle assessment, *Energy and Buildings*, 42(2): 198–202.

Utley, W.A. & Cappelen, P. 1978. The sound insulation of wood joist floors in timber frame constructions, *Applied Acoustics*, 11(2): 147–164.

Wu, D. & Huang, Q. 2002. Virtual Reality Technology and Research Status of the Development Process. *Ocean Mapping*, 22(6): 15–17.

Xin, J.G. & Zhao, J.L. 2004. *Application Analysis of the Ecological Concept on Architectural Design*. China Technology Information.

Yang, L. 2010. Computational Fluid Dynamics Technology and Its Application in Wind Environment Analysis. *Journal of Urban Technology*, 17(3): 53–67.

Yang, L. 2013a. Research of Urban Thermal Environment Based on Digital Technologies, *Nature Environment and Pollution Technology*,12(4): 645–650.

Yang, L. 2013b. Research on Building Wind Environment Based on the Compare of Wind Tunnel Experiments and Numerical Simulations, *Nature Environment and Pollution Technology*, 12(3): 375–382.

Yang, L. He, B.J. & Ye, M. 2014a. CFD Simulation Research on Residential Indoor Air Quality, *Science of the Total Environment*, 472: 1137–1144.

Yang, L. He, B.J. & Ye, M. 2014b. Application Research of ECOTECT in Residential Estate Planning, *Energy and buildings*, 72: 195–202.

Yang, L. He, B.J. & Ye, M. 2014c. The application of solar technologies in building energy efficiency: BISE design in solar-powered residential buildings, *Technology in Society*, 1–8.

Zhang, M. 2000. Virtual Reality Applications in Architectural Design. *Central Building*, 18(1): 51–52.

Zhang, Z.Q. Wang, Z.J. & Lian, L.M. 2004. Indoor thermal environment of residential buildings Numerical Simulation. *Building heat ventilation air-conditioning*, 23(5): 88–92.

Advances in Civil, Architectural, Structural and Constructional Engineering – Kim, Jung & Seo (Eds)
© 2016 Taylor & Francis Group, London, ISBN 978-1-138-02849-4

Carbon footprint reduction in building construction by less usage of ore-based materials

A. Malakahmad, J.H. Kok & S.S.S. Gardezi
Department of Civil and Environmental Engineering, Universiti Teknologi PETRONAS,
Bandar Seri Iskandar, Perak, Malaysia

ABSTRACT: The construction industry requires a fair quantity of ore-based materials. The extraction, processing and transportation of these materials require extensive amount of fossil fuel and this resulted in substantial emission of greenhouse gases such as CO_2. In this study less utilization of ore-based materials and adopting different alternate materials in the construction industry were carried out for reduction of CO_2 emissions. A double-story office building was considered as case study for the research work. Various alternative building materials such as recycled materials in concrete mixtures, wood, bamboo and straw bale were identified and opted in different construction elements of the office building. The CO_2 emission of each activity was quantified using a carbon calculator. The study has achieved tCO_2e reduction of 53% by adopting green concrete and 5.3% by adopting brick wall instead of conventional concrete. By replacing steel with bamboo, carbon footprint from reinforcement has been reduced from 13.2 to 0.11 tCO_2e. Similarly, 72% reduction in carbon footprint of flooring has been achieved by replacing the conventional concrete floor with wooden flooring. These findings indicate that alternative building materials cause lesser environmental impact. Therefore, it is suggested that those materials to be utilized in an office building after consideration of building structural requirements.

1 INTRODUCTION

The construction sector holds a very vital role in provision of buildings and facilities to satisfy human being's requirements in daily life, encourage employment opportunities directly and indirectly, and contribute towards the national economy development. However, contributing in the country's development, the construction sector and its activities also bring negative impacts to our environment, social and economy. During the construction stage, noise, dust (air pollution), traffic congestion, water pollution and waste disposal are common issues. Apart from this, a large amount of natural and human resources is required for the construction. Construction industry is one of the major consumers of natural resources and the construction materials are one of the prime resources for any construction project (Gardezi et al. 2014).

The building sector was one of the major consumers of the energy and natural resources which spend 40% of the materials entering the global economy and generated 40–50% of the global output greenhouse gases (Asif et al. 2007). Similarly, the European Commission DG ENV on March 2011 reported that the construction accounted for 24% of global raw materials removed from the earth. Unfortunately, the activities undertaken to extract, process and transport these construc-

tion materials are not environmentally friendly as Green House gas (GHG) and CO_2 emissions are generated by these activities. These processes for construction materials consume large quantities of energy and water. The use of natural resources in the construction stage of different projects is not only affecting the environment by making significant contribution in the terms of GHG emission, but also depleting them in quantity. Therefore, sustainability in the construction sector is a common concern for all. The World Commission on Environment and Development (WCED) defines the "sustainable development" as the one which meets the needs of the present without compromising the ability of future generations to meet their own needs (Glavinich & De 2008). In order to avoid the depletion of natural resources and adverse effects of construction sector, there is an urgent need that the construction sector considers alternate building materials which not only reduce the consumption of natural resources but also reduce GHG emissions. A part from the conventional material, there are several alternative building materials including green concrete, wood, bamboo and straw bale which have been adopted by different researchers in the past. Different researches have utilized alternate construction materials in their research. A summary of previous research is given in Table 1.

Table 1. Alternative construction materials in previous studies.

Researcher	Findings
Venkatarama et al. 2003	Total embodied energy of load bearing masonry buildings can be reduced by 50% when energy efficient/alternative building materials are used.
Lippke et al. 2004	A net CO_2 avoided was 55 metric tons for the wood-frame house compared to a net source of emissions of 185 metric tons for the steel-frame house.
Ghavami K 2005	Substitution of bamboo as steel reinforcement and finding that the structural elements developed could be used in different building constructions.
Gustavsson et al. 2006	Comparison wood and concrete framed building and finding that wood framed building has lower energy and CO_2 emission balances as compared to its counterpart.
Gonzalez et al. 2006	The study showed the possibility of reducing the CO_2 emissions up to 30% in the construction phase, through a careful selection of low environmental impact materials.
Monahan et al. 2011	Reduction of embodied carbon by 34% using a novel offsite panelized modular timber frame system as compared to traditional concrete frame for 3 bedroom semi-detached house in UK.
Briga-Sa et al. 2013	The application of the woven fabric waste (WFW) and woven fabric sub waste (WFS) from the textile industry as thermal insulation for building. This insulation building material in the external double wall increased its thermal behavior in 56% and 30%, respectively.
Elsayed	Indicating that straw bales have high level of energy efficiency over other building materials.

The objective of the study is to promote adoption of alternate construction materials and reduction of their carbon footprints in building construction. It also aims to reduce the materials demand of ore-based in the construction sector. Various alternative building materials are being identified, analyzed and evaluated for the construction of buildings in terms of its environmental impact to achieve low carbon footprint for a building.

2 RESEARCH METHODOLOGY

An office building was selected for the designing and working with different construction materials. After completion of designing phase, the quantification of materials was carried out and CO_2 emissions were calculated. In first case, the conventional constructions materials were adopted and in second case, alternated materials resulted from reviewing of literatures were utilized. Comparison of the building materials was performed based on the extensive review of literatures, identification of the materials availability locally and calculation of the materials quantities required for the building construction. Then, carbon calculator developed by Environment Agency has been used to quantify the greenhouse gas impacts of construction activities in terms of carbon dioxide equivalency (tCO$_2$e). The information pertinent to materials transportation distance and material quantities entered into a carbon calculator to measure the carbon dioxide equivalency of both the conventional and alternative building materials. The building materials were compared based on their embodied CO_2 emissions for the different building sections in the proposed office building, to determine the impact and carbon-friendliness of the alternative materials over the conventional options. Based on the results, the alternative materials were being recommended to be used in the office building in terms of their respective embodied CO_2 contributions.

2.1 Office building details

The office building is proposed to be constructed at Universiti Teknologi PETRONAS in Perak, Malaysia. This green resource center aims to create awareness of the impact of green movement in construction site and to provide tours for professionals, students and public to deliver the information about the environmental-friendly building construction. Figures 1 and 2 show the building in different directions.

2.2 Rationalization of selection of building materials

The conventional concrete which made up of cement, coarse aggregate and fine aggregate is proposed to be substituted with the green concrete which consists of the slag cement, recycled concrete aggregate and recycled glass. The conventional wall system which usually made up of concrete or bricks is compared with the straw bale wall system in terms of its weight and carbon emission. Then, the bamboo reinforcement which is utilized as the beam and column reinforcement

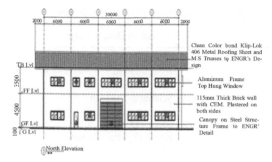

Figure 1.   North elevation of the office building.

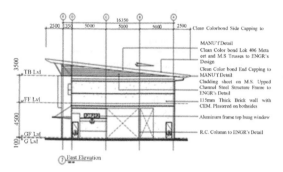

Figure 2.   East elevation of the office building.

Table 2.   The comparison between conventional and alternative building materials.

No.	Conventional Building Materials	Alternative Building Materials
1	Conventional Concrete: – Cement – Coarse Aggregate – Fine Aggregate	Use of Recycled Materials in Concrete Mixtures (Green Concrete): – Ground granulated blast furnace slag (GGBFS) / Slag Cement – Recycled Concrete Aggregate (RCA) – Recycled Glass (Crushed waste glass)
2	Concrete Wall System/ Brick Wall System Steel Reinforcement Structure	Straw Bale Wall System
3	– Steel Reinforced Concrete Beam/Column	Bamboo Reinforcement Structure – Bamboo Reinforced Concrete Beam/Column
4	Concrete Flooring	Wood Flooring/Bamboo Flooring

with proven strength and durability is proposed to replace the steel reinforcement system. Table 2 shows the comparison between conventional and alternative building materials.

3   RESULTS AND DISCUSSION

3.1   *Material quantities*

The quantities of materials required for the construction of the different component of the office building were initially calculated and are given in Table 3.

3.2   *Carbon footprint comparison*

Table 4 shows the summary of the carbon footprint for the conventional and alternative building materials for its utilization in different building sections. The graphical representations of results obtained from the calculator, in term of tCO_2e for both conventional and alternative building materials used in different building sections are shown in Figures 3 and 4.

The results show that the conventional concrete has been responsible for 61 tCO_2e for an outer wall. However, this emission is reduced to 28.9 tCO_2e and 0.93 tCO_2e if green concrete and straw bale are being adopted as alternate materials. Similarly, for inner wall green concrete produces 28.9 tCO_2e and brick wall 55.4 tCO_2e as compared to 61 tCO_2e if conventional concrete is used.

However, if conventional steel reinforcement is replaced with the alternate bamboo reinforcement, the reduction of carbon foot print is quite remarkable (more than 99%). Similarly, in case of flooring, the adoption of alternate materials reduces the carbon footprint up to 125 tCO_2e in case of wood and 3.6 tCO_2e in case of bamboo flooring as compared to 433 tCO_2e if conventional concrete is used. In an over scenario, a comparative analysis of alternatives 1 and 2 adapted to the conventional materials has been graphically represented in Figure 5.

Table 3. Materials quantities required for proposed construction site.

Description.	Unit	Quantity	Remarks
Building Volume	m³	3924	
Wall (6 inch thickness assumed)	m³	120	N & S elevation = 80, E & W elevation = 40
Flooring	m²	850	Ground floor = 400, First floor = 450

Table 4. Carbon footprint comparison of conventional and alternative building materials.

Materials	Quantities	Transport Distance (km)
A. External Wall		
Conventional Concrete	120 m³	Ipoh – 47 km
Green Concrete		Ipoh – 47 km
Straw Bale		Selangor – 220 km
B. Internal Wall		
Conventional Concrete		Ipoh – 47 km
Bricks	120 m³	Kuala Dipang – 30 km
Green Concrete		Ipoh – 47 km
C. Concrete Reinforcement		
Steel	9 tonnes	Ipoh – 47 km
Bamboo		Taiping – 94 km
D. Flooring		
Concrete	850 m³	Ipoh – 47 km
Wood		Ipoh – 47 km
Bamboo		Taiping – 94 km

Materials	CO_2 Emission Factor	Carbon Footprint (tCO$_2$e)
A. External Wall		
Conventional Concrete	0.083	61.0
Green Concrete	*varies	28.9
Straw Bale	0.04108	0.93
B. Internal Wall		
Conventional Concrete	0.083	61.0
Bricks	0.240	55.4
Green Concrete	*varies	28.9
C. Concrete Reinforcement		
Steel	1.46	13.2
Bamboo	0.0020412	0.11
D. Flooring		
Concrete	0.083	433.0
Wood	0.240	125.0
Bamboo	0.0020412	3.6

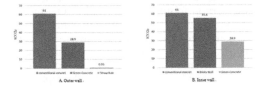

Figure 3. CO$_2$ equivalency of conventional and alternative building materials for walls.

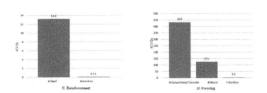

Figure 4. CO$_2$ equivalency of conventional and alternative building materials.

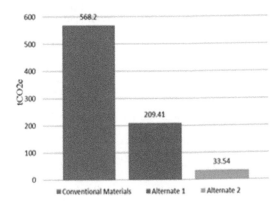

Figure 5. Comparative analysis of alternatives 1 and 2 adapted to the conventional materials.

Considerable amount of carbon footprint can be avoided using alternative 1 or alternative 2 material options as compared to conventional construction materials. The alternative 1 which consists of green concrete as outer wall, inner bricks wall, bamboo reinforcement and wood flooring, is able to reduce the total carbon footprint of the conventional building materials by 63% while alternative 2 (straw bale as outer wall, green concrete inner wall, bamboo reinforcement and flooring) can further avoid the carbon footprint up to 94% when compared to the conventional building materials.

4 CONCLUSIONS

The study has revealed that the alternative building materials can be utilized to reduce the environmental impact of the materials up to a considerable context when used in building construction. By adoption of these alternative materials, in this case study, a reduction of 53% of tCO_2e by adopting green concrete and 5.3% by adopting brick wall instead of conventional concrete was achieved. Similarly in the case of reinforcement, replacing steel with bamboo, causes carbon footprint reduction from 13.2 to 0.11 tCO_2e. 72% reduction in carbon footprint was achieved by replacing the conventional concrete floor with wooden floor. The study highlights that carbon foot-print during the manufacturing, transportation and construction phase can be reduced profusely by utilizing the alternative building materials. It is expected that the study will act as the motivation and guideline for designers/engineers in their future re-search on sustainable building construction.

ACKNOWLEDGEMENT

The authors are thankful to Ministry of Education, Malaysia for providing financial support (Grant No. 0153AB-J13) for this research under MyRA grant scheme.

REFERENCES

Asif, M. Muneer, T. & Kelley, R. 2007. Life cycle assessment: A case study of a dwelling home in Scotland. *Building and Environment*, 42: 1391–1394.

Briga-Sá, A. Nascimento, D. Teixeira, N. Pinto, J. Caldeira, F. & Varum, H. et al. 2013. Textile waste as an alternative thermal insulation building material solution. *Construction and Building Materials*, 38: 155–160.

Elsayed, M.S.G. *Straw Bale is Future House Building Material*. In Architect-Egypt (ed).

Gardezi, S.S.S. Shafiq, N. Zawawi, N.A.W.A. & Farhan, S.A. 2014. Embodied carbon potential of conventional construction materials used in typical Malaysian single storey low cost house using Building Information Modeling (BIM). *Advanced Materials Research*, 242–246.

Ghavami, K. 2005. Bamboo as reinforcement in structural concrete elements. *Cement and Concrete Composites*, 27: 637–649.

Glavinich, T.E. & De, P. 2008. BOOK TOOLS.

González, M.J. & García Navarro, J. 2006. Assessment of the decrease of CO2 emissions in the construction field through the selection of materials: Practical case study of three houses of low environmental impact. *Building and Environment*, 41: 902–909.

Gustavsson, L. & Sathre, R. 2006. Variability in energy and carbon dioxide balances of wood and concrete building materials. *Building and Environment*, 41: 940–951.

Lippke, B. Wilson, J. Perez-Garcia, J. Bowyer, J. & Meil, J. 2004. CORRIM: Life-cycle environmental performance of renewable building materials. *Forest Products Journal*, 54: 8–19.

Monahan, J. & Powell, J. 2011. An embodied carbon and energy analysis of modern methods of construction in housing: a case study using a lifecycle assessment framework. *Energy and Buildings*, 43: 179–188.

Venkatarama Reddy, B. & Jagadish, K. 2003. Embodied energy of common and alternative building materials and technologies. *Energy and buildings*, 35: 129–137.

Advances in Civil, Architectural, Structural and Constructional Engineering – Kim, Jung & Seo (Eds)
© 2016 Taylor & Francis Group, London, ISBN 978-1-138-02849-4

Comparison of in-hole electric potential method and electric resistivity imaging in Polymer Cut-off Wall quality testing

R. Wang & X. Lou
Yellow River Institute of Hydraulic Research, Zhengzhou, P.R. China

ABSTRACT: Polymer cut-off wall is a new kind of anti-seepage work made by the special grouting process and is used in hydraulic and civil engineering. Unlike traditional concrete cut-off wall, the polymer cut-off wall is made by foamed polymer and the thickness of wall is very small (less than 5 cm). So, most traditional quality test methods such as elastic wave tomography, GPR and resistivity imaging are not suitable in polymer cut-off wall quality testing because of the lack of application conditions. We design a new testing method based on DC potential measuring in borehole. Electrodes are placed in borehole to excite electric fields and measure the distribution of electric potential near the wall to analyze the quality of the wall. We finish a series of indoor model and some full-scale tests and compare the test result with electric resistivity imaging. The results prove that the method is effective and practical in polymer cut-off wall quality testing.

1 INTRODUCTION

1.1 Polymer Cut-off Wall (PCW)

Cut-off walls are widely used in hydraulic and civil engineering projects such as embankments, dams and foundation ditches. Most cut-off walls are made up of plain concrete or soil-cements underground to prevent seepage channels and protect the structures or spaces behind the wall. And some cut-off walls are strengthened by steel bars inside the ground (Stephen et al. 1999, Liu & Song 2006). The size of concrete or soil-cement cut-off walls is generally tens of centimeters wide. Polymer cut-off wall is a new kind of cut-off wall that is made up of foaming polymer material (Xu et al. 2012). When building a polymer cut-off wall, the patented equipment is used to trench and inject polymer material. The advantages of the polymer cut-off wall are obvious such as rapid construction, low cost and environmental friendly. Building a polymer cut-off wall needs several injections in turn. Each injection only creates a solid polymer sheet of about 0.5 m wide and 0.05 m thick. Such sheets compose a wall that can be hundreds meters long.

1.2 Problems in PCW quality testing

After construction, nondestructive quality testing is necessary to ensure that the polymer sheets made by injections contact each other closely and there are no apertures and holes in the wall to ensure the anti-seepage function. Considering the small thickness of walls, the high absorption coefficient and the high resistivity of polymer cut-off walls, most traditional quality testing methods are not practical. Those methods include GPR (Ground Penetrating Radar), cross-hole acoustic wave testing, elastic wave tomography, electric resistivity imaging and transient electromagnetic sounding (Leng et al. 2003). Electric resistivity imaging is a possible method but the resolution of the result is not gratifying. Thus, a new testing method is needed for evaluating the quality of the polymer cut-off wall.

2 THEORY OF THE IN-HOLE ELECTRIC POTENTIAL METHOD

The resistivity of foaming polymer is very high. It is generally much higher than $10^{10}\ \Omega \cdot m$ and can be regarded as a totally insulating material. The cracks and holes in the polymer cut-off wall will destroy the insulation of the walls. We use electrical dipole on ground to excite an electric field underground and the insulation wall will affect the distribution of the electric field. If the wall has holes or cracks, current will flow through them. The electric potential gradient near the holes and/or cracks will be different from the gradient near insulation walls. By measuring the underground electrical gradient in boreholes with another dipole, we can find the position of cracks/holes and evaluate the quality of walls.

3 EXPERIMENTAL WORK

3.1 Design and process

We build an experimental rig to simulate the testing situation. A plastic tank with saltwater inside was regarded as the ground. A piece of synthetic glass was placed in the middle of saltwater and regarded as the polymer cut-off wall. A 3 mm hole was drilled in the middle of glass 10 cm under the water surface to simulate the quality defect.

We designed 6 different combinations of exciting and measuring dipoles (Dipoles Set A ~ Dipoles Set F) and finished 6 experiments (Exp. No. 1 ~ Exp. No. 6) to evaluate the effects of different dipole sets. Figure 1 shows the dipole sets

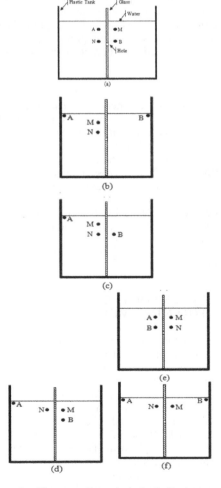

Figure 1.   The experimental rig and dipoles sets: (a) Exp. No. 1 and Dipoles Set A (b) Exp. No. 2 and Dipoles Set B (c) Exp. No. 3 and Dipoles Set C (d) Exp. No. 4 and Dipoles Set D (e) Exp. No. 5 and Dipoles Set E (f) Exp. No. 6 and Dipoles Set F.

used in these experiments. Electrodes "A" and "B" excited the dipole connected to a DC source and the exciting current was steady during testing. Electrodes "M" and "N" were measuring dipole connected to a high-precision potential meter. In the experiments, we used a DC resistivity tester to do the exciting and measuring work.

In Exp. No. 1, poles A, B, M, and N were all placed underwater. In Exp. No. 2, poles M and N were placed underwater and poles A and B were on the water surface. In Exp. No. 3, poles B, M and N were underwater poles and pole A was on the water surface. In Exp. No. 4, poles B, M and N were underwater poles and pole A was on the water surface. In Exp. No. 5, poles A, B, M and N were all underwater poles. In Exp. No. 6, poles M and B were underwater poles and poles A and N were on the water surface.

In all experiments, all poles on the water surface were steady and underwater poles were moved synchronously from the bottom to the surface. At each testing point, we charged the A and B dipoles and measured the value of the potential between MN and the source current. After the recording of the data, we moved the underwater poles synchronously 1 cm up and tested again. We continued the testing until the underwater poles reached the water surface.

3.2 Results

The source current magnitude will affect the value of testing potential. In order to eliminate the effect, we used a calculated value K by using Equation (1) as the testing result instead of the potential value

$$K = U_{MN}/I \qquad (1)$$

where U_{MN} is the potential between M and N; and I is the source current through A and B.

It should be noted that Equation (1) is similar in form of Ohm's law but the physics meaning is totally different. K means the potential value caused by a unitary current but not resistivity.

The results of the experiments are shown in Figures 2–7. These figures are K-depth graphs.

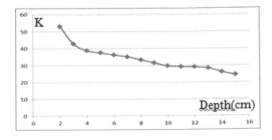

Figure 2.   K-Depth graph of Exp. No. 1.

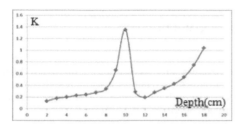

Figure 3.   K-Depth graph of Exp. No. 2.

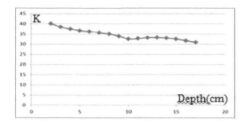

Figure 4.   K-Depth graph of Exp. No. 3.

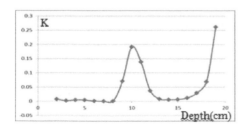

Figure 5.   K-Depth graph of Exp. No. 4.

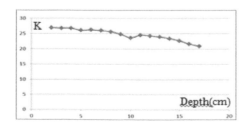

Figure 6.   K-Depth graph of Exp. No. 5.

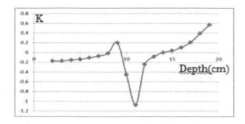

Figure 7.   K-Depth graph of Exp. No. 6.

The depth of the hole in the glass is marked by vertical lines.

We observed peaks/valleys, as shown in Figures 3, Figure 4 and Figure 5, and the position of peaks/valleys were relative to the position of the hole in glass. All peaks/valleys were located at the position of holes and lower boundary of the glass. It means that Dipoles Set B, Dipoles Set C and Dipole Set E can locate the position of holes in the wall and the lower wall's boundary. For example, in Figure 6, the curve is almost flat and the K value is near 0 from 2 cm to 8 cm in depth. Starting at 8 cm, the value of K increases rapidly and reaches the peak value of 0.2 at 10 cm in depth where the hole is located. At the depth of 13 cm, the value of K falls back to almost 0. It rises again at the depth of 17 cm. Comparing the curves in Figure 3, Figure 4 and Figure 6, the peak values of K shown in Figure 4 are largest. That means Dipoles Set C has the best noise immunity.

We changed the distances between poles and repeated the experiments (b), (c) and (e). The results indicate that the distances affect the width and height of the anomaly peak. The height of the anomaly peak is inversely proportional to the distance between the measuring dipole and the hole. The width of the anomaly is directly proportional to the electrode spacing of MN. According to the results of the experiments, the underground dipoles should be near the wall as close as possible to increase the signal-to-noise ratio and ensure the distinguishability of anomaly peaks.

4   ELECTRIC RESISTIVITY IMAGING ANALYSIS

Electric resistivity imaging is a widely used geophysical method. It can get underground electric resistivity distribution images by electrodes array and inversion software. We used the finite-difference model to simulate the testing result of electric resistivity imaging. The models and the results of electric resistivity imaging are shown in Figure 8.

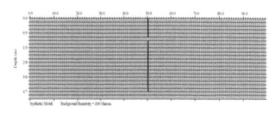

Figure 8a.   Finite-Difference Model, dark area represents high electric resistivity, simulate the polymer cut-off wall.

187

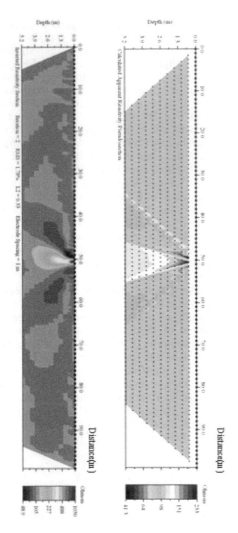

Figure 8b.   Finite-difference model.

5   CONCLUSIONS

The results of in-door experiments indicate that the electric potential testing method is useful and practical in polymer cut-off wall quality testing. Three kinds of dipoles sets are effective in testing works. The electric resistivity imaging can be used to determine the horizontal location of the polymer cut-off wall and probable depth, but it is difficult to locate quality problems such as cracks and holes.

ACKNOWLEDGMENT

The research was sponsored by China's Hydraulic Science and Technology Promotion Program "Embankments and Soil-Concrete Contact Surface Strength by Polymer Injecting Technology, Project Number: 1261420162562".

REFERENCES

Liu, H. & Song, E. 2006. Working mechanism of cutoff walls in reducing uplift of large underground structures induced by soil liquefaction. *Computers and Geotechnics*, 33: 209–221.

Leng, Y., Huang, J. & Wang, R. 2003. Research progress in safety detecting of Cut-off wall. *Progress In Geophysics*, 18(3): 404–409.

Stephen, L., Garvin, C. & Hayles, S. 1999. The chemical compatibility of cement–bentonite cut-off wall material. *Construction and Building Materials*, 13(6): 329–341.

Xu, J., Wang, F. & Zhong, Y. Stress analysis of polymer diaphragm wall for earth-rock dams under static and dynamic loads. *Chinese Journal of Geotechnical Engineering*, 34(9): 1699–1704.

Advances in Civil, Architectural, Structural and Constructional Engineering – Kim, Jung & Seo (Eds)
© 2016 Taylor & Francis Group, London, ISBN 978-1-138-02849-4

A novel Reliability Based Design Optimization method for truss and frame structural systems

C.L. Yu, Z.T. Chen & C. Chen
School of Naval Architecture and Ocean Engineering, Harbin Institute of Technology, Weihai, China

J.S. Lee
School of Naval Architecture and Ocean Engineering, University of Ulsan, Ulsan, Korea

ABSTRACT: A proper design procedure must consider various uncertainties. However, the expensive computational cost is the primary factor limiting application of RBDO on practical engineering. In this study, an efficient procedure for reliability assessment and its based optimization design of stochastic structural system are developed. First order method is adopted to calculate the safety indices since the safety margin equation is linear. Probabilistically dominate failure modes are selected by applying the branch-and-bound method. PNET is used to assess failure probability of structural system. The optimization design problem is formulated as a nonlinear programming problem to minimize weight with constraints on reliability and dimensions. The optimum vector algorithm is applied to solve the optimization problem under special convergence criterion. Two numerical examples are performed to illustrate the feasibility of proposed process for any novel truss and frame structure. The present optimization routine shows that the approach is efficient in the practical application.

1 NTRODUCTION

Due to many sources of uncertainty, inherent in structural design, probabilistic approach is more rational. However, the reliability of a real structure is usually much more difficult to be evaluated since more than one component can fail. There exists possibility of more than one failure mode for the structural system. Therefore, it is widely accepted that reliability analysis of a structural system is very important in the complex structure such as bridge, marine structure, and spacecraft and so on.

Reliability-based structural optimization is an approach to minimize values of some predetermined utility functions, such as cost or weight, under the constraint that the failure probability of the structure shall stay within a certain allowable range. The question of reliability-based optimization was raised at more than sixty years ago. Despite all these efforts, optimization algorithms combined with reliability analysis procedures are still confined to academic type of examples. It is assumed that the explicit formulations of the limit state functions are available, which is hardly possible in practice. For structures at a practical size, points at the limit state function are usually determined by using finite element procedure, bounding formulas and response surface methods. Among three optimization approaches for limit states

functions above, finite element method has been extensively applied, especially for the novel structure. Another factor limiting the development of RBDO (Reliability-Based Design Optimization) on practical structural system is expensive computational cost. The solution for RBDO problem is usually regarded as nested double optimization loops. Since both loops are solved by iterative computation, they can consume a lot of computing time.

In this study, a novel method combining SFEM (Stochastic Finite Element Method) with structural reliability theory is used to achieve reliability assessment of structural systems. The optimum vector algorithm is employed to solve the optimization problem under prescribed convergence criterion. This proposed approach is proved to be very robust and efficient for any novel truss and frame structural systems, which can be applied extendedly to practical engineering.

2 RELIABILITY ASSESSMENT OF STRUCTURAL SYSTEMS

2.1 *Limit state function*

Regardless of structural styles, the internal force of q-component can be expressed as Equation 1:

$$S_q = \sum_{j=1}^{Q} b_{qj} \boldsymbol{F}_j - \sum_{j=1}^{q-1} a_{qj} \boldsymbol{R}_j \qquad (1)$$

where, a_{qj} is resistance coefficient, i.e. load effect in j-section due to unit value of opposite resistance; b_{qj} is load coefficient for F_j, i.e. load effect due to unit value of load F_j; F_j is external load which subjects to the structure; R_j is resistance of each component.

The parameters of both a_{qj} and b_{qj} can be obtained by finite element method. Failure of structural system is assumed to be occurred once the q-component fails. Hence, the safety margin equation of i-th failure mode is expressed as Equation 2:

$$Z_i^q = R_q - S_q = \sum_{j=1}^{2n} a_{qj} R_{pj} - \sum_{j=1}^{Q} b_{qj} F_j \qquad (2)$$

where, a_{qj} is defined as following:

$$a_{qj} = \begin{cases} 1 & (j = q) \\ 0 & (j = I_c) \\ \text{otherwise, obtained by FEM} \end{cases}$$

and I_c is the set of components that do not fail in the i-th failure mode;

Because the safety margin equation is linear, the first order method is adequate to calculate the safety index of each component of structural system. The safety index is defined as the ratio between mean value and standard deviation about limit state equation, as shown in Equation 3:

$$\beta_i = \frac{\mu_{\boldsymbol{Z}_i}^q}{\sigma_{\boldsymbol{Z}_i}^q} = \frac{\sum_{j=1}^{2n} a_{qj} \mu_{\boldsymbol{R}_{pj}} - \sum_{j=1}^{Q} b_{qj} \mu_{\boldsymbol{F}_{pj}}}{\sigma_{\boldsymbol{Z}_i}^q} \qquad (3)$$

where, μ^* is mean value with respect to responding variables; σ^* is a standard deviation with respect to responding variables.

2.2 Reduced stiffness and fictitious load

For many redundant ductile and semi-brittle systems, structural collapse occurs only after several components simultaneously reach their maximum capacity. After each componential failure, the stiffness is locally modified and the residual strength of failed components is accounted by applying the fictitious load (or equivalent nodal forces) on the structure. The derivation and expression of reduced stiffness and fictitious load for truss and frame structural element are referred as An & Cai (2007).

Finally, when one component fails, the failure criterion is satisfied. Then one failure mode is formed. Usually, singularity of stiffness matrix is defined as the structural failure criterion. However, it is almost impossible that the stiffness matrix is completely singular for a complex structure. Consequently, small ratio (10^{-6}) of stiffness matrix value to initial structural stiffness matrix value is regarded as the failure criterion of structural system in this study.

3 RELIABILITY OF STRUCTURAL SYSTEMS

There are too many failure paths in a highly redundant structure to generate all of them, which necessitates a procedure for selecting only the probabilistically significant failure paths. Efficient methods using a branch-and-bound technique have been proposed and is adopted in this paper. The probability of failure of the structural system is estimated by modeling the different dominant failure modes of structural system as a series system. The joint failure probabilities of failure modes are considered to calculate the failure probability of structural system. In order to avoid complicated calculation on joint failure probabilities of failure modes as well as obtain the numerical estimation about failure probability, PNET is employed to assess the failure probability value of structural system. For PNET method, dominant failure modes are divided into several representative failure groups. It is assumed that each representative failure group is independent of others, so Equation 4 can be used to obtain the failure probability of structural system:

$$P_f = 1 - \prod_{i=1}^{G} (1 - P_i) \qquad (4)$$

where, P_i is representative failure probability of each failure mode group.

4 OPTIMIZATION ALGORITHM

4.1 Model of optimization problem

The optimum design problem is formulated as a nonlinear programming problem, which aims at minimizing weight with constraints on reliability and dimensions, as shown in Equations 5 to 7.

Minimize:

$$W(x) = \sum_{i=1}^{n} C_{bi} x_i \qquad (5)$$

where, $W(x)$ is the vector of structural weight; C_{bi} is the weight of unit cross-sectional area.

Subjected to:

$$\beta_s(x) \geq \beta_s^a \qquad (6)$$

and

$$X^L \leq X \leq X^H \qquad (7)$$

where, β_s^a prescribed target reliability index; X^L and X^H are lower and upper bound of design variables, respectively. The objective function of above problem is linear. However, the constraint function is non-linear. The optimum vector algorithm is known as one of the most appropriate methods to find the optimum design point to solve this kind of problems.

4.2 Optimum vector algorithm

The principle of optimum vector is shown in Figure 1. Let the initial design point be A_0. Firstly, move the design point to the point A_1 on the constraint boundary along the negative gradient direction of aim function. Then, move it to design point A_2 along the optimum vector direction (D direction), make the design point far from the constraint boundary to increase the safety index of structural system. Repeat moving the design point in the same direction until the convergence criterion Equation 8 is satisfied:

$$\left| \frac{W^{(K)} - W^{(K-1)}}{W^{(K)}} \right| \leq \varepsilon_W \qquad (8)$$

where, ε_W is a small positive number, usually $\varepsilon_W = 10^{-4} \sim 10^{-3}$.

It is credible that the direction of optimum vector (D direction) is most significant during the optimum design process. The expression of D direction is show as Equation 9:

$$D = \frac{\nabla g}{\nabla W^T} - \nabla W \nabla g + (\nabla g^T \nabla W)\nabla g - \nabla W \qquad (9)$$

where,

$$\nabla g = \frac{\left[-\dfrac{\partial \beta_s}{\partial x_1}, -\dfrac{\partial \beta_s}{\partial x_2}, \cdots -\dfrac{\partial \beta_s}{\partial x_n} \right]^T}{|\nabla g|}$$

and

$$\nabla W = \frac{\left[C_{b1}, C_{b2}, \cdots C_{bn} \right]^T}{|\nabla W|}$$

4.3 Procedure

The code is developed with MATLAB following the procedure. The whole system is divided into

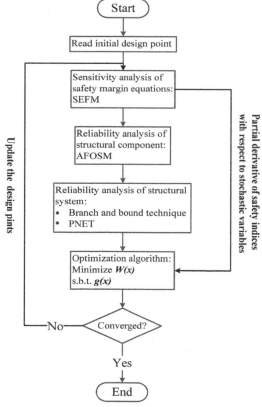

Figure 2. Flow chart of RBDO procedure.

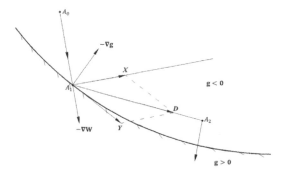

Figure 1. Optimum vector algorithm.

five parts: input module, reliability analysis module of each dominant failure mode, reliability analysis module of structural system, optimization module and output module. The flowchart of reliability assessment and RBDO is summarized in Figure 2.

5 ILLUSTRATIVE EXAMPLES

5.1 *Truss structural system*

A statically indeterminate 6-member truss with one degree of redundancy as shown in Figure 3 is selected for the present reliability analysis.

It is also carried out by Kalatjari et al., Shao and Murotsu and Kima et al. The allowable yield stress and applied loads are uncorrelated normal random variables. Mean value and coefficient of variation of yield stress of members are 270.662 MPa and 0.05, respectively. Statistics data of applied loads are given in Table 1.

All the members' cross-sectional area is 230 mm². In this example, failure due to buckling is not considered. The behavior of the material in compression and tension are considered to be identical. To optimize this structure, the prescribed allowable safety index of structural system is given as 2.5000. There is no limitation on the ultimate cross-sectional area of component.

Three most possible failure paths are chosen as dominant failure modes and their corresponding limit state functions are summarized in Table 2, safety indices are compared with those of reference as in Table 3. Positive and negative signs denote failing members in tension and compression, respectively.

Finally, correlation coefficient between failure mode 1 and another four failure modes are 0.8386, 0.1472 and 0.3102, respectively. The safety index of structural system is 2.9756, and corresponding failure probability is 1.45×10^{-3}. Comparing with referred safety index obtained by MCS (Monte-Carlo Simulation) method is 2.9859, difference is about 0.35%.

The change of structural weight with optimization step is drawn in Figure 4. Finally, the safety index of optimized structure is 2.5102 and its weight is 10.61 kg, reduced by 18.5% comparing

Table 2. Safety margin equations for most dominant failure modes.

Failure path	Limit state function
6+−2+	$Z = \sigma_y A_2 + 0.6\sigma_y A_6 - 0.0033(L_1 + L_3) A_2\, 0.0043\, L_2 A_2$
6+−1+	$Z = \sigma_y A_1 + 0.8\sigma_y A_6 - 0.0043(L_1 + L_3 + L_5)A_1$
6+−5−	$Z = \sigma_y(A_5 + A_6) - 0.0054(L_1 + L_3)A_5$

Table 3. Safety margin equations for most dominant failure modes.

Failure path	Safety index	
	Present study	Kalatjari et al.
6+−2+	3.0630	3.0528
6+−1+	3.4581	3.4480
6+−5−	4.8867	4.8771

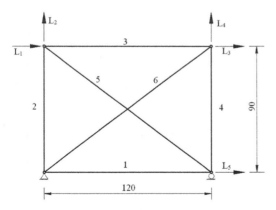

Figure 3. Plane truss structural system (Unit: mm).

Table 1. Statistic data of loads (6-member truss).

Random variable	Force KN	Coefficient of variation
L_1	49.0	0.1
L_2	29.4	0.1
L_3	19.6	0.1
L_4	29.4	0.1
L_5	19.6	0.1

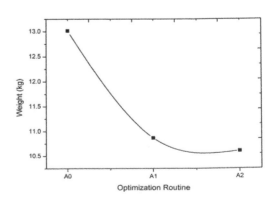

Figure 4. Variation of structural weight.

the original weight 13.02 kg. The light weight is remarkable.

5.2 Frame structural system

The plane frame model in Figure 5 has 8 possible hinges as components. It is especially sensitive and has been frequently selected for evaluation of the system reliability and sensitivity analysis. Both resistance and loading are assumed to be independent and normally distributed variables. Mean value of component properties are shown in Table 4. COV of resistance is 0.05; modulus of elasticity of material is 210 GPa, density of material is 7.85×10^3 kg/m³. Mean of P1 and P2 is 20 KN and 40 KN respectively, COV of loading is 0.3. where, $*$ is the nodal number, $(*)$ is the number of component, U is the number of element.

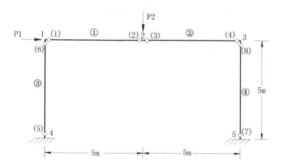

Figure 5.   Frame structure.

Table 4.   Mean value of component properties for frame structure.

Component	A_k	I_k	R_k
(1) & (2)	4.0×10^{-3}	4.77×10^{-5}	1.01×10^5
(3) & (4)	4.0×10^{-3}	4.77×10^{-5}	1.01×10^5
(5) & (6)	4.0×10^{-3}	3.58×10^{-5}	7.5×10^4
(7) & (8)	4.0×10^{-3}	3.58×10^{-5}	7.5×10^4

Table 5.   Safety index for the top five possible failure modes.

Failure mode number	Failure path	Present safety index	Referred safety index
Model 1	(8)-(2)-(6)	2.49	2.48
Model 2	(8)-(2)-(7)-(6)	2.49	2.48
Model 3	(8)-(2)-(7)-(1)	2.91	2.88
Model 4	(8)-(2)-(1)	2.91	2.88
Model 5	(8)-(2)-(7)-(5)	2.95	2.91

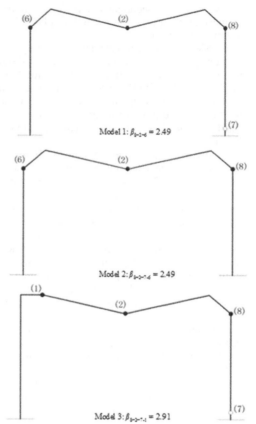
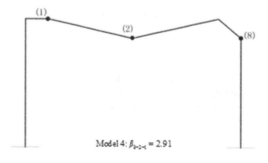
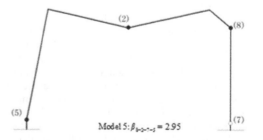

Figure 6.   Collapse forms of the top five possible failure modes for frame structure.

where, A_k = cross sectional area (m^2), I_k = moment of inertia (m^4), R_k = mean strength (= plastic bending moment, N·m).

FOSM is applied to calculate the safety index of each possible failure mode. Branch-and-bound method is adopted to find the dominant failure modes and then PNET is to calculate the failure probability. The comparison of safety indices for the top five possible failure modes is summarized in Table 5. The collapse forms are shown in Figure 6.

Figure 6 illustrates that component (7) does not contribute to the safety margin equations for failure mode 1 and 3, i.e. without it a collapse mechanism can be formed which are paths of (8)-(2)-(6) and (8)-(2)-(1). Those are one of failure modes having the same correlation as the modes for paths of (8)-(2)-(7)-(6) and (8)-(2)-(7)-(1). This component (7) at collapse is referred to as a non-active hinge, but before collapse it was being considered in the particular failure path being examined. The present results coincide with reference.

The allowable safety index of structural system is given as 3.00. There is no limitation about the ultimate cross-sectional areas of components, but its minimum value is assumed to be limited as 4.0×10^{-3} m^2. The section of beam is circular and initial value of cross-sectional area is set to be $A = [3.14159\,3.14159\,3.14159\,3.14159] \times 10^{-2}$ m^2.

After seven iteration steps, the convergence condition is satisfied. Finally, the reliability index of optimized frames is about 3.18. Comparing the weight in iteration step 7 with original structural weigh, it decreases 464.7 kg, which is by about 23%.

6 CONCLUSIONS

This paper not only develops an efficient and accurate method to identify dominant failure modes of stochastic structural system and compute the failure probabilities of the overall structural system, but also proposes a valid process to complete the optimum design. Due to the application of SFEM, the pro-posed method for RBDO can be applied to any novel stochastic structural system in practical engineering. The safety indices for each component are calculated by approximate method, thus integration of the failure probability density function is avoided to save the computation cost. PNET method is adopted to calculate the safety index of structural system with considering the correlation between different most dominant failure modes. The complicated calculation on joint failure probabilities of failure modes can be therefore avoided and numerical estimation for failure probability of structural system rather than any other bounds can be obtained. The RBDO problem is modeled

as a nonlinear programming problem that aims at minimizing weight with constraints from reliability and dimensions. The codes combining structural system reliability assessment and optimization algorithm have been developed with MATLAB.

In the two numerical examples, some most possible dominant failure modes are found and the structural system reliability indices are obtained. The rationality of present approach is indicated through comparison between the present results and those found in references. As far as the present results are concerned, weight are remarkably reduced by applying the RBDO procedure with keeping the structural safety greater than its allowable level. The variations of components' cross-sectional areas and structural weights with optimization iteration steps show that the proposed optimization algorithm can make the optimum design reach quickly. It is also proved that the proposed procedure is efficient and can be widely applied in practical engineering.

ACKNOWLEDGEMENT

This research was supported by HIT Discipline Guide Fund (Project No.: WH20140102).

REFERENCES

An, W.G. & Cai, Y.L. 2007. *Reliability analysis and optimization of stochastic structural system (in Chinese)*. Harbin: Harbin Engineering University Press.

Ang, A.H.S. & Tang, W.H. 1984. *Probability Concepts in Engineering Planning and Design Volume II*. New York: John Wiley and Sons.

Ditlevsen, O. 1979. Narrow reliability bounds for structural system. *Journal of Structural Mechanics*, 7(4): 453–472.

Feng, Y. 1989. A method for computing structural system reliability with high accuracy. *Computers & Structures*, 33(1): 1–5.

Frangopol, D.M. 1985. Sensitivity studies in reliability-based analysis of redundant structure. *Structural Safety*, 3: 13–22.

Freudenthal, A.M. 1956. Safety and the probability of structural failure. *Transaction-ASCE*, 121: 1337–1397.

Gasser, M & Schueller, G.I. 1997. Reliability-based optimization of structural systems. *Mathematic Methods of Operations Research*, 46:287–307.

Kalatjari, V. Kaveh, A. & Mansoorian, P. 2011. System reliability assessment of redundant trusses using improved algebraic force method and artificial intelligence. *Asian Journal of Civil Engineering (Building & Housing)*, 12(4): 523–550.

Kima, D.S. Okb, S.Y. Songc, J. & Kohd, H.M. 2013. System reliability analysis using dominant failure modes identified by selective searching technique. *Reliability Engineering & System Safety*, 119: 316–331.

Lee, J.S. 1989. *Reliability Analysis of Continuous Structural Systems*. Unpublished Ph.D. dissertation, University of Glasgow, UK.

Leheta, H.W. & Mansour, A.E. 1997. Reliability-based method for optimal structural design of stiffened panels. *Marine Structure*, 10: 323–352.

Lin, T.S. & Corotis, R.B. 1983. Reliability of ductile systems with random strength. *Journal of Structural Engineering-ASCE*, 109: 1585–1601.

Liu, N. & Tang, W.H. 2004. System reliability evaluation of nonlinear continuum structures—A probabilistic FEM approach. *Finite Element in Analysis and Design*, 40: 595–610.

Melchers, R.E. 1999. *Structural reliability analysis and prediction (2nd ed.)*. New York: John Wiley.

Moses, F. 1990. New directions and research needs in system reliability research. *Structure Safety*, 7: 93–100.

Park, S.Y. 2001. A new methodology for the rapid calculation of system reliability of complex structures. *Architecture Re-search*, 3(1): 71–80.

Ranganathan, R. & Deshpande, A.G. 1987. Generation of dominant modes and reliability analysis of frames. *Structural Safety*, 4: 217–228.

Shao, S. & Murotsu, Y. 1999. Approach to failure mode analysis of large structures. *Probabilistic Engineering Mechanics*, 14: 169–177.

Zhao, Y.G. & Ono, T. 1998. System reliability evaluation of ductile frame structures, *Journal of Structural Engineering*, 124(6): 678–685.

Zheng, M.B. & Chen, G.H. 2011. An effective sensitivity analysis methodology for the reliability-based structural optimization design to high temperature components. *Quality and Reliability Engineering International*, 27(8): 1211–1220.

Advances in Civil, Architectural, Structural and Constructional Engineering – Kim, Jung & Seo (Eds)
© 2016 Taylor & Francis Group, London, ISBN 978-1-138-02849-4

Stresses in the concrete cylinder subjected to radiation exposures

V.I. Andreev

Moscow State University of Civil Engineering, Moscow, Russia
Research Institute Building Physics of the Russian Academy of Architecture and Building Sciences,
Moscow, Russia

ABSTRACT: In this paper, we consider the planar problem (plane strain) theory of elasticity of the propagation of neutrons in a thick concrete cylinder. Under the influence of the neutron flux, concrete was found swollen and there are forced radiation deformations. We consider the influence of irradiation on the deformation properties of concrete. We present the solution of the quasi-stationary axisymmetric problem of radiation stresses in a thick-walled concrete cylinder when on its inner surface is given the integrated neutron flux (fluence) equal to Φ_0. The problem is solved, taking into account changes in the elasticity modulus of the material in the process irradiation.

1 INTRODUCTION

The mechanics of inhomogeneous bodies is one of the relatively new trends in solid mechanics. Various physical factors (temperature, radiation, moisture) affect the strength and deformation properties of materials. Unfortunately, there is sufficiently small fundamental literature available in this area (Olszak, Ryhlevsky & Urbanovsky 1964, Lomakin 1976, Andreev et al. 2002). This article examines one of the most urgent problems of the influence inhomogeneity due to radiation on the concrete cylindrical shell.

2 DISTRIBUTION OF NEUTRON FLUENCE

The distribution of the neutron flux is described by the Poisson equation:

$$\nabla^2\Phi - \frac{\Phi}{L^2} = 0,$$

which in this case is represented as.

$$\frac{d^2\Phi}{dr^2} + \frac{1}{r}\frac{d\Phi}{dr} - \frac{\Phi}{L^2} = 0. \tag{1}$$

Here r is the radius and L is the diffusion length depending on the neutron spectrum.

A study (Andreev & Dubrovskiy 1982, Dubrovskaya 1989) used the approximate solution of Equation (1):

$$\Phi = \Phi_0 \frac{a}{r}\exp\left(-\frac{r-a}{L}\right), \tag{2}$$

where a is the inner radius of the cylinder.

The following is an *exact* solution. By substituting $\rho = r/L$ in Equation (2), it reduces to the form.

$$\rho^2\frac{d^2\Phi}{d\rho^2} + \rho\frac{d\Phi}{d\rho} - \rho^2\Phi = 0. \tag{3}$$

This equation is one of versions of the Bessel equation (Watson 1995), the general solution of which has the form.

$$\Phi = C_1 I_0(\rho) + C_2 K_0(\rho) \tag{4}$$

Here I_0 is the modified Bessel function of the first kind of zero order and K_0 is the zero-order Macdonald function.

Assuming that the thickness of the cylinder ensures full absorption of neutrons, we can write the boundary conditions as follows:

$$r = a\,(\rho_1 = a/L), \Phi = \Phi_0;$$
$$r = b\,(\rho_2 = b/L), \Phi = 0,$$

where b is the outer radius of the cylinder.

From these conditions, we can find C_1 and C_2 as follows:

$$C_1 = \frac{\Phi_0 K_0(\rho_2)}{I_0(\rho_1)K_0(\rho_2) - I_0(\rho_2)K_0(\rho_1)};$$

$$C_2 = -\frac{\Phi_0 I_0(\rho_2)}{I_0(\rho_1)K_0(\rho_2) - I_0(\rho_2)K_0(\rho_1)}.$$

Under the influence of neutron flux in the concrete appear forced radiation deformations, $\varepsilon_f = \varepsilon_r$, similar to what occurs when heated. These deformations (Dubrovskiy 1973) are given in the empirical formula.

$$\varepsilon_r = \frac{\alpha\varepsilon_{max}\left[\exp(\beta\Phi) - 1\right]}{\varepsilon_{max} + \alpha\exp(\beta\Phi)}, \tag{5}$$

where ε_{max} is the maximum radiation strain in solution (concrete); Φ is the integrated flux (fluence) of neutrons; and α and β are empirical coefficients depending on the radiation aggregate deformability and the energy spectrum of the neutron flux.

It is known that the radiation exposure of many materials (Olszak et al. 1964) changes their mechanical properties. Dubrovskiy (1973) found the empirical dependence of the concrete's elastic modulus of fluence Φ:

$$E = E_0\left[\gamma_1 - \alpha_1 \lg(\beta_1\Phi)\right], \tag{6}$$

where E_0 is the elastic modulus of the non-irradiated material; and α_1, β_1 and γ_1 are empirical coefficients depending on the grade of concrete and the neutron energy spectrum.

The following are the results of the calculation made for the following values of the cylinder dimensions, material characteristics and constants in Equations (5) and (6): $a = 3.3$ m; $b = 4.5$ m; $\varepsilon_{max} = 0.01$; $L = 0.16$ m; $v = 0.16$; $\alpha = 1$; $\beta = 3 \cdot 10^{-24}$ m^2/neutr; $E_0 = 2 \cdot 10^4$ MPa; $\alpha_1 = 0.7$; $\beta_1 = 10^{-24}$ m^2/neutr; $\gamma_1 = 0.8$.

Note that the table values of Bessel functions $I_0(x)$ and $K_0(x)$ given in reference books (e.g. Hemming 1968) are limited to values of the argument $x = 10$. At the same time, the values of ρ lying in the range $[\rho_1 = 20{,}625; \rho_2 = 28{,}125]$ significantly exceed this value. In Bronshtein & Semendyaev (1986), asymptotic formulas for consideration Bessel functions at large values of the argument are given:

$$I_0(\rho) = \frac{e^\rho}{(2\pi\rho)^{1/2}} + O\left(\frac{1}{\rho}\right);$$

$$K_0(\rho) = \left(\frac{\pi}{2\rho}\right)^{1/2} e^{-\rho} + O\left(\frac{1}{\rho}\right). \tag{7}$$

Since the asymptotic formulas for the Bessel functions give values accurate to $O(1/\rho)$, it is of interest to determine the residual error of (3) with Formulas (7). A direct substitution of the first

term in the expression for $K_0 = (\rho)$ in (3) shows that the left side of the equation remains the term that just is the residual error of Equation (3).

$$e^{-\rho}/(4\sqrt{\rho}), \tag{8}$$

Substituting into (8) the minimum value of $\rho = 20{,}625$, we can see that this residual error is three orders smaller than the third term in (3). Using the expression in (7) for the Bessel functions, we can calculate the constants C_1 and C_2:

$$C_1 = 11{,}4\Phi_0 e^{-35{,}625}; \quad C_2 = 3{,}63\Phi_0 e^{20{,}625}.$$

Substituting these constants in (4) and using (7), we easily see that in solution (4), we can neglect the term containing the constant C_1. Then, the solution of (3) takes the form

$$\Phi(\rho) = 3.63\Phi_0\left(\frac{\pi}{2\rho}\right)^{1/2} e^{20{,}625-\rho}. \tag{9}$$

Comparing the results of the calculation of the function $\Phi(\rho)$ by Formulas (9) and (2) shows that the difference between the approximate and exact decision for consideration interval of ρ does not exceed 5%.

Figure 1 shows the graphs $E(r)$, $\Phi(r)$ and $\varepsilon_r(r)$. Functions $\varepsilon_r(r)$ and $E(r)$ are constructed according to (5) and (6). As follows from (5), the impact of radiation exposure on Young's modulus starts with a certain threshold value $\Phi = \Phi^*$, which is determined from the condition.

$$\lg(\beta_1\Phi^*) = (\gamma_1 - 1)/\alpha_1$$

If some parts of the structure integrated neutron flux are not greater than Φ^*, the modulus of

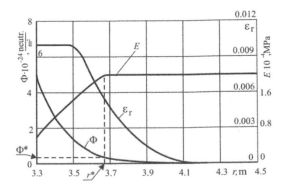

Figure 1. The distribution of the neutron flux $-\Phi$, radiation deformation ε_r and elastic modulus E on thickness of the cylinder wall.

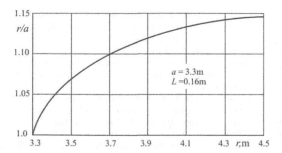

Figure 2. The dependence of area of the change of deformation properties of concrete (r^*) on the integral neutron flux on the inner wall of the cylinder.

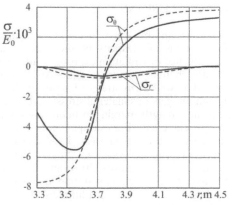

---- -homogeneous material --- inhomogeneous material

Figure 3. Diagrams of radiation stresses in the thick-walled cylinder with and without considering inhomogeneity of the material.

elasticity therein remains constant. A value Φ^* corresponds to the radius $r^* = \rho^* L$, which can be found from (9) by substituting a corresponding value $\Phi = \Phi^*$. The resulting equation for determining r^* is transcendental and solved numerically. If on the whole interval (a,b) satisfies the condition $\Phi(r) < \Phi^*$, then $r^* = a$.

Figure 2 shows an example of the dependence of $r^*(\Phi_0)$, which is determined only by two parameters, a and L, the values of which are indicated in the figure.

3   CALCULATION OF STRESS STATE

To solve the problem of determining the radiation stress in the cylinder taking into account the change in the elastic modulus, we use the equation.

$$\sigma_r'' + \varphi(r)\sigma_r' + \psi(r)\sigma_r = f(r), \tag{10}$$

The coefficients and the right-hand side of this equation when substituted in (5) and (6) take the form:

$$\varphi(r) = \frac{3}{r} + \frac{\alpha_1 \lg e \cdot \dfrac{d\Phi}{dr}}{\Phi\left[\gamma_1 - \alpha_1 \lg(\beta_1\Phi)\right]};$$

$$\psi(r) = -\frac{1-2\nu}{1-\nu} \cdot \frac{\alpha_1 \lg e \cdot \dfrac{d\Phi}{dr}}{r\Phi\left[\gamma_1 - \alpha_1 \lg(\beta_1\Phi)\right]};$$

$$f(r)\frac{\alpha\beta\varepsilon_{max}\left(\alpha + 100\varepsilon_{max}\right)e^{\beta\Phi} \cdot \dfrac{d\Phi}{dr}}{\left(100\varepsilon_{max} + \alpha e^{\beta\Phi}\right)^2}.$$

In addition to these relations, function $\Phi(r)$ is given by (9).

Figure 3 shows the stress diagrams σ_r and σ_θ, obtained by the numerical solution of Equation (10)

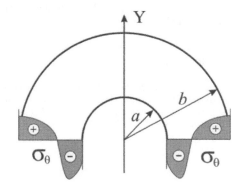

Figure 4. To the condition of the self-equilibrium diagram σ_θ.

by the sweep method with homogeneous boundary conditions for the stresses at the ends of the interval (a, b). It may be noted that taking into account the inhomogeneity of the material leads to a drastic reduction of stress concentration near the inner surface of the cylinder, reducing also stress in the stretch zone. The last fact is also supported by the need to fulfill the conditions of equilibrium.

$$\int_a^b \sigma_\theta dr = 0 \tag{11}$$

This equation indicates that the diagram should be self-equilibrium. In other words, the positive area of the diagram σ_θ in absolute value should be equal to the square of its negative side (Figure 4). For half of the cylindrical shell must be the equality $\Sigma Y = 0$, which corresponds to Equation (11).

4 CONCLUSIONS

As shown in this study, accounting changes in the modulus of elasticity of the concrete under the influence of radiation exposure leads to a significant changing the stress state. As in the compressed as well as in stretched zones of the shell, the inhomogeneous material stresses decrease, that is in the strength margin. In many cases, there is an opposite effect. For example, in the rupture of the pipeline, there is moistened soil that leads to forced swelling deformations. Studies conducted by Andreev & Avershyev (2014) and Avershyev & Andreev (2014) determined that at constrained swelling, tensile stresses in a humidified array may lead to its destruction.

ACKNOWLEDGMENT

This work was supported by the Ministry of Education and Science of Russia under grant no. 7.2122.2014/K.

REFERENCES

Andreev, V.I. & Avershyev, A.S. 2014. Two-dimensional problem of moisture elasticity of inhomogeneous spherical array with cavity. *Applied Mechanics and Materials*, 580–583: 812–815.

Andreev, V.I. & Dubrovskiy, A.V. 1982. Accounting of inhomogeneity of the material in the calculation of the dry reactor protection. *Problems of Atomic Science and Technology*. 3(13): 3–8.

Andreev, V.I. 2002. *Some problems and methods of mechanics inhomogeneous bodies*. Moscow: Publ. House ASV.

Avershyev, A.S. & Andreev, V.I. 2014. Two-dimensional problem moisture elasticity for inhomogeneous flat annular area. *Applied Mechanics and Materials*, 580–583: 2974–2977.

Bronshtein, I.N. & Semendyaev, K.A. 1986. *Handbook of mathematics for engineers and students of technical colleges*. Moscow: Science.

Dubrovskaya, E.V. 1989. *Some flat and axially symmetric equilibrium problems of an elastic body in conditions of radiation exposure*. PhD. Diss. Moscow.

Dubrovskiy, V.B. (ed.) 1973. *Radiation resistance of materials. Reference book*. Moscow: Atomizdat.

Heming, R.V. 1968. *Numerical Methods for Scientists and Engineers*. Moscow: Science.

Lomakin, V.A. 1976. *Theory of elasticity inhomogeneous bodies*. Moscow: Moscow State University.

Olszak, W. Ryhlevsky, Y. & Urbanovsky, W. 1964. *Theory of plasticity inhomogeneous bodies*. Moscow: Mir.

Watson, G.N. 1995. *A Treatise on the Theory of Bessel Functions*. Cambridge Mathematical Library. Copyrighted material. Cambridge University Press.

Advances in Civil, Architectural, Structural and Constructional Engineering – Kim, Jung & Seo (Eds)
© 2016 Taylor & Francis Group, London, ISBN 978-1-138-02849-4

Thermal and dynamic analyses of the Siraf city's wind tower: Construction of a half-scale model at Lyon university, France

G.R. Dehghan Kamaragi
GSA/ENSA Paris Malaquais, Paris-Est University, Paris, France

M. Pinon
ENSA Paris Malaquais, Paris, France

A. Hocine
Paris-Ouest University, Nanterre, France

ABSTRACT: A wind tower (Badgir in Persian) captures wind from any direction and guides it to the occupant's zones. Built on the roof, the wind tower has been used for many centuries, employed at different heights, ranging from 1 m to 33 m. These towers are a zero carbon cooling technology. This paper describes four steps: climate search in Iran; various specific tests with small-scale models at the wind tunnel to visualize the pressure drop in different conduits; construction of a half-scale model of Badgir with our students; and numerical analysis with "Fluent software".

1 INTRODUCTION

1.1 History

It is certainly the originality of this wind tower that caught our attention and led us to perform multiple analyses. The city of Siraf is an ancient port located in the central district of Kangan, province of Bu-shehr. During the Sassanid Dynasty, Siraf was a port on the north shore of the Persian Gulf. Siraf was used as a route between the Arabian Peninsula and India.

1.2 The local measurements in Siraf

Nasuri's house is located in the city of Siraf, at the foot of the mountain, at about 50 meters of the

seaside. The first day of analysis on this tower was carried out on April 17th, 2013 between 8 a.m. and 8 p.m. The second analysis commenced on April 18th, 2013 at 9 p.m. and was completed on April 19th at 7 a.m. To obtain more precision, we recorded the data for three successive days. This tower is 14.83 m high. The room located under Badgir is 12.24 m long and 2.80 m wide, with a surface area of 34.27 m² and a height under ceiling of 3.84 m. The room has a door and two small windows giving onto other rooms. The measurements are given in Table 1.

Table 1 highlights the difference between the inside and outside temperatures. The gain freshness is most notable between 3 p.m. and 4 p.m.

Figure 1. Wind tower of Nasuri's house in Siraf city.

Table 1. Measurements performed in Siraf.

Hours	Outside air Velocity (m/s)	Outside air T (°C)	Inside air Velocity (m/s)	Inside air T (°C)	ΔT
9 a.m.	2.5	33.2	1.5	27.8	5.4
10 a.m.	2.1	34.6	1.2	27.2	7.4
11 a.m.	2.6	34.9	1.5	27.1	7.8
12 a.m.	2.6	35.1	1.4	28.8	6.3
1 p.m.	2.5	35.8	1.8	28.5	7.3
2 p.m.	3.2	35.9	2.1	28.2	7.7
3 p.m.	3.1	36.2	2.1	28.2	8
4 p.m.	3.9	35.7	1.3	27.6	8.1
5 p.m.	3.8	35.1	1.2	27.3	7.8
6 p.m.	3.6	34.2	1.4	26.9	7.3

2 CONSTRUCTION OF A HALF-SCALE MODEL AT LYON UNIVERSITY

We conducted a workshop with third-year students at the University of Lyon (L'Isle-d'Abeau) over a period of 5 days, from 20 to 24 April 2015. We built a half-scale tower (6 m in height). Students learned to design the successive steps of the construction, followed by observing the use of Bernoulli's theorem and Navier-Stokes equations. To build the tower, we used lime and brick similar to the materials used in Siraf (original Badgir).

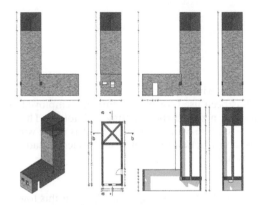

Figure 2.   Plan and section of the tower.

Figure 3.   Lime (clay, straw and water) and bricks (solid and hollow).

Figure 4.   The process of construction of a Badgir.

Figure 5.   The interior of the half-scale model.

Figure 6.   Original Badgir (left) and the half-scale Badgir (right).

Figure 7.   The interior of Nasuri's house (original wind tower).

3 CONDITION OF THE EXPERIMENT IN THE WIND TUNNEL

In this experiment, we used a subsonic Prandtl wind tunnel located in the laboratory of Ville d'Avray. This wind tunnel has the following characteristics:

The square test section (side length 450 mm) has a longitudinal length equal to 750 mm; the maximum velocity is 40 m/s; the rate of turbulence is between 1% and 2%.

We made a model to perform tests and analyses in the wind tunnel.

We installed the model in the wind tunnel. As shown in the figure, we found the data by changing the direction of the wind at different angles: 0°, 15°, 30° and 45°.

On the other side, at $\alpha = 45°$, the pressure in A and C tunnels is almost the same. The pressure in C and D tunnels is also the same. In this case, two tunnels are operated by sucking (A and B) and the two others by blowing (C and D). Everything occurs as if the building has only two tunnels A–B and C–D.

3.1   Results and analysis of wind tunnel tests

Wind tunnel tests show that when the angle $\alpha = 0°$ to 15°, there is a total lack of dynamism in the ducts, but when $\alpha = 30°$ and 45°, the four conduits support the wind and the tower begins to expel the air.

The case at $\alpha = 30°$ is an intermediate configuration: tunnels A and B are operated by suction and tunnel C by blowing. In comparison with the case $\alpha = 45°$, the contribution of tunnel B for suction is more weak.

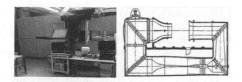

Figure 8.   View of the wind tunnel.

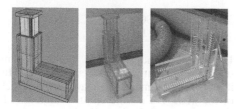

Figure 9.   View of the model.

Figure 10.   Evaluations at different wind angles.

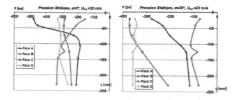

Figure 11.   Static pressure, $\alpha = 0°$ and $\alpha = 30°$
U inf = 20 m/s.

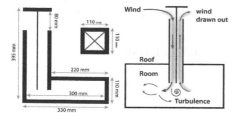

Figure 12.   Plan and section (left); a turbulent air back
to the top of the tower (right).

To verify the accuracy and validity of the results,
we compared the data obtained in parallel with the
numerical calculations with Fluent.

4   FLUENT MODELING

The geometry of a Badgir is shown in the following
figure. A 3D geometry was generated according to

Figure 13.   The 3D and the generated grid.

an experimental prototype that has been used in a
wind tunnel.

The numerical simulations were carried out at a
wind (air) speed of 20 m/s, and four different angles
for wind were considered: 0°, 15°, 30° and 45°.

To observe the Fluent results, two different
quadric surfaces are defined in Fluent, z = zmean
and x = xmean-tower = zmean (since the tower
cross section is square, so xmean-tower = zmean).
The results and contours are drawn on these two
surfaces. With the help of these planes, we can
determine whether the part of the tower is exhaling or inhaling the air flow.

4.1   0° Wind angle

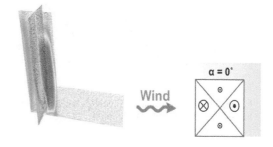

Figure 14.   The pressure is negative in canals B, C and D
Maximal velocity in canal A.

4.2   15° Wind angle

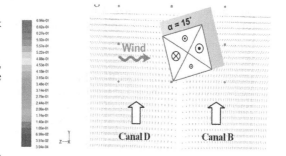

Figure 15.   The canals D and B exhale.

Figure 16. 30° Wind angle.

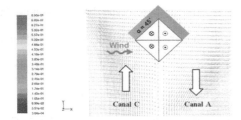

Figure 17. 45° wind angle.

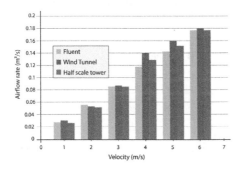

Figure 18. Comparison of Fluent, wind tunnel and the half-scale model at Lyon University.

4.3 30° Wind angle

As compared with the two previous cases, it seems that for 30° wind angle, the velocity vectors are smoother in canals B and D. Moreover, the velocity vectors in these two canals are to some extent identical with canal C. Therefore, it can be deduced for this condition that the wind enters canal A and then smoothly and uniformly exits from the three remaining canals.

4.4 45° wind angle

The air flow through canals B and D for 45° wind angle is the same with the 15° wind angle case. This result is predictable because of geometric symmetry of a Badgir. It is obvious that the geometry has a mirror symmetry with respect to the quadratic plane z = zmean. Hence there is no difference between the results of the internal air flow for 15° and 45° wind angle cases.

In summary, the smoother flow with one canal as an inhale part and three other canals as exhale parts of the Badgir tower occurs at the 30° wind angle. Since at this condition, three exhaling canals have identical velocity vector distributions. That means each of these three canals bears the same portion of air conducted to the outside.

5 FACTORS CONSIDERED IN THESE CALCULATIONS

1. Different wind speeds; 2. Different temperatures; 3. Different wind incident angles; 4. Humidity; 5. Building materials (clay soil); 6. Thickness of the walls; 7. Full and empty fountain; 8. Location openings; 9. Different surfaces.

6 CONCLUSIONS

The CFD simulations are validated with detailed wind tunnel experiments. Results show that a four-sided wind tower has great potential.

An angle of 32° remains the optimum angle for this Badgir experience.

At 32° angle, one air duct blows air into the room and three ducts expel the room air.

The air undergoes a temperature loss of 2°C upon entering the room.

When the fountain is full, we achieve 4°C in terms of cooling (the temperature drop).

The more the windows are placed in the bottom area, the more speed the output receives.

We can calculate the same parameters for adjacent rooms.

We can choose the ideal location for Badgirs in buildings with large dimensions.

ACKNOWLEDGMENT

The authors are grateful to the students of L2 and L3 for their participation. They thank Ms M. Alishahi and Mr M.R. Daneshgar.

REFERENCES

Bahadori, M.N. & Dehghani, A. 2008. *Iranian Masterpiece of Engineering*, Nashr Ketab Daneshgahi.
Elmualim, A., 2004. *Modeling of Windcatcher for Natural Ventilation*, National Renewable Energy Laboratory, Denver, Colorado. USA.
Ghobadian, V. 1998. *Climatic Analysis of the Iranian Traditional Buildings*, Tehran: Tehran University.
Roaf, S. 1990. *The significance of thermal thresholds in the performance of some traditional technologies*, Proceedings of the North Sun Conference, Reading, Pergamon.

Advances in Civil, Architectural, Structural and Constructional Engineering – Kim, Jung & Seo (Eds)
© 2016 Taylor & Francis Group, London, ISBN 978-1-138-02849-4

Energy-saving clustering algorithm based on improved PSO for WSNs

Z.Y. Sun & C. Zhou
Department of Information Engineering, Northeast Dianli University, Jilin, China

ABSTRACT: The LEACH algorithm selects the Cluster Head (CH) randomly with the same probability; however, it cannot guarantee that the number of clusters in each round is the best. In addition, non-Cluster Head node (non-CH) in the process of selecting the cluster does not take the global energy consumption into account. In this paper, considering the disadvantages of cluster's structure in LEACH, we adopt the Particle Swarm Optimization (PSO) algorithm for choosing CHs. Based on the adjustment of the inertia weight and the acceleration factor, we amend the disadvantages of PSO in terms of convergence speed in the late stage, which make the particle fall into local optimum very easily. An evaluation function is proposed to calculate which cluster the non-CH node will belong to. Simulation experiments show that the proposed algorithm is more superior than LEACH in clustering efficiency, and the lifetime of the network is significantly extended.

1 INTRODUCTION

Classical LEACH algorithm cannot guarantee that the number of clusters in each round is the best, and it does not consider the residual energy of nodes, which leads to some nodes with low energy being chosen as the CHs. The non-CH nodes take the minimum distance as the principle when selecting the CH without considering the energy consumption of transmission link. Jiang et al. (2012) proposed and analyzed the Energy-Balanced Unequal Clustering Protocol (EBUCP). By using PSO, this protocol uses unequal clustering mechanism and inter-cluster multi-hop routing, selects a set of optimal nodes as cluster heads and the network is divided into clusters of varying sizes. The distance between the cluster head and the cluster member is minimized in EBUCP to reduce the communication energy consumption within the cluster. Since the algorithm adopts the unequal clustering, the size of the cluster is only considered without taking the number of clusters into account. A dual-cluster heads clustering routing algorithm based on PSO named PSO-DH has been proposed in the literature (Han et al. 2010). The master cluster head is used for data collection and date integration. The integrated data are sent to the vice cluster head. The vice cluster head charges for the communication with the base station. However, the algorithm does not consider the shortcoming of PSO and the parameters in the algorithm without in-depth research. In addition, vice cluster heads communicate with the base station directly, which results in poor energy efficiency. The PSO is a swarm intelligence technique for global optimization (Amin et al. 2012).

However, the search efficiency of PSO decreases as the iteration process approaches to the end. In order to solve the disadvantages of PSO, this paper improves the inertia weight and acceleration factor, and then the improved algorithm is applied to the election of CH. First, we calculate the optimal number of CHs, and then we select CHs by taking full consideration of the node residual energy as well as location information. Algorithms change the cluster head nodes, taking its own minimum cost as the principle on the ownership of the communication mechanism. The algorithm determines the non-CH nodes of the optimal home cluster using the evaluation function F. The algorithm improves the attribution mechanism of the non-cluster head node based on its own minimum communication cost in the LEACH algorithm, and the optimal belonging cluster of non-cluster head is determined by the evaluation function.

2 ENERGY MODEL

The energy model used in this paper is the same as that reported in the literature (Li et al. 2011). When a node transmits l-bit data, the required energy is as follows:

$$E_{tx}(l,d) = lE_{elec} + lE_a d^n = \begin{cases} lE_{elec} + l\varepsilon_{fs}d_{to-CH}^2, d < d_0 \\ lE_{elec} + l\varepsilon_{mp}d_{to-BS}^4, d \geq d_0 \end{cases}$$

$$(1)$$

The energy required by the radio to receive an l-bit data is given by

$$E_{rx}(l) = E_{rx-e}(l) = lE_{elec} \qquad (2)$$

So, the total energy is described as

$$E_{total} = E_{tx} + E_{rx} \qquad (3)$$

where d is the distance from the source node to the destination node; n is the channel attenuation index in the range (Han et al. 2010, Guo et al. 2010); ε_{fs} is the unit power amplification; ε_{mp} is the unit power amplification of multi-path fading model; E_{elec} is the energy consumption of sending or receiving unit bit data; and E_T is the total energy consumption.

3 PARTICLE SWARM OPTIMIZATION

PSO is a kind of swarm intelligence algorithm like fish schooling. In each iteration, its velocity v_{in} and the position x_{in} in the nth dimension are updated using the following equations respectively (Wang & Cui 2014):

$$v_{in}(t+1) = wv_{in}(t) + c_1 r_1 \left(p_{in}(t) - x_{in}(t) \right)$$
$$+ c_2 r_2 \left(p_{gn}(t) - x_{in}(t) \right) \qquad (4)$$

$$x_{in}(t+1) = x_{in}(t) + v_{in}(t+1) \qquad (5)$$

where w is the inertia weight; r_1, r_2 are two different uniformly distributed random numbers in the range [0,1]; c_1, c_2 are two non-negative constants called acceleration factors, which are used to control the flight velocity of particle; and t is the current number of iterations and we always assume the max number is t_{max}.

3.1 Improvement of PSO

In the standard PSO, the inertia weight is denoted as

$$w = w_{max} - \frac{w_{max} - w_{min}}{t_{max}} \times t \qquad (6)$$

where w_{max}, w_{min} are the initial and terminal inertia weight respectively. Usually, we express w with a linear gradient strategy of 0.9 to 0.4. However, when the inertia weight is bigger, the particle moves fast. So the algorithm has strong ability of global search. With the reduction of w, the particle strengthens its ability of local search because of moving slowly. It is easy to fall into local optimum and hard to jump out. Therefore, it is beneficial to balance the global search ability and local search

ability of the particles by dynamically adjusting w. Within the suitable convergence region of particles, we suppose that particles have higher global search ability in the earlier stage to find out the right position (Wang & Cui 2014). But in the late stage, particles should have higher development ability to speed up the convergence and prevent particles from trapping in the local optimum. Therefore, in this paper, w is improved as follows:

$$w(t) = w_{min} + \left(w_{max} - w_{min} \right) \left(\frac{f' - f_{min}}{f_{avg} - f_{min}} \right)$$
$$\exp\left(-k \times \left(\frac{t}{t_{max}} \right)^2 \right) \qquad (7)$$

where f is the particle fitness function; f_{min} is the smallest fitness value; and f_{avg} is the average fitness value. It can be seen that when the target value of each particle tends to be the global optimum, the inertia weight is reduced and the velocity of the particle becomes slow. It is beneficial for searching the particle in the local area. In contrast, the particle whose fitness value is worse than the target value can move closer to the better search area because its inertia weight is larger and the velocity is faster.

The improved formula for velocity and position is defined as

$$v'_{in}(t+1) = \left(w_{min} + (w_{max} - w_{min}) \left(\frac{f - f_{min}}{f_{avg} - f_{min}} \right) \right.$$
$$\times \exp^{\left(-k\left(\frac{t}{t_{max}} \right)^2 \right)} \left. \right) v_{in}(t) + \left(c_{max} - \frac{(c_{max} - c_{min})}{t_{max}^2} \times t^2 \right)$$
$$r_1(p_{in}(t) - x_{in}(t)) + \left(c_{min} + \frac{(c_{max} - c_{min})}{t_{max}^2} \times t^2 \right)$$
$$r_2(p_{gn}(t) - x_{in}(t)) \qquad (8)$$

$$x'_{in}(t+1) = x_{in}(t) + v'_{in}(t+1) \qquad (9)$$

4 CLUSTERING BASED ON IMPROVED PSO

4.1 Determine the number of CHs

The number of the clusters has a great impact on the energy consumption of the WSN. When there are a large number of clusters, too much power of nodes will be consumed in data fusion and forwarding. So the clustering loses its meaning. If there are a small number of clusters, the burden of CHs will increase, the energy of nodes will be used up rapidly and nodes will easily die prematurely. This text assumes that N nodes are distributed in an area of $m \times m$ and they are divided into k clusters equally. So, there are $\frac{N}{k}$ nodes in each cluster, which includes one CH and $\frac{N}{k} - 1$ member nodes. When the CH and member nodes transmit or receive data, the energy consumption of CH and member nodes, in an area of r radius, can be calculated respectively according to Formulas (10) and (11):

$$E_{ch} = l \left[\left(\frac{N}{k} - 1 \right) E_{elec} + \frac{N}{k} E_{da} + E_{elec} + \varepsilon_{mp} d_{to-BS}^4 \right]$$

(10)

$$E_{non-ch} = l \times \left(E_{elec} + \varepsilon_{fs} d_{to-CH}^2 \right)$$

(11)

So, we can find out the total cost of energy of a cluster as follows:

$$E_{cluster} = E_{ch} + \left(\frac{N}{k} - 1 \right) E_{non-ch}$$

$$= l \left[\frac{2N}{k} E_{elec} + \frac{N}{k} E_{da} + \varepsilon_{mp} d_{to-BS}^4 + \frac{N}{k} \varepsilon_{fs} d_{to-CH}^2 \right]$$

(12)

And the total energy consumption of WSN is defined as

$$E_{total} = l \left[2N E_{elec} + N E_{da} + k \varepsilon_{mp} d_{to-BS}^4 + N \varepsilon_{fs} d_{to-CH}^2 \right]$$

(13)

d_{to-BS} is the distance from nodes to the base station. We suppose that the coordinate of BS is (x_0, y_0), and (x_i, y_i) is the coordinate of the node.

$$d_{to-BS} = \sqrt{(x_i - x_0)^2 + (y_i - y_0)^2}$$

(14)

If d_{to-CH} is the distance from nodes to CH, its expectation can be obtained as follows:

$$E[d_{to-CH}^2] = \iint (x^2 + y^2) \rho(x, y) dx dy$$

$$= \rho \int_0^{2\pi} \int_0^{\frac{m}{\sqrt{\pi k}}} r^3 dr d\theta = \frac{m^2}{2\pi k}$$

(15)

In order to minimize the total energy consumption of the network, $\frac{\partial E_{total}}{\partial k} = 0$, i.e.

$$\varepsilon_{mp} d_{to-BS}^4 - \varepsilon_{fs} \frac{N m^2}{2\pi k^2} = 0$$

(16)

We can draw the conclusion as follows:

$$k = \sqrt{\frac{N}{2\pi}} \sqrt{\frac{\varepsilon_{fs}}{\varepsilon_{mp}}} \frac{m}{d_{to-BS}^2}$$

(17)

It is easy to see from Formula (17) that the optimal number of CHs is determined by the network scale and the number of nodes in the network. Through the energy consumption and node location, we count the optimal number of CHs. It combines the dynamic variation process of network nodes. Compared with LEACH, the new algorithm is more helpful for reducing the overall energy consumption of WSN (Guo et al. 2010).

4.2 Clustering

When LEACH starts to select CHs, the nodes will generate a random number in the range [0,1]. If the number is less than T(n), then the node is CH.

Because we do not consider the current residual energy of nodes, some nodes with low energy can be selected as CHs in this way. Energy is consumed extensively in data fusion and forwarding, so nodes can die early because they consume energy very fast. In this paper, the improved PSO is used to select CHs. We fully consider the node residual energy and location information, then the nodes with more residual energy and nearby BS will be selected as CHs.

4.3 Algorithm process

Step 1: pre-clustering. LEACH is used to select the candidate CHs. In order to ensure that some of the nodes with lower energy are not selected as candidate CHs, the nodes must meet Formula (18):

$$E_i \geq E_{avg} = \frac{1}{N_{alive}} \sum_{i=1}^{n} E_i$$

(18)

Step 2: Particles initialization. $X = \{x_1, x_2 \mid x_i \mid x_n\}$ is selected as initialization particle swarm. Initialize the search points' position and velocity randomly. Let the current position be the individual optimal solution P_i of each particle. Find the global extreme value from the individual optimal solutions, and record the number of the particle P_g.

Step 3: Fitness function. We should calculate the fitness value of the particles f. If $f < P_i$, let f replace P_i. If the individual extreme value of total particles is better than the current global extreme value, then the global extreme P_g is updated.

Routing optimization should take the routing length, the cost of inter-cluster communication, and nodes energy consumption into consideration (Pratyay & Prasanta 2014). Therefore, the fitness function of sensor nodes is defined as

$$f = \alpha f_1 + \beta f_2 + \gamma f_3 \qquad (19)$$

$$f_1 = \frac{\sum_{i=1}^{n} E(n_i)}{\sum_{i=1}^{K} E(CH_i)} \qquad (20)$$

$$f_2 = \frac{\sum_{i \in C(i)} d(n_i, CH_i)}{|C(i)|} \qquad (21)$$

$$f_3 = \sum_{i=1}^{k} d(BS, CH_i) \qquad (22)$$

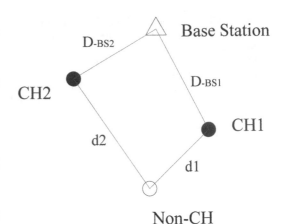

Figure 1. Different ways of selecting the CH.

where f_1 is the ratio for which the total energy of CHs accounts the total energy of the network nodes. f_2 is the ratio of average Euclidean distance between CH and member nodes. f_3 expresses the ratio of average Euclidean distance from the CH to the BS. $E(n_i)$ is the energy of network nodes. $E(CH_i)$ is the energy of CH nodes. $d(n_i, CH_i)$ is Euclidean distance between CH and member nodes. $C(i)$ is the number of non-CH. $d(BS, CH_i)$ is Euclidean distance between CH and BS. $\alpha,\ \beta,\ \gamma\ |\ [0,1]$ and $\alpha,\ \beta,\ \gamma = 1$. We can adjust the proportion of f_1, f_2, f_3 to f by the value of $\alpha,\ \beta,\ \gamma$.

Step 4: Particle status update. The particle updates its velocity and position according to the modified Formula (8) and (9), and maps it to the sensor node.

Step 5: Test whether the end condition is met. If P_g is not found or t_{max} is not reached, then it returns to Step 3. Otherwise, the algorithm ends and outputs the particle as the optimal solution in this round, which has the smallest fitness value.

4.4 The formation of clusters

In LEACH, if the node has been selected as CH, it will broadcast this new node to non-CH nodes and then the non-CH node selects its belonging-cluster and adds based on the received signal. This method follows the principle of minimum communication cost of non-CH nodes. However, it does not consider the energy consumption of the whole transport chain. In Figure 1, the non-CH node1 joins its nearest CH1. But the energy consumed by CH1 and non-CH node1 in the chain of $d1+D_{-BS1}$ may not be less than the energy consumed by CH2 and non-CH node1 in the chain of $d2+D_{-BS2}$.

Therefore, in order to balance the energy consumption, we propose an evaluation function F. According to F, the nodes determine which cluster it will join to:

$$F = \alpha \frac{d(CHi, n_i)}{\max d(CHi, n_i)}$$
$$+ \beta \frac{d(BS, CHi) - \min d(BS, CHi)}{\max d(BS, CHi) - \min d(BS, CHi)}$$
$$+ \gamma \frac{E_0 - E_{re}(CHi)}{E_0} + \mu \frac{\Delta E}{E_0} \qquad (23)$$

$\Delta E = \left| E_{re}(n_i) - \frac{E_{total}}{N} \right|$ represents the difference between the average energy of the nodes and the ideal average energy. F gives full consideration to residual energy of CH, the distance from CH to non-CH nodes and the distance from CH to BS. The non-CH nodes calculate the evaluation function of each CH respectively within its communication range, and join the cluster whose F is smallest.

5 EXPERIMENTAL RESULTS

We performed extensive experiments on the proposed algorithm using MATLAB R2012b. The experiments were performed with 200 nodes that distributed in a monitoring area of 200×200 randomly, and BS is located out of the coordinate of (100, 250), as shown in Figure 2. We used simulation parameter values, as listed in Table 1.

The numbers of CHs of the new algorithm in this paper are compared with LEACH in 50 rounds. As shown in Figure 2, the fluctuation range of the number of CHs in new algorithm is less than the range of LEACH, and the number of CH is more stable.

Figure 3 shows the survival state of the network nodes of several algorithms in different running times. In the initial operation stage of the network, all nodes are full of energy. With the operation of

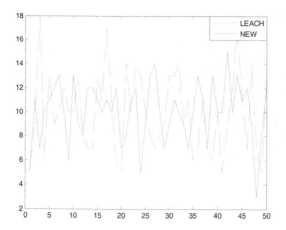

Figure 2. Comparison of the number of CHs in 50 rounds.

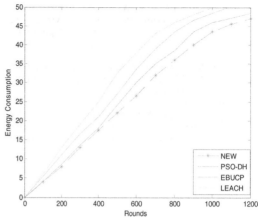

Figure 4. Comparison of energy consumption.

Table 1. Simulation parameters.

Parameter	Value
E_0	$0.5\ J$
E_{elec}	$0.5\ nJ/bit$
ε_{fs}	$10\ nJ/(bit \cdot m^{-2})$
ε_{mp}	$0.0013\ pJ/(bit \cdot m^{-4})$
E_{da}	$5\ nJ/bit$
l	$4000\ bit$

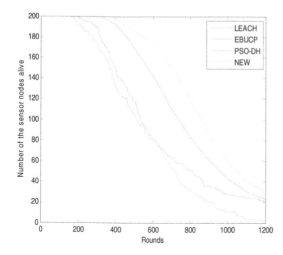

Figure 3. Comparison of active sensor nodes.

the network, the nodes consume a large amount of energy. Some nodes forward and fuse data frequently, so the energy is consumed quickly and die earlier. It can be seen from Figure 3 that in LEACH,

the first inactive node appears in 200th round. In the new algorithm proposed in this paper, the first inactive node appears in 400th round. When WSN operates in 1200th round, the nodes in Leach almost completely die. However, the new algorithm still has about 15% nodes remaining.

The relationship between the total energy consumption of the network and the running time of the network is shown in Figure 4. With the increase in running time, the energy consumption of LEACH is accelerated. Since the new algorithm takes the residual energy of the node into account during clustering, the energy consumption of each node can be reduced to some extent. It is beneficial to balance the energy consumption of the network.

6 CONCLUSIONS

In this paper, the optimal number of CHs is calculated using the relationship between energy and distance. We propose a clustering algorithm based on improved PSO, which modifies the inertia weight and the acceleration factor, to improve the convergence speed of the particle. The PSO can find CHs in the process of clustering quickly and reduce the clustering time. The new algorithm reduces the overall energy consumption of the network to a certain extent and prolongs the lifetime of the network.

REFERENCES

Amin, S. Wirawan & Gamantyo, H. 2012. Dynamic Overlapping Clustering for Wireless Sensor Networks Based-on Particle Swarm Optimization, *Journal of ICT Research and Applications*, 6(1): 43–62.

Guo, J., Sun, L.J. & Wang, R.C. 2010. A Particle Swarm Clustering Protocol for Wireless Sensor Networks Based on Particle Swarm Optimization of Clusters, *Journal of Nanjing University of Posts and Telecommunications*, 4(30): 36–40.

Han, D.X., Zhang, R.H. & Liu, D.H. 2010. PSO-based Double Cluster-heads Clustering Algorithm for Wireless Sensor Network, *Computer Engineering*, 10(36): 100–102.

Jiang, C.J., Tang, X.L. & Xiang, M. 2012. Unequal clustering routing protocal for wireless sensor networks based on PSO algorithm, *Application Research of Computers*, 8(29): 3074–3078.

Li, H.B., Yu, C.B., Yan, J.H. & Li, Y.L. 2011. Clustering strategy for energy balance of wireless sensor network based on improved particle swarm optimization clustering algorithm, *Application Research of Computers*, 28(2): 657–660.

Pratyay, K. & Prasanta, K.J. 2014. Energy efficient clustering and routing algorithms for wireless sensor networks: Particle swarm optimization approach. *Engineering Applications of Artificial Intelligence*, 33: 127–140.

Wang, D.D. & Cui, B.T. 2014. Research of WSN Clustering Routing Based on Adaptive Particle Swarm Optimization, *Computer Technology and Development*, 3(24): 82–85.

Advances in Civil, Architectural, Structural and Constructional Engineering – Kim, Jung & Seo (Eds)
© 2016 Taylor & Francis Group, London, ISBN 978-1-138-02849-4

Investigation of the shell's stability using the mixed finite element method

L.U. Stupishin & K.E. Nikitin
Urban and Road Construction and Structural Mechanic Department, South-West State University, Kursk, Russia

ABSTRACT: This paper proposes a methodology and algorithm for determining critical loads based on the mixed finite element method. It considers axisymmetric geometrically nonlinear shallow shells made of orthotropic material. Finally it presents the results of numerical investigations of stability by changing the shape of shells, ratio of elastic modulus of the material and parameters of the support contour.

1 INTRODUTION

The developed methodology is based on the Galerkin method and finite element method in a mixed formulation. This approach has been described in the literature (Stupishin and Nikitin 2014a, 2014b, Stupishin and Kolesnikov 2014, Andreev and Duborovskiy 2013), which has several advantages over the classical formulation of the displacements finite element method. The solution of the nonlinear problem is determined by using one of the algorithm continuation methods.

2 DESCRIPTION OF THE METHODOLOGY

We used an algorithm continuation method (Grigolyuk and Shalashilin 1988), which is as follows:

$$\xi_k^0 = t \cdot dX_{k-1}, X_k^0 = X_{k-1} + \xi_k^0,$$

$$\begin{cases} J\left(X_k^i\right)\Delta X_k^{i+1} = -F(X_k^i) \\ (\xi_k^i, \Delta X_k^{i+1}) = \frac{1}{2}(t^2 - (\xi_k^i, \xi_k^i)) \end{cases},$$

$$X_k^{i+1} = X_{k-1} + \Delta X_k^{i+1}, \xi_k^{i+1} = X_k^{i+1} - X_{k-1}, \qquad (1)$$

where $F(X)$ is the nonlinear system of n equations with n+1 unknown variables that is solved; X is the vector of unknown variables. In our problem, it includes a system of nodal degrees of freedom and the parameter of loading q. The system of equations $F(X)$ has a set of solutions that can be represented by curve solutions. Here $k = 0$,

1, 2 … is the number of steps along the curve solutions; $i = 0, 1, 2$ … is the number of refinement iterations at ach step k; $F(X_k^i)$ and X_k^i are respectively the system of equations and vector of its solutions at step k and refinement iteration i; ξ_k^0 is the initial direction vector for searching a solution X_k^i at step k; t is the step value; dX_k is the unit vector tangent to the curve solutions at current step k; dX_{k-1} is the unit vector tangent to the curve solutions at previous step; X_k^0 is the initial vector of solution of the system equation $F(X_k^i)$ at step k; X_{k-1} is the last refined vector of solution of the system equation $F(X_k^i)$ at previous step; $J(X_k^i)$ is the Jacobi matrix system of equations $F(X_k^i)$ at step k and refinement iteration i; ξ_k^i is the direction vector for searching a solution X_k^i at step k and refinement iteration i; ΔX_k^i is the vector of iterative refinement of the solution vector X_k^i at step k and refinement iteration i; X_k^{i+1} is the refined vector of solution of the system equation $F(X_k^i)$ at step k and refinement iteration i; and ξ_k^i is the refined direction vector for searching a solution X_k^i at step k and refinement iteration i.

Refinement continues until the following conditions hold:

$$\|\Delta X_k^{i+1}\| > \varepsilon, \qquad (2)$$

where ε is a constant that specifies the search accuracy solutions.

The value of dX_k is determined by solving the system of equations of the form

$$\begin{cases} J\left(X_k\right)dX_k = 0 \\ (dX_k, dX_k) = 1 \end{cases} \qquad (3)$$

The stress-strain state of the shell is determined by using the finite element method, and matrices and vectors that are derived using the Galerkin method. An expression describing the state of the jth finite element is of the form

$$F_j(X_{k,j}^i) = \{r_1 \quad r_2 \quad r_3 \quad r_4\}^T, \tag{4}$$

where

$$r_1 = \left(\frac{\eta - \chi}{3}\bar{C}_2 + \chi\bar{C}_{12} - e_1\bar{C}\right)\Phi_1 + e_3\theta_1 +$$
$$+ \left(e_1\bar{C}_1 + \frac{\eta - \chi}{6}\bar{C}_2\right)\Phi_2 + e_4\theta_2$$
$$+ (\eta^2 + 3\chi\eta - 4\chi^2)\frac{\theta_1^2}{40}$$
$$+ (2\eta^2 + \chi\eta - 3\chi^2)\frac{\theta_1\theta_2}{60}$$
$$+ (3\eta^2 - \chi\eta - 2\chi^2)\frac{\theta_2^2}{120}, \tag{5}$$

$$r_2 = -e_3\Phi_1 + \left(\frac{\eta - \chi}{3}\bar{D}_2 - \chi\bar{D}_{12} - e_1\bar{D}_1\right)\theta_1$$
$$+ e_4\Phi_2 + + \left(e_1\bar{D}_1 + \frac{\eta - \chi}{6}\bar{D}_2\right)\theta_2$$
$$- \frac{qhC\sqrt{\mu}}{120}(-8\chi^4 + 7\eta^2\chi^2$$
$$+ 7\eta\chi^3 - 3\eta^3\chi - 3\eta^4)$$
$$- \frac{s_i}{6}(\eta^2 + \eta\chi - 2\chi^2)$$
$$+ (4\chi^2 - 3\chi\eta - \eta^2)\frac{\theta_1\Phi_1}{20}$$
$$+ (3\chi^2 - \chi\eta - 2\eta^2)\frac{\theta_2\Phi_1}{60}$$
$$+ (3\chi^2 - \chi\eta - 2\eta^2)\frac{\theta_1\Phi_2}{60}$$
$$+ (2\chi^2 + \chi\eta - 3\eta^2)\frac{\theta_2\Phi_2}{60}, \tag{6}$$

$$r_3 = \left(e_2\bar{C}_1 + \frac{\eta - \chi}{6}\bar{C}_2\right)\Phi_1 + e_4\theta_1$$
$$+ \left(\frac{\eta - \chi}{3}\bar{C}_2 + \eta\bar{C}_{12}\right.$$
$$\left. - e_2\bar{C}_1\right)\Phi_2 - e_3\theta_2$$
$$+ (4\eta^2 - 3\chi\eta - \chi^2)\frac{\theta_2^2}{40}$$
$$+ (3\eta^2 - \chi\eta - 2\chi^2)\frac{\theta_1\theta_2}{60}$$
$$+ (2\eta^2 + \chi\eta - 3\chi^2)\frac{\theta_1^2}{120}, \tag{7}$$

$$r_4 = e_4\Phi_1 + \left(e_2\bar{D}_1 + \frac{\eta - \chi}{6}\bar{D}_2\right)\theta_1 + e_3\Phi_2$$
$$+ \left(\frac{\eta - \chi}{3}\bar{D}_2 + \eta\bar{D}_{12} - e_2\bar{D}_1\right)\theta_2$$
$$- \frac{qhC\sqrt{\mu}}{120}(-7\chi^4$$
$$+ 23\eta^2\chi^2 - 7\eta\chi^3 - 7\eta^3\chi - 12\eta^4)$$
$$- \frac{s_i}{6}(2\eta^2 - \eta\chi - \chi^2)$$
$$+ (\chi^2 + 3\chi\eta - 4\eta^2)\frac{\theta_2\Phi_2}{20}$$
$$+ (2\chi^2 + \chi\eta - 3\eta^2)\frac{\theta_1\Phi_2}{60}$$
$$+ (3\chi^2 - \chi\eta - 2\eta^2)\frac{\theta_1\Phi_1}{60}$$
$$+ (2\chi^2 + \chi\eta - 3\eta^2)\frac{\theta_2\Phi_1}{60}, \tag{8}$$

$$e_1 = \frac{\eta^2 + \eta\chi + 4\chi^2}{6(\chi - \eta)},$$
$$e_2 = \frac{4\eta^2 + \eta\chi + \chi^2}{6(\chi - \eta)},$$
$$e_3 = \frac{K}{12}(\eta^2 + 2\eta\chi - 3\chi^2),$$
$$e_4 = \frac{K}{12}(\eta^2 - \chi^2), \tag{9}$$

$$X_{k,j}^i = \{\Phi_1, \theta_1, \Phi_2, \theta_2, q\}^T, \tag{10}$$

All notations in Formulas (5)–(10), assembly algorithm for the generalized vector $F(X_k^i)$, have been described elsewhere (Stupishin and Nikitin 2014a, 2014b).

Designed calculation algorithm determines the stress-strain state of the shell at each step of the method of continuation on the parameter. In the process of calculating, the deformation analysis changes and fixes their extreme values. Based on the above calculation, it determines the values of the upper and lower critical loads corresponding to the upper and lower extreme points on the graph of the deformation of the load (i.e. the upper and lower levels at the turning point).

3 COMPARISON OF THE RESULTS

The results of the calculations were compared for the values of the upper and lower critical loads (Valishvili 1976). The difference in the values is found to be not more than 1.2%.

To establish the convergence of the algorithm, test calculations were made for different finite element mesh densities in the radial direction.

The results showed convergence forces, moments and displacements with increasing density of the grid. Acceptable practical calculation results were obtained by partitioning the shell at least 20 elements in the radial direction.

4 INVESTIGATION OF STABILITY

The developed methodology was used to study the influence on the values of the upper and lower critical loads of the following parameters:

– parameter that specifies the shape of the shell (parameter Z);
– parameters of the shell's elastic fastening support contour (parameters n and m);
– the ratio of the elastic modulus of the shell material in the radial and circumferential direction (parameter Θ).

Shape generatrix of a shell of revolution is defined by the function parameter Z:

$$f(\overline{\rho}) = f_{max}\rho^Z, \tag{11}$$

where f_{max} is the rise in the center of the shell and ρ is the dimensionless coordinate, which changes from 0 in the center of the shell to the value 1 on the support contour.

Forms of generatrix of the shell for different values of Z are shown in Figure 1.

The ratio of the elastic modulus in the radial direction (E_1) of the elastic modulus in the circumferential direction (E_2) can be calculated as follows:

$$\Theta = E_1/E_2. \tag{12}$$

It is assumed that the shell is based on the elastic support contour, which specifies the parameters of the flexibility n and m. Parameter n describes the flexibility of the support contour in the radial direction, and may range from -0.3 (if displacement in the radial direction is restricted) to infinity (if displacement in the radial direction is not restricted). The parameter m describes the flexibility of the support contour in its torsion and may

range from $+0.3$ (if the reference contour is a hinge and rotation is not restricted) to infinity (if rotation of the support contour is restricted).

The investigations revealed that the parameters varied the shell in the following ranges: Z = from 1 to 2.5 (see Figure 1); n = from -0.3 to 100; m = from 0.3 to 100. We considered the shallow shell of revolution with base radius a = 1 m and height f_{max} = 1 m; thickness h = 4 mm, Poisson's ratio of the shell material v1 = v2 = 0.3. The shell has a uniformly distributed load of intensity p.

Some results are shown in Figures 2, 4, 6 and 8, which present the graphs of the dimensionless deflection depending on the values of the load on the shell and the values of variable parameters. Figures 3, 5, 7 and 9 show the graphs of the upper and lower critical loads, depending on the values of the variable parameters of the shell.

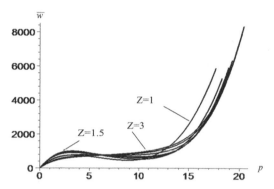

Figure 2. The graph change dimensionless displacements in the center of the shell for the values of the load p for different values of Z (for the shell with $E_1 = E_2 = 2.1 \cdot 10^{11}$ Pa and fixed support contour).

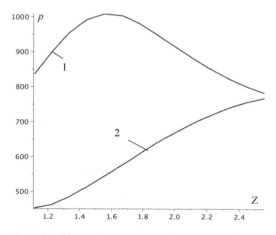

Figure 3. The graphs change the values of critical loads p, depending on the value of Z (1–upper critical load 2–lower critical load).

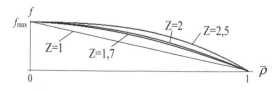

Figure 1. Forms of generatrix of the shell for different values of Z.

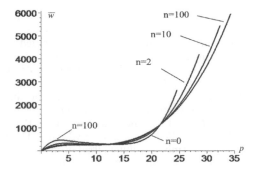

Figure 4. The graphs change dimensionless displacements in the center of the shell for the values of the load p for different values of n (for the shell with $E_1 = E_2 = 2.1 \cdot 10^{11}$ Pa; Z = 2 (spherical shell); m = 100).

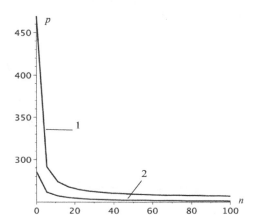

Figure 5. The graphs change the values of critical loads, depending on the value of n (1–upper critical load 2–lower critical load).

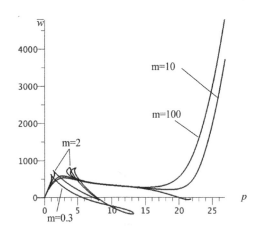

Figure 6. The graphs change dimensionless displacements in the center of the shell for the values of the load p for different values of m (for the shell with $E_1 = E_2 = 2.1 \cdot 10^{11}$ Pa; Z = 2 (spherical shell); n = –0.3).

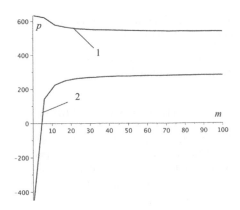

Figure 7. The graphs change the values of critical loads, depending on the value of m (1–upper critical load 2–lower critical load).

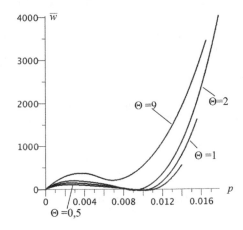

Figure 8. The graph change dimensionless displacements in the center of the shell for the values of the load p for different values of Θ (for the shell with Z = 2 (spherical shell) and hinge-fixed support contour).

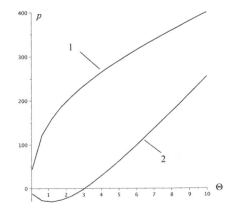

Figure 9. The graphs change the values of critical loads, depending on the value of Θ (1–upper critical load 2–lower critical load).

We change the Z observed extreme values of the upper critical load, which is located near $Z = 1.6$ (see Figure 3). We can assume that the shell of such form has the highest safety factor than other forms of the shell under the influence of uniformly distributed loads. At the same time, the extremum for the lower critical load is absent. By increasing the shape parameter Z, we observe only an increase in the values of the lower critical load (Figure 3). The largest gap between the values of the upper and lower critical loads occurs precisely at the point of the extremum upper critical load.

Changing the parameters n and m shows the growth of critical load with an increase in the stiffness of the support contour in the radial direction and increase flexibility with respect to the rotation (Figures 4–6). The highest values of critical loads correspond to the hinge-fixed support contour.

Increasing the parameter Θ results in an increase of the upper critical load (Figure 7). At the same time, with $\Theta = 1$, there is an extreme value of the lower critical load.

5 CONCLUSIONS

The proposed methodology based on the mixed finite element method shows good accuracy and convergence of the results. It is more effective than finite elements in other formulations.

Applying the methodology of investigating, the shell's stability allows us to find more optimal combination of size and shape of the shell and its parameters of the support contour.

REFERENCES

Andreev, V.I. and Dubrovskiy, I.A. 2013. Stress state of the hemispherical shell at front movement radiating field. *Applied Mechanics and Materials*, 405–408: 1073–1076.

Grigolyuk, E.I. and Shalashilin, V.I. 1988. *Problems of nonlinear deformation*. Moscow: Nauka Publ.

Stupishin, L.U. and Kolesnikov, A.G. 2014. Geometric Nonlinear Orthotropic Shallow Shells Investigation. *Applied Mechanics and Materials*, 501–504: 766–769.

Stupishin, L.U. and Nikitin, K.E. 2014. Mixed finite element for geometrically nonlinear orthotropic shallow shells of revolution. *Applied Mechanics and Materials*, 919–921: 1299–1302.

Stupishin, L.U. and Nikitin, K.E. 2014. Mixed finite element of geometrically nonlinear shallow shells of revolution. *Applied Mechanics and Materials*, 501–504: 514–517.

Valishvili N.V. 1976. *Methods of analyzing shells of revolution on electronic digital computers*. Moscow: Mashinostroenie Publ.

Advances in Civil, Architectural, Structural and Constructional Engineering – Kim, Jung & Seo (Eds)
© 2016 Taylor & Francis Group, London, ISBN 978-1-138-02849-4

Mode II stress intensity factor determination for cracked timber beams

L. Stupishin, V. Kabanov & A. Masalov
South-West State University (SWSU), Kursk, Russia

ABSTRACT: The mode II (forward shear) fracture of wood beams is studied. The Stress-Intensity Factor (SIF) is determined by the section method. The dependencies of mode II SIF from the geometry of the wood beam loaded by the concentrated load are presented. These results demonstrate a good agreement with the results of Barret-Foschi and Murphy, obtained by using the finite element method and the boundary value collocation method. The parameters on crack growth steps and growth of crack surface displacement and critical stress intensity factor of timber member adhesive joints on resorcinol glues are presented in this paper.

1 INTRODUCTION

For development of design methods for timber structures with flaws, we need a simple method for SIF calculation and normalized experimental data of SIF values (Stupishin et al. 2014).

Until now, the theoretical values of SIFs have been determined by using the finite element method and the boundary value collocation method (Barret & Foschi 1977, Murphy 1979). The solution based on the section method is known for mode I SIF (cleavage) (Morozov 1969). In the first part of our paper, the method for determination of mode II SIF is presented, which is also based on the section method. In the second part of the paper, some results of timber beam tests are presented.

2 RESULTS AND DISCUSSION OF SIF VALUE CALCULATIONS

Let us consider an end cracked cantilever beam loaded by force on the beam end (Figure 1a). Its cross-section is rectangular. This problem may be presented as a superposition of two others: in the first one, a solid beam with a span $L-c$ is loaded by two bending moments $M_c/2$, where $M_c = F \cdot c$ is the bending moment at the crack tip cross section (Figure 1b); in the second one, the same beam is loaded by two forces F/2 (Figure 1c).

Problem 1 (Figure 1b). Let us cut the lower part of the beam and substitute its effect by some tangential stresses τ, distributed by some law. Normal stress σ_M is distributed accordingly to the equilibrium condition in the support cross-section. Its maximum value is $\sigma_M = M_c/W_c$, where W_z is the moment of resistance in the cross-section. Since the sum of all forces projected upon the X-axis is

equal to zero, the equivalent force T_M, from normal stresses σ_M, is equal to the equivalent force T_Q, from tangential stresses τ_M. Let us hypothetically assume that the distribution of stresses τ along the beam or the X-axis is

$$\tau = \frac{KIIM}{\sqrt{\pi}}\left(\frac{1}{\sqrt{x}} - \frac{1}{\sqrt{h}}\right) \qquad (1)$$

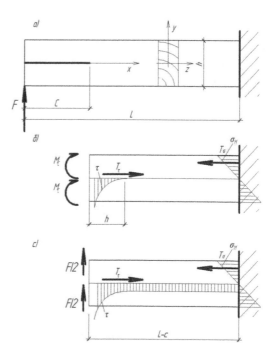

Figure 1. Design schemes of a cracked cantilever wood beam.

where K_{IIM} is the SIF of the bending moment Mc effect at the crack tip and x is the horizontal coordinate that originates at the crack tip.

Let us also assume that tangential stresses τ can have values of the same sign.

Figure 2 shows the distribution of stresses τ given in Equation (1). The value of $x = h$, where $\tau = 0$, is approximately chosen according to the Sent-Venant principle. From Equation (1), the resulting force T_τ of τ is given by

$$T_\tau = \int_0^h \tau dx = K_{IIM}\sqrt{\frac{h}{\pi}} \tag{2}$$

The resulting force T_σ of σ in the upper beam part is given by

$$T_\sigma = \frac{3}{2}\frac{M_c}{h} \tag{3}$$

From the equilibrium condition $\Sigma X = T_\tau\text{-}T_\sigma = 0$, and Equations (2) and (3), it follows that

$$K_{IIM} = \frac{3}{2}\frac{M_c}{h}\sqrt{\frac{h}{\pi}} \tag{4}$$

It may be written as

$$K_{IIM} = \sqrt{\pi c}Y_M \tag{5}$$

where

$$\tau = \frac{3}{2}\frac{F}{h} \tag{6}$$

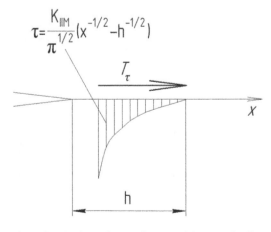

Figure 2. Design scheme of tangential stress distribution at the crack tip.

is the nominal tangential stress in the neutral plane of the beam, calculated by Zhuravsky's formula for unity of width, where

$$Y_M = \sqrt{\frac{c}{\beta h}} \tag{7}$$

is the parameter, depending on the dimensions of the beam (geometry factor); β is the parameter whose exact value needs to be determined (according to the Sent-Venant principle β may be taken equal to unity).

Problem 2 (Figure 1c). Let us cut the lower part of the beam and substitute its effect by tangential stresses τ. Normal stress σ_M is distributed according to the equilibrium condition at the support cross-section. Its maximum value is

$$\sigma_M = \frac{F(L-C)}{W_z} \tag{8}$$

Let us hypothetically assume that the distribution of τ along the beam (Parton & Morozov 1985) is

$$\tau = \tau_K \frac{x}{\sqrt{x^2 - c^2}} \tag{9}$$

$$\tau_K \frac{K_{IIQ}}{\sqrt{\pi c}} \tag{10}$$

where K_{IIQ} is the SIF of the lateral force effect.

From Equation (9), the resulting force of the tangential stresses T_τ is given by

$$T_\tau = \int_0^h \tau dx = 2K_{IIM}\sqrt{\frac{L^2 - c^2}{\pi c}} \tag{11}$$

From the equilibrium condition $\Sigma X = T_\tau\text{-}T_\sigma = 0$ and Equation (11), it follows that

$$K_{IIQ} = \frac{3}{4}\frac{F(L-C)}{h}\sqrt{\frac{\pi c}{L^2 - c^2}} \tag{12}$$

or

$$K_{IIQ} = \tau\sqrt{\pi c}Y_Q \tag{13}$$

where

$$Y_Q = \frac{L-c}{2\sqrt{L^2 - c^2}} \tag{14}$$

is the parameter that depends on the dimensions of the beam (geometry factor).

Thus, K_{II} may be determined as

$$K_{II} = K_{IIM} K_{IIQ} = \tau \sqrt{\pi c} \left(Y_M + Y_Q \right) \tag{15}$$

where

$$Y_M + Y_Q = \sqrt{\frac{c}{\beta h} + \frac{L - c}{2\sqrt{L^2 - c^2}}} \tag{16}$$

Let us graphically compare the solution of this problem by using the finite element method (Barret & Foschi 1977), the boundary value collocation method (Murphy 1979) and the section method. Let us consider a cantilever beam with the length of $L = 8$ m and the beam depth of $h = 0.5, 1.0, 1.5, 2.0$ m.

Figure 3a, b, c, d show the results of the comparison.

On the abscissa axis, the ratios of the crack length to beam span, c/L, are shown; on the ordinate axis, the values of the geometry factors are shown. The results for the beam with 0.5 m, 1 m, 1.5 m and 2 m depth are shown in Figure 3a, b, c, and d respectively. Comparative diagrams of the geometry factor are derived by using the section method (MC), the finite element method (BF) and the boundary value collocation method (Mer).

Let us calculate an admissible value of end through crack length cl for the design scheme of timber beam, as shown in Figure 1. Let us assume the following short-term characteristics of glued laminated timber: bending shear strength $R_s = 4,5$ MPa, bending tensile strength R = 45 MPa, critical

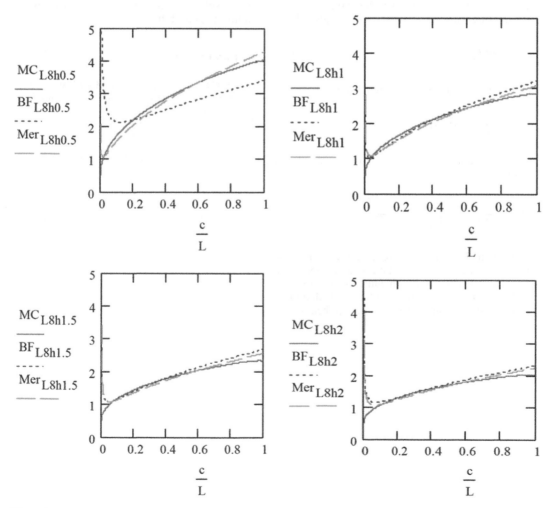

Figure 3. Comparative diagrams of the geometry factor, derived by using the section method (MC), finite element method (BF) (Barret &Foschi 1977) and boundary collocation method (Mer) (Murphy 1979).

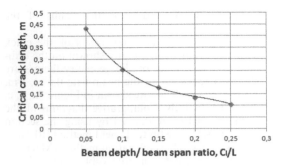

Figure 4. The dependence of critical crack length on the beam depth-to-beam span ratio.

SIF $K_{IIc} = 1,4$ MPa·m$^{1/2}$, beam span $L = 3$ m, beam depth $h = 0,6$ m and beam width $b = 0,2$ m.

Fracture load when bending shear is given by

$$F = \frac{2}{3} bh R_s = 270 \, kN \qquad (17)$$

Fracture load when bending tension is given by

$$F = \frac{Rbh^2}{6L} = 135 \, kH \qquad (18)$$

From Equations (15) and (16), we can derive Equation (19):

$$K_{IIC} = \frac{3}{2} \frac{F}{bh} \sqrt{\pi c_1} \left(\sqrt{\frac{c_1}{h}} + \frac{L - c_1}{2\sqrt{L^2 - c_1^2}} \right) \qquad (19)$$

By using Equation (19), we can estimate the critical crack length. From the data obtained above, we find the critical crack length equal to $c_l = 0.136$ m.

Figure 4 shows the results of the calculation of critical crack length for the cantilever beam with the same parameters as mentioned above, except the beam depth h varying from 0.15 m to 0.75 m.

3 CONCLUSIONS

All the three solutions show good agreement with each other except on the area approximately up to $c/L = 0.1$. But as Figure 4 shows, those crack lengths in most cases were lower than their critical lengths. It is important to note that the Barret & Foschi (1977) and Murphy (1979) analyzed the geometry factor for the end-cracked beams with the loading scheme of three—and four-point bending. Thus, it may be concluded that the derived Equation (16) can be applied to design the structures of this type with a similar load scheme.

Future research will take into account the material anisotropy to determine the boundaries of the solution applicability.

REFERENCES

Barret, J.D. & Foschi, R.O. 1977. Mode II stress-intensity factors for cracked wood beams. *Engineering Fracture Mechanics.* 9: 371–378.
Morozov, E.M. 1969. Section method in theory of cracks. *Proceeding of Universities. Construction.* 3: 22–25.
Murphy, J.F. 1979. Strength of wood beams with end splits. *Forest Products Laboratory Research Paper FPL, 347.*
Parton, V.V. & Morozov, E.M. 1985. *Mechanics of elastic-plastic fracture.* Moscow. Science.
Stupishin, L. Kabanov, V. & Masalov, A. 2014. Fracture resistance of bended glued timber elements with flaws. *Advanced Materials Research.* 988: 363–366.

Advances in Civil, Architectural, Structural and Constructional Engineering – Kim, Jung & Seo (Eds)
© 2016 Taylor & Francis Group, London, ISBN 978-1-138-02849-4

Modular eco-wall solution for african rehousing programs

M.P. Amado
CITAD—Centro de Investigação em Território, Arquitectura e Design, Lisboa, Portugal
FCT UNL—Faculdade de Ciências e Tecnologia da Universidade NOVA de Lisboa, Libsoa, Portugal

I. Ramalhete
GEOTPU—Gabinete de Estudos de Ordenamento do Território e Planeamento Urbano, Caparica, Portugal
CIAUD—Centro de Investigação em Arquitectura, Urbanismo e Design, Lisboa, Portugal

ABSTRACT: This paper discusses the research on eco-modular wall solution for rehousing programs in Africa where pre-fabrication can be possible if adapted to the social, economic and environmental reality. The solution for using partial concrete panels in association with local material like earth, which provides a reduction in embodied energy, is available. Also, the modular solution developed for housing programs can offer an adaptive solution to the family dynamic typology and different uses in parts of the house. The concept of a thin optimal panel made up of reinforced concrete, and parameters as dimensions, materials and constructive processes are a good solution and available to be adopted to the reality of Cape Verde, Angola, Sao Tomé Principe, Guinea-Bissau and Mozambique.

1 INTRODUCTION

African developing countries are currently experiencing a rapid population growth due to fast economic development. This context promotes the necessity of a huge number of houses, where the absence of response results in the formation of urban slums due to massive migrations from rural areas. These situations result in precarious housing conditions and public health problems (Amado et al., 2014). Owing to these problems, governments are compelling to apply imported housing models, non-adequate to environmental, social and economic contexts. The absence of an integrated method to build material selection and consequent construction process compromises housing affordability due to building processes that require specialized labor force but also due to the application of imported materials, more costly because of transportation costs and non-adequate to local climatic conditions (Habib et al. 2009). This has economic impacts on housing stakeholders and also on users, besides environmental consequences (Mao et al. 2013).

Considering the self-construction potential of slum inhabitants, which is already institutionalized in several countries (Greene e Rojas, 2008, Ntema, 2011), the implementation of prefabricated modular solutions combined with local materials and local building techniques show an opportunity to address the housing problem, besides economic and environmental advantages. Economically, prefabricated solutions are more viable and therefore affordable by low-income populations due

to their cheaper production with a reduction of 30% compared with common solutions (Stallen & Chabannes, 1994). The usage of standard components also helps the assembly process in the context of assisted self-construction. For manufacturers, prefabrication results in 52% less waste materials and 50% (Stallen & Chabannes, 1994) less energy and water consumption due to the rationalization process of resources management and consumption, leading to cost reduction and also environmental implications (Tam et al. 2007).

Besides the environmental and economic aspects, application of local materials has impacts on social matters, which refers to population acceptance by a tectonic that they are familiar with in a rehousing and relocation scenario.

Therefore, this research focuses on a modular eco-wall solution for housing that combines the application of a prefabricated structural module complemented with local materials, considering social, economic and environmental parameters.

This paper presents the material selection process that takes into account social, economic and environmental aspects.

2 METHODOLOGY

The research methodology raises the development of an eco-wall solution for low-cost housing to the reality of Cape Verde, Angola, Guinea-Bissau, São Tomé and Principe and Mozambique, which is adaptable to each different social, economic and environmental context (Amado et al. 2013). The methodology refers to an interactive process with three stages, in which the first establishes the main parameters for adaptability and the others corresponds to the consequent conception process (prefabricated structural module) and local adjustment (through complemented local materials) in order to get an optimal modular-wall solution for housing.

The first step considers the main bases for module adaptability: housing stock analysis; legal requirements; and transportation conditions.

Housing stock analysis integrates the identification of prevailing layouts, construction methods and applied materials. This leads to the identification of needs and opportunities for an optimal solution according to the geographical context. This also aims at the identification of currently applied materials as well as common and traditional building processes that will enable the best options corresponding to the local adjustment phase.

Legal requirements and transportation costs define the minimum dimensions to consider in modular solution based on costs and environmental impacts (energy consumption and resource management).

The second step corresponds to module optimal dimensioning through a parametric framework based on the research made in the first stage: legal requirements establish the minimum dimension for housing (area and ceiling height) that is related to module composition (quantity and joints). These parameters are combined with transportation conditions in order to transport maximum modules in a sea container. Those parameters are selected from a comparative score action in order to achieve the optimal solution in terms of the module height and length. Thickness is reached through mechanical tests to compression and traction made with models.

The third step refers to local adjustment and adequacy based on the housing stock analysis previously made. It discusses the application of local materials (external and internal) to complement the structural module solution in order to achieve the optimal wall solution for housing according to the local place.

The four steps are the material selection of the wall element that compound the eco-wall solution adapt to local reality and that combine social, economic and environmental criteria for a durable and sustainable solution. The solution is also able to apply the solution of affordable housing and self-construction solutions.

3 RESULTS AND DISCUSSION

Material selection for eco-wall solution considers a set of principles that combine social, economic and environmental criteria and local reality. These

Figure 1. First sample with common concrete C25/30.

criteria are interrelated, and an analysis between the entire positive and constraints of each decision is considered.

The eco-wall is composed of a prefabricated structural module that is complemented *in situ* with local materials, using traditional techniques, in an assisted self-building scenario.

It is important to consider, at first moment, the structural function of the module in order to ensure structural safety and durability requirements and therefore less maintenance costs and more quality. Thus, the structural model corresponds to a concrete panel with 7 cm thickness. The first sample considered concrete type commonly used in construction in order to be easily available and affordable: C25/30; structure class A500, type Nq50; steel profiles class S280GD+Z, Ω shape. The environmental impacts of concrete, namely its embodied energy and carbon emissions due to cement percentage, are pondered because of its durability and safety aspect to the house solution.

Durability has important consequences on cost reduction but also important impacts on the environment due to less maintenance and substitution of modules along time. Durability is also related to quality, which is an important principle for low-cost housing in social terms (acceptance and social inclusion).

The referenced values considered for embodied energy and carbon emissions are those from ICE Inventory of Carbon & Energy 2011 from University of Bath project for Carbon Vision Buildings.

Concerning concrete values, ICE considered a Cradle-to-Gate approach (energy consumption since extraction to factory gate before its transport to consumers). ICE shows an embodied energy for concrete C25/30 MPa of 0.78 MJ/kg and carbon emissions of 0.106 kg CO_2/Kg (Hammond & Jones, 2011).

In order to reduce its embodied energy and carbon emissions and thus environmental impacts, several module samples were developed and tested

Figure 2. Module sample type A 1:3:1 (cement, soil, gravel).

by adding soil to concrete mixture in order to reduce cement percentage but maintaining its performance.

The structural concrete module is prefabricated and then transported to the building site, where the population will be able to proceed to its assemblage through an assisted self-building process. The complemented materials are applied *in situ*, corresponding to the local adjustment phase. These completed materials refer to external and internal cover materials whose selection takes into account social, economic and environmental criteria.

The usage of local materials considers the contribution that its production has to local economy, but also the reduction of environmental impacts due to less harmful emissions associated with transportation and manufacturing. Social aspects were also observed, i.e. the population identifies itself with a familiar architectural imagery, reinforcing collective identity, social cohesion and acceptance.

The housing stock analysis previously made in the first phase identifies several available and adequate materials that ensure wall performance and sustainability. Thus, the analysis considers the application of compressed earth block (CEB) as external covering and wallpaper as internal finish.

CEB are commonly used in the African case studies previously mentioned, and meet economic viability, low environmental impacts in terms of energy consumption and resource management, thermal performance and cultural expression. Its natural and ecological proprieties, besides its raw material being inexhaustible, presents low embodied energy and a good thermal performance in climatic countries where cooling is essential for thermal comfort. This aspect is also important for users, not only in terms of comfort and quality of housing environment, but also in terms of costs: energy-efficient housing represents less energy

costs and consumption for cooling and ventilation that reverts directly to household budget.

According to ICE data, in a Cradle-to-Site approach (energy consumption since extraction to building site), CEB show an average embodied energy of 0.45 MJ/kg, where its lowest value show only 0.15 MJ/Kg. In terms of carbon emissions, the average values are 0.023 kg CO_2/kg.

However, CEB performance depends on soil type and characteristics, which leads to the addiction of cement or other stabilizers in some situation. Thus, ICE data show an average embodied energy of 0.68 MJ/Kg if 5% of cement is added, where the lowest value observed was 0.48 MJ/kg. Carbon emissions values show an average of 0.06 kg CO_2/kg, being much higher compared with non-cement CEB. As observed in concrete analysis, the durability factor is the main principle for sustainability. Even with higher values of embodied energy and carbon emissions, if the soil is not able to ensure durability and quality, improvement should be made in a cost benefit consideration.

Aiming thermal improvement of the wall solution, the addition of a straw or canes layer between the CEB and the structural module is considered. These two raw materials are commonly used in traditional housing in the four studied territories and, besides being applied at its natural state (neither embodied energy nor carbon emissions), are also inexhaustible.

As internal finish, the research focused on a low-cost ecological material that simultaneously does not compromise housing affordability, presents low environmental impacts and should be appellative to occupants. Therefore, the application of recycled wallpaper composed of agricultural wastes is proposed, which is also found in the four case studies that are originated locally, and also improves the wallpaper blending aggregation and its mechanical performance. Nevertheless, it is recognized that paper production sustainability should be attached to a strategy for production forests in order to be an inexhaustible resource.

The sustainability of wallpaper should not only consider the blending itself but also its application into inner walls, where glue, pastes or other kinds of chemicals can compromise the ecological purpose of the material. These chemicals are also usually present in pigments and paper binders that compromise its full recycling capacity. Thus, the research considers the application of a natural-based glue made with wood pulp and water (Methyl Cellulose) in order to provide a full recycling process and keep ecological principles (Bogati, 2011). This product is a natural powder that dissolves in cold water (between 40°C and 50°C) and has adhesive properties, i.e. energy consumption for its production is low, reverting in fewer costs and

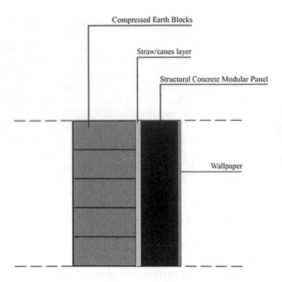

Figure 3. Section of the eco-wall solution.

emissions. It also can be used as an emulsifier for paper blending, improving its water absorption resistance (Rabi et al. 2009).

Thus, ICE data are not considered for wallpaper. The reference values refer to common wallpaper with chemical glue, showing high values. Data consider neither the recycling properties nor the production forest factor, aspects that are envisage in present research.

Therefore, eco-wall solution is formed by a structural concrete panel, a straw/canes layer, CEB masonry as external cover and recycled wallpaper as internal cover.

4 CONCLUSIONS

Developing countries in Africa are facing a fast development that is leading to massive migrations to urban areas, and consequently slum formation and expansion are emerging. This requires an urgent response by the housing sector that should promote a role of social, economic and environmental integration for a sustainable future.

The research focus on the concept of an eco-wall solution for Cape Verde, Angola, Mozambique, São Tome and Principe and Guinea-Bissau. The methodology is formed by an adequacy approach to these territories that combines social, economic and environmental local criteria. This paper also focuses on the material selection that is considered

in the sustainability of the wall composition in terms of costs for users and for promoters, social acceptance by the population and environmental concerns, namely embodied energy and carbon emissions.

The research of the eco-wall solution combines a prefabricated module in concrete that ensures durability, quality and structural safety, with local materials (CEB and wallpaper). The local adjustment through natural and locally produced materials shows almost no transportation efforts as well as low embodied energy and carbon emissions, leading to economic and environmental benefits. The approach towards the social impacts focuses on the use of local materials familiar to the population for a better acceptance process easier to be applied in the rehousing programs.

REFERENCES

Amado, M.P. Pinho, F. Faria, P. & Ramalhete, I. 2014. *Eco-wall modular solutions for buildings.* In 9th International Masonry Conference Proc. Guimarães, 7–9.

Amado, M.P. Lopes, T. & Ramalhete, I. 2014. *Eco-Wall: Modular Solution for Low-Cost Houses.* In CISBAT 2013 Proc. Lausanne, 4–6.

Bogati, D. 2011. *Cellulose Based Biochemicals.* Saimaa: University of Applied Sciences.

Greene, M. & Rojas, E. 2008. Incremental construction: A strategy to facilitate access to housing. *In Environment and Urbanization,* 20(1): 89–108.

Habib, R.R. Mahfoud, Z. Fawaz, M. Basma S.H. & Yeretzian J.S. 2009. Housing quality and ill health in a dis-advantaged urban community. *In Public Health,* 123(2): 174–181.

Hammond, G. & Jones, C. 2011. *Inventory of Carbon & Energy (ICE) Version 2.0.* Bath: University of Bath.

Mao, C. Shen, Q. Shen, L. & Tang, L. 2013. Comparative study of greenhouse gas emissions between off-site prefabrication and conventional construction methods: Two case studies of residential projects. *In Energy and Buildings,* 66: 165–176.

Ntema, L.J. 2011. *Self-help housing policy in South Africa: Paradigms, policy and practice.* Bloemfontein: University of Free State.

Rabi, J. Santos, S. Tonoli, G. & Savastano, H. 2009. *Agricultural Wastes as Building Materials: Properties,* Performance and Applications. New York: Nova Science Publishers Inc.

Stallen, M. & Chabannes, Y. 1994. Potentials of prefabrication for self-help and mutual-aid housing in developing countries. *In Habitat International,* 18(2): 13–39.

Tam, V.W. & Tam, C.M. & Zeng, S.X. & Ng, W.C.Y. 2007. Towards adoption of prefabrication in construction. *In Buildings and Environment,* 42(10): 3642–3654.

Advances in Civil, Architectural, Structural and Constructional Engineering – Kim, Jung & Seo (Eds)
© 2016 Taylor & Francis Group, London, ISBN 978-1-138-02849-4

Effect of patching thickness on the flexural performance of patched reinforced concrete beams

S.A. Kristiawan, A. Supriyadi & M.K. Muktamirin
SMARTCRete Research Group, Department of Civil Engineering, Sebelas Maret University, Indonesia

ABSTRACT: The aim of this study is to investigate the effect of Unsaturated Polyester Resin (UPR)-mortar as patch repair material on the flexural behaviour of patched reinforced concrete beams. The variable studied is the patching thickness. Based on the observed experimental works, it is shown that an increase in patching thickness tends to reduce cracking density and improve the maximum load capacity but it is not likely to alter the stiffness and ductility factor of reinforced concrete beams.

1 INTRODUCTION

Reinforced concrete is a construction material that has become the primary options in the development and construction of various infrastructures. When it is designed properly, reinforced concrete tends to have a good strength and durability. However, as those of other construction materials such as steel and wood structures reinforced concrete is not risk-free from deterioration. Several factors could raise the deterioration process in reinforced concrete; mostly due to exposure to aggresive agents from the environment. As an example, penetration of chloride ion will trigger corrosion of the reinforcement which eventually damages the concrete cover. The forms of damage include cracking, spalling and delamination of concrete in the corrosion zone. In addition, corrosion also reduces the cross sectional area of reinforcing bars as well as destroying the bonding between reinforcing bars and concrete (Sahamitmongkol et al. 2008).

The occurence of damages in the corrosion zone influences the performance of reinforced concrete in term of capacity and serviceability. In such situation, it is necessary to stop the deterioration progress before its capacity and serviceability reaching the limit state. There are many repair options that could be applied to deal with such problem. Patching technique offers simple and convinient method to repair the damage of concrete due to reinforcement corrosion (JSCE, 2001). This technique includes the following actions: removing the deteriorated concrete in the corrosion zone, cleaning the corroded reinforcement and embedding new reinforcements if necessary, and finally patching the damaged area with suitable repair material. Repair of damaged concrete with this technique is expected to recover the size and appearence, protect the reinforcement from continuos corrosion,

and regain the structural performance of the reinforced concrete (Jumaat et al. 2006).

The structural parameters that may be employed to characterize the flexural performance of reinforced concrete beam are capacity, flexural rigidity (stiffness), deflection, ductility and cracking pattern (Rio et al. 2008, Haddad et al. 2008). The effectiveness of patching material and method to restore the performance of deteriorated reinforced concrete beams may be assessed by comparing the above structural parameters between the patched and non-patched (control) reinforced concrete beams (Supriyadi et al. 2015). This research is aimed to investigate the effect of patching on the flexural performance of reinforced concrete beams. The patch repair materials used in the research are Unsaturated Polyester Resin (UPR) mortars. The variable investigated is the patching thickness.

2 MATERIALS AND METHOD

2.1 Materials

The concrete materials used for fabricating reinforced concrete specimens were proportioned following the method proposed by Building Research Establishment (Teychenne et al. 1997) to attain a target strength of 25 MPa. Meanwhile patch repair materials (UPR mortar) developed in this research were composed of Unsaturated Polyester Resin (UPR), sand, cement and fly ash. UPR used in this investigation was YUKALAC® 157 BQTN-EX, an unsaturated orthophtalic type resin made from the polymerization of di-carboxylic acids with glycols. The thermo-setting of this material is initiated by curing agent (hardener) which in this investigation proportioned at 3% by weight of UPR. Sand used in the development of this UPR mortar was conformed to the requirements for grading and quality of fine aggregate as specified by ASTM C33, with the specific gravity of 2.56 and fineness modulus of 2.4. Cement and fly ash were incorporated in the mixture as filler. Transformation of the mixture into the hardened state was relied upon the polymerization of the UPR which bond the ingredients into a solid material. No water was added to the mixture. Hence, hydration reaction did not take place. Table 1 shows the composition of UPR mortar.

2.2 Mechanical properties

Both concrete and UPR mortar used for fabricating the beam specimens have been tested to determine their mechanical characteristics. Table 2 shows these characteristics. It is clear from this table that UPR mortar has a high strength and a low modular ratio to concrete.

2.3 Specimens

Four types of reinforced concrete beams were cast and tested in this research. The beam dimension and reinforcements lay out for all the specimens were similar but each specimen had different patching depth. The first beam was identified as BN indicating reinforced concrete beam with no patching which would be used as a control beam. The other three beams were identified as BR7, BR9 and BR11 representing reinforced concrete beam repaired with a patching thickness of 7, 9 and 11 cm, respectively. The zone and length of patching were made similar for all these three beams. Figure 1 illustrates the reinforced concrete beams. UPR mortars were applied to patch the reinforced concrete after 90 days after casting of the beam specimens. On the following day after

Table 1. Proportion of UPR mortar.

Cement (kg/m³)	Fly ash (kg/m³)	Sand (kg/m³)	UPR (kg/m³)	Hardener (kg/m³)
808	143	950	475	24

Table 2. Mechanical properties of concrete and UPR mortar.

Materials	Strength (MPa)	Elastic Modulus (MPa)
Concrete*	30.35	24,410.
UPR mortar**	76.88	13,334

*determined at 28 days; **determined at 1 days.

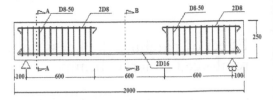

(a) Control reinforced concrete beam

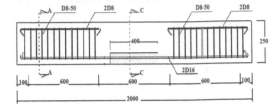

(b) Patched reinforced concrete beam

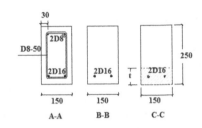

(c) Cross sections of the beams

Figure 1. Illustration of beam specimens.

Figure 2. Loading of the specimen.

being repaired, the patched reinforced concrete beams together with the control beam were tested under flexural loading.

2.4 Flexural testing

All four beams were tested in a four-point bending where the loading span was one-third of the support span. Loading was applied at an increment of 200 kilograms and the corresponding deflection at the mid-span was measured (Figure 2). When the first crack appeared, the corresponding load was noted to calculate the cracking moment. Continues loading after first cracking will trigger new cracks. The evolution and intensity of cracks were noted and mapped continuously until failure of the specimens.

3 RESULTS AND DISCUSSION

3.1 *Cracking behavior*

The configuration of loading creates a maximum and pure bending moment in the area between applied point loads. In turn, it is expected the first crack will take place in this zone. Observation on the Control Beam (BN) confirms this where the first crack occurred at a loading about 2 ton or equals to a bending moment of 0.6 ton-meter. Similar findings were observed on the patched reinforced concrete beams where the first cracks occurred at loading in the range of 1.8–2.2 ton (or bending moment of 0.54–0.66 ton-meter). It was noticed that these cracks took place beyond the repair zone even though they are still within the area of maximum bending moments. It can be deduced that repairing (patching) the reinforced concrete beam with UPR-mortar could increase the cracking moment in the repair zone. Theoretically, a higher tensile stress will be distributed in the repair zone since this zone has a lower modular ratio compared to the zone of the beam without repair. However, first crack does not take place in the repair zone since the strength of UPR mortar is superior in comparison to that of concrete; which in turn, contribute to the increase of cracking moment in the repair zone. Thus, a loading which is equal to a cracking moment in the zone without repair has not reached the corresponding cracking moment in the repair zone of patched reinforced concrete beam.

An increase of loading after first crack will create another cracks in the beams. It is interesting to note that more cracks were observed in the control beam (BN) when compared to those of patched reinforced concrete beams. It is also noteworthy that cracks were evenly distributed in the control beam (BN) while cracks in the patched reinforced concrete beams tend to spread beyond the repair zone. Up to failure loads only a maximum of two cracks were observed in the UPR mortar of patched reinforced concrete beams. These cracks were generated at a higher load as the patching thickness is increased. The corresponding loads that trigger first cracks in the UPR mortar of patched reinforced concrete beams are 6.2, 7.2 and 10.2 ton for BR7, BR9 and BR11, respectively. The final evolution of cracks for all beams is presented in Figure 3. It was also noted that strong bond exists between UPR mortar and concrete and so no early cracks are initiated in this interface.

3.2 *Load-deflection behavior*

The loads and the corresponding deflections of the beams under a four-point bending are given in Figure 4. It is obvious from this figure that at the beginning of loading a linear relationship between load and deflection is obtained for all the beams. There is no significant different in the load-deflection behaviour between the control beam (BN) with those of patched reinforced concrete beams at this stage of loading. Theoretically, the flexural stiffness of beam will decrease after cracking occurred. Hence, it is expected that an increase in loading beyond cracking moment would change the linear relationship of load-deflection i.e. the gradient of the linier relationship tends to lower.

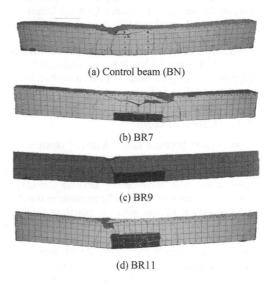

(a) Control beam (BN)

(b) BR7

(c) BR9

(d) BR11

Figure 3.   Final cracking pattern of the beams.

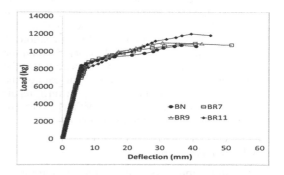

Figure 4.   Load-deflection behaviour of the beams.

However, it was recorded in this investigation that only few cracks were observed in the beams when the loading did not yet cause yielding of the reinforcements. Consequently, there is no significant reduction in the flexural stiffness of the beams at this stage of loading. Furthermore, tension stiffening could also play important role in maintaining the flexural stiffness of the beams. In conclusion, both the presence of few cracks only and tension stiffening could be responsible to the occurrence of linear behaviour of load-deflection at this stage.

At a loading of about 8.2 ton, the linear behaviour of the beams is vanished. The beams start to develop significant deflection at a little increment of load. This behaviour is related to the yielding of reinforcements. When loading has attained the yield load, an increase in loading will initiate elongation of the reinforcement at considerable rate especially in the cracking zone. As a result, the beams show large deflection rate.

There is a slightly different rate of deflection at post-yield behaviour between Control Beam (BN) and patched reinforced concrete beams. Beams patched with UPR mortar are likely to require a higher load to attain similar deflection with that of control beam (BN). It is also indicated that a higher patching thickness causes a higher load to produce similar deflection.

3.3 Flexural capacity

Comparison between failure loads of control beam (BN) and patched reinforced concrete beams suggests that utilization of UPR mortar for patch repair application will enhance the flexural capacity of reinforced concrete beams. The increase in the flexural capacity is in line with the patching thickness. A higher patching thickness yields a greater flexural capacity. This investigation showed an increase in flexural capacity by 2.29%, 2.61% and 13.81% when the patching thickness of 7, 9 and 11 cm were applied, respectively. The increase in the flexural capacity could be related to tension stiffening effect. It is suggested that tension stiffening does not only affect stiffness of reinforced concrete beam but it can also be taken into account in the calculation of flexural capacity (Behfarnia, 2009). Utilization of UPR mortar tends to spread the cracks beyond the repair zone as previously shown. Consequently, the effect of tension stiffening is more significant on the repair zone than on unrepair zone. In turn, repair zone provides more contribution with regard to tension stiffening effect on flexural capacity.

3.4 Ductility factor

Ductility represents the ability of reinforced concrete to show inelastic deformation prior to collapse. To measure ductility, it is generally accepted to use ductility factor (μ) which represents the ratio of rotation (θ), curvature (φ), deflection (Δ), or absorbed energy (E) at ultimate load to the corresponding property when the reinforcement starts yielding (Afefy & Mahmoud, 2014). The inelastic behaviour of the reinforced concrete is dictated by the property of reinforcement; not by the concrete which is a brittle material. However, concrete does influence the inelastic behaviour of reinforced concrete by providing tension stiffening effect.

Inelastic behaviour of reinforced concrete starts at a point of yield stress/load. In this investigation, the yield load of all reinforced concrete beams is about similar at 8.2 t. The post-yield deformations of the reinforced concrete beams, which characterize the inelastic behaviour, feature dissimilar tendency. It is likely that utilization of UPR mortar slightly increases the stiffening of reinforced concrete in the inelastic zone. Additionally, the UPR mortar affects the extent of deflection at ultimate load. To account for the inelastic behaviour of reinforced concrete beams investigated in this research, the ductility factors based on the calculations of ratio of deflections at ultimate and yield loads are quantified. The obtained values are 7.00, 7.05, 6.05 and 7.06 for BN, BR7, BR9 and BR11, respectively. There is no specific influence of patching thickness that can be deduced from these values. A similar value of ductility factor between control and patched reinforced concrete beams could be expected due to similar reinforcement ratio.

4 CONCLUSIONS

This paper presents the results of experimental work to investigate the effect of patching thickness on the flexural performance of patched reinforced concrete beams with UPR mortar. It is found that the superior strength of UPR mortar offers an advantage in term of reducing cracking tendency in the repair zone. The cracks are spread beyond the repair zone and the number or density of cracks is reduced. With the use of UPR mortar, it is discovered that the cracking moment and flexural capacity of patched reinforced concrete beams is greater as the patching thickness increases. The reduction of cracking density (thus the increase in tension stiffening) is also accountable for maintaining the stiffness of patched reinforced concrete to compensate for the reduction caused by lower modular ratio. Meanwhile, the ductility factor of reinforced concrete beam is not likely to be affected when UPR mortar is being applied to patch the reinforced concrete.

ACKNOWLEDGEMENT

This research was made possible due to financial support by Directorate General of Higher Education through Hibah Kompetensi Research Scheme 2015 (Contract No. 339/UN.27.11/PL/2015).

REFERENCES

Afefy, H.M.E & Mahmoud, M.H. 2014. Structural performance of RC slabs provided by pre-cast ECC strips in tension cover zone. *Construction and Building Materials*, 65: 103–113.

Behfarnia, K. 2009. The effect of tension stiffening on the be-haviour of R/C beams. *Asian Journal of Civil Engineering (Building and Housing)*, 10(3): 243–255.

Haddad, R.H. Shannag, M.J. & Al-Hambouth, M.T. Repair of reinforced concrete beams damaged by alkali silica reactions. *ACI Structural Journal*, 105(2): 145–153.

JSCE. 2001. *Standard specifications for concrete structures-maintenance*. Tokyo, Japan Society of Civil Engineers.

Jumaat, M.Z. Kabir, M.H. & Obaydullah, M.A. 2006. A review of the repair of reinforced concrete beams. *Journal of Applied Science Research*, 2(6): 317–326.

Rio, O. Andrade, C. Izquierdo, D. & Alonso, C. 2008. Behaviour of patch-repaired concrete structural elements under increasing static loads to flexural failure. *Journal of Material in Civil Engineering*, 17(2): 168–177.

Sahamitmongkol, R. Suwathanangkul, S. Phoothong, P. & Kato, Y. 2008. Flexural behaviour of corroded RC members—experiments and simulations. *Advanced Concrete Technology*, 6(2): 317–336.

Supriyadi, A. Kristiawan, S.A. & Raditya, S.B. 2015. Experimental investigation on the flexural behaviour of patched reinforced concrete with unsaturated polyester resin mortar. *Applied Mechanics and Materials*, 754–755: 457–462.

Teychenne, D.C. Franklin, R.E. & Entroy, H.C. 1997. *Design of normal mixes concrete*. Second Edition, Building Re-search Establishment Ltd, Garston, Watford, UK.

Advances in Civil, Architectural, Structural and Constructional Engineering – Kim, Jung & Seo (Eds)
© 2016 Taylor & Francis Group, London, ISBN 978-1-138-02849-4

Analysis of reliable project overheads during the construction stage

N.M. Jaya

Department of Civil Engineering, The University of Udayana, Denpasar, Bali, Indonesia

ABSTRACT: Project overheads are commonly used for supporting multiple cost objects; however, they cannot be directly assigned to individual construction activities. Construction project overheads must be identified and distributed to every activity for improving the Cost Management and Controlling Practices (CMCPs). The literature documented 47 items of project overheads and categorized into four hierarchies (Unit, Batch, Project and Facility levels). Questionnaire data were administered among 250 project professionals with 107 responses. Descriptive statistics were used to analyze the identification of construction project overheads. The mean statistic value was subtracted by standard deviation to determine the lowest point value of 256.28. It measured the availability of project overheads during the construction stage. This statistical measure eliminated 8 items of project overheads. Therefore, 39 construction project overheads were considerably reliable having greater values than 256.28. This measurement should enable an improvement to the CMCPs in construction projects.

1 INTRODUCTION

Project overheads have been identified previously and documented through the literature review and field project observations. About 47 items of project overheads are categorized into four (4) hierarchies: Unit level, Batch level, Project sustaining and Facility sustaining overheads (Jaya & Frederika 2015), as presented in Table 1.

Project overheads reported in scientific literature are slightly different from real construction activities in practice. Construction projects implemented various items and categories of overheads in different projects depending on the complexity of particular projects, such as site characteristics, remote positions, availability of required local resources, weather and unforeseen ground conditions.

Every project must investigate and create overheads and maintain relevant costs to the specific project. However, it would be easier when the list of project overheads is available and implemented successfully during similar projects. Therefore, identification of construction project overheads presented in Table 1 was analyzed and carefully selected for future projects, in order to improve sustainable Cost Management and Controlling Practices (CMCPs).

Table 1. Identification and documentation of construction project overheads.

Four categories	47 items of construction project overheads
Unit-level overheads	Equipment depreciations, Direct tool sets, Safeguards
Batch-level overheads	General inspections, Mobilization and setup equipment, Demobilization materials and equipment, Drawing reviews, Change orders, Sample of materials, Material tests, Placing purchase orders, Materials deliveries, Receiving materials, Paying suppliers, Moving materials, Quality inspections, Intermediate project release
Project-sustaining overheads	General planning, Scheduling projects, Planning resources, Planning costs, Engineering costs, Controlling costs, Project reporting, Soft drawing, As built drawing
Facility-sustaining overheads	Site-office & project storage, Site-project administration, Site-project supervision, Site-project labor, First aids, Project insurance, Legal expenses, Rental plant and equipment, Rental land, and base camp for workers, Scaffolding, Hoarding screen, Temporary building, Water supply, Power and lighting, Telephones and communications, Security services, Cleaning services, Transport and haulage, Managing contract conditions, Project's working conditions, Project sundries

2 DATA ANALYSIS AND DISCUSSION

2.1 Data findings

Academic or professional research normally involves two important activities: literature review and field research. A literature review is the requirement to make judgments related to both valuing and organizing the references for creating ideas and findings (Saunders et al. 2007 & 2009). Field research is the essential procedures to collect data and create findings from individuals and institutions that do not have any controls, sanctions and structured limitations (Yin 2009). It would be possible to derive research findings from both data sources, such as the literature and field research. This section presents field data findings and analysis.

A total of 250 questionnaires were delivered to project professionals, and 107 responses were received. Of these, 42.8% response rates were considered as adequate data, and appropriate data analysis for amounts greater than 100 data sets (Fellow & Liu 2008).

Project professionals are associated with their positions in construction projects, which include senior management (e.g. management directors, operational managers, accounting department managers, and management representatives), office management staff (e.g. project managers, procurement, marketing and resource development managers, cost controls, central logistics, quality control and safety managers), and the project management teams (e.g. site managers, engineers, quantity surveyors, site officers, site logistics, architects, and supervisors). An ethical purpose prevents the presentation of participants by the name. However, a detail explanation of the respondents is provided in Table 2.

Project professionals were asked to complete the questionnaire with five Likert scales, which contain a neutral value of 3, and ranked between 1 (strongly disagree) and 5 (strongly agree), related to the occurrences of project overheads during the construction stage. The weighted scores of project overheads were calculated in spreadsheets based on the 107 responses related to the 47 variables of project overheads. The weighted scores of the project overheads are shown in Figure 1.

In order to compare their relative positions, the weighted scores of the project overheads can be classified ranging from the least score of 'paying suppliers' (score = 322) to the highest score of 'mobilization and setup equipment' (score = 446). The weighted scores of other project overheads within this range are also shown in Figure 2.

The top five scores of project overheads are represented by mobilization and setup equipment,

Table 2. The list of respondents.

Management levels	Current jobs/ positions	Number of respondents	
Senior management	Management director	1	1%
	Operational manager	2	2%
	Accounting department manager	2	2%
	Project sponsor	0	0%
	Management representative	1	1%
Office management staff	Project manager	4	4%
	Procurement manager	9	8%
	Marketing/hr development	4	4%
	Cost control manager	2	2%
	Central logistic manager	6	6%
	Quality control/ safety manger	5	5%
Project management teams	Site manager	23	21%
	Engineering manager	8	7%
	Quantity surveyor	11	10%
	Site-office and logistic	4	4%
	Architect/drawing	2	2%
	Supervisor	23	21%
Total responses		**107**	100%

planning cost, general planning, demobilization material and equipment, and controlling costs. On the other hand, the bottom five fall into the categories of paying suppliers, project sundries, hoarding screen, placing order and receiving material.

2.2 Descriptive statistic analyses

Descriptive statistics can be used as the basis of analyzing central tendencies of series data set. Central tendency statistics include mode, median, mean, variance and standard deviation.

The most basic form of statistical analyses may use the range of weighted scores of the data set to calculate a basis of assessment for examining the availability and reliability of project overheads during the construction stage. The Statistical Package for Social Sciences (SPSS) helps analyze the central

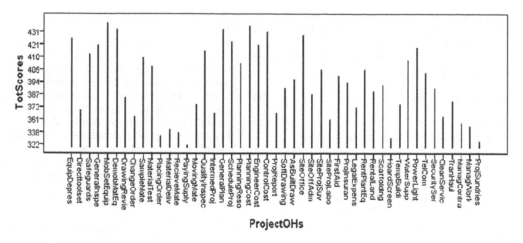

Figure 1. Weight of the 47 items of project overheads.

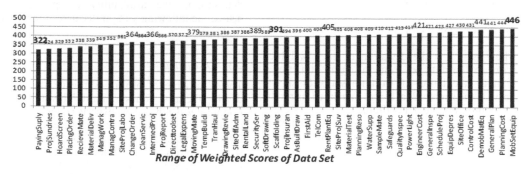

Figure 2. Range of the 47 items of project overheads.

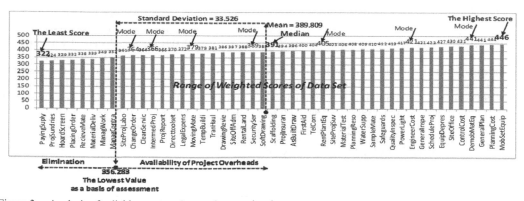

Figure 3. Analysis of reliable construction project overheads.

tendency statistics based on the weighted scores of the data set. However, the section of analysis also provides statistical analyses using the statistic equations, and the results are shown in Figure 3.

Modal statistics are determined by the most frequent scores that occur within a range of data, and are represented by seven different modes of project overheads, such as the lowest score (364) that occurs twice and the highest mode that also occurs twice with a score of 441. The other five modal statistics are shown in Figure 3.

A median statistic represented by the range-ordered score of the middle number of data set falls to the weighted score of 391.

The mean statistic can be calculated by using Equation (1):

$$\text{Mean } (\bar{x}) = (\Sigma x)/n \tag{1}$$

where

$(\bar{x})$	the mean statistic of the data set;
Σx	the sum of weighted scores of the data set; and
n	the number in the data set

The sum value of the data set (18,321) and the number in the data set (47) are calculated and the resulting mean statistic is 389.809. This is a similar value with the mean statistic calculated using the SPSS program (refer to Table 3).

A variance statistic can be analyzed through Equation (2) as follows:

$$\text{Variance} = \Sigma(x-\bar{x})^2/(n-1) \tag{2}$$

where $(x-\bar{x})$ is the deviation of each weighted score in the data set from the mean statistic. A manual calculation of variance statistics should provide the same value. It was calculated using the SPSS program, i.e. 1,124 (Table 3).

The Standard Deviation (SD) can therefore be calculated by the square-root of the variance, which is given in Equation (3):

$$\text{SD} = \sqrt{\left\{ \Sigma(x-\bar{x})^2/(n-1) \right\}} \tag{3}$$

Resulting,

$$\text{SD} = \sqrt{1,124} = 33.526$$

Range statistics analysis provides the facility to range-order the distribution of the data set between the minimum and maximum values of weighted scores, and examines the relationship between every point value of all elements of central tendency statistics. The lowest point value of assessment basis for the availability and reliability of project overheads can be determined by subtracting the mean value with the standard deviation, which is as follows: 389.809–33.526 = 356.28. The project overheads that are positioned at lower values than this point value of assessment basis would be disqualified and eliminated because they are considered to be unreliable overheads in construction projects. There are eight items of project overheads that fall into this category: paying suppliers, project sundries, hoarding screen, placing order, receiving materials, material deliveries, managing work schedules and managing contract conditions (refer to Figure 3). The reason for excluding these eight item overheads is because they no longer directly support construction project activities. It may be more appropriate to include them into general-office rather than site-project expenses.

Having analyzed the descriptive statistics for determining the availability of project overheads, the following section discusses the research findings.

2.3 Discussion

The 47 variables of project overheads were used to develop the survey questionnaire. This survey questionnaire was advised by experts (refer to Table 1) during the consultation in their offices prior to the 250 questionnaires sent to the professionals in construction projects. The choice of 250 construction project professionals was considered to be an appropriate number of potential respondents, which were 'greater than five times' of the 47 question variables, and up to 107 responses were received (refer to Table 2). The response rate of 42.8% is considered to be adequate for analyzing the data set. Fellow and Liu (2008) suggested that at least 100 usable data sets are appropriate for data analysis.

The degree of agreement of the respondents for the availability of project overheads was measured using five values of Likert scales to determine weighted scores of the 47 variables of project overheads. The lowest value 1 represents 'strongly disagree' and the value 2 means 'disagree'. A 'neutral' value is represented by score 3. A higher value (4)

Table 3. Descriptive statistic analysis.

Descriptive statistics

	N statistic	Range statistic	Minimum statistic	Maximum statistic	Sum statistic	Mean Statistic	Std error	Std deviation statistic	Variance statistic	Skewness Statistic	Std error
TotScores	47	124	322	446	183.21	389 81	4 891	33.529	1.124E3	–0.325	0.347
Valid N (listwise)	47										

indicates 'agree' and the highest score (5) indicates 'strongly agree'. The Likert scale does not represent the exact or precise values of personal judgments attached on the variables to measure their absolute scales. However, ordinal numbers (1 to 5) of Likert scales can provide a relative difference of values inherent between the variables of project overheads.

Figure 2 shows the accumulated scores of 107 respondents for each of the 47 variable overheads. Weighted scores of variable overheads range between the lowest total score of 322 for 'paying supplies' and the highest total score of 446 for 'mobilization and setup equipment'. The relative weighted scores and range statistics were used to examine the occurrence of project overheads during the construction stage through descriptive statistics analysis.

Descriptive statistic techniques provide the basic statistics of central tendency and spread statistic measures. The central tendency analysis may include mode statistics, median and mean, while spread statistics include the range statistic, variance and standard deviation. The first three characteristics of the central tendency analysis (e.g. mode, median, and mean statistics) provided three different alternatives of central point values within the range statistic in order to measure the lowest value of 'basis point' for determining the occurrence of project overheads during the construction stage (refer to Figure 3). The mode statistics did not provide an absolute or central point value by seven different point values (i.e. 364, 366, 379, 389, 405, 421 and 441). However, the median value was represented by 'scaffolding' (with a total score of 391), which provided an approximate result with a mean statistic of 389.809 (~391). The mean statistic is the most suitable value to be used as a central point to determine the basis point of measurement through shifting this central point to the left side of Figure 3, by the value of standard deviation (33.526). The basis point of measurement is now positioned at the point value of 356.28 (389.809–33.526).

Eight items of project overheads are disqualified because their values positioned at lower values than the basis point value of measurement (each point value of the eight overheads is less than 356.28). Therefore, descriptive statistics have revealed the 39 items of construction project overheads that occur most often in construction projects and are categorized into four hierarchies: unit-level, batch-level, project-sustaining and facility-sustaining overheads (refer to Table 1).

The eight items of disqualified overheads are as follows: paying suppliers, project sundries, hoarding screens, placing purchase orders, receiving materials, material deliveries, managing work and contract conditions. These project overheads are disqualified because they are not related to supporting activities during the construction stage.

They may be more appropriate to support general office expenditures. For example, paying suppliers, hoarding screen and placing orders are normally included in the marketing department, while managing schedule and contract conditions cannot be categorized into site-project overheads, due to the fact that project management teams have more technical activities than administrative activities.

3 CONCLUSIONS

Project costs include overheads on the percentage basis and arbitrarily allocated to multiple cost objects; however, it cannot be directly distributed to maintain particular construction activities. Project overheads are required for supporting most of the construction activities. Construction project overheads are expected to be slightly higher in real projects than that reported in the literature.

The literature review identified 47 items of project overheads in construction. Descriptive statistics analyses determined the lowest value of assessment basis as 356.28, for the availability of construction project overheads. Of these project overheads, 8 were disqualified due to weights less than 356.28 and inconsistent occurrences in construction projects, which include: paying suppliers, project sundries, hoarding screens, placing purchase orders, receiving materials, material deliveries, managing work and contract conditions. In contrast, the remaining 39 project overheads are most often present in construction projects, and they are categorized into four hierarchies: unit level, batch level, project sustaining and facility sustaining overheads. These should be able to sustain and improve the Cost Management and Controlling Practices—the CMCPs of construction project overheads.

REFERENCES

Fellows, R. & Liu, A. 2008. *Research Methods for Construction*. Oxford, United Kingdom. Blackwell Publishing Ltd.

Jaya, N.M. & Frederika, A. 2015. *An Identification of Construction Project Overheads for Sustainable Cost Management and Controlling Practices (CMCPs)*. Recent Decisions in Technologies for Sustainable Development. International Journal of Applied Mechanics and Materials. 776: 121–126.

Saunders, M., Lewis, P. & Thornhill, A. 2007. *Research Methods for Business Students. Fourth Edition*. Essex, England: Pearson Education Limited.

Saunders, M., Lewis, P. & Thornhill, A. 2009. *Research Methods for Business Students. Fifth Edition*. Essex, England: Pearson Education Limited.

Yin, R.K. 2009. *Case Study Research Design and Methods. Fourth Edition*. London, United Kingdom: SAGE Inc.

Advances in Civil, Architectural, Structural and Constructional Engineering – Kim, Jung & Seo (Eds)
© 2016 Taylor & Francis Group, London, ISBN 978-1-138-02849-4

Experimental evaluation on UHPFRC-wide beams under bending loads

Y. Hor, T. Wee & D.K. Kim
Department of Civil and Environmental Engineering, Universiti Teknologi PETRONAS, Perak, Malaysia

ABSTRACT: Ultra-High Performance Fibre Reinforced Concrete (UHPFRC) is a relatively new type of High Strength Concrete (HSC) in the construction industry. UHPFRC offers excellent structural properties, such as ultra-high strength of 150–200 MPa, high fracture toughness, high ductility and better durability. However, testing data on UHPFRC structural members were very limited. This paper extends experimental investigation on UHPFRC-wide beams. Three beams were tested and compared to Normal Strength Concrete (NSC). Results show that beam UHPFRC with reinforcement bars gave much higher strength amongst all beams. Furthermore, the recent standard design, *fib* Model Code 2010 was used to validate the experimental results. More information on the results were discussed in detail.

1 INTRODUCTION

Ultra-High Performance Concrete (UHPC) would be one of the solutions for huge architectural structural features, durable components exposed to aggressive environments, non-penetrable cover and protective members subjected to impact or blasts rather than Normal Strength Concrete (NSC). UHPC known as Reactive Powder Concrete (RPC) was developed by (Richard & Cheyrezy 1995). Recently, UHPC offers a ultra-high compressive strength in excess of 150 and up to 200 MPa without heat curing (Wille et al. 2011). In addition, incorporation of steel fibers in UHPC, so-called Ultra-High Performance Fibre Reinforced Concrete (UHPFRC) exhibits excellent mechanical and durability properties, extremely low permeability and even strain hardening behaviour in tension (Habel et al. 2006, Tayeh et al. 2012).

Based on its exceptional properties, UHPFRC structural members have been carried out by several researchers (Voo et al. 2010). Brühwiler & Denarie (2008) suggested a basic concept to use UHPFRC as hardening parts of existing Reinforced Concrete (RC) structures where the outstanding properties of UHPFRC could be fully exploited (Brühwiler & Denarie 2008). The combination of UHPFRC layer and mother RC members enhances the structural response including maximum loads, stiffness and cracking behaviour (Alaee & Karihaloo 2003, Habel et al. 2007, Noshiravani & Bruhwiler 2013). However, despite the fact that UHPFRC has been used to benefit its superior properties, it still lacks experimental data on UHPFRC structural elements for quantifying assessment of the material.

In this study, the investigation of experimental testing on UHPFRC-wide beams is addressed. The study focuses on the results of the maximum loads, deflection and strain behaviour.

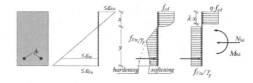

Figure 1. Stress-strain relationship (CEB-FIP, 2012).

Furthermore, current design code for Fibre Reinforced Concrete (FRC) is used to predict the failure capacity of the beams.

The design provisions such as Italian recommendation (CNR-DT/204, 2006) and fib Model Code (CEB-FIP, 2012) adopted the principle of stress-strain for predicting the bending moment resistance (M_{Rd}). The calculation can be obtained by equilibrium equations through the sectional analysis. Instead of ignoring the tensile strength of normal strength concrete (NSC), the FRC was accounted for in the calculation.

The principle assumptions are shown in Figure 1. The simplified stress-strain relationship has been adopted similar to EN-1992-1-1 (CEN 2004). The maximum compressive and post-peak tensile stress may be transformed to simply rectangular stresses. Further detail of the results is discussed.

2 EXPERIMENTAL PROGRAM

2.1 Overview

The experimental program consists of three wide beams with no shear steels. The RC beam with normal strength concrete is denoted as NSC. The UHPC beams without and with reinforcement bars are named as UHPFRC and UHPFRC-R (R refers to reinforcement), respectively.

A same amount of steel bars (10T12) is used in NSC and UHPFRC-R beams. Steel fibre content of 3% by volume is added in UHPC mix matrix. The beam specimens are tested based on simply supports of the three point bending with a ratio of load span-depth a/d = 8.11. The supports are placed in 200 mm distance from end edges of the beams (see the Figures 1, 2). It should be noted that the transverse steels (R6) are utilised in the end edges of the beams purposely to prevent anchorage failure of longitudinal steel bars. The details of cross-sections and reinforcement bars are depicted in Figure 2.

2.2 Material properties

As for the normal concrete, the commercial ready-mixed concrete is used for the NSC RC beam. The cylinders (100 × 200 mm) and cube specimens (100 × 100 × 100 mm) are also prepared to determine the compressive strength of concrete. This involves the average of six cylinders and six cubes for NSC.

UHPC mixing casted for UHPFRC beams have been done by trail mixes. Selected proportion of mix matrix is shown in Table 1. Average of four cylinders and four cubes are used for the determination of

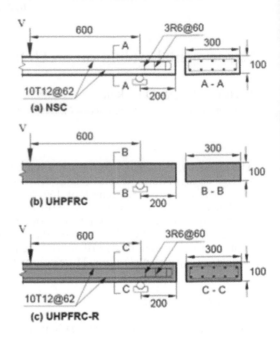

Figure 2. Beam layouts for experimental examination.

Table 1. Average values of material properties.

Normal strength concrete (NSC)

$f_{c,cube}$, MPa	$f_{c,cyl}$, MPa	f_{ct}, MPa	E_c, GPa
32.5	22.7	2.4[†]	26.9[§]

UHPFRC

Proportion by weight		Spread, mm	$f_{F,cube}$, MPa	$f_{F,cyl}$, MPa	$f_{F,fl}$, MPa	E_{cF}, GPa
Cement	1.00					
Silica Fume	0.25					
Quartz Powder	0.25					
Quartz Sand	0.48	223.7	168.6	152.8	27.4	50.8[§]
River Sand	0.80					
S.P[*]	0.05					
W/C	0.20					
Steel fibres%Vol.	3%					

Steel bars

f_y, MPa	D_s, mm	Surface	E_s, GPa
460	12	Ribbed	200

Steel fibres

σ_{fu}, MPa	d_f, mm	Shape	l_f, mm	E_{fu}, GPa
2300	0.2	Straight fibres	13	200

[*]S.P = superplasticiser (Sika ViscoCrete 2044).
[†]Concrete tensile strength: $f_{ct} = 0.3(f_{cm})^{2/3}$.
[§]Young's modulus: $E_c = 9500(f_{cm})^{1/3}$.

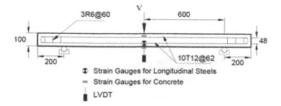

Figure 3. Details of experimental setup and instruments.

compressive strength. As for the flexural strength, standard three point bending tests on prims specimen (100 × 100 × 400 mm) were conducted. The workability of UHPC is evaluated by the spread diameters in accordance to the table slump flow test given by the instruction of the standard EN 1015–3 (EN 1015–3, 1999). Table 1 indicates the mean values of the concrete properties tested at 28 days and also includes the steel and fibre properties used.

2.3 Test setup

The details of specimen setup are illustrated in Figure 3. Three simply supported beams with 1600 mm long and 300 × 100 mm cross-section were tested. All beams were reported in three point loading. The roller supports were used to minimise the friction and to allow the free rotation and translation.

Vertical mid-span deflection and stain of longitudinal steels were measured respectively by LVDTs (linear variable displacement transducers) and strain gauges positioned under the applied loading. Concrete strain gauges were also attached to the top concrete surface of the mid-span (compression zone) and to the underneath concrete surface (tension) in order to record the strain responses of all beams.

3 RESULTS AND DISCUSSION

3.1 Load-deflection of UHPFRC beams

The load-mid span deflection relationships of the beams are shown in Figure 4. Both UHPFRC and UHPFRC-R were significantly stiffer than NSC. Unlike beam NSC failed in brittle, beam UHPFRC shows a good ductility due to the steel fibres bridging the cracks. Beam UHPFRC-R with the present of reinforcement bars gives much higher failure loads among all beams and performed ductile failure. The ultimate load of failure of beam UHPFRC-R was 46% and 63% higher in comparison to beams NSC and UHPFRC, respectively.

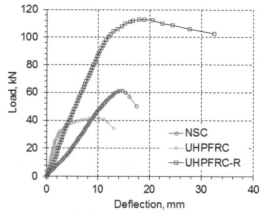

Figure 4. Comparison of load versus mid-span deflection relationship of the beams.

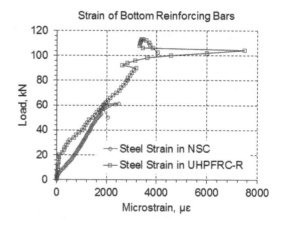

Figure 5. Load-strain curves of bottom longitudinal steel bars.

3.2 Internal strain of longitudinal bars

Strain of longitudinal steels performed is given in the load-strain curves as shown in Figure 5. According to the yield strain of steel bars ($\varepsilon_y = f_y / E_s$), the strain performed in the tested beam NSC does not reach the yield while for beam UHPFRC-R, the strain developed and reached to the yielding strain.

3.3 External strain evaluation of UHPFRC

Strain values measured by gauges attached at the underneath for tension and at the top chord for compression of the beams (see the Figure 3 for the instruments) are illustrated in Figures 6, 7, respectively in load-strain relationships.

From the illustration in Figure 6, the gauges could record only a small displacement during

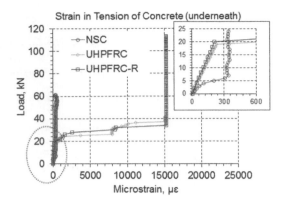

Figure 6. Load-strain curves of the beams in tension.

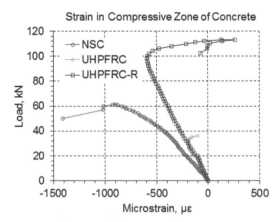

Figure 7. Load-strain curves of the beams in compression.

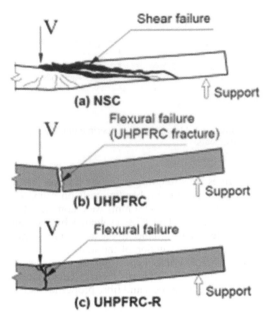

Figure 8. Crack patterns of failure modes in tested beams.

the micro cracks at underneath appeared. However, it is worth to see that strain of the beams UHPFRC and UHPFRC-R shows a very similar trend.

For the strain in the compression zone of concrete, Figure 7 shows that apart from NSC, beams UHPFRC and UHPFRC-R give stiffness not much difference until about 90% of failure loads. Both beams UHPFRC and UHPFRC-R show the strain values not only in compression but also turn the performance in tension after the remaining 10% of the failure loads.

3.4 Modes of failure

All the beams showed the different failure mechanism as seen in Figure 8. The following descriptions of the failure are based on experiment observed.

Apparently, the normal strength concrete beam NSC failed in shear, although micro flexural cracks occurred earlier than diagonal cracks of the failure. This flexural cracks may be because of the concrete weakness in deform while the strong reinforcements have not acted in yielding strength. It can also be evident from load-deflection curves (Figure 4) that the NSC beam shows a small ductile before the failure load becomes increasingly brittle.

Beam UHPFRC with no reinforcement bars failed in flexure by fracture capacity of UHPFRC. Meanwhile, beam UHPFRC-R having the same reinforcements as beam NSC gave a flexural failure. Because the present of steel, fibres contributes to the shear capacity and the UHPFRC had a high compressive strength, this led the concentrated loads transform in bending manner.

3.5 Validation of experimental results

Table 2 shows the experimental results including maximum loads (V_u), modes of failure and moment resistance with the values of $M_{u,exp}/M_{R,pre}$ ratios. The moment resistance ($M_{R,pre}$) was calculated according to current standard design, *fib* Model Code 2010 and the ultimate moment ($M_{u,exp} = V_u a$) is derived from V_u multiply by the distance between applied load and support (a). In this Table 2, the high ratio of $M_{u,exp}/M_{R,pre}$ demonstrates the poor prediction from the design code. This implies the

Table 2. Test results and moment resistance.

Beams	Experimental results		Moment resistance	
	Failure modes	V_u, kN	$M_{R,pre}$, kN.m	$M_{u,exp}/M_{R,pre}$
NSC	Shear	61.0	21.34	1.72
UHPFRC	Flexure	41.5	15.25	1.63

further research on analytical method is needed for such wide-beams.

4 CONCLUSIONS

As shown in experimental results, the following conclusions could be made:

Due to the superior properties of UHPC material, the beam UHPFRC-R carried a high load capacity of 113.0 kN and granted 46% greater than beam NSC. Moreover, UHPFRC-R performed a very good deflection as it is evident in Figure 7 that about 90% of failure loads the strain turned into tension manner.

The strong UHPC material gave beams UHP-FRC and UHPFRC-R demonstrating a ductile failure rather than brittle failure as beam NSC.

Prediction from the current design code failed in such wide-beams with significantly lower than experiment performed. This led the further research in this area.

ACKNOWLEDGEMENT

This work presented a part of the research project of YUTP Fundamental Research Grant (YUTP-FRG). Authors would like to thank and express their gratitude and sincere appreciation to the Universiti Teknologi PETRONAS (UTP) for providing financial support for this research work.

REFERENCES

Alaee, F. J. & Karihaloo, B. L. 2003. Retrofitting of reinforced concrete beams with CARDIFRC. *Journal of Composites for Construction*, 7: 174–186.

Brühwiler, E. & Denarie, E. 2008. *Rehabilitation of concrete structures using ultra-high performance fibre reinforced concrete*. UHPC-2008: The Second International Symposium on Ultra High Performance Concrete, Citeseer, 05–07.

CEB-FIP, M. 2012. *Model code 2010: final draft*, Lausanne, Switzerland, International Federation for Structural Concrete (fib).

CEN 2004. *Design of Concrete Structures: Part 1–1: General Rules and Rules for Buildings*. Eurocode 2. British Standards Institution.

EN 1015–3, B. 1999. Methods of test for mortar for masonry part 3: Determination of consistence of fresh mortar (by flow table).

Habel, K. Denarié, E. & Brühwiler, E. 2006. Experimental investigation of composite ultra-high-performance fiber-reinforced concrete and conventional concrete members. *ACI Structural Journal*, 104(1): 93–101.

Habel, K. Viviani, M. Denarié, E. & Brühwiler, E. 2006. Development of the mechanical properties of an ultra-high performance fiber reinforced concrete (UHP-FRC). *Cement and Concrete Research*, 36: 1362–1370.

Noshiravani, T. & Bruhwiler, E. 2013. Experimental investigation on reinforced ultra-high-performance fiber-reinforced concrete composite beams subjected to combined bending and shear. *ACI Structural Journal*, 110(2): 251–262.

Richard, P. & Cheyrezy, M. 1995. The composition of reactive powder concretes. *Cement and concrete research*, 25: 1501–1511.

Tayeh, B.A. Bakar, B.A. Johari, M.M. & Voo, Y.L. 2012. Mechanical and permeability properties of the interface between normal concrete substrate and ultra-high performance fiber concrete overlay. *Construction and Building Materials*, 36: 538–548.

Voo, Y.L. Poon, W.K. & Foster, S.J. 2010. Shear strength of steel fiber-reinforced ultrahigh-performance concrete beams without stirrups. *Journal of structural engineering*, 136: 1393–1400.

Wille, K. Naaman, A.E. & Parra-Montesinos, G.J. 2011. Ultra-high performance concrete with compressive strength exceeding 150 MPa (22 ksi): a simpler way. *ACI Materials Journal*, 108(1): 46–54.

Pullout strength of ring-shaped waste bottle fiber concrete

J.M. Irwan & S.K. Faisal
Jamilus Research Centre for Sustainable Construction (JRC), Faculty of Civil and Environmental Engineering, Universiti Tun Hussein Onn Malaysia, Malaysia

N. Othman
Micropollutant Research Centre (MPRC), Faculty of Civil and Environmental Engineering, University Tun Hussein Onn Malaysia, Malaysia
Jamilus Research Centre for Sustainable Construction (JRC), Faculty of Civil and Environmental Engineering, Universiti Tun Hussein Onn Malaysia, Malaysia

M.H. Wan Ibrahim
Jamilus Research Centre for Sustainable Construction (JRC), Faculty of Civil and Environmental Engineering, Universiti Tun Hussein Onn Malaysia, Malaysia

ABSTRACT: Polyethylene Terephthalate (PET) is one of the most commonly used plastic containers that are normally discarded by people, which results in environmental pollution. One of the best ways to reduce the environmental pollution is by recycling PET as a ring-shaped PET (RPET) fiber in concrete. Therefore, it is appropriate to investigate the RPET fiber in term of pullout strength of RPET fiber concrete. A total of 30 cubes for the fiber pullout test were made with three embedded fiber length sizes, namely 15 mm, 20 mm and 25 mm. The RPET-10 fiber exhibited the highest pullout load of 103.6% compared with RPET-5 fiber at 15 mm embedded length. This finding shows a similar pattern for 20 mm and 25 mm embedded lengths. RPET-10 fiber obtained 77.2% and 13.7% increases in pullout strength compared with RPET-5 fiber, at 20 to 25 mm embedded lengths. In conclusion, this result for the pullout load strength is important as new inputs to the technical properties of RPET FC during the pullout load test.

1 INTRODUCTION

Waste Polyethylene Terephthalate (PET) bottles can be used in various applications. Efforts have been made to explore their use in concrete. The development of new construction materials using recycled PET fibers is important in the construction and PET recycling industries. In the field of civil engineering, the recycled PET has begun to be adopted in the concrete.

Foti (2011) found that the possibility of using fibers from Polyethylene Terephthalate (PET) in bottles increased the ductility of the concrete. The author determined the tensile strength of ring-shaped PET with 0.20 mm thickness, 335 total length, and 5.0 mm width of fiber. The author showed that ring '0' shape PET obtained an average value of tensile strength equal to 180 MPa. Shannon (2011) studied the melt extrusion and tensile strength of different materials, namely High Density Polyethylene, HDPE and PP fiber with 1.76 mm of width fibers. The author found that fiber with different materials, but with similar size of width and embedded length would tend to show differences in pull-out

load. It can be pointed out that HDPE of 1.76 mm of width obtains higher load strength compared with PP pf 1.76 mm of width fiber at both 15 mm and 20 mm embedded length.

Richardson et al. (2009) examined the behavior of polypropylene fiber with 0.19 mm in diameter with a different embedded length of fiber. The authors found that the load was increased as the embedded length of fiber increased. The authors confirmed that the increases were on average of 39.3% to 48.1% strength with an embedded length ranging from 45 mm to 55 mm. Singh et al. (2010) investigated the pull-out behavior of polypropylene fiber from an embedded length on the pull-out fiber characteristics with matrix concrete. Singh et al. (2010) found that the embedded length of fiber in concrete shows significant effects on the pull-out characteristic. The authors indicated that the increase in the embedded length from 19 mm to 38 mm would increase 39% to 68.2% of load strength. Therefore, in this experiment, the pullout strength test of ring PET (RPET) fiber was performed. The aims of this experiment are to determine the bond mechanisms of recycled PET in different embedded lengths and sizes of RPET fiber.

2 EXPERIMENTAL WORK

The materials used in this study were ordinary PC, class F fly ash, coarse aggregates, sand, water and superplasticizer. Two sizes of fibers were used in the experiment: RPET-5 and RPET-10 fibers. The diameter of the fiber was 60 ± 5 mm cross section. The bottles were first cleaned to remove impurities and then cut into dog bone-shaped rings, as shown in Figures 1(a) and 1(b). The bottom of the dog bone-shaped rings was cut to confirm that fiber slipped during the pullout test. This is due to the fact that the full ring-shaped RPET fiber was failed by tensile load. Figure 1(a) and 1(b) illustrates the RPET-5 and RPET-10 fibers used in the pullout test.

A total of 30 cubes with the embedment length of 15, 20 and 25 mm were used according to previous research (Singh et al. 2004 & Shannon et al. 2011).

Concrete was inserted into the mold at certain levels, as shown in Figure 2(a) to 2(c). Half-round polystyrene was placed at the center of the cube mold. This polystyrene maintained the form of RPET

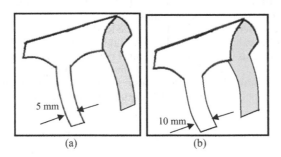

Figure 1.   (a): RPET-5 and (b): RPET-10 fiber.

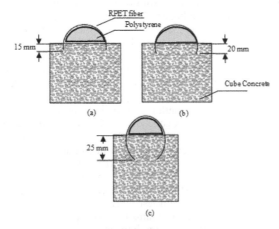

Figure 2.   (a): Embedment 15 mm of RPET, (b): Embedment 20 mm of RPET, and (c): Embedment 25 mm of RPET.

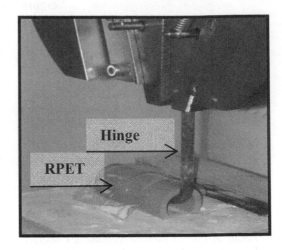

Figure 3.   Hinge support for the pullout strength test.

fibers during the insertion of RPET into the concrete mold. Then, the process was continued by pushing RPET fibers into the desired embedment length. The concrete was placed around the fiber until the top level of the mold was reached. Pullout test was conducted using a tensile test machine (Instron 3369) with a pull loading capability of 50 kN. The end of RPET fibers at the top was restrained using the customized fixture, as shown in Figure 3. This fixture reduced local stresses around the clamped section of the fibers. The threaded rod was adjusted, such that the specimen satisfied the upper restraining plate. Adjustments were made carefully to avoid tightness, which induces compression. The fixture was designed to restrain the specimen with minimal addition of lateral compressive forces around fibers.

Specimens were handled cautiously to ensure that no twisting or stretching would occur during the set up. An initial tension was applied to ensure that no slack or kink was present in the sample. Pullout test was conducted at a constant displacement rate of 2.5 mm/min (Shannon 2011). Each test was conducted until complete pullout of the fibers was achieved or fiber ruptured.

3 RESULTS AND DISCUSSION OF FIBER PULLOUT TEST

The strength behavior of FRC depends on the pullout behavior of a single fiber in the concrete matrix.

3.1 *Load-end displacement response*

The typical load-end displacement responses of RPET-5 and RPET-10 fibers are shown in Figures 4 and 5, respectively (load-displacement).

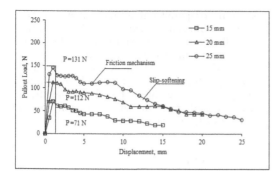

Figure 4. Load-displacement curves for RPET-5 fiber (Specimen 1).

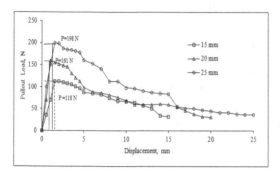

Figure 5. Load-displacement curves for RPET-10 fiber (Specimen 1).

In the experiment, comparisons were made between the RPET FC sizes with respect to peak and post-strength after peak response, which differentiated each size of the fiber used. The pullout load pattern initially increased nearly linearly with the slip at the beginning of the load until it achieved the critical load (linear region).

The nonlinear region indicates the start of debonding of a fiber from the matrix. The pullout curves presented a distinct linear initial portion that terminated at a point (P_{max}), in which nonlinear behavior was exhibited until the peak load was reached, followed by fiber/matrix debonding and frictional sliding. The initial incline of the fibers pulled from the concrete matrix was excessively steep. It was observed that the load decreased gradually after reaching the P_{max} point. It was also observed there was not a complete loss of bond, and the frictional resistance mechanism was activated. However, the part of the curve that corresponded to frictional sliding demonstrated an increase in the pullout load with increasing embedded length. This increase can be attributed to the increase in friction between the fiber and the concrete matrix

Table 1. Analysis of different lengths of RPET fiber.

| Type of fiber | Length of fiber (mm) | P_{max} (N) | Difference percentage compared | | |
			15 mm, (%)	20 mm, (%)	25 mm, (%)
RPET-5	15	75.30	–	−25.9	−42.0
	20	101.60	34.9	–	−21.7
	25	129.75	72.3	27.7	–
RPET-10	15	153.30	–	−21.3	−31.4
	20	194.80	27.1	–	−12.8
	25	223.40	72.2	14.7	–

because of fiber abrasion, which occurs when the fiber slides out of the matrix (Choi et al. 2003, Singh et al. 2004).

3.2 Effect of embedded length of fiber on pullout load

The experiment determined the pullout behavior pattern, and the results are summarized in Table 1.

Hence, the part of the curve that corresponds to frictional sliding indicates an increase in the pullout load. This increase can be attributed to the increase in friction between the fiber and the concrete matrix. The abrasion effect increases with the increase in the embedded length of fiber. The embedded length of 20 mm for RPET-5 and RPET-10 fibers exhibited increases in pullout of 34.9% and 27.1%, respectively, compared with 15 mm embedded length. Meanwhile, the embedded length of 25 mm for RPET-5 and RPET-10 fibers presented increases of 72.3% and 72.2%, respectively, compared with 15 mm embedded length. Nevertheless, these findings showed that the embedded length of 25 mm for RPET-5 and RPET-10 fibers demonstrated increases in pullout load of 27.7% and 14.7%, respectively, compared with the 20 mm embedded length.

This finding confirmed that an embedded length of 25 mm exhibited improved performance compared with the 15 mm and 20 mm embedded lengths. An increase in the embedded length of fiber leads to increase in pullout strength that contacted with matrix concrete. This finding is also in agreement with that reported in Singh et al. (2004) and Richardson et al. (2009).

3.3 Effect of RPET fiber size on pullout load

The effect of RPET fiber was determined to measure RPET-5 and RPET-10 fibers in the pullout test. The findings confirmed that the pullout load at embedded lengths of 15, 20, and 25 mm obtained

Table 2. Analysis of different sizes of RPET fiber.

| Type of fiber | Length of fiber (mm) | P_{max} (N) | Difference percentage | |
			RPET-5, (%)	RPET-10, (%)
RPET-5	15	75.30	–	–50.9
	20	101.60	–	–47.8
	25	129.75	–	41.9
RPET-10	15	153.30	103.6	–
	20	194.80	91.7	–
	25	223.40	72.2	–

for RPET-10 fiber is higher than that for RPET-5 fiber. The experiment indicated that RPET-10 fiber exhibited the highest pullout load because of higher loads of 103.6% compared with RPET-5 fibers, respectively, at 15 mm embedded length, as presented in Table 2. This finding shows a similar pattern to that for 20 mm and 25 mm embedded lengths. RPET-10 fiber obtained 91.7% increases in pullout strength compared with RPET-5 fiber, respectively, at 20 mm embedded length. Meanwhile, RPET-10 fiber obtained 77.2% increases in pullout strength compared with RPET-5 fiber, respectively, at 25 mm embedded length.

Based on the analysis, the surface area connected to concrete is a prior contribution of significant pullout load results with regard to embedded length. RPET-10 fiber has a higher surface area connected to the concrete matrix compared with RPET-5 fiber. The result also showed that RPET-10 fiber has surface area contacts with averages of 300 mm^2 to 500 mm^2, which are high compared with those of RPET-5 fiber with averages of only 150 mm^2 to 250 mm^2 in the surface area. Thus, the highest number of the surface area significantly contributes to friction energy and interfacial bond strength between the fiber and the concrete matrix during the pullout test (Sammer et al. 2010, Shannon et al. 2011).

4 CONCLUSIONS

The RPET-10 fiber exhibited the highest pullout load of 103.6% and 33.6% compared with RPET-5 fiber, at 15 mm embedded length. The 20 mm and 25 mm embedded lengths of RPET-10 fiber obtained 77.2% and 13.7% increases in pullout

strength compared with RPET-5 fiber, respectively, at 20 to 25 mm embedded lengths. It was confirmed that fibers with a large width will exhibit improvement in pullout load energy results compared with fibers with a small width. A high surface area of fibers produces high friction and slip hardening energy during the pullout test. Therefore, the load-displacement area is higher than that in fibers with a small surface area.

ACKNOWLEDGMENT

This work was financially supported by the University of Tun Hussein Onn Malaysia (UTHM) and Minister of Education Malaysia (KPM) through Exploratory Research Grant Scheme (ERGS) Vot. No. E-002.

REFERENCES

Ali, M. Xioyang, Li. & Nawawi C. 2013. Experimental investigations on bond strength between coconut fiber and concrete. *Material & Design*, 44: 596–605.
Choi, O.C. & Lee, C. 2003. Flexural performance of ring type steel fibre reinforced concrete. *Cement and Concrete Research*, 33: 841–849.
Foti, D. 2012. Use of recycled waste PET bottles fibers for the reinforcement of concrete. *Composite Structure*, 96: 396–404.
Hamoush, S. Heard, W. & Zornig, B. 2010. Effect of matrix strength on pullout behavior of steel fiber reinforced very-high strength concrete composites. *Construction and Building Materials*, 25(1): 39–46.
Irwan, J.M. Othman, N. Koh, H.B. Asyraf, R.M. Faisal, S.K. & Annas, M.M.K. 2013. The Mechanical Properties of PET Fiber Reinforced Concrete from Recycled Bottle Wastes, *Advanced Materials Research*, 795: 347–351.
Liao, W. Chao, S. & Naaman, A. 2010. Experience self-consolidating high performance fiber reinforced mortar and concrete. *American Concrete Institution*. 274(6): 79–94.
O'Connell, S. 2011. *Development of a new high performance synthetic fiber for concrete reinforcement*, July 2011, Dalhousie University Hailfax, Nova Scotia.
Richardson, A.E. & Sean L. 2009. Synthetic fibers and steel fibers in concrete with regard to bond strength and toughness, *Built Environment Research*, 2(2): 128–140.
Singh, S. Shukla, A. & Brown, R. 2004. Pullout behaviour of polypropylene fibers from cementitious matrix. *Cement and Concrete Research*, 34: 1919–1925.

Advances in Civil, Architectural, Structural and Constructional Engineering – Kim, Jung & Seo (Eds)
© 2016 Taylor & Francis Group, London, ISBN 978-1-138-02849-4

Empirical filtering method for natural frequencies evaluation of low rise reinforced concrete school building using ambient vibration approach

A.F. Kamarudin & M.E. Daud
Faculty of Civil and Environmental Engineering, Universiti Tun Hussein Onn Malaysia, Malaysia

A. Ibrahim
Faculty of Civil Engineering, Universiti Teknologi MARA, Malaysia

Z. Ibrahim
Department of Civil Engineering, Faculty of Engineering, Universiti Malaya, Malaysia

ABSTRACT: Ambient vibrations are composed by various sources of seismic noise wave fields such as nature and human activities. These sources may disrupt the origin prediction of natural frequencies of site-structure if without special attention is taken care during field measurements and data processing. A simple empirical filtering protocol has been used with application of Fourier Amplitude Spectra (FAS) and Horizontal to Vertical Spectral Ratio (HVSR) methods, to determine the origin natural frequencies of a 4-storey reinforced concrete primary school building in Johor Malaysia. Three predominant modes of building frequencies at 4.21 Hz and 6.58 Hz in the transverse axis and 4.35 Hz in the longitudinal axis significantly observed from the computed spectral. The results are in good agreement against two ratios applied from the filtering protocol used, with relative deviation less than 7.13% and 18.85% in respective axes.

1 INTRODUCTION

Advances in sensor recording systems nowadays have proven that weak motion instrument can measure rather than strong motion, and the strong motion sensors are almost as sensitive as the weak motion sensors (Havskov & Alguacil, 2004). However, interpretation of natural frequencies from ambient vibration measurements for buildings is not straightforward as for the free-field case (ground) (Gosar 2010). The main difficulties are to detect and eliminate the effects of fundamental frequencies of the nearby free-field and other buildings in the vicinity (Gosar 2010). Strong ground amplification may also contaminate the building responses and causes inaccurate estimation of the building natural frequencies (Kamarudin et al. 2014). In this study, significant peaks of FAS and HVSR curves were distinguished and evaluated in order to predict the originality of predominant frequencies (f_o) of the building and its fundamental ground frequencies (F_o) from ambient vibration sources, on a low rise existing reinforced concrete school building of SK Sri Molek.

2 MICROTREMOR METHODOLOGY

2.1 Instruments and fieldwork parameters

Ambient vibration measurements were measured using three units of Lennartz portable tri-axial seismometer sensors (S1, S2, and S3) with 1 Hz eigenfrequency, 400 V/m/s output voltage, 136 dB of dynamic ranges and CityShark II data logger with 1 GB memory card (see Figure 1). The ground and structure measurements fieldwork were carried out under good weather condition. Short period of natural or artificial noises perturbation were strictly monitored and controlled according to recommendations by SESAME (2004) guideline. The data logger setup was based on 100 Hz sampling rate frequency at optimum gain level within 15-minutes of recording length. Ambient vibration signals were measured in three main components i.e. vertical direction (UD) and horizontal directions (North-South: NS and East-West: EW). All recorded measurements were automatically stored into a flash card and manually transferred to a notebook for data extraction and processing.

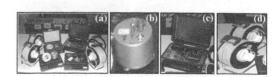

Figure 1. (a) Microtremor equipment, (b) Lennartz seismometer sensor, (c) CityShark II datalogger and (d) reinforced cable.

2.2 Fieldworks on ground surfaces and building

All sensors were aligned to the True North direction which perpendicularly to the longitudinal axis of the school building (see Figure 3). Simultaneous three unit of seismometer sensors were used and labeled as S1, S2, and S3. S1 acting as a reference sensor and permanently placed at specific reference point in every measurement taken.

In building measurements, the sensors are lying along the school corridor and positioned closer to the structure main frame joints. According to Oliveira and Navarro (2010), a single instrument located at the top of building can in most cases accurately determine the building fundamental period (its inverse gives the fundamental frequency). S1 was positioned on the upper floor and mid-span (C11) of the building in both horizontal and vertical directions of sensor's alignments. Details illustration and breakdown of sensor positions used in ambient vibration measurement carried out are given in Figure 2 and Table 1.

For the ground ambient vibration measurements, sensors were placed along lines B and C, which only discussed in this paper (other lines were also carried out outside of the school compound). S1 was placed at the center whereas S2 and S3 are positioned respectively to the West and East regions as shown in Figure 3. Approximate 25 meter grid spacing was used between measurement points. Details sensor locations for every line and measurement performed are given in Table 2.

2.3 Processing tool and analysis procedures

The recorded microtremor wave fields were transformed into FAS then HVSR curves using opensource software of GEOPSY developed by the

Table 1. Measurements breakdown of ambient vibration on building (refer to Figure 2).

Orientations of Sensor Alignments	Sensor 1 (S1). Ref. Point	Sensor 2 (S2)	Sensor 3 (S3)
Upper floor	C11	C21	C1
measurements	C11	C3	C19
(horizontally)	C11	C6	C16
3rd Floor	C11	C9	C13
Vertical	3rd floor, 3F	2nd floor, 2F	1st floor, 1F
measurements	3rd floor, 3F	2nd floor, 2F	Ground
at Column 11			floor, GF

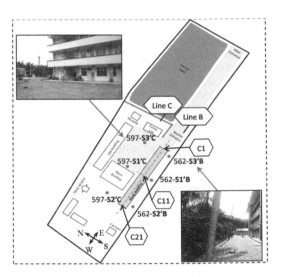

Figure 3. Positions of seismometer sensors on the ground surfaces in SK Sri Molek compound.

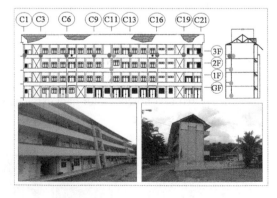

Figure 2. Sensor alignments on the upper floor in horizontal measurements, and column 11 (C11) in vertical measurements.

Table 2. Measurements breakdown of ambient vibration on ground (refer to Figure 3).

Measurement Lines & File Index No.	Sensor 1 (S1'), Ref. Point	Sensor 2 (S2')	Sensor 3 (S3')
Line B & 562	Parallel to C11	Between C19 and C20 (25 m from S1')	Between C2 and C3 (25 m from S1')
Line C & 597	Parallel to C11	Between C19 and C20 (25 m from S1')	Between C2 and C3 (25 m from S1')

SESAME group. Fourier spectrum from three components of ambient vibration were computed with 15-sec automatic window length, anti-triggering algorithm, 5% cosine taper and smoothing constant 40 by Konno Ohmachi. Generated FAS and HVSR from GEOPSY were filtered to determine the F_o and f_o according to the protocol as illustrated in Figure 4. In the first filtering stage, the peaks spectra within $FAS_{NS, EW, UD}$, and FAS vs HVSR were compared, to distinguish the origin of every peak frequency and its amplitude consistency patterns, to enable preliminary evaluation against a natural frequency mode of ground or building that strongly presence to be made. Two ratios were introduced as expressed in Equation 1 and 2, to further verify the dominating building frequencies that attribute to translational, torsional or rocking modes when there has significant clear peak bump on the spectrum. Figure 5 shows application of Ratio 2 computed from selected normalized FAS column against HVSR ground spectral.

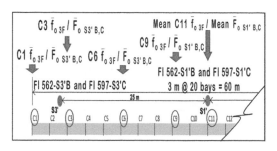

Figure 5. Distribution method of normalized FAS curve to respective column against normalized HVSR curves to respective sensor (refer to Ratio 2).

Ratio 1
$(BC_{ref} : BC_i)$

Normalized average FAS amplitudes at C11 from all four measurements against the max.

$$= \frac{\text{amplitude value for "Mean C11} \bar{f_o}\text{"}}{\text{Normalized FAS amplitudes against the}}$$

max. amplitude value at respective column

for "$C_{i, i=1-21}, \bar{f_o}$"

(1)

Ratio 2
$(BC_i : GS_i)$

$$= \frac{C_{i,i=1-21}, \bar{f_o}}{}$$

Normalized average HVSR of line B and C

amplitudes at respective sensor no.

against the maximum amplitude

for "$F_o S_i, B,C, {i=1,2,3}$"

(2)

3 RESULTS AND DISCUSSIONS

Three main components of $FAS_{NS, EW \text{ and } UD}$ curves measured at each joint on the upper floor building are shown in Figures 6 and 7 (c-d). The mean FAS curve peaks show at 3.09 Hz, 4.21 Hz, and 6.58 Hz in the NS direction, two peaks in the EW (2.98 Hz and 4.35 Hz) and two peaks as well (4.21 Hz and 7.29 Hz) in the UD directions. Meanwhile, only a single sharp peak of HVSR curves was obtained from lines B and C as shown in Figures 6 and 7 (e-f), at similar spectral mean HVSR curves frequency of 2.79 Hz.

The distributions of ground fundamental frequencies values from S3 (in the East), S1, to S2 (in the West) HVSR curves were extracted as in Table 3,

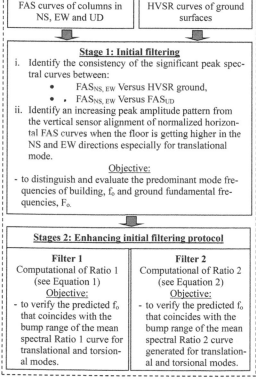

Figure 4. Filtering protocol for determination of predominant frequencies of building, f_o, and fundamental frequencies of ground, F_o.

The flowchart content:

FAS curves of columns in NS, EW and UD | HVSR curves of ground surfaces

Stage 1: Initial filtering
i. Identify the consistency of the significant peak spectral curves between:
- $FAS_{NS, EW}$ Versus HVSR ground,
- . $FAS_{NS, EW}$ Versus FAS_{UD}

ii. Identify an increasing peak amplitude pattern from the vertical sensor alignment of normalized horizontal FAS curves when the floor is getting higher in the NS and EW directions especially for translational mode.

Objective:
- to distinguish and evaluate the predominant mode frequencies of building, f_o and ground fundamental frequencies, F_o.

Stages 2: Enhancing initial filtering protocol

Filter 1	Filter 2
Computational of Ratio 1 (see Equation 1)	Computational of Ratio 2 (see Equation 2)
Objective:	Objective:
- to verify the predicted f_o that coincides with the bump range of the mean spectral Ratio 1 curve for translational and torsional modes.	- to verify the predicted f_o that coincides with the bump range of the mean spectral Ratio 2 curve generated for translational and torsional modes.

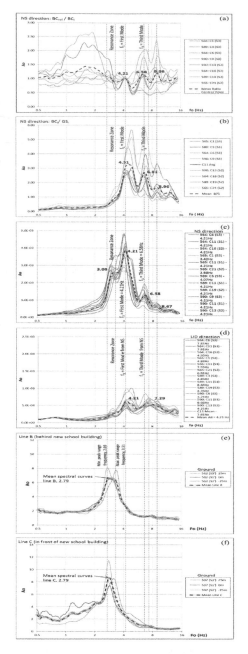

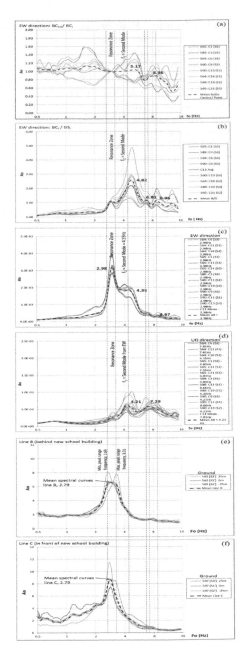

Figure 6. (a) NS direction: Ratio 1 as in Equation (1), (b) NS direction: Ratio 2 as in Equation (2), (c) NS direction: FAS curves, (d) UD direction: FAS curves, (e) Ground HVSR curves at line B and (f) Ground HVSR curves at line C.

Figure 7. (a) EW direction: Ratio 1 as in Equation (1), (b) EW direction: Ratio 2 as in Equation (2), (c) EW direction: FAS curves, (d) UD direction: FAS curves, (e) Ground HVSR curves at line B and (f) Ground HVSR curves at line C.

have shown 0.62 Hz difference between the bottom frequency value of 2.69 Hz to the top of 3.31 Hz frequency value. This range is also classified as

resonance zone that could encourage to structural destruction under seismic activity, if the building predominant frequency is found to be in this region.

Table 3. Peak frequencies picking from HVSR curves.

File Index No. and Measurement Points	HVSR, F_o First Peak (Hz)
562 S3'B	3.31
597 S3'C	3.09
562 S1'B	2.98
597 S1'C	2.88
562 S2'B	2.69
597 S2'C	2.79
Mean line B	2.79
Mean line C	2.79

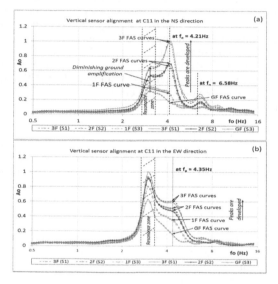

Figure 8. Vertical sensor alignment in (a) NS direction and (b) EW direction.

Clearly observed from Figure 8, the first peak frequency depicted from FAS curves in NS and EW directions in Figures 6 and 7 (c) at 3.09 Hz and 2.98 Hz have been contaminated due to strong ground amplification and shorter number of building stories. This effect is diminishing and easily detected in the weak axis of the building (NS direction) when the amplitudes of the first peak f_o in the resonance zone is reducing when the number of floor is increased. It is noted that, elimination of this peak region frequencies from the FAS curves in NS and EW directions show the peaks frequencies of building as depicted in Table 4, can be represented by three significant mean modes at 4.21 Hz, 4.35 Hz, and 6.58 Hz. However, none of corresponding vertical frequency (7.29 Hz) is significant appeared in both FAS horizontal curves which can be interpreted to additional building mode due to the rocking effect.

Reliability of three mean modes of predominant frequencies initially determined from the overlapping process are successfully agreed with the peak mean spectral ratio curves computed from Ratio 1 and Ratio 2 in both NS and EW directions, as shown in Figures 6 and 7 (a–b). These ratios are able to eliminate the contamination of strong ground amplification and at the same time magnify the dominant building vibration modes effectively. Very weak building vibration mode at the last peak mean (8.36 to 8.96 Hz) had risen from these spectral Ratio 1 and Ratio 2 coincide with very low amplitude indicated at similar mean frequency observed in Figures 6 and 7 (c).

Relative deviations are calculated to measure the percentage of difference between the mean predicted building mode of frequencies against

Table 4. First and second peaks frequency picking from both FAS curves in the NS and EW directions.

Column No.	NS direction, f_o		EW direction, f_o	
	First Peak (Hz)	Second Peak (Hz)	First Peak (Hz)	Second Peak (Hz)
C1	4.21	6.81	4.07	–
C3	4.07	7.04	4.35	6.81
C6	4.21	6.58	3.93	6.58
C9	4.21	6.58	4.66	6.81
C11	4.21	6.81	4.51	6.58
C11	4.21	6.58	4.51	7.04
C11	4.21	6.58	4.07	–
C11	4.21	6.58	4.51	7.04
C13	4.21	6.58	4.66	–
C16	4.21	6.58	4.07	–
C19	4.21	6.58	4.51	6.81
C21	4.21	6.81	4.35	6.58
Overall ranges	4.07 to 4.21	6.58 to 7.04	3.93 to 4.66	6.58 to 7.04

Table 5. Calculation of relative deviation for the first peak frequency in (a) the NS direction and (b) the EW direction, and for the second peak frequency in (c) the NS direction and (d) the EW direction.

(a)	First peak frequency	NS direction	Deviation (Hz)	% Relative Deviation
Mean		4.21		
Mean Ratio 1		4.21	0	0.00
Mean Ratio 2		4.51	0.30	7.13
Min ground frequency range		2.69	1.52	36.10
Max ground frequency range		3.31	0.90	21.38

(b)	First peak frequency	EW direction	Deviation (Hz)	% Relative Deviation
Mean		4.35		
Mean Ratio 1		5.17	0.82	18.85
Mean Ratio 2		4.82	0.47	10.80
Min ground frequency range		2.69	1.66	38.16
Max ground frequency range		3.31	1.04	23.91

(c)	Second peak frequency	NS direction	Deviation (Hz)	% Relative Deviation
Mean		6.58		
Mean Ratio 1		6.58	0	0.00
Mean Ratio 2		6.81	0.23	3.50
Min ground frequency range		2.69	3.89	59.12
Max ground frequency range		3.31	3.27	49.70

(d)	Second peak frequency	EW direction	Deviation (Hz)	% Relative Deviation
Mean		-		
Mean Ratio 1		8.36	None of a clear second peak was observed from the FAS curves on the top floor of this building in the EW direction	
Mean Ratio 2		6.81		
Min ground frequency range		2.69		
Max ground frequency range		3.31		

the resonance frequencies and the mean peaks frequencies of Ratio 1 and Ratio 2 (see Table 5). The filtering approach protocols work efficiently in separating the building and ground frequencies at very low percentage of difference, which only less than 7.13% in NS direction and 18.85% in the EW direction.

The level of soil-structure resonance can be determined by taking the ratio between building frequencies which is closer to the site frequency (Gosar 2010). When the difference is within $\pm15\%$, the danger of soil-structure resonance is high, if it is within ±15 to 25% it is medium, while if it is higher than $\pm25\%$, then it is low (Gosar 2010). It means, the lowest percentage of relative deviation may encourage to the highest risk of resonance level. From the calculation of relative deviations, the resonance level was found to be in the medium zone level at 21.38%.

4 CONCLUSIONS

Prediction of natural frequencies from ambient vibration measurements for site-structure is widely accepted today. However the interpretation of reliable vibration modes frequency of a building is not straightforward and could lead to misinterpretation due to misidentified origin frequencies of microtremor (artificially) and microseismic (naturally) nearby activities, including strong local site application which may perturbs the accuracy of prediction. It can be concluded that, identification of three modes of predominant building frequencies, f_o, and fundamental resonant frequencies, F_o, in this study have showed a reliable verification process based on the filtering protocols used in this study. Building vibration modes are also successfully separated from BC_{ref}: BC_i (Ratio 1) and BC_i: GS_i (Ratio 2) ratios used, without significant rocking mode interfered from both FAS NS and EW horizontal spectral curves.

ACKNOWLEDGMENT

The authors would like to acknowledge the financial support provided by Ministry of Education Malaysia (Higher Education) under ERGS grant 011 and Universiti Tun Hussein Onn Malaysia for the study sponsorship and the publication of this paper. Great appreciation to Faculty of Civil and Environmental Engineering, UTHM for the instrumentations, the school administration and colleagues for their co-operation along the period of this research being conducted.

REFERENCES

Gosar, A. 2010. Site effects and soil-structure resonance study in the Kobarid basin (NW Slovenia) using microtremors. *Natural Hazards Earth System Sciences*, 10: 761–772.

Havsjov, J. & Alguacil, G. 2004. *Instrumentation in Earthquake Seismology*. Springer, Netherlands.

Kamarudin, A.F. Ibrahim, A. Ibrahim, Z. Madun, A. & Daud, M.E. 2014. Vulnerability assessment of existing low-rise reinforced concrete school buildings in low seismic region using ambient noise method. *Advanced Materials Research*, 931–932: 483–489.

Oliveira, C.S. & Navarro, M. 2010. Fundamental periods of vibration of RC buildings in Portugal from in-situ experimental and numerical techniques. *Bull Earthquake Engineering*, 8: 609–642.

SESAME. 2004. *Guidelines for the implementation of the H/V spectral ratio technique on ambient vibrations: measurements, processing and interpretation*, European Commission—Research General Directorate Project No. EVG1-CT-2000-00026.

Advances in Civil, Architectural, Structural and Constructional Engineering – Kim, Jung & Seo (Eds)
© 2016 Taylor & Francis Group, London, ISBN 978-1-138-02849-4

Dynamic behavior of kart bodies on curved sections

Y.J. Choi
Institute of Industrial Technology, Kyonggi University, Suwon, Kyonggi-do, South Korea

J.S. Lee
Department of Mechanical System Engineering, Kyonggi University, Suwon, Kyonggi-do, South Korea

ABSTRACT: Elastic deformation and fatigue damage can cause permanent deformation of a kart's frame in turn affecting the kart's driving performance. To analyze the dynamic behavior of a kart along curved section, the GPS trajectory of the kart is obtained and the torsional stress acting on the kart-frame is measured in real time. The mechanical properties of leisure and racing karts are investigated by analyzing material properties and conducting a tensile test. The torsional stress concentration and frame distortion are investigated through a stress analysis of the frame on the basis of the obtained results. Leisure and racing karts are tested in each driving condition using driving analysis equipment. The behavior of a kart when being driven along a curved section is investigated through this test. Because load movement occurs owing to centrifugal force when driving along a curve, torsional stress acts on the kart's steel frame.

1 INTRODUCTION

A kart is a medium-sized and high-speed vehicle that is smaller than a sports utility vehicle; karts are used for both leisure and racing. This cart has moderate construction costs and operating expenses, an important factor in future utility vehicle as well as excellent punctuality and environment-friendliness compared with other road leisure sports utility systems. However, it is necessary to improve the fatigue strength of a kart while maintaining low weight. In structures subjected to constantly fluctuating loads, the fatigue strength is influenced more by the stress and its direction than by the stiffness. In recent years, fatigue strength has been evaluated by experiments and by simulations using a finite element analysis program.

Jang (2010) constructed a test system capable of applying various design loads in order to evaluate the structural safety of weak points of a kart while driving on a curve and analyzed the characteristics and shifts in load of the steering system. Kim (2010) studied the vehicle behavior by analyzing the dynamic driving characteristic in which the minimum ground clearance of the kart is changed by load shift and driving mode shift of the frame when driving on curve. However, previous-studies did not consider the effect of changes in speed as well as the effects of various structural materials so that these studies did not consider the complex effects of the dynamic load occurring when driving on a track. Therefore, in this study, the fatigue strength of a kart frame was evaluated by varying the actual dynamic stress at various driving speeds.

2 TYPE OF KARTS

A kart consists of engine, chassis, and frame, as shown in Figure 1. The structure of frames is typically made of Fe and Mn; that of a racing kart may additionally consist of Si and Cr, the static strength and stiffness of which has been evaluated in previous studies.

The fatigue strength of a frame has also been analyzed because it is subjected to complex fatigue loads of amplitudes and frequencies, such as the static load from the weight of the kart and the dynamic load during curving, braking, etc.

The leisure kart's frame is typically made of Mn alloy, which affords improved elongation rate with strength. The racing kart's frame is typically made of Si and Cr, which improved hardness and elastic limit. Table 1 lists the component of kart frames. Also, the results of tensile test of each frame of karts are listed in Table 2.

(a) Leisure kart

(b) Racing kart

Figure 1.   Type of karts.

Table 1.   Component of karts frames.

[%]	Fe	Mn	Si	Cr
Leisure kart	98.30	1.70	–	–
Racing kart	97.37	1.04	0.48	1.15

Table 2.   Mechanical properties of materials.

Material	Yield strength [MPa]	Ultimate strength [MPa]	Elongation [%]
Steel	431.3	447.2	6.4
Leisure kart	471.9	558.0	27.0
Racing kart	598.9	754.0	13.1

3   MEASUREMENT AND ANALYSIS

3.1   *Load condition and stress measurement points*

Before the actual dynamic stress measurement, the stress was measured under zero and full load conditions as shown in Tables 3 and 4. The weight of a leisure kart was 102 kg$_f$, 161 kg$_f$ and that of a racing kart was 131 kg$_f$, 167 kg$_f$.

Figure 2 shows the 26 and 30 measurement points selected in the leisure and racing karts, respectively, for structural analysis and static load measurements.

Table 3.   Weight of a leisure kart.

[kgf]	FL	FR	RL	RR	Total
Tare	15.93	13.55	29.98	42.6	102.06
Load	32.47	23.3	45	60.4	161.57

Table 4.   Weight of a racing kart.

[kgf]	FL	FR	RL	RR	Total
Tare	15.5	29	36.5	50.5	131.5
Load	29	30.5	50.5	57	167

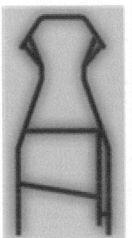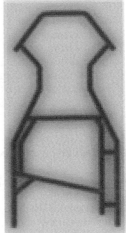

(a) Leisure kart   (b) Racing kart

Figure 2.   Type of kart's frames.

3.2   *Measurement and analysis considering driving conditions*

The torsional stress in each kart was measured at various driving speeds. Owing to the limitation of the test track, we measured the torsional deformations at various speeds while driving on the track. A constant speed was maintained when driving on the curve section. After eliminating the effects of noise, when driving one lap, the torsional deformation of the frame was measured and analyzed based on the following factors.

1. The GPS trajectory, change in steering angle, and time required for completing the lap and to reach certain points.
2. Torsional change analysis along the curve with time.
3. Torsional change analysis along the curved with speed.

Figure 3 shows the relationship between the speed and the steering angle while driving. To detect

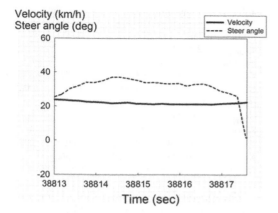

Figure 3.   Speed-steering angle in 1-lap.

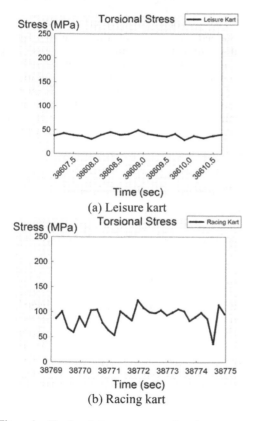

(a) Leisure kart

(b) Racing kart

Figure 4.   Torsional stresses on curved section.

the curved sections, the point at which the steering angle began changing was selected as the starting point of the curve. The speed hardly changes with the steering angle, suggesting that the kart maintains constant speed along the curved section.

Figure 4 shows the torsional stresses of the frame with time while driving along an arbitrary curved section. The torsional stress of a leisure kart is smaller than and varies less than that of a racing kart when driving along a curve.

Table 5 shows the torsional stress distributions acting on a racing kart's frame on curved sections. Figure 5 shows the torsional stress acting on both types of karts; stress acting on the racing kart is more than two times that acting on the leisure kart; the thick solid line indicates the maximum torsional stress of each type of kart at a certain speed. The dotted line indicates torsional stress data compensated after the measurement; the long and short dotted lines respectively indicate the racing and leisure karts. In both cases, the torsional stress increases with speed.

Table 5.   Torsional stress distributions.

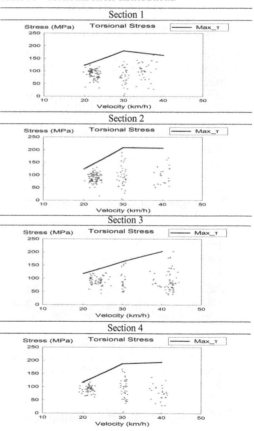

[MPa]	Section 1	Section 2	Section 3	Section 4
20 km/h	36.746 ~ 123.603	61.472 ~ 118.115	49.366 ~ 124.297	58.949 ~ 118.368
30 km/h	3.063 ~ 180.562	4.759 ~ 187.689	5.108 ~ 208.946	48.352 ~ 163.783
40 km/h	27.915 ~ 162.774	29.682 ~ 193.492	51.632 ~ 206.928	38.008 ~ 202.449

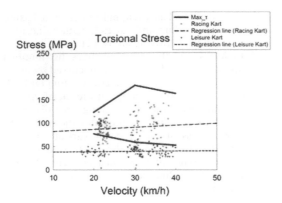

Figure 5. Torsional stress changing for the speed.

4 FATIGUE STRENGTH ASSESSMENT

4.1 Assessment method for fatigue strength

The stress-life, curve represents the relationship between the alternating stress (S) and the number of iterations (N) until a part is broken. A rotating bending test is used most commonly to obtain the data plotted in an S-N curve (Bannantine et al. 1990). Log-log coordinates are used for the average S-N test data plotted on the S-N curve.

Materials have a certain endurance limit or fatigue limit (S_e), below which they have infinite life. For industrial purposes, the limit has generally been considered as 10^7 cycles.

The fatigue limit can be determined based on the tensile strength.

$$S_e(ksi) \approx 0.5 \times S_u, \quad S_u \leq 200 \text{ ksi}$$
$$S_e \approx 100 \text{ ksi}, \quad S_u > 200 \text{ ksi} \quad (1)$$

The volume concept can be applied to fatigue test data obtained using a similar specimen. An axial load specimen has a more highly stressed volume because it does not have stress gradients. A fatigue limit ratio of 0.6–0.9 is obtained from the axial load and the rotation bending test. This test data may contain errors related to the eccentricity of the axial load.

$$Se(axial) = 0.7 \, Se(bending) \quad (2)$$

Therefore, the following safety assessment is used. A Fatigue limit ratio of 0.5–0.6 is obtained from the axial load and the rotation bending test. A theoretical value of 0.577 is given as the von Mises failure criterion. A reasonable estimate is obtained as follows.

$$Se(torsion) = 0.577 \, Se(bending) \quad (3)$$

Table 6. Fatigue limit based on the stress (40 km/h).

(a) Leisure kart

$\tau_{e\,(MPa)}$	229.97			
Section	1	2	3	4
τ	52.391	68.138	58.172	59.169
K	0.235	0.306	0.261	0.265

(b) Racing kart

$\tau_{e\,(MPa)}$	310.75			
Section	1	2	3	4
τ	162.774	193.492	206.928	202.449
K	0.524	0.623	0.666	0.651

The torsional fatigue limit of kart frame can be obtained by applying these load effects. To predict the torsional stresses that affect the kart's frame, the torsional fatigue limit was calculated based on the tensile test result and the load effect.

The torsional fatigue limit coefficient (K) is obtained as a ratio of the material's torsional fatigue limit τe to the measured torsional fatigue limit τ. For $K \leq 1$, the actual acting torsional stress occurs in a smaller range than the endurance limit of the material. This does not caused fatigue deformation and damage to the material, as a result of which its life can be infinite. On the other hand, for $K > 1$, damage is caused to the material by fatigue.

$$K = \tau_e / \tau \quad (4)$$

4.2 Fatigue strength evaluation

Table 6 shows the torsional stresses, measured torsional fatigue limit, and torsional fatigue limit coefficient (K) when driving at a speed of 40 km/h.

5 CONCLUSIONS

Fatigue strength assessment was conducted based on measured results obtained when a kart was being driven on a track. Both leisure and racing karts showed lower torsional stress than the torsional endurance limit. Therefore, plastic deformation and fatigue damage did not occur in both karts when they were being driven along a curved section.

REFERENCES

Bannantine, J.A., Comer, J.J. & James L. 1990. *Fundamentals of Metal Fatigue Analysis,* Prentice Hall.
Jang, H.T. 2010. *A Kinematic Analysis of a Racing Kart Steering Mechanism,* Proc. of 1st KAIS Conference, 1156–1158.
Kim, Y.H. 2010. *A Study on the Twisting Characteristic of the Body at Cornering of a Kart,* Doctoral Thesis, Kyonggi University.

Advances in Civil, Architectural, Structural and Constructional Engineering – Kim, Jung & Seo (Eds)
© 2016 Taylor & Francis Group, London, ISBN 978-1-138-02849-4

Diagonal compression behavior of wallets constructed using wood-wool cement composite panel

M.S. Md Noh
Faculty of Civil and Environmental Engineering, University Tun Hussein Onn Malaysia,
Batu Pahat, Johor, Malaysia

Z. Ahmad
Institute for Infrastructure Engineering and Sustainable Management, University Teknologi Mara,
Shah Alam, Selangor, Malaysia

A. Ibrahim
Faculty of Civil Engineering, University Technology Mara, Shah Alam, Selangor, Malaysia

P. Walker
Department of Architecture and Civil Engineering, University of Bath, Bath, UK

ABSTRACT: Wood-Wool Cement Composite Panel (WWCP) is a wood based product which is produced by mixing the wood-wool strand with Portland cement paste and bonded together under pressure to form a panel. In Malaysia, the application of WWCP as a wall system in building construction is still new and thus, the structural behavior and the installation technique of this composite wall are still not well established. Therefore, in this study two types of installation techniques denoted as W1 and W2 have been considered in the fabrication of 600 mm × 600 mm wallettes and tested under diagonal compression load in accordance to ASTM E 519–02. For W1, 100 mm thickness of WWCP has been used and cut into 300 mm width and 600 mm length and stacked vertically in two layers. The top and bottom panels were connected together with 10 mm thick mortar paste and three vertical steel bars were inserted between panels. For W2, a new panel arrangement technique has been proposed by cross laminating two layers of 50 mm thickness of WWCP. The front and back side panels were bonded together using either adhesive or mortar mix with different thicknesses to form approximately 100 mm (±15 mm) thick wallettes. The results indicated that the wallettes W2 which bonded with 15 mm thick mortar mix shows a highest diagonal loading capacity compared to all tested wallettes.

1 INTRODUCTION

Nowadays, sustainable construction methods are becoming more important as people look at ways to reduce environmental destruction. It is clear that the development levels of construction activity keeps continuing and increasing from time to time. If trend is continuing, these will intensify environmental deterioration and resources stress. To solve this problem construction industry have to find right ways as well as use the materials that are produced from renewable resources, lower energy consumption and less carbon emissions during its life cycle (Goverse et al. 2001).

Wood-Wool Cement Composite Panel (WWCP) is a wood composite product that is produced from renewable resources, lightweight and available for the construction industry to replace less

eco-friendly structural and non-structural materials for housing and construction (Soffi et al. 2014). WWCP are manufactured from long wood-wool strands which shredded from local fast grown timber species known as Kelampayan. Wood-wool then coated with Portland cement paste and bonded together under pressure. When cured, the resulting WWCP are stable and have a density range from 300 kg/m³ to 500 kg/m³ (Ahmad et al. 2011).

In Malaysia, the application of WWCP as a wall system in building construction is still low due to lack of knowledge and limited study conducted on the structural performance of this construction system under imposed loads. Besides, the installation technique of this composite panel is also not well established and required further investigation.

Fatihah (2011) experimentally investigated the axial compression behavior of wall fabricated using 50 mm and 75 mm thicknesses of the wood-wool cement composite board which was vertically stacked at running bond pattern. The results indicated that the buckling failures at the panel joint were observed for the wall without surface plaster whereas the highest loading capacity was reported for 75 mm thick walls with surface plaster. In another study, Firdaus (2012) experimentally investigated the performance of a timber frame wall in-filled with 50 mm thick wood-wool cement composite board under in-plane lateral load. The result revealed that the used of timber frame and diagonal corner bracing effectively increased the lateral load capacity of the walls. However, the frame failure mode was observed in this study due to the weakness of the frame and the weakness of the joint between wall panel and frame. Soffi et al. (2014) previously studied the axial compression behavior of wallettes fabricated using wood-wool cement composite panel at different panel arrangement and joint techniques denoted as W1 and W2. The results indicated that a new proposed cross-laminated WWCP wallettes with mortar mix as an adhesive, W2 showed a 100% increase of load carrying capacity for each increment of 5 mm mortar thickness and the wallettes W2 with 15 mm mortar thickness showed a higher loading capacity compared against all tested wallettes. This present paper is a part of the research work reported in the previous paper.

The diagonal compression load test, also known as in-plane shear resistance test. This test is useful in order to evaluate the shear stiffness and capacity of wallettes when subjected to an applied load in-plane direction. This test also conducted to determine the resistance of wall system against lateral wind load (Manalo, 2012). Therefore, in this paper, the diagonal compression behavior of wallettes is explored. The results obtained will be used as guidance for future research work which related to the structural behavior of the prefabricated wall constructed using wood-wool cement composite panel.

2 MATERIALS AND METHODS

2.1 Wood-Wool Cement Composite Panel (WWCP)

WWCP used in this study is a local factory product which is manufactured by Duralite (M) Sdn. Bhd. This composite panel is produced by mixing the wood-wool, Portland cement and water in a specific mixing ratio. The local fast grown timber species known as Kelampayan are cut

into 500 mm long timber log and shredded using a professional shredding machine to produce 1.5 mm – 3.5 mm wide and 0.5 mm thick wood-wool as shown in Figure 1c. The uniformly mixture of wood-wool, cement, and water were spread into a wooden mould, and then it will be pressed under the concrete block for 24 hours (Figure 1e). When hardened, it will result a dimension stable panel with standard panel size of 2400 mm length, 600 mm wide and varies from 25 mm to 100 mm thicknesses (Figure 1f). The strength properties of the WWCP used in this study is shown in Table 1.

Figure 1. The production process of WWCP.

Table 1. Strength properties of WWCP (Soffi et al. 2014).

Panel Thickness	Density	Bending Properties		Comp. Strength	Tensile Strength
		MOR	MOE		
mm	kg/m³	N/mm²	N/mm²	N/mm²	N/mm²
50	328	1.15	444	0.84	0.060
100	272	0.40	239	0.30	0.018

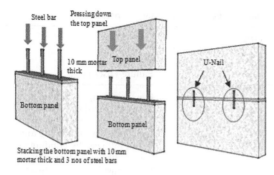

Figure 2. Fabrication of wallettes W1.

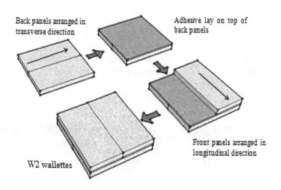

Figure 3. Fabrication of wallettes W2.

2.2 Fabrication of wallettes specimens

In this study, two types of panel arrangement techniques were considered in the fabrication of wallettes and denoted as W1 and W2. The W1 is the currently practiced technique of wall construction using WWCP and W2 is a new technique proposed for wall construction using WWCP.

For W1, the 100 mm thick WWCP was used and cut into the size of 300 mm width and 600 mm length. The cut panels were then stacked vertically (two layer) in running bond to form a 600 mm × 600 mm wallettes. The top and bottom panels were connected together with 10 mm thick mortar paste and three vertical steel bars were inserted between panels (Figure 2). Two U-Nail have been fixed at the connection area on each side in order to increase the stability of the connection. The W1 without and with U-Nail was refereed as W1-NON and W1-WIN respectively.

For W2, a new panel arrangement technique has been proposed by a cross-laminated two layers of 50 mm thickness of wood-wool panels (cut into size of 300 mm width and 600 mm length) with different orientation of panel arrangement to form a 600 mm × 600 mm wallettes as shown in Figure 3.

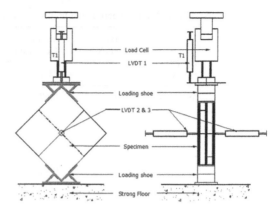

Figure 4. Illustrated diagonal compression load testing set-up of WWCP wallettes.

The front and back layers of the panels were bonded together using adhesive to form 100 mm (±15 mm) thick wall. For wallettes W2-S3, Sikadur 30 has been used as an adhesive with 3 mm thick glue line, whereas W2-M5, W2-M10 and W2-M15 the Emaco R1 mortar mix with thicknesses of 5 mm, 10 mm, and 15 mm respectively have been used as an adhesive.

2.3 Testing set-up and procedures

In this study, a three replicates of each type of wallettes were tested under diagonal compression load in accordance to ASTM E 519–02. The load was applied using 250 kN hydraulic jack through loading shoes positioned on two diagonally opposite corners of the wallettes specimen. The testing set-up was instrumented with three LVDTs, where LVDT 1 is used to measure the vertical displacement and LVDT 2 and 3 are used to measure the horizontal displacement of the wallettes as shown in Figure 4. The straightness of the specimens was checked using a spirit level to avoid load eccentricities. The load was then applied at a uniform rate of 0.005 mm/min up to failure. The failure load, maximum displacement at peak and the failure mode was observed.

3 RESULTS AND DISCUSSION

A summary of maximum load, displacement at peak load and failure description of each type wallettes are presented in Table 2. Load-displacement curves selected from each wallettes specimen also presented in Figure 5.

From the testing results of wallettes W1, it can be seen that the diagonal compression load capacity of W1-NON exhibited the lowest among tested

Table 2. Maximum load, deflection and failure mode of wallettes under the diagonal compression load.

Wallettes types	Specimen references	Maximum load (Mean) (kN)	Standard Deviation (kN)	Vertical displacement (Mean) (mm)	Failure Discription
W1	W1 - NON	2.01	0.67	7.61	Panel crushing at loading area and followed by bonding failure within wood-wool panel
	W1 - WIN	3.25	0.12	14.87	Panel crushing at loading area and followed by bonding failure within wood-wool panel
W2	W2 – S3	17.18	0.64	8.46	Panel crushing at loading area and followed by vertical cracking on the web of wallettes near to panel joint
	W2 – M5	18.20	2.42	11.25	Panel crushing at loading area and followed by vertical cracking on the web of wallettes near to panel joint
	W2 – M10	34.10	6.62	7.10	Panel crushing at loading area
	W2 – M15	48.13	9.60	14.05	Panel crushing at loading area

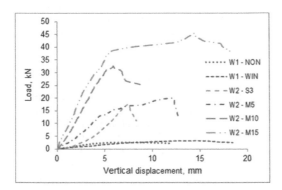

Figure 5. Load-displacement curves of selected from each types of wallettes.

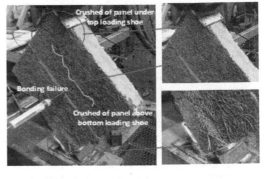

Figure 6. Failure mode of wallettes W1-WON.

wallettes which failed at maximum load of 2.01 kN at vertical displacement of 7.61 mm. Whereas, for wallettes W1 with U-Nail (W1-WIN) the loading capacity was recorded 62% improved to a maximum load of 3.25 kN and the displacement also increased to 14.87 mm. Since, there are improvements in term of loading capacity and displacement, but the results recorded were much lower compared to wallettes W2. This condition was similarly as reported in Soffi et al. (2014) where the porosity of 100 mm thick panels and weak wood-wool bond strength are the major contribution to the lower loading capacity of wallettes W1. The load-displacement curves of wallettes W1 exhibited similar shear response where the load gradually increases up to the maximum load and after

this load, the drop in an applied load was observed but the wallettes was seen continued to carry the load before its final failure. In terms of failure modes, both W1 wallettes showed the similar failure pattern where the crushing of the panel under loading shoe is the main failure cause of these wallettes. There are no failure mechanisms detected at the panel joint, however, the loss in bonding within the wood-wool matrix were contributed to the drop in of diagonal load (Figures 6 and 7).

For wallettes W2, the results indicated that the new proposed fabrication technique significantly increased the loading capacity of wallettes. A higher failure load and vertical displacement were observed for this type of wallettes. Wallettes W2–M15 recorded as a highest diagonal load

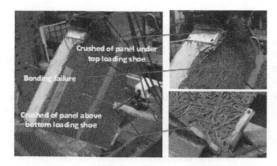

Figure 7.   Failure mode of wallettes W1-WIN.

Figure 8.   Failure mode of wallettes W2.

Figure 9.   Vertical crack on the web of wallettes W2-S3 and W2-M5.

carrying capacity and reaches its peak shear resistance at a load of 48.13 kN and a vertical displacement of 14.05 mm.

For wallettes W2-M10, the results showed that the reduction from 15 mm to 10 mm mortar thickness significantly reduced the loading capacity about 30% to the failure load of 34.10 kN and displacement of 7.10 mm. The drop in of loading capacity was continued about 60% for wallettes W2-M5 when the mortar thickness was further reduced to 5 mm to an applied failure load of 18.20 kN. It is evident that, the shear capacity of wallettes W2 is governed by the mortar thickness provided as an adhesive. This was similar findings reported in the previous paper (Soffi et al. 2014) which studied the axial compression capacity of wallettes W2. On the other hand, the wallettes W2-S3 identically failed as W2-M5 with a little different to the failure load of 17.18 kN. For load—displacement curves, the results showed that all wallettes W2 also exhibited similar shear response where the loads linearly increased up to the maximum load. The drops in after maximum load were observed for wallettes W2-S3, W2-M5 and W2-M10 however, for wallettes W2-M15 it can be seen that the wallettes continued to resist the load before its final failure. In terms of failure modes,

the crushing panels at diagonal corners under the loading shoes were observed as the main failure mechanism of wallettes W2 (Figure 8). There is no vertical crack that appeared on the surface of the panel due to the high porosity of wood-wool panel.

The vertical crack on the web of the panel was seen to occur for wallettes W2-S3 and W2-M5 but not for W2-M10 and W2-M15 as shown in Figure 9. This probably due to insufficient thickness of adhesives to laminate two layers of the panel.

4   CONCLUSIONS

Based on experimental results, the several conclusions can be drawn as follows:

The panel arrangement technique significantly contributed to the stability as well as increased the load carrying capacity of wallettes. The results showed that the new panel arrangement technique proposed in this study (W2) performed better compared against wallettes that currently constructed using WWCP at the construction industry (W1).

The shear capacity of wallettes (W2) depends on the strength of WWCP and the mortar thickness used as an adhesive. It can be seen that, the wallettes made from cross laminated with two layers of 50 mm thick panels bonded with 15 mm thick mortar was recorded as the highest diagonal compression load capacity of 48.13 kN with vertical displacement of 14.05 mm.

The failure mechanism of wallettes mainly due to the crushing the diagonal corner of wallettes and it was followed by the bonding failure of wood-wool matrix without damaging the connection.

ACKNOWLEDGMENT

This research is financially supported by Exploratory Research Grant Scheme (ERGS) Ministry of Higher Education Malaysia and laboratory testing is supported by Faculty of Civil Engineering, Universiti Teknologi Mara, Malaysia which are greatly acknowledged.

REFERENCES

Ahmad, Z. Wee, L.S. & Fauzi, M.A. 2011. Mechanical Properties of Wood-wool Cement Composite Board Manufactured Using Selected Malaysian Fast Grown Timber Species. *ASM Science Journal*, 5(1): 27–35.

ASTM (2002). ASTM E 519–02, Standard Test Method for Di-agonal Tension (Shear) in Masonry Assemblages. ASTM International, West Conshohocken, PA.

Fatihah, W.W.M. 2011. *Behaviour of Wall Constructed Using Wood-Wool Cement Board*, Master Dessertation, Universiti Teknologi Mara.

Firdaus, M.B. 2012. *Behaviour of Wall Using Wood-Wool Cement Panel*, Master Dessertation, Universiti Teknologi Mara.

Goverse, T. Hekkert, M.P. Groenewegen, P. Worrell, E. & Smits, R.E.H. 2001. Wood innovation in the residential construction sector; opportunities and constraints. Resources, *Conservation and Recycling*, 34(1): 53–74.

Manalo, A. 2013. Structural behaviour of a prefabricated composite wall system made from rigid polyurethane foam and Magnesium Oxide board. *Construction and Building Materials*, 41: 642–653.

Md Noh, Ahmad, Z. & Ibrahim, A. 2014. Axial compression behaviour of wallettes constructed using wood-wool cement composite panel. *Advanced Materials Research*, 1051: 671–677.

Conceptual design automation: Consideration of building materials impact at early stages of AEC design

S. Abrishami

University of Portsmouth, Portsmouth, UK

ABSTRACT: Advance consideration of building materials at early design stages can enhance sustainability and energy consumption, as well as overall performance of the building. Proposed method and conceptual prototype within this research, enable designer to analyze building performance by considering building materials (with different values) at conceptual design level. The prototype using genetic algorithm for generation of design alternatives, and BIM integration allow designers to study and analyze different building materials (using the model) by exploiting different BIM simulation capabilities (i.e. energy analysis, cost, schedule, etc.). Amongst a range of benefits (using evolutionary systems), integration of building materials is capable of having a notable impact on the construction process lifecycle if considered correctly. This paper present a conceptual framework which exploits genetic algorithm for generation of design alternatives by reflecting materials at the early design stages (through parameterization). Therefore, material parameters will be added to the evolutionary system as design requirements and constraints for generation.

1 INTRODUCTION

Application of the evolutionary design could bring about some benefits in Architecture, Engineering, and Construction (AEC) domain. Moreover, building materials analysis can be advanced using Building Information Modelling (BIM) within their integration to early and mid-level design levels. However, BIM tools are often lacking in their ability to incorporate conceptual design requirements at early design stages. Amongst the emerging design automation systems for conceptual design, generative systems have been assisting designers to rapidly explore design solutions and can enhance design process by saving time and effort and assess more possible alternatives to the design requirements (Narahara & Terzidis 2006). Whilst generative tools assist AEC designers in their design projects, it fails to meet the very basic principles of information modelling and data management requirements. In order to address this problem, 'Generative BIM workspace' (G-BIM) developed by the researcher (Abrishami et al. 2014a), enables design creativity, fluidity, and flexibility by the adoption of generative design approach. Using such an integrated platform, relevant information to the design requirements can form the system input, and the design algorithms can generate the design output, which can assist designers to solve complex multi-criteria design problems.

2 BACKGROUND: EVOLUTIONARY DESIGN

Formerly, the parametric approach used to be entirely successful to optimize a single aspect of a design model at a late design stage. However, possessing the ability to evolve the overall design of a building at an early stage in the design process would engender an extremely wider range of potential benefits. For this to be possible, the design models being evolved would need to vary significantly from one another, and as a result, the generative approach can enhance the design process.

The evaluation step is responsible for the relative assessment of the design models in the population at any given time. This further signifies that all design models can be meaningfully compared and contrasted to one another. When there is a significant restriction in the variability of design models, three main problems related to evaluation may arise; in the first place, when variability is not restricted, almost all of the generated models turn out to be chaotic forms that are uninterpretable, either by the designer or the computer, as a design. One possible choice for the evolutionary system is to determine the models that can be interpreted as designs and discard any chaotic models. Secondly, to compare and contrast the designs employed by various architectural concepts and languages requires an intuitive reasoning to be made. In other words, the betterment and priority of any given design in comparison with another one would be very much due to personal taste. As a consequence, such judgements cannot be performed by the computer. One possibility is to grant the designer the permission to stay in contact with the evolutionary system. Typically, programs used for analysis and simulation of designs require designs to be specified as complicated representations that adopt high-level semantic concepts to describe a design. For instance, representations generally include spaces, walls and floors in specific arrangements. Conversely, generative programs that produce unlimited variability describe designs through making use of basic low-level geometric primitives. Some researchers, in spite of the aforementioned problems, have seriously followed the development of evolutionary systems which are able to evolve designs that vary in exceedingly unlimited ways. The major motivation lying behind this remarkably generic approach is the tendency to avoid restrictive processes that are likely to eliminate the best designs.

The evaluation of a design solution is implemented through the fitness score that can be defined by any user or system. Furthermore, the fitter the solutions are, the more probable are the solutions to be selected for reproducing the next generation of design solutions. Consequently, there is constant optimisation after each generation through imitating natural evolution. On the other hand, the situation with the parametric associative model is that it provides hardly noticeable function for cogent exploration of design alternatives; as a consequence, users have to manually tune the design by adjusting the parameters to reach a final design. The quality of the design is still extremely centred on the number of design alternatives which have been explored to bring about the final design. Whereas, BIM tools provide parametric design and modelling, lacking function to hone cogent exploration of design space is a regularly consistent matter.

2.1 Examples of developed systems

There had been a number of experimental generative evolutionary design systems developed, which have yielded successful outcomes. These systems employ countless number of generative techniques to generate designs, including Genetic Algorithm (GA), shape grammars, L-systems, and cellular automata and the forth. Some examples of generative design systems are provided below:

Voxels (Baron et al. 1999): Evolve forms containing aggregations of elementary particles. A grid of voxels is initially defined and the state of each voxel in the grid is encoded then by the genotype.

Two-dimensional orthogonal generation: Rosenman and Gero (1999) have developed several systems to evolve two-dimensional plans for various buildings. The aforementioned plans are generated by making use of a set of shape grammar growth rules that exert small alterations to an existing plan. What the genotype does afterwards is to encode the selection and application of rules.

Evolving structural trusses (von Buelow 2007): A connectivity matrix defines the topology of the truss, and the positions of the joints define the geometry. The genotype, afterwards, encodes both the topology matrix and the joint positions.

Space frame structures (Shea 1997): Generation of the structures is accomplished by using a set of shape grammar growth rules that are able to add, replace and modify structural members using simulated annealing.

Form Generation: Frazer (2002), Coates et al. (1999), Jackson (2002), and Bentley (1999) have developed evolutionary systems which are mainly used for evolving forms and solid object designs. A set of evaluation routines is employed by these systems to assess the designs.

On the positive side, a myriad of the aforementioned systems are capable of generating complex forms that are dissimilar in overall make-up and configuration. However, the downside to these systems is that none is able to evolve three-dimensional forms that simulate buildings with embedded information which can be used to analyse the overall performance of the building.

3 FRAMEWORK DEVELOPMENT

A comprehensive research study has been carried out in developing interactive design models for three-dimensional graphics design by Takekata et al. (2005), Wakayama et al. (2009), and Akase et al. (2012) parallel to other research groups' work in evolutionary design. The primary focus has been on applying interactive evolutionary computation to three-dimensional digital models. The basic idea is to use evolutionary computing method to automatically produce a set of design alternatives by making use of an evolutionary computing method. Furthermore, each design fitness will be evaluated through different categories by the designer. Afterwards, the evaluation scores of the designs will be transferred to the evolutionary computing method to evolve the next generation of designs. The repetition of the process can take as long as desired to obtain a final design population.

A glimpse through the existing literature in the area of evolutionary design for digital graphics, architectural and geometry designs indicates that there is clearly mounting interest in creating intelligent design tools to carry out effective exploration of the design space. More specifically, as put forth by Séquin (2007), BIM tools are considered the most helpful in the final phases of design nowadays. However, a remarkable amount of the validation depends on much detailed, tedious computation, which are offloaded to machines by human beings. What seems to be the case is that today's computational tools are probably the least contributory when it comes to the commencement of the design process in the initial creative phase of the conceptual design. Consequently, for the purpose of developing the AEC computational support tools function to support the initial creative design, it is extremely vital to conduct research in developing a general framework and a tool that fulfils the exploration of the design space. The proposed framework presents a valuable set of rubrics in order to support the early design process, specially:

Creation of models with relevant links to all required information and details for the development process;
Creating a generative process is capable of controlling the variability of design outcomes, and generation of designs with required level of complexity. Moreover, generate alternatives that differ significantly in terms of overall organization and configuration;
Creating an innovative collaborative environment which enables designers to communicate in an efficient way through conceptual design phases (enable both short-term asynchrony and long-term asynchrony);

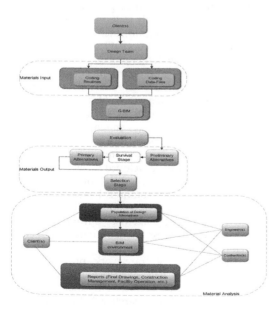

Figure 1.   Conceptual framework.

Enable designers to analyze building materials at early stages of design;
Allow designers to simulate the energy consumption of different design from the population of alternatives;

Enable designers to analyze the overall performance of any design using BIM advance capabilities (i.e. energy analysis, cost, schedule, etc.) by applying BIM integration.
Figure 1 presents the proposed conceptual framework.

4   PROTOTYPE DEVELOPMENT

During the development of the theoretical foundations of this study, content analysis as a qualitative approach (Creswell 2003) was adopted in order to uncover a deep understanding of the current state of computational support during the conceptual architectural design phase (Abrishami et al. 2015). The main issues focused on identifying the theoretical framework (for adopting generative design) as a method of automation for conceptual design.

Existing evolutionary systems have ostensibly been formed based on source-code libraries or as programming toolkits (Alba et al. 2002). Contrarily, they differ from each other based on their architecture and the utilized methods for data creation and retrieval. Existing evolutionary tools are rarely implemented as ready-made menu-driven systems. G-BIM proposes ready-made menu-driven support tool consisting of the integration of evolutionary design with Autodesk Revit. The users are not envisaged to be programmers, nor experts in GA. Therefore, the proposed architecture uses an approach similar to some current tools such as Grasshopper in terms of Graphical User Interface (GUI). Moreover, the BIM integration allows the tool to exploit exiting Revit material functions at the generation and fitness step.

The proposed architecture of the generative system uses an approach that is a generalized version of the concept-seeding approach by Frazer (2002). In particular, the architecture divides the evolutionary system into two parts: a generic core and a set of specialized components. The generic core provides the main infrastructure for creating an evolutionary design system. The specialized components, on the other hand, must be defined by the design team and will include a set of rules and representations that define the transformations performed by the evolution steps. During the evolutionary process, the generic core will invoke the specialized components, which may invoke other specialized components.

The prototype developed by Abrishami et al. (2013) fulfils all four conditions by means of careful controlling of the variability of the design models. In the first place, the design encompasses the typical kinds of constructs that characterize buildings, like spaces, walls, floors, windows and doors. Secondly, all of the designs share an identical design. Thirdly, the designs are somewhat varied in terms of the overall form, space organization, and the treatment of the facades. Lastly, all designs are shown by high-level semantic constructs that are constrained so as to make sure that merely valid building designs can be generated.

In this research, Autodesk Revit was selected according to a survey as the central BIM tool for G-BIM (Abrishami et al. 2014b), largely due to the fact that it is featured by having a well-documented Application Programmer's Interface (API) and Revit's advance materials function. A glance through a development perspective indicates that a tremendous merit of Autodesk is the pre-existing modelling capabilities used for visualization and user interaction. When it comes to G-BIM's users, it is highly advantageous in the sense that learning how to use the tool is way easier compared to other tools since it is part of the Revit environment. G-BIM's data object level integration within Revit gives an indication to the fact that regular addressable Revit information models, which are highly propitious, allow the user to manipulate it in the usual Revit ways.

All in all, when it comes to the process of generating new designs, G-BIM implemented works with Revit environment in an interactive manner. This application functions basically as follows: receiving the input design data which is to be optimized; equipping the user with a GUI which provides them with the opportunity to select the design parameters including requirements, constraints, and site data; enhancing the design parameters through using GA engine, and generating new design files; assessing the designs and proceed to the next generation; Repeating last two steps to the point where decent or most favorable solutions are obtained.

In order to employ GA to hammer out a reasonable solution, Revit API which provide the functionality for managing Revit engine were used. Technically, the GA is initiated with the input data provided by the user. Thereafter, to proceed to the next generation; doing so makes a new population of solution to be generated while GA will continue to progress with the next generation. Finally, to pause and stop the engine, in this stage, GA data processing will be either paused or discontinued.

There are three kinds of data used by G-BIM engine which are the design requirements, the solution generation and the fitness score, the description of each is provided as follows:
Design requirements: The input data comprises mainly of design variables and the parameters that are used by the engine for the purpose of optimization. It is worth knowing that GUI is responsible for generating the input data. The population size needed at the beginning, and the other engine parameters are further specified by the user through using G-BIM GUI, which are all sent to the engine as input.

Solution generation: Once a generation of GA optimization is obtained, GA engine records population of design solution. Every single solution is made use of to create a design, which will be assessed by a fitness score (using Revit engine) to measure the merit of the design solution in accordance with the predefined criteria.

Survival stage: after the initial population is created, the fitness of each solution ought to be computed and transferred into a score file. The engine makes use of the fitness values to continue producing the next generation of solutions. This process has to be done repeatedly up to the time the desirable or near-optimal solutions are obtained.

5 CONCLUSIONS

This paper presented a G-BIM framework with the capability of integrating building materials analyzing and optimizing design solutions at the conceptual design stage to: provide techniques for exploring and generating design solutions; create models with appropriate information and details needed for the development process; analyze materials impact at early design stages; create a generative process capable of controlling the variability of design outcomes (i.e., generating alternatives that differ significantly in terms of overall organization and configuration), including the generation of designs with the required level of complexity at each stage of the design process.

The proposed framework/prototype adopts the same approach as conventional and existing design process. Even though it enables design creativity, fluidity, and flexibility by the adoption of generative design, it makes minimal changes to the common design process. Therefore, relevant information to the design requirements forms the system input, and the developed system generates the design output within the BIM context. The system will provide design solutions based on input data such as site data, constraints, building materials, and requirements; likewise, during the conventional design process, the same data will be considered by the designer.

The conceptual framework provides for the exploitation of new concepts in computational design and architecture. The theoretical basis underpinning this research will be used to integrate building materials input into the G-BIM prototype. From a research limitation perspective, it is acknowledged that whilst this work is relatively in its infancy, primary data and empirical evidence presented in this paper support these findings. That being said, it is equally important to acknowledge that further studies are needed to develop and validate this framework, using domain experts and focus-groups (development iteration), in order to capture the precise rubrics and parameters needed to shape and further refine this model as a part of the holistic development process.

REFERENCES

Abrishami, S. Goulding, J. Pour Rahimian, F. & Ganah, A. 2015. Virtual generative BIM workspace for maximising AEC conceptual design innovation: A paradigm of future opportunities. *Construction Innovation*, 15(1): 24–41.

Abrishami, S. Goulding, J.S. Ganah, A. & Pour Rahimian, F. 2013. Exploiting modern opportunities in AEC industry: A paradigm of future opportunities. *American Society of Civil Engineers*, AIE 2013: 321–333.

Abrishami, S. Goulding, J. Rahimian, F.P. Ganah, A. & Sawhney, A. 2014a. G-BIM framework and development process for integrated AEC design automation. *Procedia Engineering*, 85(0):10–17.

Abrishami, S. Goulding, J.S. Pour-Rahimian, F. & Ganah, A. 2014b. Integration of BIM and generative design to exploit AEC conceptual design innovation, *Journal of Information Technology in Construction*, 19:350–359.

Akase, R. Nishino, H. Kagawa, T. Utsumiya, K. & Okada, Y. 2012. *An avatar motion generation method based on inverse kinematics and interactive evolutionary computation*. In Sixth International Conference on Complex, Intelligent, and Software Intensive Systems, CISIS 2012, Palermo, Italy, July 4–6, 2012: 741–746.

Alba, E. Nebro, A.J. & Troya, J.M. 2002. Heterogeneous computing and parallel genetic algorithms. *Journal of Parallel and Distributed Computing*, 62(9): 1362–1385.

Baron, P. Fisher, R. Tuson, A. & Mill, F. 1999. A voxel-based representation for evolutionary shape optimization. *Artificial Intelligence for Engineering Design, Analysis and Manufacturing*, 13(3): 145–156.

Bentley, P. 1999. From coffee tables to hospitals: Generic evolutionary design. *Evolutionary Design by Computers*, 18: 405–423.

Coates, P. Broughton, T. & Jackson, H. 1999. *Exploring three-dimensional design worlds using Lindenmayer Systems and Genetic Programming*, Evolutionary design using computers, 323–341.

Creswell, J. 2003. *Research Design: Qualitative*, Quantitative, and Mixed Methods Approaches. SAGE Publications.

Frazer, J.H. 2002. *Creative Evolutionary Systems, chapter Creative design and the generative evolutionary paradigm*, 253–274. Evolutionary Computation Series. Morgan Kaufmann, Academic Press, London, UK.

Jackson, H. 2002. *Toward a symbiotic coevolutionary approach to architecture*. Creative Evolutionary Systems, Morgam Kaufimann, San Francisco, 299–312.

Narahara, T. & Terzidis, K. 2006. *Multiple-constraint genetic algorithm in housing design*. In International Conference, Synthetic Landscapes, Digital Exchange, Louisville, USA. 12–15 October.

Rosenman, M.A. & Gero, J.S. 1999. *Evolving designs by generating useful complex gene structures*. Evolutionary Design by Computers, Morgan Kaufmann, San Francisco, 345–364.

Séquin, C.H. 2007. Computer-aided design and realization of geometrical sculptures. *Computer-Aided Design and Applications*, 4(5): 671–681.

Shea, K. 1997. *Essays of Discrete Structures: Purposeful De-sign of Grammatical Structures by Directed Stochastic Search*. Ph.D. thesis, Carnegie Mellon, University Pittsburgh, PA.

Takekata, K. Azmi, M.S.B. Nishino, H. Kagawa, T. & Utsumiya, K. 2005. *An intuitive optimization method of haptic rendering using interactive evolutionary computation*. In Proceedings of the IEEE International Conference on Systems, Man and Cybernetics, Waikoloa, Hawaii, USA, October 10–12, 2005: 1896–1901.

von Buelow, P. 2007. Advantages of evolutionary computation used for exploration in the creative design process. *Journal of Integrated Design and Process Science*, 11(3): 5–18.

Wakayama, Y. Takano, S. Okada, Y. & Nishino, H. 2009. *Motion generation system using interactive evolutionary computation and signal processing*. In NBiS 2009, 12th International Conference on Network-Based Information Systems, Indianapolis, Indiana, USA, 19–21 August 2009: 492–498.

Advances in Civil, Architectural, Structural and Constructional Engineering – Kim, Jung & Seo (Eds)
© 2016 Taylor & Francis Group, London, ISBN 978-1-138-02849-4

Measurement of the static and dynamic characteristics of the cable-stayed crossing gas pipe bridge over the Fu-Jiang river

C.J. Liu
State Key Laboratory of Geohazard Prevention and Geoenvironment Protection, Chengdu University of Technology, Chengdu, China

Z.L. Zheng
College of Civil Engineering, Chongqing University, Chongqing, China

Y.X. Zeng & X. Zheng
College of Environment and Civil Engineering, Chengdu University of Technology, Chengdu, China

ABSTRACT: Vibration acceleration method is employed to measure the vibration frequency, damping ratio, vibration mode, acceleration, velocity and displacement at different points of the Fu-Jiang pipe bridge, and parameters relating to vibration displacement and tension of stayed cables of the pipe bridge were measured in different pigging conditions. In the course of field measurement, acceleration sensors are fixed by magnetic supports at 1/2, 1/4 and 1/8 span, two pedestals and two compensators, respectively. The detected signals are transmitted to an INV-306 intelligent signal acquisition and processing analyzer after being amplified by a charge amplifier, and then stored in a computer to be treated by DASP2000 data processing program to obtain the desired results.

1 GENERAL SITUATION OF THE PIPE BRIDGE

Fu-Jiang crossing gas pipe bridge is close to Jiangyou city, Sichuan Province, China. It is a three-span cable stayed pipe bridge structure with two towers and vertical compensator. The length of main span length is 320 m. The main gas pipe is made of double-sided submerged arc welding spiral steel; its size is $\varphi 720 \times 14$ mm. Simple rail is employed as the safety measures of maintenance, but without walkway plates on the gas pipe. The tower frame of the bridge is conical fully welding steel frame structure. The height of the two towers equals 50 m. The cable-stayed pulley block was installed on the overhead of the tower to adapt deformation and internal force adjustment of the stayed cable. Each tower is the symmetric axis of the stayed cable. Eight pairs of stayed cable were installed on the two sides of each tower. The stayed cable is galvanized steel wire rope, and the sizes of the stayed cable are $\varphi 43.5$ mm, $\varphi 34$ mm, $\varphi 28$ mm and $\varphi 22$ mm. The structural schematic diagram of the pipe bridge is shown in Figure 1.

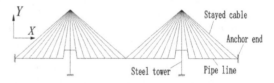

Figure 1. The structural schematic diagram of the Fu-Jiang crossing gas pipe bridge.

2 ITEMS OF THE MEASUREMENT

According to the request of the project contract of "Pigging Dynamic Study of the Wide or Medium Span Crossing Pipeline" in the engineering technology study of the "West-to-East Gas Transmission" project, we did on-line test on the Fu-Jiang crossing gas pipe bridge. The items of measurement are listed as follows.

1. Vibration frequency, damping ratio and vibration mode of the pipe bridge.
2. Vibration acceleration, velocity, and displacement of the pipe bridge.

3. Vibration displacement of the pipe bridge under different pigging working condition.
4. Pretension of the stayed cable of the pipe bridge.
5. Positions of measurement: 1/2, 1/4 and 1/8 span of the pipe bridge, the pedestal and compensator of the pipe bridge.

3 METHOD AND THEORY OF THE MEASUREMENT

We measured the vibration acceleration of each measuring point of the pipe bridge to achieve the measurement of related parameters. This method is readily available, but it usually aims at one component (stayed cable, pipe bridge, tower frame and so on) (Wang et al. 2014, Park et al. 2013, Carmelo & Alberto 2006, Carmelo et al. 2001, Conte et al. 2008, Cai et al. 2009, Brownjohn et al. 1999). The purpose of this project was to measure the whole large flexible crossing structure. The layout of the measuring points and the interaction of each part of the structure could cause the overrun of the measured values and so on, which are all challenging problems. In addition, the noise deletion in the measuring process and the new situation under the pigging state would increase the difficulty of the problem.

Because the input excitation is not easy to be accurately measured and simulated, the input response signal is the only signal that used in the pulsation experiment (Zhang et al. 1996, De Miranda et al. 2010, Brownjohn & Xia 2000). Also since modal parameters of the structure are inherent characteristics of the structure, they must appear in output signals every time. Therefore, it would be very easy to eliminate noise in the signal analysis (Chen et al. 2002, Nakamura 2007), thus recognizing modal parameters of the structure correctly. For large flexible structures, this method is valid. The natural frequency can be obtained from the peak value of the response spectrum.

The acceleration sensor was installed on the corresponding positions of the structure by magnetic supports in field measurement (Zhu et al. 2003). The signal was amplified by charge amplifier and input into the INV-306 intelligent signal acquisition and process analyzer, and stored in computer for all kinds of processing and analysis. Generally, the analysis includes time and frequency domain. The contents of the time domain analysis include amplitude, low frequency phase characteristics and damp of the structure while those of the frequency domain analysis consist of spectrum characteristics, phase characteristics, vibration mode and damp of each order and so on (Salawu & Williams 1995, Sokol & Flesch 2005, Du et al. 2002). For the concrete analysis process, in order to obtain the natural frequency accurately, it is necessary to pretreat and refine the analysis steps according to the complex degree of waveform record, analysis precision and so on. Meanwhile, camera watch, human judgment and other comprehensive measures must be used to solve this problem.

3.1 Vibration frequency and damping ratio

There are always some interference signals in the input and output signals of the practical measurement system. The auto-power spectrum of the measurement value of the system output response STT(f) is made up of the output spectrum obtained by input excitation through linear system SVV(f) and the output spectrum caused by uncertain factors SNN(f). Especially, because the value of ambient excitation amplitude is too low, or measurement point of the output response is close to the node of mode shape, the output spectrum obtained by input excitation SVV(f) will be too small, then the influence of the output caused by uncertain factors will be more obvious. In view of this situation, we used coherence function to determine the influence degree of SNN(f). In addition, for the large flexible structure, there is coupling phenomenon in each mode shape, so we used the phase of the measurement value of output to determine the coupling vibration frequency of the structure.

The data of output response spectrum and half-power bandwidth method was employed to estimate modal damping ratio,

$$\xi_i = B_i / 2 f_i \qquad (1)$$

where, ξ_i is the damping ratio of the i order mode shape; B_i is the half-power point bandwidth of the i order mode shape response peak value; f_i is the recognition frequency of the i order mode shape.

3.2 Determination of the mode shape

We used the amplitude value and phase of the measurement points and the transfer function of the corresponding reference points to determine the mode shape approximately. That is the ratio of the cross and self spectrum between arbitrary measurement point and reference point, including the magnitude and positive or negative sign of the amplitude value.

3.3 Acceleration, velocity and displacement

We used acceleration sensor to measure the vibration acceleration, and integrate the vibration acceleration to obtain the velocity and displacement of the corresponding point.

3.4 Tension of stayed cable

The length of stayed cable is generally several tens meters, comparing to the length of the whole bridge cable and the tension. The stayed cable could be regarded as a suspended chord fixed on the two ends and its dead weight can be neglected.

Basic formula of the stayed cable tension is,

$$T = \frac{4l^2 f_n^2 \rho}{n^2} = 4l^2 f_1^2 \rho \qquad (2)$$

where, T is the tension; l is the length of stayed cable; f_n is the natural frequency of n order of the stayed cable; f_1 is the basic frequency of the stayed cable; ρ is the line density of stayed cable; n is the order number.

Therefore, if the basic frequency of the stayed cable is measured, the theoretical tension of the stayed cable can be calculated (Zhang & Jiang 1996). In order to measure the basic frequency of the stayed cable, the acceleration sensor was fixed on the stayed cable, and it is on the vertical plane. The measured actual vibration signals of the stayed cable include the basic and high order frequency, and other kinds of interference. Because the stayed cable is very long; the middle point of the stayed cable is very high above the ground; the condition restriction in the actual measurement, the acceleration sensor only can be installed on the stayed cable near the pipe bridge. The basic frequency part is relatively small in the signals obtained by the acceleration sensor. Using several resonance peak values to get the corresponding basic frequency; and the average value of them is the basic frequency of the stayed cable.

The acceleration sensor was fixed on the corresponding positions of the structure by magnetic supports in field measurement. Signals were amplified by charge amplifier and input into the INV-306 intelligent signal acquisition and processing analyzer, and stored in computer. The DASP2000 data processing and analysis program was used to process these signals. The distribution of measurement instrument is shown in Figure 2.

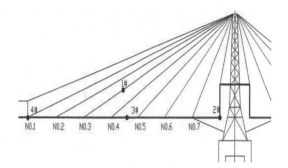

Figure 3. The distribution of measuring points.

Table 1. The measuring results of natural vibration characteristic of Fu-Jiang crossing project.

Direction	Frequency (Hz)	Period (s)	Amplitude spectrum (m/s²)	Damping ratio
Transversal	0.0821	12.180	35.242	0.098
	0.107	9.346	16.261	0.080
	0.150	6.667	9.902	0.078
	0.193	5.181	9.720	0.062
	0.257	3.891	6.203	0.057
	0.281	4.762	4.887	0.018
	0.386	2.591	5.101	0.013
Vertical	0.3281	2.618	0.855	0.048
	0.468	2.137	0.0261	0.040
	0.656	1.524	0.0212	0.038
	0.843	1.186	0.0520	0.034

The measuring points were set on the pipe bridge according to the request of the measuring positions. The distribution of these measuring points is shown in Figure 3.

4 ANALYSIS OF THE MEASURING RESULTS

4.1 Results of natural vibration characteristic

The natural vibration characteristics of Fu-Jiang pipe bridge are shown in Table 1.

4.2 Results of the cable tension

We measured the tension of the stayed cable of the Fu-Jiang crossing project under the normal working condition. The measuring results were shown in Table 2. The results in Table 2 basically tallies with the results in Ref. (Zhu & Wang 2003) (Yao et al. 1999)

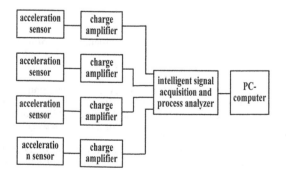

Figure 2. The measurement instrument and equipment of the crossing gas pipe project.

Table 2. The measuring results of natural vibration characteristic of Fu-Jiang crossing project.

Cable No.	Length (m)	Diameter (mm)	Line density (kg/m)	Basic frequency (Hz)	Tension (kN)
1	155.45	45.0	8.260	0.393	123.7
2	136.25	45.0	8.260	0.450	124.6
3	117.32	42.0	7.203	0.529	123.7
4	98.816	42.0	7.203	0.665	124.6
5	81.022	36.0	5.292	0.837	97.5
6	64.530	36.0	5.292	1.046	96.5
7	50.642	25.5	2.659	1.517	62.8

Table 3. The measuring results of the vertical vibration displacement of Fu-Jiang crossing project.

Working condition		Measuring position		
		1/2 span point	1/4 span point	Compensator point
1	Transmission pressure, MPa	1.23	0.021 m	−0.074 m	0.053 m
	Rate of flow, × 104 m³/d	77.00			
	Injected water volume, m³	0.00			
	Average velocity of pigging, m/s	1.84			
2	Transmission pressure, MPa	1.13	0.094 m	−0.412 m	−0.102 m
	Rate of flow, × 104 m³/d	73.00			
	Injected water volume, m³	3.00			
	Average velocity of pigging, m/s	2.01			
3	Transmission pressure, MPa	1.12	0.161 m	−0.762 m	−0.616 m
	Rate of flow, × 104 m³/d	77.00			
	Injected water volume, m³	6.00			
	Average velocity of pigging, m/s	4.08			

4.3 Results of the pigging vibration displacement

We measured the vibration displacement of Fu-Jiang crossing project at the points of 1/2, 1/4 span and compensator. The results were shown in Table 3.

5 CONCLUSIONS

This work has to lead the following results:

1. The natural frequencies of Fu-Jiang cable-stayed pipe bridge in each direction are relatively concentrated. The transversal basic frequency is 0.061 Hz, and the others are 0.107, 0.150, 0.193, 0.257, 0.281, 0.323, 0.386 (Hz). The vertical basic frequency is 0.3281Hz, and the others are 0.468, 0.656, 0.843 (Hz).
2. It is known from Table 1, the difference between the transversal and vertical damping ratio of Fu-Jiang cable-stayed pipe bridge is little. The value of damping ratio of corresponding frequency of each order will decrease with the increase of the frequency.
3. The maximum vibration displacement of Fu-Jiang cable-stayed pipe bridge didn't appear at mid-span, but appear at the section from 1/4 span to compensator. This provide basis for guiding field pigging and making damping measures effectively.

ACKNOWLEDGEMENT

This work was supported by the sub-project of the "West-to-East Gas Transmission" engineering technology study. The project number is 010201-01.

REFERENCES

Brownjohn, J.M.W. & Xia, P.Q. 2000. Dynamic assessment of curved cable-stayed bridge by model updating. *Journal of structural engineering New York, N.Y.*, 126(2): 252–260.

Brownjohn, J.M.W., Lee, J. & Cheong, B. 1999. Dynamic performance of a curved cable-stayed bridge. *Engineering Structures*, 21(11): 1015–1027.

Carmelo, G. & Alberto, G.S. 2006. Dynamic testing and modeling of a 30-years' old cable-stayed bridge. *Structural Engineering International: Journal of the International Association for Bridge and Structural Engineering (IABSE)*, 16(1): 39–43.

Carmelo, G. & Francesco, M.Y.C. 2001. Dynamic assessment of a curved cable-stayed bridge at the Malpensa 2000 Air-port, Milan, Italy. *Structural Engineering International: Journal of the International Association for Bridge and Structural Engineering (IABSE)*, 11(1): 52–58.

Cai, C.S., Araujo, M. & Nair, A. et al. 2009. Static and dynamic performance evaluation of a pre-stressed concrete bridge through field testing and monitoring. *International Journal of Structural Stability and Dynamics*, 9(4): 711–728.

Chen, Z.W. & Wang S.C. 2002. Realization of Digital Filter in Signal Processing System. *Journal of Vibration and Shock*, 21(3): 18–21.

Conte, J.P., He, X.F. & Moaveni, B. et al. 2008. Dynamic testing of Alfred Zampa Memorial Bridge. *Journal of Structural Engineering*, 134(6): 1006–1015.

De Miranda, M., De Palma, A. & Zanchettin A. 2010. "Ponte del mare": Conceptual design and realization of a long span cable-stayed footbridge in Pescara, Italy. *Structural Engineering International: Journal of the International Association for Bridge and Structural Engineering*, 20(1): 21–25.

Du, J.C., Pan, J.J. & Zhuang, B.Z. et al. 2002. Measuring the tension of stayed cable by vibration method. *Journal of Vibration and Shock*, 21(1): 98.

Nakamura, S.I. 2007. Static and aero-dynamic studies on cable-stayed bridges using steel pipe-girders. *Structural Engineering International: Journal of the International Association for Bridge and Structural Engineering*, 17(1): 68–71.

Park, J.W., Sim, S.H. & Jung, H.J. et al. 2013. Development of a wireless displacement measurement system using acceleration responses. *Sensors (Switzerland)*, 13(7): 8377–8392.

Salawu, O.S. & Williams, C. 1995. Bridge assessment using forced-vibration testing. *Journal of structural engineering New York, N.Y.*, 121(2): 161–173.

Sokol, M. & Flesch, R. 2005. Assessment of soil stiffness properties by dynamic tests on bridges. *Journal of Bridge Engineering*, 10(1): 77–86.

Wang, H., Cheng, H.Y. & Li, A.Q. et al. 2014. Whole-process measurement of buffeting response of Sutong Bridge under action of Typhoon Haikui. *Bridge Construction*, 44(4): 15–21.

Yao, J.F., Xue, D.J. & Yue, Q. 1999. A study of the online measurement on diagonal pulling wire rope tension of aerial crossing gas pipeline bridge over the Fu-Jiang River. *Natural Gas Industry*, 19(3): 81–83.

Zhang, J.X. & Jiang D.F. 1996. The Device and Measurement System of Three Dimensional Acceleration Sensor. *Journal of Vibration and Shock*, 15(2): 81–91.

Zhu, H.P. & Wang D.H. 2003. Parametric Study of dynamic characteristics of cable-stayed bridges. *Journal of Vibration and Shock*, 3(22): 12–15.

Advances in Civil, Architectural, Structural and Constructional Engineering – Kim, Jung & Seo (Eds)
© 2016 Taylor & Francis Group, London, ISBN 978-1-138-02849-4

Experimental testing of Solidification Products containing hazardous waste-Neutralization Sludge with aim of material utilization

Jakub Hodul, Rostislav Drochytka & Božena Dohnálková

AdMaS Centre, Faculty of Civil Engineering, Brno University of Technology, Brno, Czech Republic

ABSTRACT: In order to reduce the amount of hazardous waste-Neutralization Sludge (NS) this work deals with the possibility of using solidification and stabilization (S/S) technology to transform the hazardous waste with the aim to achieve material utilization of Solidification Product (SP). NS have been characterized as waste arising after neutralization of waste acids from various industrial processes that contain hazardous substances, mainly heavy metals. Solidification Products (SPs) prepared according to five different solidification formulas were tested according to the optimal testing methodology. Physical and mechanical properties-compressive strength, bulk density, permeability were determined with the Solidification Products (SPs) and within environmental demands the leaching test was carried out. Solidification Products (SPs) have to meet required properties to ensure its future safe use, e.g. as material for base layer of roads or material for reclamation of landfills.

1 INTRODUCTION

The permanent production of hazardous waste, especially Neutralization Sludge (NS) as waste products during galvanic metal plating and metal surface treatment in the Czech Republic has a negative impact on the environment—extension of inconvenient and unsecured landfills. The neutralization sludge contains huge amounts of heavy metals and it is difficult to find a technology using which these contaminants will be firmly incorporated into a new and further efficiently usable product. Stabilization and solidification (S/S) appears as the most appropriate technology of the treatment of sludge containing heavy metals. Inorganic binders, especially cement and fly ashes are the most commonly used solidification agents for S/S of Neutralization Sludge (NS) and also for soils contaminated by heavy metals. Current economic and environmental conditions require the highest amount of NS and the lowest content of binders in Solidification Product (SP) while meeting all the requirements. For the safe and effective future utilization of SP it is necessary to meet physical and mechanical and ecological demands result from a future optimal use of SP in the building industry or as the technological material in remediation processes.

2 IDENTIFICATION OF INPUT RAW-MATERIALS

2.1 Hazardous waste—neutralization sludge

Neutralization Sludge (NS) with the designation NS-X and NS-Y were selected as input hazardous dangerous wastes. (Dohnálková et al. 2014) characterize NS as waste generated after neutralization of waste acids from various industrial processes containing dangerous substances. Selected NS are waste products produced in consequence of waste water treatments from galvanic processes and they are produced in huge amounts (in the Czech Republic even hundreds of tons per year). (Rossini & Bernardes 2005) confirm that many heavy metals can be found in this sludge. Table 1 shows pollutant concentrations in NS-X and NS-Y in the dry matter and it is clear that both of the sludge are characterized by high content of chromium (Cr), lead (Pb) and NS-X contains also high concentration of carbohydrates C_{10}-C_{40}.

2.2 Solidification agents

Mainly secondary raw materials were used as solidification agents. Two types of fly ashes were used in all five formulas—Fly ash from Circulating

Table 1. Results of concentration of pollutants in dry matter of the NS.

Component	NS-X [mg/kg dry*]	NS-Y [mg/kg dry**]
As	12.3	1.49
Cd	0.10	0.38
Cr	854	159
Hg	0.078	0.009
Ni	68.6	77.8
Pb	247	1870
V	32.2	6.99
Carbohydrates $(C_{10} - C_{40})$	5360	23.1

*NS-X contains 45.40% dry.
**NS-Y contains 42.89% dry.

Table 2. Chemical analysis of the fly ashes and the cement.

Component	CFBC [%]	PCC [%]	CEM* [%]
SiO_2	36.90	50.16	19
Al_2O_3	13.60	27.54	5
Fe_2O_3	4.92	13.08	3
SO_3	6.12	0.07	2.6
CaO	20.10	2.51	62
MgO	1.31	1.46	2
K_2O	4.92	1.35	0.78
Na_2O	0.65	0.35	0.13
P_2O_5	0.66	0.19	–

*Cement (CEM II/B-M (S-LL) 32,5R).

Fluidized Bed Combustion (CFBC fly ash) and fly ash from Pulverized Coal Combustion (PCC fly ash). Though both of these fly ashes are produced as by-products of combustion of lignite in thermal power plant, the CFBC fly ash can be classified according to ASTM C618 in class "C" and PCC fly ash can be classified in class "F". Chemical composition of CFBC and PCC fly ash is stated in the Table 2 above. (Dermatas & Meng) demonstrated that by CFBC fly ash heavy metals like chromium (Cr) and lead (Pb) can be successfully immobilized. As other solidification agent used for preparation of Solidification Products (SPs) Portland mixed cement with granulated blast furnace slag and limestone (CEM II/B-M (S-LL) 32,5R) was used and in one formula even carbide lime (by-product in acetylene production) was applied. (Chen et al. 2009) state that essential principle of heavy metals fixation during S/S technology using cement is represented mainly by chemical reactions between heavy metals and cement hydration products with production of low soluble metal compounds.

3 PREPARATION OF SOLIDIFICATION PRODUCTS

3.1 Solidification formulas

Within this work 5 different formulas were experimentally tested. Solidification formulas with designation S1 and S4 contained neutralization sludge NS-Y and solidification formulas with designation S2, S3, and S5 formulas contained NS-X. The composition of all tested solidification formulas is shown in Table 3. Three formulas contained 50% NS by weight and two others 40%. Especially due to evaluations of cement amount effect on physical and mechanical properties of Solidification Products (SPs) higher cement amount was proposed at solidification formulas (Solidification Products (SPs) S4 and S5, by up to 15% by weight. (Qian et al. 2005) stated in his work that in order to obtain the minimal compressive strength 0.3 MPa at least 5–15% of cement must be used at technology of S/S to incorporate the heavy metals to the matrix of SP.

3.2 Preparation of test specimens

Homogenization of all solidification mixtures was carried out using manual electrical mixer. All components of solidification formulas were dosed by weight while firstly the Neutralization Sludge (NS) was weighed which was subsequently crushed in smaller pieces to provide better homogenization. Subsequently the solidification agents were added to the mixer vessel and mixing of the mixture started. Water was gradually added to provide the mixture required consistency. To produce samples of Solidification Products (SPs) for examination of compressive strength the mixture was fulfilled into the prepared metal triple molds of dimensions $100 \times 100 \times 100$ mm, and for production of the samples for permeability test High Density

Table 3. Composition of solidification formulas.

Component	S1 [%]	S2 [%]	S3 [%]	S4 [%]	S5 [%]
NS-X	–	50	50	–	40
NS-Y	50	–	–	40	–
FA	20	20	15	20	20
FFA	25	25	25	25	25
CEM*	5	5	–	15	15
Carbide lime	–	–	10	–	–

*Cement (CEM II/B-M (S-LL) 32,5R).

Polymer (HDPE) circular molds with diameter 120 mm were used. The samples in cubic molds were taken out after 2 days and then stored in laboratory conditions (temperature 20°C, relative humidity (R.H.) = 60%). The samples for permeability tests were stored in environment with 90% R.H., after 7 days they were taken out of the molds and remained in damp conditions till its further testing. Bulk density of SPs was determined before the determination of compressive strength with the same samples (cubes 100 × 100 × 100 mm). (Stegemann & Yhou 2009) mentioned in their work that the compressive strength of SPs containing cement and fly ash matrix should be higher when they are stored in damp conditions due to better formation of hydration products and more efficient incorporation of heavy metals in the solidification product matrix.

4 ANALYTICAL METHODS

4.1 Compressive strength

According to the study prepared by (Barth et al. 1989) untreated waste generally does not show sufficient compressive and tensile strength, however when the waste is incorporated in the cement matrix after solidification and stabilization, the strength significantly increase. (Hills & Pollard 1997) stated that Environmental Protection Agency in the USA (EPA) requires the Solidification Product (SP) to show the minimal compressive strength 0.35 MPa. This minimal set value is required for storage of SP on landfill. Study of (Zivica 1997) proved that compressive strength of Solidification Products (SPs) with cement matrix depends mainly on hydration products of the matrix, components occurring in the waste and their mutual reaction, but above all it depends on those products which cause porous structure of SP matrices. SPs including Neutralization Sludge (NS) containing heavy metals like Ni, Cr, Cd and Hg show better strength characteristics with fly ash matrix than with the cement one (Roy et al. 1992).

Mechanical properties of solidification products (SPs) prepared according designed solidification formulas were tested according to EN 12390–3: Testing of hardened concrete—Part 3: Compressive strength of specimens. The specimens, cubes with dimensions 100 × 100 × 100 mm, were loaded on test press meeting EN 12390–4 requirements. Maximum load (power in [N]) when the specimen was crushed was recorded and compressive strength of SP was calculated. Figure 1 shows the results of compressive strength of SPs depending on their maturing time. Each of the samples was stored in laboratory environment till the test.

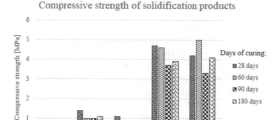

Compressive strength of solidification products

Figure 1. Results of compressive strength of Solidification Products (SPs) in different ages.

4.2 Bulk density

Bulk density is one of the physical properties of Solidification Products (SPs). (Malviya & Chaudhary 2005) stated that by increasing binder content (fly ash, cement) the pores of SP are more filled and the bulk density grows—the matrix is more compact. Bulk density is generally higher with SPs with cement matrix at the beginning of hydration (fresh state) as with completely hardened SPs, mainly due to the evaporation of excess water from the matrix.

Bulk density was determined according to EN 12390–7 with 100 × 100 × 100 mm cubes (same like with the compressive strength tests). SPs with higher content of cement in the matrix (S4, S5) had higher bulk density than other SPs.

4.3 Permeability

Permeability is expressed as permeability coefficient or hydraulic conductivity while speed during which the water is able to flow through permeable materials is measured in [m/s]. This parameter expresses ability of material to let leak liquids through its structure. Cement based SPs have certain ability to maintain water inside its structure and thus prevent excessive leaching of incorporated pollutants. Determination of permeability is important mainly for realization of S/S of contaminated soils to exclude releasing of pollutant to subsoil layers and underground water. According to (Dohnálková et al. 2014) the permeability of SP is influenced mainly by distribution of particle sizes, Liquid/Solid (L/S) ratio, compacting rate, hydration products, material homogeneity and in case of consistent monolithic SP also by the amount and redistribution of cracks. Considering the future optimal application of SP the lowest permeability is required.

The permeability was determined according to the standard EN ISO/TS 17892-11 with cylindrical shape specimens (Ø 120 mm, h = 70 mm) after 28 days of curing.

4.4 *Leaching test*

Leaching tests belong to the basic tests according to which national legislations formed and still form criteria for further safe utilization of industrial treated waste. Leachability may be used to simulate natural attenuation, evaluate the risk of contaminant leaching and reaching the natural ground water. Czech legislation specifies leaching test according the standard EN 12457–4—Characterization of waste—Leaching—Compliance test for leaching of granular waste materials and sludge—Part 4: One stage batch test. This standard specifies a compliance test providing information on leaching of granular wastes and sludge under the experimental conditions and particularly a liquid (water) to solid ratio of 10 l/kg dry matter. In the work (Song et al. 2013), it was demonstrated that leachability of heavy metals from Solidification Products (SPs) is higher with lower pH and increased temperature within semi-dynamic tests. (Sweeney et al. 1998) studied the dependence of heavy metal leaching from SPs on carbonation and found that chromium leachability is higher at SP where no carbonation was performed than at carbonated ones. It was also found that copper leaching does not depend on the carbonation phase of SP.

5 RESULTS AND DISCUSSION

5.1 *Compressive strength*

SPs containing less Neutralization Sludge (NS) and more cement (S4, S5) in the matrix showed higher compressive strength, approximately by 3 to 4 MPa. It is clear from Figure 1 that highest compressive strengths had SP prepared according to the solidification formula S5 after 60 days of curing (5.0 MPa). Compressive strength of SPs were declining in time, however a slight increase after 90 days curing was recorded at all SPs apart from S3, where the carbide lime was used instead of the cement.

5.2 *Bulk density*

It can be seen from Figure 2 that the highest bulk density had SP after 28-day curing where residual water was not fully evaporated and hydration reactions had not carried out completely yet. SPs with NS-Y content showed little bit higher bulk density, probably due to physical properties and chemical composition of this NS.

5.3 *Permeability*

Solidification Products (SPs) with higher cement content (S4, S5) have the lowest permeability as it

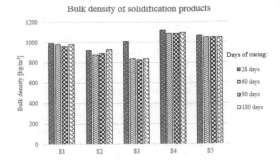

Figure 2. Results of bulk density of SPs in different ages.

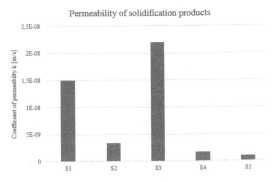

Figure 3. Results of permeability of SPs in different ages.

is clear from Figure 3. The highest permeability content was determined with SP prepared according to the solidification formula S3, where the coefficient of permeability was 2.2×10^{-8} m/s.

5.4 *Leaching tests*

Leaching tests was carried out according to EN 12457–4. On the basis of results shown in the Table 4 it can be assessed that all five solidification formulas met the limits for classification in leaching class IIb stated in particular Regulation no. 294/2005 in Czech Republic.

Compared to the leaching tests of input neutralization sludge (NS-X, NS-Y) there is in the leachates of SPs lower concentration of some parameters, especially sulphates and Dissolved Solid Substances (DSS). Water leachates of SPs with designation S4 and S5 contained less DSS, sulphates, As and Cu compared to others mainly due to higher cement content in the matrix of SPs which is related to higher pH and better incorporation and stability of heavy metals in matrix. None of the SPs showed excessive leaching of heavy metals.

Table 4. Results of leaching tests.

Parameter	S1 [mg/l]	S2 [mg/l]	S3 [mg/l]	S4 [mg/l]	S5 [mg/l]
pH*	11.3	9.8	9.7	11.5	11.6
Dissolved solid substances	2450	2980	2990	1840	866
Chlorides	904	45.9	27.8	598	46.1
Sulphates	122	1760	1430	87.6	466
As	<0.001	0.004	0.007	<0.001	0.003
Ba	0.152	0.127	0.081	0.122	0.008
Cr	<0.03	<0.03	<0.03	<0.03	0.1
Cu	0.025	0.111	0.374	0.023	0.14
Ni	<0.02	<0.02	0.017	<0.02	<0.02
Pb	<0.001	<0.001	<0.001	<0.001	<0.001
Se	0.002	0.007	0.007	<0.001	0.004
V	<0.01	0.015	<0.01	<0.01	0.039
DOC**	23	54	65	16	33

*Value of pH has not units [–].
**Dissolved organic carbon [mg/l].

6 CONCLUSIONS

Within this work, the possibility of the technology of S/S was examined mainly for the purpose of reduction of negative impact of hazardous waste—Neutralization Sludge (NS) on the environment. To obtain such properties of Solidification Product (SP), as product of this technology, in order it could be further efficiently used was another important aim of this work. Physical and mechanical properties of SPs were examined by determination of compressive strength, bulk density, and permeability. It was found that the SPs containing 15% of cement, solidification formulas S4 and S5, showed higher compressive strength than the samples with 5% or none content of the cement. Bulk density of hardened SPs varied around 900–1000 kg/m³. The Solidification Products (SPs) with highest compressive strength (S4, S5) showed the lowest permeability, therefore it can be said that the higher cement content in the matrix of the SPs is closely related to the higher compressive strength and lower permeability. However, the application of cement as solidification agent is financially and environmentally demanding and on that ground there is an effort to substitute it with CFBC and PCC fly ash. For evaluating the risk of contaminant leaching of SPs to environment the leaching test according EN 12457–4 was performed. Upon the results, it can state that heavy metals occurring in input Neutralization Sludge (NS) are firmly incorporated in the matrix of SP and they should not be released into the environment, where SP will be used.

The optimal utilization of SPs could be backfill material, material for base layers of roads, material

shaping the terrain and mainly as material for landfill reclamation and remediation of old ecological burdens. The use of the SPs as a building material in common earthwork is unlikely due to the possible high pollutant content in dry matter, which is not permitted by applicable strict national legislation. However SPs prepared according to the formulas S4 and S5 could probably be used as backfilling materials for land restoration after mining if the relevant ecological limits will be met. Nevertheless for this kind of application also other properties of the SPs must be examined, in particular, those related to their long-term durability.

ACKNOWLEDGMENT

This paper has been worked out under the project No. LO1408 "AdMaS UP - Advanced Materials, Structures and Technologies", supported by Ministry of Education, Youth and Sports under the "National Sustainability Programme I" and under the project No. TA01021418 "Technology of neutralization sludge application in the reclamation process and in the building industry" supported by The Technology Agency of The Czech Republic.

REFERENCES

Barth, E., Percin, P., Arozarena, M., Zieleinswski, J., Dosani, M., Maxey, H., Hokanson, S., Pryately, C., Whipple, T., Kravitz, R., Cullinane, M., Jones, L. & Malone, P. 1989. *Stabilization and Solidification of Hazardous Waste.* New Jersey: Noyes Data Corporation.

Chen, Q.Y., Tyrer, M., Hills, C.D., Yang, X.M. & Carey, P. 2009. Immobilisation of heavy metal in cement-based solidification/stabilisation: A review. *Waste Management,* 29: 390–403.

Dermatas, D. & Meng, X. 2003. Utilization of fly ash for stabilization/solidification of heavy metal contaminated soils. *Engineering Geology,* 70: 377–394.

Hills, C.D. & Pollard, S.J.T. 1997. Influence of interferences effect on the mechanical, microstructural and fixation characteristics of cement solidified hazardous waste forms. *Journal of Hazardous Materials,* 52: 171–191.

Hodul, J. 2013. *Development of a new methodology for evaluation durability of solidification products made of hazardous waste.* Brno: University of technology, Institute of Technology of Building Materials and Components, Faculty of Civil Engineering. Bachelor Thesis.

Hodul, J. 2015. *Solidification product made of hazardous waste and possibilities of its launch to the market.* Brno: University of technology, Institute of Technology of Building Materials and Components, Faculty of Civil Engineering. Diploma Thesis.

Malviya, R. & Chaudhary, R. 2006. Factors affecting hazardous waste solidification/stabilization: A review. *Journal of Hazardous Materials B,* 137: 267–276.

Qian, G., Cao, Y., Chui, P. & Tay, J. 2005. Utilization of MSWI fly ash for stabilization/solidification of industrial waste sludge. *Journal of Hazardous Materials B*, 129: 274–281.

Rossini, G. & Bernardes, A.M. 2005. Galvanic sludge metals recovery by pyrometallurgical and hydrometallurgical treatment. *Journal of Hazardous Materials*, 131: 210–216.

Roy, A., Harvill, E.C., Carteledge, F.K. & Tittlebaum, M.E. 1992. The effect of sodium sulphate on solidification/stabilization of synthetic electroplating sludge in cementious binders. *Journal of Hazardous Materials*, 30. 297–316.

Song, F., Gu, L., Zhu, N. & Yuan, H. 2013. Leaching behavior of heavy metals from sewage sludge solidified by cement-based binders. *Chemosphere*, 92: 344–350.

Stegemann, J.A. & Zhou, Q. 2008. Screening tests for assessing treatability of inorganic industrial wastes by stabilisation / solidification with cement. *Journal of Hazardous Materials*, 161: 300–306.

Sweeney, R.E.H., Hills, C.D. & Buenfuld, N.R. 1998. Investigation into the carbonation of stabilised/solidified synthetic waste. *Environmental Technology*, 19: 893–905.

Zivica, V. 1997. Hardening and properties of cement-based materials incorporating heavy metals oxides. *Bulletin of Materials Science*, 20: 677–683.

Advances in Civil, Architectural, Structural and Constructional Engineering – Kim, Jung & Seo (Eds)
© *2016 Taylor & Francis Group, London, ISBN 978-1-138-02849-4*

Exo-type damping system with isotropic dampers

Y.S. Chun
Land and Housing Institute, Korea Land and Housing Corporation, Daejeon, Korea

M.W. Hur
Department of Architectural Engineering, Dan-Kook University, Gyeonggi-do, Korea

ABSTRACT: This study was conducted to address the problem of reduction in damping effect due to the torsional and out-of-plane behavior of a unidirectional damping system. This study is a part of a R&D project aimed at implementing an efficient damping structure. This paper presents the Exo-type Kagome Damping System, which is a new type of damping system with isotropic dampers, applicable to mid/low-rise structures (≤20 stories). Further, it presents a method for achieving the target damping performance; this method was established after performing numerical analyses with variables such as the difference in stiffness between the main and support structures, size of the damping device, and the number of stories of the main structure.

1 INTRODUCTION

A structure with damping systems is a structure that is designed to improve the earthquake-resistance of a building or structure by installing dampers that mechanically control the seismic vibration inside or outside the structure, and making them absorb most of the vibration energy. The effectiveness of such a structure with damping systems is well proven and it has been widely applied to construction and retrofitting of buildings in earthquake-prone countries such as Japan, the U.S., China, and New Zealand (Constantinou et al. 1993, Housner et al. 1993, Soong & Spencer Jr. 2002, Nagarajaiah et al. 2003).

In South Korea, various types of dampers (steel-hysteretic, frictional, visco-elastic, and viscous types) have recently been developed, and are used with increasing popularity (Lee 2013, Baek et al. 2014, Han et al. 2014, Chun et al. 2014). However, most of these dampers are designed to operate under in-plane loading in uniaxial direction, and hence, they are likely to fail when they operate under torsional and out-of-plane loading.

This paper presents the Exo-type Kagome Damping System (EKDS), which can be applied to mid/low-rise buildings and structures. EKDS can overcome to a great extent the aforementioned drawback of conventional dampers since Kagome dampers are isotropic dampers that exhibit equal damping characteristics in biaxial directions. The effectiveness of the EKDS was tested by performing a cyclic loading test on a scaled down frame specimen of a 20-story framed apartment building.

2 EXO-TYPE KAGOME DAMPING SYSTEM

2.1 *Characteristics of kagome damper*

Kagome damper is a type of steel damper that makes use of its capacity to dissipate the energy by inducing shear deformation of the Kagome truss structure. Unlike other steel dampers, it has a wire-wounded texture characterized by high resistance against fatigue or repeated loading, thus maintaining the overall damping function, even in cases that it is partially damaged. Its other advantages include a high strength-to-weight ratio, high energy absorption through large shear deformation, and modularized production in accordance with the column cross-section and size as well as wall thickness, thus enabling a parallel arrangement and providing capacity to easily satisfy the strength requirements. Moreover, because of its porous structure with large empty interior space, it is small in size and light with respect to its volume, and thus easy to install and handle. It absorbs the vibration energy, irrespective of the direction of seismic loading, because of its isotropy (Hwang 2013). Figure 1 shows the shape of the Kagome damper and its hysteretic behavior under shear deformation.

2.2 *Structure of EKDS*

EKDS involves constructing an independent cantilever-type support structure outside the main structure and installing dampers between the floor level

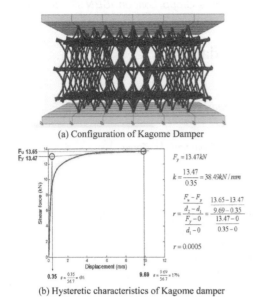

(a) Configuration of Kagome Damper

$F_y = 13.47 kN$

$k = \dfrac{13.47}{0.35} = 38.49 kN/mm$

$r = \dfrac{\dfrac{F_u - F_y}{d_2 - d_1}}{\dfrac{F_y - 0}{d_1 - 0}} = \dfrac{\dfrac{13.65 - 13.47}{9.69 - 0.35}}{\dfrac{13.47 - 0}{0.35 - 0}}$

$r = 0.0005$

(b) Hysteretic characteristics of Kagome damper

Figure 1. Configuration and hysteretic characteristics of Kagome damper.

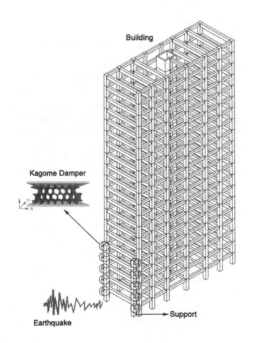

Figure 2. Concept diagram of Exo-type system.

of the main structure and the support structure. It is designed to cause deformation of the Kagome damper as a result of the relative deformation arising out of the difference in stiffness between the main and support structures. Figure 2 shows a schematic arrangement of EKDS.

3 ANALYTICAL STUDY

3.1 Building for study

The building type investigated is an apartment building with four units per floor on a rectangular plan, as shown in the floor plan in Figure 3. The staircase and the elevator core are constructed as shear walls. Columns with modules of length 4.0 m, 4.6 m, and 3.5 m are arranged along the longitudinal axis, and two columns, approximately 10.0 m apart, support the slab along the transverse axis. We analyzed two buildings having 15 and 20 stories, respectively, with the story height of 3.0 m. Its seismic design variables as per Korean Building Code (KBC2009) are as follows: zone factor = 0.22, soil profile = S_d, and importance factor = 1.5. The fundamental period of the building obtained from the eigenvalue analysis is 2.30 s (transverse axis) and 2.25 s (longitudinal axis) for the 15-story building, and 3.12 s (transverse axis) and 2.90 s (longitudinal axis) for the 20-story building.

The target reduction rate of the base shear due to the damping effect was set at 20% of the base shear, for easiness of practical application. To achieve this target, it was assumed that EKDS support structures are installed at four positions, corresponding to both edges of the lateral walls of the building. The installation heights of the support structures were set at three and five floor heights, which can be considered as the appropriate height and stiffness applicable for the support structure.

3.2 Analysis variables and analysis method

The support-to-building stiffness ratio (the total stiffness ratio of the structure for the height considered for damping) was controlled by changing the modulus of elasticity of the support structure, thereby setting six cases as variables. For the yield load of the Kagome damper, a total of eight capacities were considered as variables with reference to the basic module (100×100 mm). As a result, the number of cases to be analyzed was 194, consisting of two different numbers of stories, two different installation heights of the support structure, eight volumetric

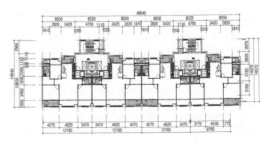

Figure 3. Model building structure (Architectural plan).

capacities of dampers, and six levels of the modulus of elasticity, including the undamped structures for comparison purposes. The base shear strength in each case was analyzed for cross-comparisons among individual cases. In order to verify the reduction effects of the base shear, the base shear of the damping structure and that of the undamped structure were compared in terms of relative ratio. The base shear of the main structure of structure with damping system was calculated by subtracting the base shear increased by the installation of the support structure from the total base shear. Table 1 presents the specifications of the Kagome damper. Analysis was performed using MIDAS GEN 2015.

For the analysis, three seismic waves were chosen out of the recorded waves observed in relatively stable soil (El-Centro 1940, Taft 1952, Hachinohe 1968). Spectrum characteristics of these seismic waves are compatible with KBC (2009) and ASCE /SEI (2010) design spectrum. The soil conditions were assumed to be S_d.

3.3 Results of analysis

Figures 4–7 show the reduction of the base shear of the main structure under each seismic wave, expressed in terms of relative ratios with respect to the base shear of the undamped structure. The calculation results were used to determine the damping effect according to the applied height of the support structure, support-to-building stiffness ratio, and the yield strength ratio of the damper to the base shear of the main building. The figures indicate that despite slight changes depending on the seismic waves, the increase in the stiffness ratio of the support structure to the building and the yield strength ratio of the damper to the base shear of the main building has good correlation with the increase in the reduction rate of the base shear. One aspect to be noted is that, when the EKDS with damper size of 200×200 mm and height of support structure of three stories was applied, the base shear increased instead of decreasing, regardless of the stiffness ratio between the main and support structures. However, when it was applied to five stories, the base shear decreased to a greater extent for the same stiffness ratio, when compared to the case of three stories. From this result, it can be inferred that to secure the effectiveness of EKDS, choosing the correct height required for EKDS is of vital importance.

The analysis results revealed that the target reduction of the base shear can be achieved by ensuring the threshold values of the influencing variables. In the case of application of EKDS to three stories, the threshold values were analyzed to be 7.0 for the support-to-building stiffness ratio, and 8.0% of the base shear for the yield strength ratio of the damper to the base shear of the main building ($V_{damper}/V_{b,shear}$); in the case of application to 5 stories, these values were 2.5 and 3.5%, respectively.

Table 1. Specifications of Kagome damper.

Damper hight (mm)	Yield strain (%)	Ultimate stain (%)	Yield stress (MPa)	Ultimate stress (MPa)
200	0.23	15.0	0.79	1.31

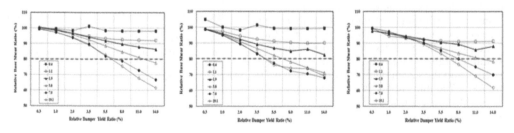

Figure 4. Distribution of the base shear with stiffness ratio and damper yield ratio (15F-3Story).

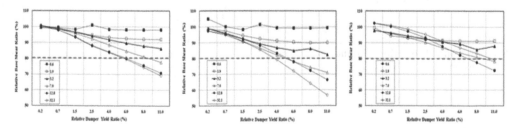

Figure 5. Distribution of the base shear with stiffness ratio and damper yield ratio (20F-3Story).

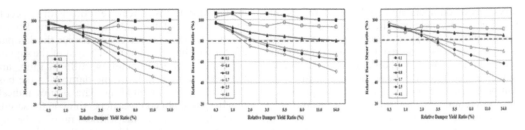

Figure 6. Distribution of the base shear with stiffness ratio and damper yield ratio (15F-5Story).

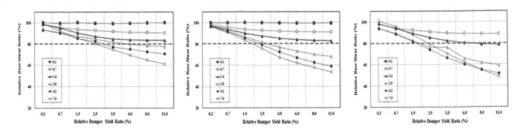

Figure 7. Distribution of the base shear with stiffness ratio and damper yield ratio (20F-5Story).

4 CONCLUSIONS

Within the framework of a R&D project for implementing an efficient damping structure, we proposed the Exo-type Kagome Damping System (EKDS) characterized by isotropic damping, and presented its application method for achieving targeted damping effects, after performing numerical analyses with variables such as the difference in stiffness between the main and support structures, size of the damping device, and the number of stories of the structure. Analytical results show that in order to achieve a damping performance of about 20% in terms of base shear, when a support structure is applied for the height of three stories in a 20-story building, the stiffness ratio between the building and support structure and the size of dampers should exceed 7.0 and 700 × 700 mm, respectively; when applied for the height of five stories, these values are 2.8 and 400 × 400 mm, respectively.

ACKNOWLEDGEMENT

This research was supported by a grant (R2013 01005) from Residential Housing Research Program funded by Korea Land and Housing Corporation.

REFERENCES

ASCE, 2010. Minimum Design Loads for Buildings and Other Structures (ASCE 7-10).

Baek, E.L., Oh, S.H. & Lee, S.H. 2014. Seismic Performance of an Existing Low-Rise Reinforced Concrete Piloti Building Retrofitted by Steel Rod Damper, Journal of EESK, 18(5): 241–251.

Chun, Y.S. et al. 2014. Development of Damping System for improvement of Seismic Performance of Apartment Buildings (2): Application Plan of Damping System for Apartment Buildings, Land & Housing Institute.

Constantinou, M.C. & Symans, M.D. 1993. Seismic Response of Structures with Supplemental Damping, The Structural Design of Tall Buildings, 2: 77–92.

Han, S.W. et al. 2014. Seismic Behavior of Reinforced Concrete Moment Frames Retrofitted by Toggle Bracing System with High Density Friction Damper, Journal of EESK, 18(3): 133–140.

Housner, G.W. et al. 1997. Structural Control: Past, Present, and Future, Journal of Engineering Mechanics, 123(9): 897–971.

Hwang, J.S., Park, S.C. & Kang, K.J. 2013. A study on the hysteresis properties and mathematical model of Kagome truss damper, Journal of Architectural Institute of Korea, 29(9): 21–29.

KBC, Korean Building and Commentary, 2009. Archtectural Institute of Korea.

Lee, H.H. 2013. Displacement and Velocity Dependence of Clamped Shape Metallic Dampers, Proceeding of Korea Institute for Structural Maintenance Inspection, 17(2): 62–70.

Soong, T.T. & Spencer, Jr. B.F. 2002. Supplemental energy dissipation: state-of-the-art and state-of-the practice, Engineering Structures, 24: 243–259.

Spencer, Jr. B.F. & Nagarajaiah, S. 2003. State of the Art of Structural Control, Journal of Structural Engineering, ASCE, 2003: 845–856.

Advances in Civil, Architectural, Structural and Constructional Engineering – Kim, Jung & Seo (Eds)
© 2016 Taylor & Francis Group, London, ISBN 978-1-138-02849-4

Behaviour of concrete cylinders confined by Steel-BFRP Hybrid Stirrup (SBHS)

A. Ishag, W. Gang, Z. Sun & Z. Wu
*Key Laboratory of Concrete and Prestressed Concrete Structures of the Ministry of Education,
Southeast University, Nanjing, China*

ABSTRACT: This paper presents experimental results for evaluating the behavior of axially loaded of concrete cylinder confined with Steel-BFRP Hybrid Stirrup (SBHS). SBHS is a kind of new stirrup made up of externally basalt fiber reinforced polymer and inner steel wire, which is characterized as high modulus, high strength, and anti-corrosion performance. 18 specimens of SBHS confined concrete cylinders with different SBHSs were tested under axial compressive loading, and the parameters include: steel/BFRP ratio, SBHS diameter and stirrup volume ratio. To validate the model, three-dimensional (3D) Finite Element Analysis (FEA) is used and compared with experimental values. The comparison between experimental and FEA results show that the model provides satisfactory predictions of the stress-strain response of the cylinders reinforced with SBHS. The experimental results indicated that the smaller spacing gives an average strength of concrete core 18% higher than that with larger spacing.

1 INTRODUCTION

The confinement effects on concrete behavior have been thoroughly studied in the past decades. Previous researchers conducted and demonstrated that lateral confinement in concrete columns increases the ductility, energy absorption and compressive strength of concrete capacity (Richart et al. 1929). Currently, FRP's is widely used as a new material for confining concrete structural due to FRP's chemical and mechanical properties, capacity of confinement and easiness of application. In recent years, many research have been published in confined concrete behavior, such as internal and external confined by transverse reinforcement such as steel jackets and fiber composites (Mander et al. 1988). Several studies were conducted on the behavior of concrete confined by FRP and applied to the cylindrical specimens (Samaan et al. 1998, Toutanji 1999). The effect of FRP spiral reinforcement of the strength and deformation capacity of concrete has been also studied (Afifi et al. 2013). Many researchers studied the confining pressure under earthquake, loading type, ductility and straining capacity of the RC concrete which controls the inelastic response of the structure (Ahmad et al. 1991, Wang et al. 2007).

This study is conducted on concrete cylinder made of steel-BFRP hybrid stirrup (SBHS) in order to obtain the stress-strain of the SBHS and spacing effects, as shown in Figure 1. Therefore, this paper is directed towards this endeavor. Strain

Inner steel wires   Outer FRP

Figure 1. Steel-BFRP Hybrid Stirrup (SBHS).

gauges were attached in SBHS hoop with 20 mm concrete cover. SBHS is characterized by light weight, high-tensile capacity and anti-corrosion and rust in aggressive environment due to BFRP. The steel wire's properties were described as non-magnetic, corrosion resistance, and tensile strength of 42 kN/mm². All the specimens were subjected to a compressive axial load, and axial stress-strain, ultimate failure load, and failure modes were measured along the cylinder.

2 EXPERIMENTAL INVESTIGATION

2.1 Cylinder design

The experimental investigations was designed as per ASTM (ASTM Standard C31/C31M-12. 2010) provisions. It comprised of testing of a total number of 18 cylindrical specimens having a diameter of 150 mm and height of 300 mm, and divided into 6 groups; every group had 3 specimens based to stirrup spacing and subjected to axial compressive load after 28 days until failure as shown in Table 1 and Figure 2, respectively. The basic manufacturing characteristics and mechanical properties of fiber-reinforced polymers to produce the stirrup were shown in Table 2 and Figure 1.

2.2 Concrete casting

The specimens were casted using the concrete mix proportions of material per weight as 1: 2: 4: 0.45 (cement: sand: coarse aggregate: water). The properties of Basalt Fiber Reinforced Polymer (BFRP) were used in the present investigation of outer steel wires as shown in Table 2 and Figure 1.

Table 1. Details of concrete cylinders.

Spacemen's type	D (mm)	H (mm)	Stirrup Space	Cover	Amount
C0	150	300	0	20	3
C1	150	300	40	20	3
C2	150	300	50	20	3
C3	150	300	60	20	3
C4	150	300	80	20	3
C5	150	300	100	20	3

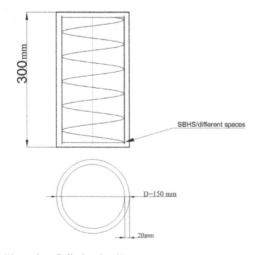

Figure 2. Cylinder details.

Table 2. Material properties supplied by manufacturer.

Type	E (GPa)	Tensile strength (Mpa)	Density (g/cm³)	Diameter (µm)	Elongation rate (%)
BFRP	90	2,250	2.63	13	2.5
Resin	3.6	95	1.06	–	6.1

3 FINITE ELEMENT MODEL ANSYS

3.1 Concrete properties

The finite element program ANSYS was used in this analysis. Solid 65 elements (ANSYS 12 2009) were used to model the concrete. It is worth mention that Linear and multi-linear is-tropics material properties are require to predefine when using ANSYS, as well as some additional concrete material properties, to simulate real concrete behavior. The shear transfer coefficient of an open and closed crack, $\beta t = 0.2$, and $\beta c = 0.8$ (Kachlakev et al. 2001, Werasak & Meng 2008). The concrete modulus of elasticity can be calculated from the equation $E_c = 4700\sqrt{f'_c}$, and the tensile strength $f_r = 0.62\sqrt{f'_c}$, and 0.2 Poisson's ratio of concrete were used. The compressive uniaxial stress-strain values for the concrete model were obtained from the following equations for the multi-linear isotropic stress-strain curve for the concrete:

$$E_c = f_{el}/\varepsilon_{el}, \varepsilon_0 = \frac{2f'_c}{E_c} \quad \text{and} \quad f = \frac{E_c\varepsilon}{1+(\varepsilon/\varepsilon_0)^2} \quad (1)$$

where, f_{el} is the stress at the elastic strain ε_{el} in the elastic range ($f_{el} = 0.30f'_c$), ε_0 is the strain at the ultimate compressive strength, and f is the stress at any strain ε.

3.2 Hybrid stirrup

Link 8 elements were used to model the hybrid stirrup (SBHS). This element is a 3D spar element with two nodes and three degrees of freedom, and capable of plastic deformation. In the finite element model, the rebar was assumed to be a bilinear, isotropic, elastic and perfectly plastic material that behaves identically in tension and compression, with an elastic modulus equal to 210 GPa and a 0.3 Poisson's ratio. A yield stress of 256 MPa is directly obtained from SBHS mechanical tests.

3.3 Test setup

The sand was placed on top and bottom in all the specimens to ensure a flat surface and to distribution the load uniformly, and axial load was applied through the hydraulic machine with load rate 0.5 kN/min which is kept constant during the test until specimen's failure.

4 TEST RESULTS AND DISCUSSION

The specimens were tested after 28 days of casting by applying axial compression. Stressed versus strains are demonstrated in Figures 3–6. To identify the behavior of the cylinder reinforced by SBHS Stirrup, some parameters such as load, displacement, stress-strain and spacing effects were investigated.

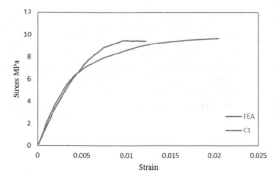

Figure 3. Axial stress Vs. strain for cylinders type C1.

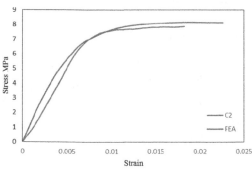

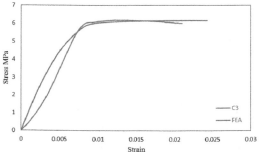

Figure 4. Axial stress-strain for cylinders type C2 and C3.

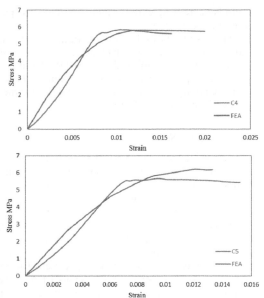

Figure 5. Axial stress Vs. strain for cylinders type C4 and C5.

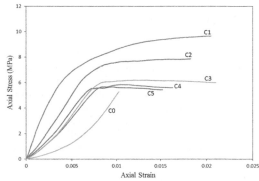

Figure 6. Axial stress-strain behavior for cylinders.

Table 3. Cylinders average stress and load.

Specimens	Spaces	Average load	f_{cc} MPa	f'_c MPa
C0	0	93.30	5.28	–
C1	40	170.83	–	9.67
C2	50	139.13	–	7.87
C3	60	109.30	–	6.19
C4	80	103.21	–	5.84
C5	100	100.05	–	5.66

4.1 Ultimate load capacity

The cylinders load carrying capacity are shown in Table 3. From the experiment results, it is obviously that the cylinders ability to carrying axial load is decreasing with the increase in SBHS spacing in the cylinders.

The results showed that the variation in spacing gives different value of axial load carrying capacity for all cylinders reinforced with SBHS. Axial load carrying capacity increases by 45% and 32.94% in load carrying capacity for C1 and C2 specimen, respectively compared with C5.

4.2 Stress—Strain

To examine the reliability and validity of the experimental program an extensive verification is carried out using a FEA results data, and the average values indicated by the FEA are very close to the experimental results. The SBHS and spacing effects of FRP as well as effects of BFRP and steel wires on stress—strain behavior for cylinders were exhibited in Figure 5. Minor difference in stress behavior has been observed for specimens C4 and C5. The average stress—strain versus FEA curves for axial loaded of concrete reinforced with SBHS is shown in Figure 6. The curve represents the axial stress (f_c', f_{cc}) versus axial strain (ε_c) response for the specimens with different stirrup spacing.

4.3 Axial Stress-Strain comparison

Average axial stress-strain is evaluated from average axial deformation for the specimens. Figure 6, shows the average axial stress and strain comparison for all specimens, and shows that the cylinders reinforced with SBHS has higher axial stress-strain compared to control cylinder C0 (un-confined), the results show that the specimens C4, C5 have less effectiveness in improving the axial strength compared to C1, C2, which has small spaces between stirrups.

4.4 Failure modes

The specimens reinforced with SHBS failed due to failure of concrete under axial loading, and the

Figure 7. Failure modes of cylinders.

specimen C0 has failed after reaching its ultimate compressive strength due to non-stirrup. And cracking the concrete at ultimate load led to the spalled out of the concrete cover and cracked as shown in Figure 7.

5 CONCLUSIONS

In this paper, the experimental and FEA simulation was adopted to predict the contribution of confining behavior, stress-strain, and SBHS stirrup to the concrete cylinders capacity. According to results, the following conclusions can be drawn:

1. The confined concrete strength is basically dependent on the maximum confining pressure where the Steel-BFRP Hybrid Stirrup (SBHS) is applied. All the resulting stress versus strain curves presented an increasing axial load behavior of cylinders. The better confinement was achieved when concrete cylinders were confined with Hybrid Stirrup than without confined specimens.
2. The SBHS stirrup of small diameter and close spacing is better than the one using SBHS of wide spacing. Compared with different spacing of stirrups, the cylinder capacity is increased due to small stirrup spacing.
3. Based on the comparisons of FEA with the experimental results, the FEA more accurate in predicting the confining behavior effective to the concrete.

REFERENCES

Afifi, M.Z., Mohamed, H.M. & Benmokrane, B. 2013. Strength and Axial Behavior of Circular Concrete Columns Reinforced with CFRP Bars and Spirals. *Journal of Composites for Construction*, 18(2): 04013035.

Ahmad, S., Khaloot, A. & Irshaid, A. 1991. Behaviour of concrete spirally confined by fibreglass filaments. *Magazine of Concrete Research*, 43(156): 143–148.

ANSYS 12. 2009. *Swanson Analysis Systems*. Computer software user's manual, Canonsburg, Pennsylvania, USA.

ASTM Standard C31/C31M-12. 2010. *Standard practice for making and curing concrete test specimens in the field*. West Conshohocken: PA: ASTM International.

Kachlakev, D., Miller, T., Yim, S., Chansawat, K. & Potisuk, T. 2001. *Finite element modeling of concrete structures strengthened with FRP laminates: final report*. SPR 316.

Mander, J.B., Priestley, M.J. & Park, R. 1988. Theoretical stress-strain model for confined concrete. *Journal of structural engineering*, 114(8): 1804–1826.

Werasak, R. & Meng, J. 2008. *Finite element analysis on lightweight reinforced concrete shear walls with different web reinforcement*. The Sixth Prince of Songkla Univ. Engng. Conf., Hat Yai, Thailand: 61–67.

Advances in Civil, Architectural, Structural and Constructional Engineering – Kim, Jung & Seo (Eds)
© 2016 Taylor & Francis Group, London, ISBN 978-1-138-02849-4

Influence of span-length on seismic vulnerability of reinforced concrete buildings based on their fragility curves

M.P. Cripstyani
Department of Civil Engineering, Sebelas Maret University, Surakarta, Indonesia

S.A. Kristiawan & E. Purwanto
SMARTCrete Research Group, Department of Civil Engineering, Sebelas Maret University, Surakarta, Indonesia

ABSTRACT: This study discusses about the influence of span-length on seismic vulnerability of reinforced concrete buildings based on their fragility curve. Models are developed in a typical structural system with variation of span-length in one direction. Non-linear static pushover analysis is carried out to obtain seismic structural response. Damage states are defined according to drift and strength values upon analyzing the structural response. Fragility function is then used to develop curves that relate spectral displacements to the probability of reaching structural damage levels. Based on the obtained fragility curves, it can be concluded that span-length affect the probability of exceedance failure at extensive and complete damage state.

1 INTRODUCTION

Asessment of vulnerability of multi-story buildings in the seismic zone is an important part of the effort to reduce the risk of loss in the event of earthquake. Proper loss assessment methods as means of predicting the likelihood of damage that will occur in the buildings have been developed by various researchers in the field of seismic resistance buildings. One of such methods is the development of seismic fragility curve which represents the conditional probability of failure of a structure or component for a given seismic input motion parameter. This method (i.e. HAZUS) was first introduced in USA and later adapted by many countries in an approach to assess the damage of buildings for a given ground motion level (Bilgin 2013, Hamid & Mohamad 2013, Bakhshi & Asadi 2013). Application of seismic fragility curve to evaluate the vulnerability of various buildings offers a possibility to identify a variety of structural parameters which affect the probability of structural failures. This study investigates the effect of span-length in similar structural system of reinforced concrete building on the seismic vulnerability of the building based on the development of fragility curve.

2 FRAGILITY CURVE DEVELOPMENT

2.1 Non-linear static pushover analysis

The essential step in the development of fragility curve is performing seismic analysis to obtain the structural response. The structural response will be used to identify the level of damage that occurs in the structure due to specific earthquake ground motions. A non-linear static pushover analysis was employed for this case. This analysis consists of two phases. At first gravity load, which is a combination of dead load and reduced live load, is applied to the structure. Secondly, an inverted triangular distribution of lateral load from the static equivalent analysis is applied and monotonically increased causing the weakest element of the structure to deform firstly. Repeating the application of this lateral load will initiate yieding of other elements and finally the structural collapse occurs. From non-linear static analysis, the expected plastic joints in the structural elements could also be identified (ATC-40 1996).

2.2 Hazus damage states

The next step in the development of fragility curve is setting the criteria of various damage states. Damage states could be defined according to drift and strength values determined upon analyzing the response of the structure. These damage states (levels) in HAZUS are identified from the first yield component (refers to slight damage), the yield-point of equivalent elastoplastic system with equal energy absorption (moderate damage), the median point of peak plateau (extensive damage), and the ultimate load capacity (complete damage). Different scales of limit states/damage states can be adopted depending on the methodology used to compute the fragility functions and depending

Table 1. HAZUS average inter-story drift ratio of reinforced concrete structures (Moderate Code Design Level).

Building type	Structural damage states			
	Slight	Moderate	Extensive	Complete
Low Rise	0.0050	0.0090	0.0230	0.0600
Mid Rise	0.0030	0.0060	0.0150	0.0400
High Rise	0.0025	0.0045	0.0115	0.0300

on the choices of the researchers. HAZUS Average Inter-Story Ratio of Reinforced Concrete Structure presented in Table 1 is an example of such limit states based on FEMA, HAZUS-MH MR5 Advanced Engineering Building Module (AEBM) 2010. It should be noted that there are some studies that do not refer to any of the typical damage scales as set out by HAZUS, but they define and develop their own criteria of specific damage state scales.

2.3 Fragility function

The fragility function to develop the relationships between exceedance probabilities failures with structural response in terms of spectral parameters are given by the Equation 1 (Frankie 2010):

$$P\left(Exceedences_i \mid S_d\right) = \Phi\left[\frac{1}{\beta_{tot}}\ln\left(\frac{S_d}{LS_i}\right)\right] \quad (1)$$

where, S_d = structural response variable; LS_i = threshold value for the ith limit state; and β_{tot} = uncertainity response parameter.

2.4 Uncertainity response parameter

Papailia (2011) stated that the dispersion (β) of the fragility curve took into account explicitly the model uncertainty for the estimation of the damage measure for a given intensity and the uncertainty of the capacity in terms of the damage measure. This latter uncertainty response parameter (β_{tot}) included both the model uncertainty and the dispersion of material and geometric properties. Frankie (2010) describe the uncertainty response parameter (β_{tot}) as in Equation 2 below:

$$\beta_{tot} = \sqrt{\left[\left(CONV[\beta_C, \beta_D]\right)_i\right]^2 + \left[\left(\beta_{LS}\right)_i\right]^2} \quad (2)$$

where, β_{LSi} = the uncertainty associated with the limit state threshold values (taken to be 0.4 in HAZUS); β_C = the uncertainty associated with the building capacity; and β_D = the uncertainty associated with imposed earthquake demand (taken to be 0.45 at short periods and 0.5 at long periods in HAZUS). Finally, CONV [β_C, β_D] is the combination of uncertainty associated with the capacity and demand, and is obtained by convolution.

3 METHODOLOGY

3.1 Building model

The non-linear static analysis must appropriently represent the building response and capture the damage pattern of it's elements to develop fragility functions. Building models are required for loss estimation and could be developed for horizontal direction of the building response of interest to judge the mode of failure.

Four moment resisting frames system plan views as presented in Figure 1, and are chosen to study the seismic fragility response of the building with span-length as the variable paramaters. The variation of structural models are shown in Table 2. All of the buildings are regular both in plan and in elevation but with variation of span-length in the X direction. These seven-story buildings with four

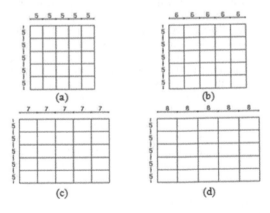

Figure 1. Plan view of structural models (a) SL-5; (b) SL-6; (c) SL-7; (d) SL-8.

Table 2. The variation of structural model.

Model code	Span-length	
	in X direction m	in Y direction m
SL-5	5	5
SL-6	6	5
SL-7	7	5
SL-8	8	5

meter inter-story heights are designed for seismic, gravity and live loads.

3.2 *Developing fragility curve*

According to Frankie (2010) fragility curves are generated by probabilistic analysis of structural response data. They are presented both as a function of ground-motion intensity and as structural demand to comply with the format requirements of different loss-assessment. Developing fragility curve of a building need a statistical analysis performed on the results obtained from the structural response assessment. The data must represent the building capacity variance under numerous ground motions. It is critical to select some appropriate ground motion records for structural response assessment. For this purpose building capacity curve should be introduced from analysis of the variations of demands considered. Range of Peak Ground Acceleration (PGA) is selected from 0.2 g to 0.6 g at intervals of 0.05 g.

4 RESULT

Non-linear static pushover analysis is carried out to obtain capacity curve of each structural model in this study. Based on the capacity curve, median value of damage states are identified for developing fragility curve using a procedure introduced by Duan & Pappin (2008). Table 3 shows limit state values resulted from each structural model.

As shown in Table 3, the different span-length in the same type of RC frame structure produces variation values in limit states. For the slight and moderate state, the increase in span-length generates the increase of limit states value. At the more critical limit state, the extensive and complete state, the effect of span-length has influence on the limit state value.

Using Equation 2, the total uncertainity response parameter is generated to draw the fagility curve of each structure. Table 4 shows that the variation of span-length affects the uncertainity value (β).

Table 3. Limit state values based on the non-linear static pushover analysis.

| Model code | Limit states | | | |
	Slight m	Moderate m	Extensive m	Complete m
SL-5	0.01589	0.02384	0.14274	0.30932
SL-6	0.01592	0.02388	0.12782	0.27951
SL-7	0.01594	0.02391	0.12497	0.27385
SL-8	0.01596	0.02394	0.14155	0.30705

Table 4. Uncertainty response parameter (β) from the non-linear static pushover analysis.

Model Code	β_c	β_{tot}
SL-5	0.3029	0.4226
SL-6	0.4057	0.4397
SL-7	0.3585	0.4313
SL-8	0.4905	0.4569

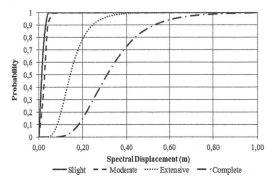

Figure 2. Fragility curve of a five meters span length seven-storey RC frame structure (SL-5).

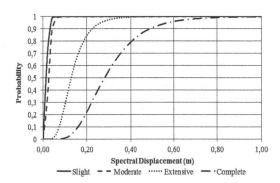

Figure 3. Fragility curve of a six meters span length seven-storey RC frame structure (SL-6).

After all required parameter are available, the fragility function can be determined. Figures 2–5 illustrates the fragility curves of each structural model in this study. The variations in β generates different curves.

HAZUS average inter-story drift ratio in the moderate code design level as shown in Table 1 is used as parameter, and then coverted to damage state threshold value by multiplying the drift ratios by the pushover modal factor values (α) from the analysis as shown in Table 5 and height of the building at the roof level.

The results of damage state threshold value for the structural system considered in this study are presented in Table 6 below.

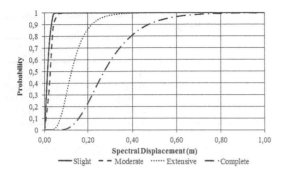

Figure 4. Fragility curve of a seven meters span length seven-storey RC frame structure (SL-7).

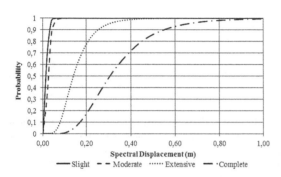

Figure 5. Fragility curve of a eight meters span length seven-storey RC frame structure (SL-8).

Table 5. Pushover modal factor of each model structure.

Model code	α
SL-5	0.2001
SL-6	0.2241
SL-7	0.2431
SL-8	0.3026

Table 6. Damage state treshold value.

Damage state	S_d (m)
Slight	0.0168
Moderate	0.0376
Extensive	0.1021
Complete	0.3389

In order to be able to conduct comparison and provide the means to determine the conclusion, probability of each model structure to reach the threshold value at a specific spectral intensity parameter are calculated. For this purpose, a value of ground motion level is chosen at 0.45 g as a parameter to see the influence of the span-length

Table 7. Probability of failure according to HAZUS damage states.

Model code	Limit states			
	Slight	Moderate	Extensive	Complete
SL-5	0.9976	0.9682	0.6769	0.3398
SL-6	0.9984	0.9785	0.7528	0.4384
SL-7	0.9994	0.9888	0.8042	0.4920
SL-8	0.9994	0.9905	0.7652	0.4216

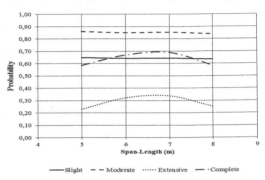

Figure 6. Comparion of probability at HAZUS damage state of seven-story RC frame structure with different span length.

on the seismic vulnerability of RC buildings. Probability of each model to reach the threshold value for a given spectral intensity for each limit state are shown in Table 7 and Figure 6 below.

5 CONCLUSIONS

The fragility curves generated in this study are based on variation of peak ground acceleration from 0.2 g to 0.6 g at intervals of 0.05 g. This study points out that the fragility curves are influenced by the span-length. The probability of reinforced concrete structure to reach certain threshold values of extensive and complete damage states are influenced by the span-length. However, the span-length does not seems affect the probability of exceedance failure at slight and moderate damage state.

REFERENCES

Applied Technology Council (ATC). 1996. *Seismic evaluation and retrofit of concrete buildings.* Rep. ATC-40. Redwood City, CA: Applied Technology Council.
Bakhshi, A. & Asadi, P. 2013. Probabilistic evaluation of seismic design parameters of RC frames based on fragility curve. *Scientia Iranica Transaction A: Civil Engineering*, 20(2): 231–241.

Bilgin, H. 2013. Fragility-based assessment of public buildings, *Engineering Structures*, 56: 1283–1294.

Duan, X.N. & Jack W.P. 2008. *A Procedure for Establishing Fragility Functions for Seismic Loss Estimate of Existing Buildings Based on Nonlinear Pushover Analysis*. The 14th World Conference on Earthquake Engineering. Beijing: WCEE.

FEMA. 2010. *HAZUS-MH MR5 Advanced Engineering Building Module (AEBM)*. Washington, DC: Federal Emergency Management Agency.

Frankie, T.M. 2010. *Simulation-based Fragility Relationships for Unreinforced Masonry Buildings*. M.S. thesis. Urbana, IL: Dept. of Civil and Environmental Engineering, Univ. of Illinois at Urbana-Champaign.

Hamid, N.H.A. & Mohamad, N.M. 2013. Seismic assessment of a full-scale double-storey residential house using fragility curve. *Procedia Engineering*, 54: 207–221.

Papailia, A. 2011. *Seismic Fragility Curves for Reinforced Concrete Buildings*. M.S. thesis. Patra: University of Patras.

Advances in Civil, Architectural, Structural and Constructional Engineering – Kim, Jung & Seo (Eds)
© 2016 Taylor & Francis Group, London, ISBN 978-1-138-02849-4

Influence of story number on the Seismic Performance Factors (SPFs) of reinforced concrete frame buildings calculated using FEMA P695 procedure

J.Y. Mattovani
Department of Civil Engineering, Sebelas Maret University, Surakarta, Indonesia

S.A. Kristiawan & A. Supriyadi
SMARTCRete Research Group, Department of Civil Engineering, Sebelas Maret University, Surakarta, Indonesia

ABSTRACT: Seismic Performance Factors (SPFs) in the structural system consist of the response modification coefficient (R), the system over-strength factor (Ω_0), and the deflection amplification factor (C_d). Quantification of these factors has been proposed in FEMA P695. This paper is aimed to identify the influence of story number on the SPFs of reinforced concrete frame building as determined using FEMA P695 procedure. The structural responses of the buildings under consideration are determined by non-linear static pushover analysis. The obtained capacity curves from this analysis are converted into capacity spectrum from which the SPFs are calculated. The results confirm that there is an influence of story number on the SPFs of reinforced concrete buildings.

1 INTRODUCTION

Seismic Performance Factors (SPFs) are used in the design codes to estimate strength and deformation demands on seismic-force-resisting systems that are designed using linear methods of analysis, but responding in the non-linear range. SPFs include the response modification coefficient (R), the system over-strength factor (Ω_0), and the deflection amplification factor (C_d). Most of the codes determine these factors based on qualitative and comparative considerations with known response capabilities of other structural systems. In the last three decades, significant work has been carried out to determine the value of the response modification coefficient (R) (Mondal et al. 2013) and the system over-strength factor (Ω_0) (Mohd et al. 2012). However, many of the seismic-force-resisting systemshave not been subjected to any significant level of earthquake ground motion. Consequently, the seismic response of many of these structural systems and the ability to meet the design performance objectives are untested and unknown. For this reason, FEMA P695 (2009) was released as a recommended methodology for reliably quantfying building system performance and response parameters for use in seismic design. The work presented in this paper focuses on determination of SPFs for reinforced concrete frame buildings based on FEMA P695 procedure. The number of story is investigated as a varying

parameter. The significant contribution of this paper is to show the influence of story number on the non-linear seismic response of reinforced concrete frame building, which in turn, confirming the quantitative different values of SPFs even for one type of seismic-force-resisting system.

2 FEMA P695 FOR QUANTIFYING SPFs

SPFs can be determined with the use of capacity curve obtained from pushover analysis (ATC-40 1996). SPFs defined by FEMA P695 are parallel with the pushover concept, but they require spectral coordinates rather than base shear and roof displacement. Therefore, this capacity curve obtained from pushover analysis has to be converted into spectral coordinates with the assumptions that 100% of the effective seismic weight of structure participates in the fundamental mode at period, T. The conversion results in a spectrum capacity from which SPFs are determined. The steps for determining SPFs from the capacity spectrum include defining the Maximum Considered Earthquake (MCE) spectral acceleration at the period of the system (S_{MT}), the maximum strength of the fully-yielded system (S_{max}) and the seismic response coefficient (C_s).

The ratio of the MCE spectral acceleration to the seismic response coefficient, which is the design-level acceleration, is equal to 1.5 times the R:

$$1.5R = \frac{S_{MT}}{Cs} \qquad (1)$$

The 1.5 factor accounts for the definition of design earthquake ground motion in ASCE/SEI 7–05 (2006), which is two-thirds of MCE ground motions.

The over-strength parameter, (Ω_0), is defined as the ratio of the maximum strength of the fully-yielded system, S_{max}, to the seismic response coefficient, C_s:

$$\Omega_0 = \frac{S_{max}}{Cs} \qquad (2)$$

Inelastic system displacement at the MCE level is defined as $1.5C_d$ times the displacement corresponding to the design seismic response coefficient, C_s, and set equal to the MCE elastic system displacement, effectively redefining the C_d:

$$C_d = R \qquad (3)$$

The same assumption of displacement may be used for most conventional systems if or when the effective damping is more or less the same as the nominal rate of 5% used to define the response spectral acceleration and displacement. The structural system with higher (or smaller) levels of damping would have significantly smaller (or larger) displacements compared to the elastic response of 5% damping.

3 DESCRIPTION OF THE STRUCTURAL SYSTEM UNDER STUDY

The structural systems considered for this study are typical symmetric-in-plan reinforced concrete frame buildings having five-, seven-, and nine-story configurations, intended for regular office building. The value of Ca (Peak Ground Acceleration) refers to the Earthquake Zoning Map issued by Ministry of Public Works in 2010. The structural model is simulated in the Padang Region, Indonesia with 'medium' soil condition, SD ($S_s = 1.35$, $S_1 = 0.6$). All structures have the same plan arrangement with five numbers of bays in the X direction (7.0 m each) and three numbers of bays in the Y direction (5.0 m each) as shown in Figure 4.

The height of inter-story is 4.0 m. A typical elevation is shown in Figure 1–3. Further details on these planar frames, such as total height (from the foundation level), fundamental period, total seismic weight, and design base shear, are provided in Table 1. The fundamental periods of the structures, presented in Table 1, are calculated based

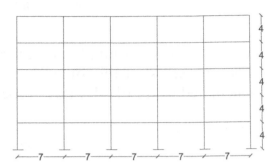

Figure 1. Elevation of the five-story reinforced concrete frame structure.

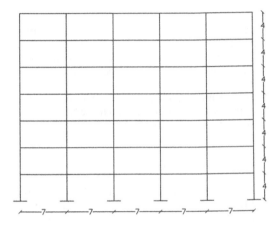

Figure 2. Elevation of the seven-story reinforced concrete frame structure.

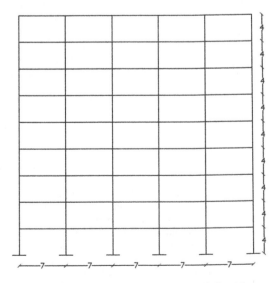

Figure 3. Elevation of the nine-story reinforced concrete frame structure.

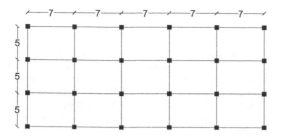

Figure 4. Structural arrangement in plan.

Table 1. Details of the reinforced concrete frames considered for the case study.

Frame	Height (m)	T (s)	W (kN)	V_d (kN)
5-story	20.0	0.6907	39004.99	4941.01
7-story	28.0	0.9350	50917.35	4764.81
9-story	36.0	1.1724	65919.86	4920.81

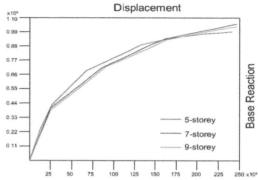

Figure 5. Capacity curves of the structural systems considered in this study.

on the empirical formula recommended in SNI 1726:2012.

The design of reinforced concrete selected for these buildings are based on common practices adopted by design engineers. The strong-column-weak-beam requirement is considered in these designs. The dimensions beams and columns for all frames are same. The dimension beams and column are 500 mm × 700 mm and 600 mm × 600 mm, respectively. The concrete properties are as follows: density is 2400 kg/m³; compressive strength (f'c) is 35 MPa, and the elastic modulus (E) is estimated by E equals to 4700√f'c.

4 NON-LINEAR STATIC PUSHOVER ANALYSIS OF REINFORCED CONCRETE FRAMES

Non-linear static pushover analysis are performed to generate capacity spectrums of the reinforced concrete structural frames under consideration, which are required for computing the response modification coefficient (R), the system over-strength factor (Ω_0), and the deflection amplification factor (C_d). The presence of the rigid floor diaphragm in every floor and the symmetric-in-plan configuration which eliminates any torsional motions, only a two-dimensional pushover analysis is performed for these evaluations.

The non-linear static pushover analysis consists of two phases i.e. at first phase, gravity loads are applied to the structure which are a combination of dead load and reduced live load. At the second

stage, lateral monotonic load is applied. Lateral load intensity at the second stage is incrementally increased to resulting in yielding of the weakest structural component in the system. The process is repeated to cause other elements to yield, until the structure finally collapses. The recorded base shear and roof displacement are plotted to obtain capacity curves. The capacity curves of the structural systems considered in this study are presented in Figure 5. These capacity curves are then converted into capacity spectrum that will be used to compute the response modification coefficient (R), the system over-strength factor (Ω_0), and the deflection amplification factor (C_d) for each frame.

5 COMPUTATION SPFs FOR THE STRUCTURAL SYSTEM CONSIDERED IN THIS STUDY

The capacity spectrum of the structural system is plotted as a relationship between the spectral acceleration and the corresponding spectral displacements, as shown in Figure 6. The MCE spectral response is also drawn in the same figure. The figure illustrates the definition of S_{MT}, S_{max}, and C_s from which SPFs will be determined. These values are then used to compute the SPFs using Equations 1–3. The resume values of SPFs for all the structural systems considered in this study are given in Table 2 and Figures 7–9.

The results clearly indicate that reinforced concrete frame building with a higher story tends to have a lower R and C_d, but slightly increase Ω_0. This finding confirms that a single type of seismic-force-resisting-system could have different value of SPFs depending on the number of story.

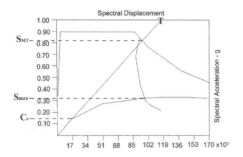

Figure 6. Capacity spectrum for the 5-story reinforced concrete frame structure.

Table 2. Seismic Performance Factors (SPFs) of reinforced concrete frame buildings based on FEMA P695.

Frame	S_{MT}	C_s	S_{max}	R	Ω_0	C_d
5-storey	0.83	0.13	0.32	6.26	2.44	4.17
7-storey	0.56	0.10	0.24	5.69	2.44	3.79
9-storey	0.43	0.08	0.19	5.52	2.45	3.68

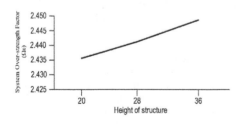

Figure 7. Influence of story number on R.

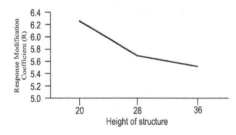

Figure 8. Influence of story number on Ω_0.

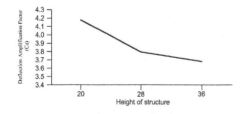

Figure 9. Influence of story number on C_d.

6 CONCLUDING REMARKS

The work presented here has considered three reinforced concrete moment frame buildings, with fundamental vibration periods covering a large spectrum. The design and details following the Indonesian code SNI 1726:2012. The values of SPFs determined using FEMA P695 procedure indicate the influence of story number on the SPFs of seismic-force-resisting-reinforced concrete frame. The values of R and Cd tend to decrease in line with the addition of in story number. Meanwhile, the value of Ω_0 tends to be linearly increased as the story number increased.

The conclusions of the present study are limited by the facts that only a single plan configuration in one single seismic zone has been considered. In addition, the structural behavior has not been validated by any non-linear response-/time-history analysis.

REFERENCES

ATC-40. 1996. *Seismic Evaluation and Retrofit of Concrete Buildings Volume 1*. California: Applied Technology Council.

ASCE. 2006. *Minimum Design Loads for Buildings and Other Structures* (ASCE/SEI 7–05). Virginia: American Society of Civil Engineers.

Elkady, A. & Lignos, D. 2015. Effect of gravity framing on the overstrength and collapse capacity of steel frame buildings with perimeter special moment frames. *Earthquake Engineering & Structural Dynamics*, 44(8): 1289–1307.

FEMA. 2009. *Quantification of Building Seismic Performance Factors* (FEMA P695). Washington, D.C.: Federal Emergency Management Agency.

Mohd, Z.A.M.Z., Debbie R. & Fatehah, S. 2013. An Evaluation of Overstrength Factor of Seismic Designed Low Rise RC Buildings. *Procedia Engineering*, 53: 48–51.

Mondal A., Ghosh, S. & Reddy, G.R. 2013.Performance-based Evaluation of The Response Reduction Factor for Ductile RC Frames. *Engineering Structures*, 56: 1808–1819.

SNI 03-1726-2012. *Tata Cara Perencaan Ketahanan Gempa untuk Gedung*, BSN, Jakarta, Indonesia.

Whittaker, A., Hart, G. & Rojahn, C. 1999. Seismic Response Modification Factors. *Journal of Structural Engineering*, 125(4): 438–444.

Advances in Civil, Architectural, Structural and Constructional Engineering – Kim, Jung & Seo (Eds)
© 2016 Taylor & Francis Group, London, ISBN 978-1-138-02849-4

Methods to improve BIM-based estimations of major building material quantities in Life Cycle Assessment

S. Roh
Architectural Engineering, Hanyang University, Ansan, Korea

S. Tae & S. Shin
School of Architecture and Architectural Engineering, Hanyang University, Ansan, Korea

C. Chae
Department of Building Research, Korea Institute of Civil Engineering and Building Technology, Goyang, Korea

S. Suk & G. Ford
Department of Construction Management, Western Carolina University, Cullowhee, USA

ABSTRACT: The purpose of this study is to propose correction methods for building information modeling (BIM)-based estimations of major building material quantities in Life Cycle Assessment (LCA). Major building materials were analyzed that could be modeled using Levels of Detail (LODs) via BIM and defined the LODs that were to be used in LCA. Different correction methods for the supply quantity of each of the major building materials were proposed using the waste factor method, reinforcement factor method, and standard quantity method, and their applicability were examined in a case study. As a result, the values of all major building materials quantities to which correction methods were applied were closer to the supply quantities in actual structures than were the BIM-based values.

1 INTRODUCTION

With more emphasis in the building industry being placed on the importance of reducing environmental loads, there has been growing interest in Life Cycle Assessment (LCA), which quantitatively evaluates the potential influence a building can have on its environment, from its production phase to its disposal phase (Roh 2012). To conduct LCA on a building, it is essential to collect information regarding all building materials and energy sources invested in the structure during its service life. However, the process of collecting this kind of information leads to excessive time and resource costs and hinders the application of LCA in practice (Tae et al. 2011).

To compensate for these problems, recently, there have been developments in LCA technology that automatically calculate building material quantities supported by a Building Information Modeling (BIM) authoring tool (Kulahcioglu et al. 2011). However, because material take-off in BIM mainly provides simple information such as the surface area (in m²) or volume (in m³) of a building material modeled in BIM (Stadel et al. 2011), it must be supplemented in the following three ways

to be applied in building LCA. (1) The waste factor applied at construction sites must be considered, because an LCA requires information regarding building material quantities actually invested in structures. (2) Rebar modeling is not supported in the BIM authoring tool for building design; it must be taken into account. (3) Materials that have high environmental loads, but whose model is usually omitted in BIM families, require consideration.

This study proposes correction methods to improve BIM-based estimations of major building material quantities in LCA.

2 LOD ANALYSIS FOR THE LCA OF BIM-BASED BUILDINGS

In BIM, a building can be represented from its basic mass form up to the level of completion document, and Levels of Detail (LODs) are generally divided in-to a total of five ranks according to the level of modelling, as shown in Table 1 (Bae 2011).

LOD 100 is the level of planning design in which the ground where the building is to be located and the building's mass form are modeled. LOD 200 is the level of interim design, where the building's

Table 1. LODs of BIM components.

LOD	Example	Remarks
LOD 100		• Conceptual design • Mass form of building
LOD 200		• Schematic design • Main outline of building
LOD 300		• Construction documents • Building material information
LOD 400		• For fabrication and assembly • Information in sharp drawing
LOD 500		• For as-built conditions • Level of completion document

Table 2. LOD of major building materials in a BIM.

Building materials	LOD				
	100	200	300	400	500
Ready-mixed concrete	■	■	■	■	■
Rebar				■	■
Glass		■	■	■	■
Paint			■	■	■
Insulation material			■	■	■
Concrete product			■	■	■

main outline is modelled, and LOD 300 is the level of working design, in which the building's detailed form and construction material information are presented. LOD 400 displays detailed information needed for the construction. LOD 500 is modeling at the same level as the actual completed building.

To perform the LCA on a BIM-based building, LODs must be defined beforehand in a BIM capable of calculating material quantities, because, as a criterion of determining the modeling level of the BIM plan, the LODs of BIM are the standards needed for a consistent LCA of the building. In this research, the calculated data on the major building material quantities (ready-mixed concrete, rebar, glass, paint, insulation material, and concrete products) of the LCA drawn from a previous study (Roh et al. 2014) was analyzed according to the LODs of BIM. Results showed that at LOD 100, the quantity information of only the ready-mixed concrete could be analyzed, while, at LOD 200, that of both the ready-mixed concrete and glass were included. At LOD 300, which usually applies to work in the turnkey system of a public building and design orders, not only is the quantity information regarding both ready-mixed concrete and glass modeled, but also most building materials such as paint, insulation material, and concrete products are modeled. In addition, at LOD 400 and LOD 500, the quantity information of rebar is added. Table 2 represents the modeling information on major building materials based on the LODs of the BIM.

To simplify LCA and obtain accurate evaluation results, it is advantageous to choose LOD 400 or LOD 500, which provide all of the quantity information regarding major building materials. However, BIM plans at the levels of LOD 400 and LOD 500 are seldom compiled during the design phase, where most of the building's overall performance in environmental load reduction is determined (Kim 2014). Therefore, it is difficult

to obtain specific information regarding the structure, and time and costs are limited. Therefore, in this study, LOD 300, the most commonly applied level in BIM design, was defined as the standard for LCA, and selected the data produced from this level to propose correction methods for supply quantities.

3 PROPOSAL OF METHODS TO IMPROVE QUANTITY ESTIMATIONS OF MAJOR BUILDING MATERIALS

In this section, correction methods were proposed for supply quantities for the data produced at LOD 300 using BIM-based LCA. Here, supply quantity correction methods were divided into the waste factor method, reinforcement factor method, and standard quantity method, according to the characteristics of the quantity information regarding major building materials produced from the BIM, as shown in Figure 1.

3.1 Waste factor method

The waste factor method involves applying a waste factor to the net supply quantity calculated using the BIM's material take-off function, as in Equation 1. The method corrects the net quantity to the actual quantity used at a construction site and is applied to ready-mixed concrete, glass, and insulation material.

$$Q_1 = (1 + a_1) \times B_1 \qquad (1)$$

where, Q_i = quantity (m³ or m²) of material (i) for the building; α_i = waste factor of building material (i); and B_i = net quantity (m³ or m²) of building material (i), obtained via BIM.

For the application of the waste factor, the actual waste factors were measured for two apartments (EA1, EA2) built in Korea, and their average value was proposed as the waste factor. As for glass, whose waste factors for the two apartments

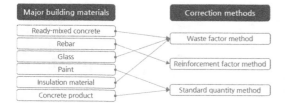

Figure 1. Improvement methods for quantity estimations of major building materials.

Table 3. Waste factors of ready-mixed concrete, glass, and insulation material.

| Building materials | Waste factors | | | |
	Proposed	EA1	EA2	AIJ
Ready-mixed concrete	2.0%	4.0%	0.3%	0.0%
Glass	2.0%	15.0%	1.0%	2.0%
Insulation material	5.0%	–	5.0%	5.0%

were considerably different, a value presented by the AIJ (Architectural Institute of Japan) was used. Table 3 presents waste factors for materials suggested in this study, which include ready-mixed concrete, glass, and insulation material.

3.2 Reinforcement factor method

The reinforcement factor method computes the quantity of rebar, whose modeling is not supported in the BIM authoring tool for building planning.

This method is typically used to estimate rebar quantities in BIM-related studies. However, the reinforcement factor method applied in existing studies only provides the rebar quantity as a simple relation to the total quantity of ready-mixed concrete. Therefore, the values are uniform and are limited in that they show large discrepancies with the actual quantities of rebar used. The reason for this phenomenon is that rebar quantities vary drastically with the structure and section of the building.

Therefore, in this research, using the previous literature (Lee 2013) as a reference, the rebar quantity calculation equation was selected as shown in Equation 2, which uses ready-mixed concrete quantities for each structure element of the building as variables, as a correction method for the supply quantities of rebar.

$$Q_R = 150B_{R,W} + 100B_{R,B} + 115B_{R,G} + 180B_{R,C} + 115B_{R,R} + 120B_{R,F}$$

(2)

where, Q_R = rebar quantity (kg) used in the LCA; $B_{R,W}$, $B_{R,B}$, $B_{R,G}$, $B_{R,C}$, $B_{R,R}$, $B_{R,F}$ = quantities (m³) of ready-mixed concrete used as a material in wall, beam, ground, column, roof, and foundation structures, respectively.

3.3 Standard quantity method

The standard quantity method is used to estimate the quantities of major building materials that have high environmental loads but are not included in BIM families, such as paint and concrete products. Paints include those that are actually used in construction sites, such as water-based paint, oil-based paint, enamel paint, and epoxy paint, and concrete products include bricks and cement.

The total volume of paint is directly proportional to the finished surface area, and the quantity used in a unit of finished surface area varies with the type and amount used in each stroke. In this study, a correction method for supply quantities, that use the finished surface area obtained from BIM and the standard quantity of paint, was proposed as shown in Equation 3. Table 4 represents the standard quantity per unit of paint.

$$Q_P = S_P \times B_{P,A}$$

(3)

where, Q_P = quantity (ℓ) of paint needed for the building LCA, S_P = standard quantity per unit of paint, $B_{P,A}$ = paint's finished surface area (m²) obtained from BIM.

Additionally, the quantities of bricks and cement must be assessed in units of numbers and bags, respectively. However, BIM usually calculates the quantity of bricks and mortar as the volume sum of concrete products. In this study, a correction method was proposed for the quantities of bricks and cement using standard brick (19 cm × 9 cm × 5.7 cm) and mortar mix ratio [cement (kg): sand (m³) = 1:3] provided by the standard quantity. That is, as seen in Equation 4, the quantity of brick infused into concrete products can be calculated from two variables: the true volume of one brick [20 cm × 10 cm × 6.7 cm = 1,340 cm³ = 0.00134 m³]

Table 4. Standard quantity per unit of paint.

| Division | Surface | Number of painting | | |
		1 time	2 times	3 times
Oil-based paint	Wooden	0.094	0.176	0.248
	Metal	0.081	0.166	0.246
Water-based paint	Wooden	0.094	0.176	0.248
	Metal	0.081	0.166	0.246
Enamel paint	Wooden	0.093	0.176	0.249
	Metal	0.082	0.165	0.238

accounting for both its mortar joint width (5 mm per brick face) and the waste factor of bricks. for the quantity of cement, as seen in Equation 5, can be computed from the amount of mortar used on each standard brick [(20 cm × 10 cm × 6.7 cm) − (19 cm × 9 cm × 5.7 cm) = 365 cm³ = 0.000365 m³] as its joint, and the amount of cement used in manufacturing 1 m³ of mortar.

$$Q_B = \frac{B_{C,V}}{V_B} \times (1 + \beta_B) \tag{4}$$

$$Q_B = \frac{B_{C,V}}{V_B} \times V_M \times \frac{\chi c}{\delta_c} \tag{5}$$

where, Q_B = brick quantity (EA) for LCA; $B_{C,V}$ = volume of concrete product (m³) calculated from BIM; V_B = unit volume of a standard brick including its mortar joint (= 0.00134 m³); β_B = brick's waste factor (3% for red brick and 5% for cement brick); Q_C = cement quantity (bags) used for LCA; V_M = mortar joint volume (= 0.000365 m³) for each standard brick; γ_C = cement input (= 510 kg) into 1 m³ of standard mortar; and δ_C = unit weight of cement (generally 40kg/bag).

4 CASE STUDY

To examine the applicability of the BIM-based correction methods for supply quantities proposed in this study, the quantity obtained from the BIM authoring tool (Q-BIM), the quantity to which the correction methods in this study is applied (Q-QCM), and the actual quantity presented in the bill of quantities (Q-BOQ) were compared. The target of evaluation was a studio apartment designed in Revit, a BIM authoring tool, at LOD 300. The building was constructed in Korea, with a reinforced concrete structure; it has 50 stories and 498 rooms.

Figure 2 presents the evaluation results of this case study. Figure 2 indicates that the quantities of ready-mixed concrete obtained were 36,642 m³, 37,375 m³, and 38,035 m³, for Q-BIM, Q-QCM, and Q-BOQ, respectively. An error rates of Q-BIM and Q-QCM with respect to Q-BOQ were 3.66%, 1.74% each, confirming that the application of the correction methods proposed in this research reduced the error by approximately 2%. The obtained rebar quantities were 5,933 tons and 6,053 tons for Q-QCM and Q-BOQ, respectively. Our correction method yielded quantity information results with an error rate of approximately 2%. Quantities for glass were 16,548 m², 16,879 m², and 17,248 m², for Q-BIM, Q-QCM, and Q-BOQ, respectively. That is, the error rates of Q-BIM and Q-QCM with respect to Q-BOQ were 4.06% and 2.14%, confirming that

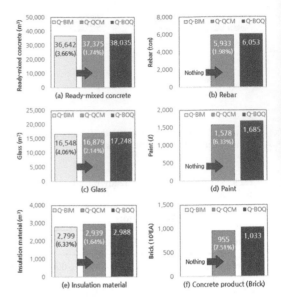

Figure 2. Evaluation results.

the error rate was reduced by approximately 2% with the application of the correction methods proposed in this study. The finished surface area obtained from Q-BIM for paint was 8,968 m², and Q-QCM 1,578ℓ. This value had an error rate of 6.33% with respect to the Q-BOQ value of 1,685ℓ. For insulation materials, Q-BIM, Q-QCM, Q-BOQ volumes were found to be 2,799 m³, 2,939 m³, 2,988 m³, respectively. An error rates of Q-BIM and Q-QCM with respect to Q-BOQ were 6.33% and 1.64%, respectively, which showed that the application of the correction methods proposed in this study reduced errors by 5%. The quantity of concrete products obtained from BIM was 1,243 m³. 955,441 bricks were obtained from Q-QCM using our correction methods. An error of 7.51% was found between this quantity and the Q-BOQ value of 1,033,000 bricks.

5 CONCLUSIONS

This study proposed correction methods for BIM-based estimations of major building material quantities in LCA, and the conclusions below were drawn.

1. The modeling information of major building materials were analyzed according to the LOD of BIM and set LOD 300 as the level to be used for LCA.
2. The waste factor method, the reinforcement factor method, and the standard quantity method were proposed as the correction methods of supply quantities for BIM-based LCA.

302

3. Through a case study to evaluate the applicability our correction methods, all major building material quantities to which these methods were applied were found to be closer to those of actual supply quantities than were the values drawn from BIM.
4. In particular, for ready-mixed concrete, glass, and insulation material, the application of our correction methods reduced the error from 4~6% to less than 2%. Furthermore, for rebar, paint, and concrete products, it is possible to generate quantity predictions with errors within 2~8%.

ACKNOWLEDGEMENT

This work was supported by the National Research Foundation of Korea (NRF) grant funded by the Korean government (MSIP) (No. 20110028794).

REFERENCES

Bae, G. 2011. *Study on the criterion establishment of LOD (level of detail) for BIM model*. Hanyang University.

Lee, S. 2013. *Modelling for estimation of RC structural mate-rial by BIM*. Hoseo University.

Kim, K. 2014. *Conceptual building information modelling framework for whole-house refurbishment based on LCC and LCA*. Aston University.

Kulahcioglu, T. et al. 2012. A 3D analyzer for BIM-enabled Life Cycle Assessment of the whole process of construction. *HVAC&R Research,* 18(1–2): 283–293.

Roh, S. 2012. *Development of simplified life cycle CO_2 emissions assessment program for low carbon building design*. Hanyang University.

Roh, S. et al. 2014. Development of building materials embodied greenhouse gases assessment criteria and system (BE-GAS) in the newly revised Korea Green Building Certification System (G-SEED). *Renewable and Sustainable Energy Reviews,* 35: 410–421.

Stadel, A. et al. 2011. Intelligent sustainable design: Integration of carbon accounting and building information modeling. *Journal of Professional Issues in Engineering Education and Practice,* 137: 51–54.

Tae, S. et al. 2011. The development of apartment house life cycle CO_2 simple assessment system using standard apartment houses of South Korea. *Renewable and Sustainable Energy Reviews,* 15(3): 1454–1467.

Development of logic for building energy consumption calculation program to match operation schedule with individual room per purpose of building

J. Lee, Y. Kim & S. Lee

Korea Institute of Civil Engineering and Building Technology, Building and Urban Research Institute, Gyeonggi-do, Korea

ABSTRACT: This study aims to develop the logic for connecting the rooms and operation schedule for each room which are categorized per purpose of the building. The logic is also utilized for energy consumption calculation program for a building which equals to the energy consumption simulation of a building. The approach has been made from the architectural planning viewpoint while scope of research encompasses schools (elementary, middle, and high schools) of educational institutes. It can be determined that the logic developed from this study can improve convenience of user who runs building energy consumption calculation program as it enables the connection to the operation schedule of each room of the target building even though no other information than the address, the most fundamental information of a building, is available.

1 INTRODUCTION

Climate change due to greenhouse effect requires reduction of energy consumption, and the Korean government confirmed the goal of energy consumption reduction by 37% until the year 2030 compared to expected emission (BAU, Business As Usual). As it is expected that buildings will play a significant part in the reduction effort, much consideration would be needed on how to reduce energy consumption of buildings.

In order to efficiently review and manage building energy demand consumption, Ministry of Land, Infrastructure and Transport and Ministry of Trade, Industry & Energy have enacted "Regulations on certification of energy efficiency level of building" in 2013. The regulation can be considered as rather practical as it provides incentives to the construction company such as bigger floor area ratio if a certain level of energy efficiency has been met. Currently, the ECO2 program is utilized for simulated analysis of building energy consumption for certification of energy efficiency level.

At the moment, this study involves development of 'building energy consumption calculation program' through simplification of entries and usage in ECO2 for the convenience of laypeople such as housewives even if they have insufficient information of a building. Awareness and identification of energy consumption for detailed functions of the building that the user resides in such as heating, air conditioning, and hot water has its own importance as it would be the first step of achieving efficient energy reduction. Values that the user failed to enter will be provided as default values, which first require development of connection logic between entered data and default values. Thus, this study aims to cover its development of connection logic for the rooms of schools (elementary, middle and high schools) and operation schedule of each room.

2 BUILDING ENERGY CONSUMPTION CALCULATION PROGRAM

Building energy consumption calculation program in this study means the energy consumption simulation program currently under development based on the ECO2 program. Number of data entry and method will have to be simplified as it targets general public.

This development can contribute to the improvement of efficiency of nationwide building energy performance, because estimating energy consumption for functions of the building such as heating, air conditioning, and hot water will lead to the wider distribution and expansion of low-energy buildings due to the increase of demand for new or remodeled buildings. Also, the program will have great practicality especially in the context of building energy policy going through reinforcement. This study defines the aforementioned simulation program as "building energy consumption calculation program".

3 METHOD OF RESEARCH

3.1 *Logic*

As the building energy consumption calculation program defines purpose of use per room rather than the whole building, purpose of the building is categorized primarily based on the entered address, then the space of each room is calculated through development logic according to the gross area and the room's capacity. The calculated area per room will be linked to the operation schedule so that the building energy consumption calculation program can be operated effectively.

3.2 *Categorization of purpose of building*

'Seumteo', or Electronic Architectural Administration Information System operated by the Korean government, has 29 categories of main purpose and 412 categories of detailed purpose for buildings in Korea in accordance with the Attached Table 1 of Enforcement Decree of the Building Act, which serve as basis for services such as issuing a building register.

The building energy consumption calculation program of this study is based on the categorization of Seumteo as it targets all types of buildings in Korea. This study involves development of matching logic for each room of the schools (elementary, middle and high schools) and their operation schedule. Table 1 shows the categories of educational and research facilities.

Table 2. Purpose of room.

Room type	Example
00 Residence	Independent/Shared house
01 Small office (over 30 m²)	Executive/personal office and etc
02 Large office (under 30 m²)	Public service/consultation room and etc
03 Meeting & seminar room	Small/medium meeting room and etc
04 Auditorium	Auditorium/lecturing room and etc
05 Cafeteria	Restaurant
06 Restroom	Restroom, and shower room and etc
07 Other rooms	Break room, changing room, reading room, snack bar and etc.
08 Auxiliary room	Hall, lobby, corridor, staircase and etc
09 Warehouse/equipment/ document room	Warehouse, archive and etc
10 Computer room	Computer room and etc (excluding the ones serving for building maintenance
11 Kitchen and cooking area	Kitchen and cooking area and etc
12 Hospital room	Patient room of hospital and etc
13 Guestroom	Guestroom of accommodation and etc
14 Classroom (elementary, middle and high schools)	Classroom of school
15 Lecture room (university)	Classroom of university
16 Store (general/ department store)	Daily life facility such as department store
17 Showroom (exhibit hall/museum)	Showroom at museum or art gallery
18 Reading room (library)	Reading room of library
19 Physical training facility	Basketball court, volleyball court, fitness center of gym in a sport facility and etc

Table 1. Categories of purpose of building.

Main purpose	Detailed purpose -1	Detailed purpose -2
Educational and research facility.		
	Job training facility	
	Private school	
	Library	
	School	
		Elementary school
		Middle school
		High school
		University
		Junior college
		College
		Kindergarten
		Other schools
	Education center	
		Education center
		Training institute
		Other education centers
	Laboratory	
		Laboratory
		Testing laboratory
		Measurement laboratory
		Other laboratories
	Other educational and research facilities	

3.3 Operation schedule per room

The building energy consumption calculation program follows the building operation schedule based on operation regulations of building energy efficiency level certification system of Korea Energy Agency, which defines the schedule including information such as starting and ending times, minimal outdoor air to be introduced indoor, time of lighting, condition of indoor heating source for purpose of 20 different rooms including residence, small office, meeting or seminar room. Table 2 shows the rooms defined by the regulation.

3.4 Development logic

The development logic to match operation schedule of individual room and building's purpose targeted educational facilities (elementary, middle, and high schools). Educational facilities are significant as they take up 4.5% of Korean buildings and take up about 50% of total area of educational, medical, and cultural facilities according to statistics of Ministry of Land, Infrastructure, and Transport.

Those values are not a small value in situation of buildings is divided into 412 categories.

Figure 1 shows the algorithm of connection logic linking operation schedule of elementary, middle, and high schools, which require minimal entry of information such as an address and number of classes for calculation of area demand per room based on its capacity and the area ratio of each room per gross area.

After entering the address and number of classes, the program can estimate the appropriate class-room size of '14 classes (elementary, middle, and high schools)' according to construction plan, by multiplying 7 m × 9 m (9 m × 9 m for lower grades) by the number of classes. In case of middle and high school, the area of a music room, an art room and a physical/chemistry laboratory are included in the class room area. Also, the address entered by the user can be referred to match information on the building register to get the lot area. To calculate school area per student of elementary, middle, and high schools, additional information has to be required to determine whether the number of student exceeds or is lower than 480 for a middle school. In case of a high school, additional information whether the class is humanities/business course or agriculture/industry course to calculate school area per student. Elementary school does not require additional information as school area is calculated with number of classes. After school area per student is calculated, total number of students can be calculated by dividing whole school area by the area per student.

Areas of '04 Auditorium' can be calculated by multiplying area demand per student and total number of students. '05 Cafeteria' and '19 Gymnasium' can be calculated by multiplying area demand per student and number of students per class, while '02 Large office' area, which means the teacher's room, is estimated by multiplying the ratio of the number of students and teachers

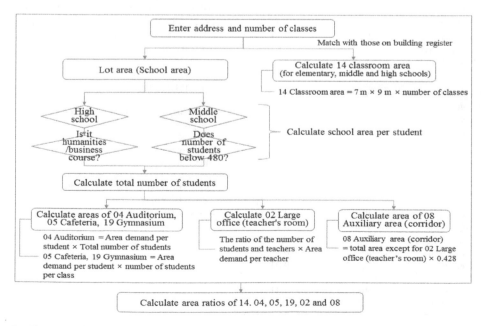

Figure 1. Sequence of algorithm calculation of room area ratio per gross area of educational facility (elementary, middle and high schools).

and area demand per teacher. Number of teachers required per student has been converted from the statistics on the KOSIS (Korean Statistical Information Service). Lastly the corridor, stair cases and entrance areas, '08 Auxiliary area', takes up 30% of all area used by students. Abovementioned approaches have been made with reference to the recommendations from the construction planning.

To represent the room area calculated with this logic in percentage per gross area, for an elementary school, classrooms are found to be taken up 58% per gross area, while auditorium is 7%, cafeteria is 1%, large office (teacher's room) is 2%, gymnasium is 2% and auxiliary area (corridor, etc) is 30%. For a middle school, classrooms are found to be taken up 52% per gross area, while auditorium is 11%, cafeteria is 1%, large office (teacher's room) is 2%, gymnasium is 3% and auxiliary area (corridor, etc) is 30%. For a high school, classrooms are found to be take up 50% per gross area, while auditorium is 13%, cafeteria is 1%, large office (teacher's room) is 3%, gymnasium is 2% and auxiliary area (corridor, etc) is 30%. All these calculations are in Table 3.

Table 3. Room area per gross area and capacity.

	Area/gross area (%)	m²/student
Elementary school		
14 Classroom	58	3.13
02 Large office (teacher's room)	2	0.11
04 Auditorium	7	0.38
05 Cafeteria	1	0.03
19 Gymnasium	2	0.13
08 Auxiliary area (corridor, etc)	30	1.62
Middle school		
14 Classroom	52	2.28
02 Large office (teacher's room)	2	0.11
04 Auditorium	11	0.48
05 Cafeteria	1	0.03
19 Gymnasium	3	0.16
08 Auxiliary area (corridor, etc)	30	1.31
High school		
14 Classroom	50	2.23
02 Large office (teacher's room)	3	0.12
04 Auditorium	13	0.59
05 Cafeteria	1	0.03
19 Gymnasium	2	0.13
08 Auxiliary area (corridor, etc)	30	1.33

4 CONCLUSIONS

This study has aimed to develop the connection logic for individual room and its operation schedule for educational facility (elementary, middle and high schools) for building energy consumption calculation. The viewpoint of construction plan has been employed for this aim, and the program is expected to have increased popularity as it improves convenience by requiring only the address, the most fundamental information to provide operation schedule of each room of the building.

Actual area per capacity and the room's percentage per gross area may vary depending on the actual utilization of the building, but development of logic for elementary, middle and high schools from the viewpoint of construction plan resulted in the numbers in Table 3.

Future studies will have to aim to develop connection logic for each room's operation schedule for building categories in the Enforcement Decree of the Building Act. The logic provided herein has been developed according to construction plan methodology, but verification has to be made whether the actual building is planned and designed according to the area suggested for the capacity before the logic is applied to the building energy consumption calculation program.

ACKNOWLEDGMENT

This research was supported by a grant (15AUDP-B079104–02) from Architecture & Urban Development Research Program funded by Ministry of Land, Infrastructure and Transport of Korean government.

REFERENCES

Kim, C.E. Kim, J.H. & Kim, J.W. 2008. *Construction Plan•Design Theory*, Seowoobooks.
Kim, G.S. 2009. *Construction Plan*, Yeamoonsa.
Korea Energy Agency, 2013. Operation regulations of building energy efficiency level certification system.
Korean Statistical Information Service, http://kosis.kr/statisticsList/statisticsList_01List.jsp?vwcd=MT_ZTITLE&parentId=O#SubCont.
Seumteo, www.eais.go.kr.

Advances in Civil, Architectural, Structural and Constructional Engineering – Kim, Jung & Seo (Eds)
© 2016 Taylor & Francis Group, London, ISBN 978-1-138-02849-4

Durability evaluation of ternary blended concrete based on low-heat cement

J.S. Mun & K.H. Lee
Department of Architectural Engineering, Graduate School, Kyonggi University, Suwon-si, Gyeonggi-do, South Korea

K.H. Yang
Department of Plant and Architectural Engineering, Kyonggi University, Suwon-si, Gyeonggi-do, South Korea

ABSTRACT: The purpose of this study is to evaluate the durability of ternary blended concrete based on low-heat cement for reduction of hydration heat and improvement of compressive strength at early ages. The main parameters were water-to-binder ratio and curing temperature for target compressive strength 40 MPa. The composition of binder are low-heat cement, modified fly-ash and lime stone powder. As a result, the compressive strength at the 28 days was satisfied on the all the mixtures. The relative dynamic modulus of elasticity by freezing-thawing decreased with decreasing of the curing temperature. But, all the mixtures were superior to the relative dynamic modulus of elasticity ratio of over 96% at 300 cycles. Similarly, the depth of carbonation and the resistance of sulfuric acid increased with increasing its age and temperature.

1 INTRODUCTION

In the construction industry, the structure of reinforced concrete is widely used because of its simple workability and excellent durability. In case of the reactor containment walls where the safety of inner/outer parts, as well as the structural performance, is very important, it is necessary to seriously consider not only the dynamic properties of concrete materials but also its durability in the design and maintenance aspects (Nevill 1995). Meanwhile, the containment walls of reactor as a special building generate high hydration heats in them due to the volume increase of material. But, as the concrete has a low heat conductivity, the difference of temperature is generated between surface and core of the material, causing the temperature stress and raising the probability of cracks (Wang & Lee 2010).

To control this hydration heat, there have been diverse studies in the field of concrete materials. For example, the materials such as fly-ash and blast furnace slag used to be mixed into the binder to control the hydration heat. When fly ash and blast furnace slag are mixed, it is excellent in reducing the hydration heat, but it may cause the problem to delay the manifestation of compressive strength at the early age, increasing the construction period (Zákoutský 2012). The problem to delay the manifestation of compressive strength in the early age can be solved by accelerating the hydration reaction of the binding material. But the acceleration of hydration reaction causes the increase of hydration heat in general. Furthermore, the concrete is sensitively influenced by the curing environment. When the concrete is cured in the very hot or cold environment, its result is very different from that in the normal environment even with the same blend.

Therefore, the curing temperature should be considered in order to improve the development of compressive strength in the early age and to reduce the hydration heat of the concrete to be developed. This study has the purpose to evaluate the durability of the ternary blended concrete based on the low-heat cement. The durability evaluation of the concrete was carried out through the tests of freezing-thawing resistance, carbonation resistance, and sulfuric acid resistance.

2 EXPERIMENTAL PROGRAM

2.1 Properties of binder

The binder of this test was made of low-heat cement, modified fly ash and lime stone powder. The chemical composition of used binder is shown in Table 1. The low-heat cement used here was the Type 4 in ASTM C 150 (ASTM, 2010), in which the percentages of C_3S, C_2S and C_3A were 31%, 48%, and 3% respectively. Its specific gravity and specific

Table 1. Chemical composition of cementitious materials (% by weight).

Materials	SiO$_2$	Al$_2$O$_3$	Fe$_2$O$_3$	CaO	MgO	K$_2$O	Na$_2$O	TiO$_2$	SO$_3$	LOI
LHC	25.3	3.1	3.4	62.5	1.7	0.57	0.10	0.09	1.9	0.8
MFA	64.0	21.9	5.5	3.8	1.2	1.1	1.0	1.5	–	2.23
LP	17.7	8.2	0.6	47.5	2.1	–	–	–	0.3	22.3

surface were 3.18 and 3,440 cm^2/g respectively. In the mass concrete, the fly-ash is usually used to reduce the heat of hydration, but it tends to lower the compressive strength. The modified fly-ash used to complement it is the one whose specific surface was enhanced with a mechanical method (milling). In the modified fly-ash, the glass coating on the surface of fly-ash is peeled to promote the early reactivity for the manifestation of compressive strength in the early age (Yang 2014). Its specific gravity and specific surface were 2.25 and 5,510 cm^2/g respectively. The fine lime stone powder is the material that has no early reactivity. It has been reported to have the effect to reduce the heat of hydration. But, when it is mixed in a large amount, it can reduce the compressive strength. Therefore, its ratio of mixture should be small. Specific gravity and specific surface of the lime stone powder used here were 2.81, 3,420 cm^2/g respectively.

2.2 Concrete mixtures

The mixture proportions of concrete to reduce the hydration heat and to development the compressive strength at the early age are shown in Table 2. Each mixture was set in consideration of the curing temperature to target compressive strength of 28 days age to 40 MPa. The curing temperatures were set as 5, 20 and 40°C in consideration of cold, reference, and hot period respectively. According to this, the water-binder ratio (W/B) was set for each climate. The durability is much influenced by the air volume. The aimed air volume was set as $4.0 \pm 1.5\%$.

2.3 Casting, curing, and testing

Before the test, the specimens were cured in the water at 5, 20 and 40 °C for the cold, reference, and hot conditions until the curing age. For the compressive strength evaluation of the concrete, cylindrical specimens of $\varnothing100 \times 200$ mm were tested with the 100 ton compressive strength tester. The freezing-thawing resistance evaluation was carried out in the way of air-freezing and water-thawing on the basis of ASTM C 666 (ASTM, 2011). The specimen was a square pillar of $100 \times 100 \times 400$ mm size, for which the relative dynamic modulus of elasticity was measured up at every 100 cycles up to

Table 2. Mixture proportions.

Curing	W/B (%)	S/a (%)	Unit weight, (kg/m^3)					
			W	LC	MFA	LP	S	G
Cold	27.5	45	148	432	81	27	730	896
Reference	30	45	162	432	81	27	715	877
Hot	32.5	45	175	432	81	27	699	857

*LC: low-heat cement, MFA: modified fly-ash, LP: lime stone powder, S: sand and G: gravel

300 cycles. For the carbonation evaluation, a cylindrical specimen of $\varnothing100 \times 200$ mm was prepared and the accelerated corrosion test was made after 28 days of curing at 5% of carbon dioxide concentration. On the sectional surface of the cylindrical specimen, the 1% phenolphthalein solution was sprayed to measure the carbonation depth.

For the sulfuric acid resistance test, a cylindrical specimen of $\varnothing100 \times 200$ mm was immersed in the 5% sulfuric acid solution after 28 days. After immersion, the specimen's mass change rate, appearance change, and compressive strength were checked on 28th and 91st days.

3 TEST RESULTS AND DISCUSSION

The compressive strength and durability evaluation of the low-heat cement based ternary concrete was carried out after curing at 5, 20 and 40°C until the age required for each test.

3.1 Compressive strength at 28 days

The compressive strength after 28 days of curing in consideration of W/B ratio and curing temperature is shown in Figure 1. In general, the early manifestation of compressive strength gets lower as the curing temperature gets lower but, in case of longer curing age, the compressive strength gets higher at the lower curing temperature. When the curing age is 28 days, 5% increase of curing temperature brings out about 2 MPa strength reductions. By the way, when the compressive strengths of cold, reference, and hot mixtures were compared, the cold specimen had the compressive strength about 1MPa lower than other ones. It seems because the

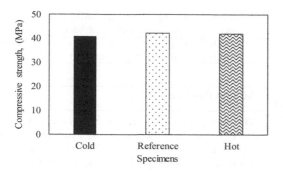

Figure 1. Compressive strength at 28 days.

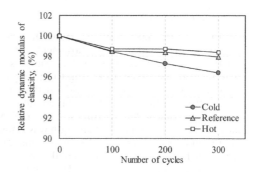

Figure 2. Relative dynamic modulus of elasticity by freezing-thawing.

low-heat cement based ternary concrete had such a low early reactivity that the reactivity was not recovered till 28 th day of curing.

3.2 *Freezing-thawing resistance*

Figure 2 shows the relative dynamic modulus of elasticity for each blend in the freezing-thawing resistance test. The freezing-thawing is a phenomenon that the water in the capillary crevices of the concrete suffers the repetitive freezing-melting process to make the crack or exfoliation, degrading the durability of the concrete. This phenomenon is greatly influenced by W/B and air volume (Joan et al. 1996). As the reduction of W/B increases the compressive strength and decreases the capillary crevices, the freezing-thawing resistance gets greater. As a result of the test, the cold mixture had the lowest freezing-thawing resistance in spite of the lowest W/B. It seems because the lower curing temperature made the compressive strength lower. However, all the specimens showed the excellent results that the relative dynamic modulus of elasticity was kept as high as 96% till 300 cycles.

3.3 *Carbonation resistance*

Figure 3 shows the carbonation depth of low-heat cement based ternary concrete after 1, 3, 4, 8 and 12 weeks from the 28th day of curing. In all the mixtures, the carbonation depth got deeper as the curing age got longer. But, it had the opposite tendency with the freezing-thawing resistance. That is, the carbonation depth got deeper as the curing temperature got higher. This result seems to reflect the tendency that the higher curing temperature decreases the compressive strength (Chang & Chen 2006), for the carbonation test was carried out after 28 days, while the freezing-thawing resistance test was carried out after 14 days of curing.

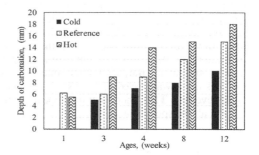

Figure 3. Depth of carbonation at differential ages.

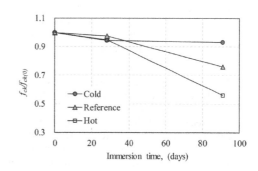

Figure 4. Residual compressive strength by immersion time.

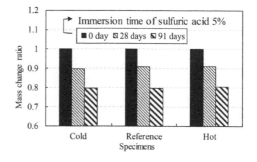

Figure 5. Mass change ratio at differential mixtures.

311

3.4 *Sulfuric acid resistance*

For the sulfuric acid resistance test, the specimens were immersed in 5% sulfuric acid solution and their mass change and residual compressive strength were measured according to the days of age (Figure 3, 4). In the sulfuric acid resistance test, the residual compressive strength turned out to get higher as the curing temperature got lower. Especially, though the compressive strength was similar at early immersion ages, the difference of manifestation rate for the compressive strength got greater as the immersion age got longer. This tendency is similar to that of carbonation resistance. That is, at the lower curing temperature, the manifestation rate of compressive strength gets higher as the curing age gets longer. The mass reduction rate was similar in all the mixtures when the immersion age was the same. This result was also shown in the appearance test of the specimens. In the early age of immersion, there was no change in the appearance of specimens. But, after some time, the paste on the surface was peeled off and some aggregates were revealed. This phenomenon appeared similarly in all the specimens.

4 CONCLUSIONS

As a basic study to develop the low-heat cement based ternary blended concrete that can lower the hydration heat and improve the early compressive strength, the evaluation of its compressive strength and durability was carried out to get the following conclusions:

1. The aimed compressive strength of 40 MPa was satisfied by all the mixtures set by W/B ratio and curing temperature after the age of 28 days.
2. The freezing-thawing resistance got lower as the curing temperature got lower. But all the mixtures had the durability index as high as 96%.
3. The carbonation depth tended to go deeper as the age got longer. This tendency got stronger in the mixtures with higher curing temperature.

4. The mass change with the sulfuric acid resistance was similar regardless of mixture. But the residual compressive strength was more excellent as the curing temperature got decreased.

ACKNOWLEDGEMENT

This work was supported by the Nuclear Power Core Technology Development Program of the Korea Institute of Energy Technology Evaluation and Planning (KETEP), granted financial resource from the Ministry of Trade, Industry & Energy, Republic of Korea (No. 20131520100750).

REFERENCES

ASTM C 150, 666 2011. *Annual book of ASTM standards*, ASTM international, West Conshohocken, Pa, USA.

Chang, C.F. & Chen, J.W. 2006. The experimental investigation of concrete carbonation depth, *Cement and Concrete Research*, 36(9): 1760–1767.

Joan, D.B. Gray, L.K. & Maher, K.T. 1996. Freeze-Thaw durability of high-performance concrete masonry unit, *ACI Materials Journal*, 93(4): 386–394.

Neville, A.M. 1995. *Properties of concrete*, Addison Wesley Longman Limited, New York, NY, USA.

Yang, K.H. 2014. *Development of unit system comprising of formwork with hight above 4m and low-heat concrete to strengthen export competition of nuclear power plant*. Technical Report (1st), Department of Plant•Architectural Engineering, Kyonggi University (in Korea).

Wang, X.T. & Lee, H.S. 2010. Modeling the hydration of concrete incorporating fly and or slag. *Cement and Concrete Research*, 40: 984–996.

Zákoutský, J. Tydlitát, V. & Černý, R. 2012. Effect of temperature on the early-stage hydration characteristics of Portland cement: a large-volume calorimetric study. *Cement and Concrete Research*, 36: 969–976.

Advances in Civil, Architectural, Structural and Constructional Engineering – Kim, Jung & Seo (Eds)
© 2016 Taylor & Francis Group, London, ISBN 978-1-138-02849-4

Suggestion for new internal transverse ties arrangement method in reinforced concrete columns

W.W. Kim
Department of Architectural Engineering, Graduate School, Kyonggi University, Suwon-si, Gyeonggi-do, South Korea

K.H. Yang
Department of Plant and Architectural Engineering, Kyonggi University, Suwon-si, Gyeonggi-do, South Korea

ABSTRACT: The purpose of this study is to evaluate the arrangement method for V-ties that can replace the cross-tie arrangement with internal transverse ties in RC columns. To this end, central axial load experiments with a compressive strength of 78 MPa were conducted on an internal ties arrangement style and on embedded V-ties. The results of the tests met all the strength formula requirements of ACI 318-11, but early buckling of the longitudinal bar occurred in the cross-tie arranged columns due to the loosening of the 90° hooks. On the other hand, this pulling-out of V-ties did not occur in the V-tie arranged columns. The V-tie arrangement method met the ultimate strength specified in ACI 318-11 and met or exceeded the ultimate strength of the cross-tie arranged columns. Furthermore, its axial ductility ratio was approx. 82% higher than that of cross-tie arrangement. Therefore, it can be concluded that the V-tie arrangement method is effective for confining core concrete and for preventing buckling of the longitudinal bar.

1 INTRODUCTION

It is very critical to secure the ductility in the seismic design of a reinforced concrete column. Internal transverse ties confining the core concrete are needed to improve the ductility. The ACI 318-11 stipulate the amount of internal transverse ties and spacing in the section of the potential plastic hinge. Furthermore, they also stipulate that because the longitudinal bar in the central section of the column is vulnerable to buckling, the internal transverse ties should be arranged, if the horizontal spacing is greater than 150 mm. The method that is the most frequently used for internal transverse ties in the confined core concrete is of cross-tie type, where the hooks are crossed perpendicularly in the manner of 90° for one hook, 135° for other hooks. However, a 90° loosening of the hook in the large deformation of the column, when cross-tie arrangement was carried out causes buckling of the longitudinal bar and lowering of ductility.

Therefore, this study suggests the V-tie arrangement method as a countermeasure to prevent the early buckling of the longitudinal bar and lowering of ductility. The V-tie arrangement method is a method of adhering bar bent in the form of symmetrical V-shape using one-touch fixing clips, in order to fix them to the longitudinal bar. Because this method does not cause loosening of the hook, it is not only effective for preventing the longitudinal bar from buckling but also has a high efficiency in confining inner core concrete. In this study, a central axial load experiment was performed on the column member to assess the V-tie reinforcement method that can replace the cross-tie reinforcement method.

2 EXPERIMENTAL METHOD AND USED MATERIALS

2.1 *Specimen details*

Experimental parameters used in this study are shown in Table 1 and Figure 1. Cross-section of the column specimen is 300 × 300 mm, and the filling thickness is 25 mm. To induce failure in the experimental section, the specimen was reinforced with carbon fiber sheet in 2-ply at its upper part and lower part. The main variables are cross-tie arrangement method and V-tie arrangement method to verify the confining effect of V-tie as the internal transverse tie. Reinforcement spacing was determined by the amount of transverse reinforcement ($A_{sh, aci}$) presented in the seismic design of ACI 318-11 (Figure 2). Cross ties were designed in the same manner as in the conventional reinforcement method, and the embedded length of V-tie

Table 1. Detail of specimens.

| Specimens | Supplementary ties | | Amount of transverse reinforcement | |
	Type	S(mm)	l_{db}	ρ_{sh}
C-78	Crosstie	85	78($6d_b$)	2.94
V-78	V-tie	85	78($6d_b$)	2.54
V-97.5	V-tie	85	97.5($7.5d_b$)	2.69

Note: S = spacing of transverse reinforcement, l_{db} = embedment length of V-Tie leg into core concrete, ρ_{sh} = volumetric ratio of transverse reinforcement.

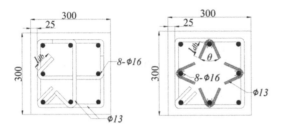

Figure 1. Detail of specimens of section.

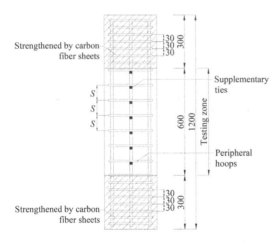

Figure 2. Detail of arrange of reinforcement.

hook was determined to be the mini-mum length of hook [Max ($6d_b$, 75mm)], which is a reference value of ACI 318-11. Also, changes were applied to the embedded length of V-tie to evaluate it. One-touch fixing clip used in the reinforcement of the V- ties was decided as the open type for easy combining to the longitudinal bar. For V-tie to be combined to the longitudinal bar, was used one-touch fixing clip (Figure 3). Reinforced plastic was

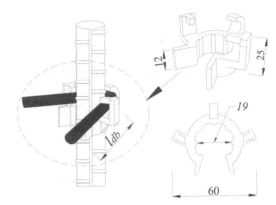

Figure 3. Detail of connector joining V-tie.

Table 2. Mechanical properties of steel bars and polypropylene.

Type	f_y MPa	f_s MPa	E_c GPa	ε_y	ε_u %
Steel	ϕ16 518	619	189.6	0.0026	0.165
Bars	ϕ16 518	691	180.2	0.0025	0.162
Polypropylene	37.2	105	2.4	0.0154	$\geq$4.5

selected because it is an elastic material that allows easy insertion.

2.2 Used materials

High strength concrete was decided as a concrete to be used in this study because it has a serious problem in brittleness fracture. The result of 78 MPa was obtained by measuring the compressive strength using a Φ100 cylinder at the same time of adding pressure to column specimen. As longitudinal bar and internal transverse ties, Φ16 and Φ13 were used. Their yield strengths are 518 MPa and 508 MPa, and elasticity coefficients are 199.7 GPa and 200.4 GPa, respectively. As one-touch fixing clips that were used for arranging V-ties, impact polypropylene of reinforced plastics was used. Its yield strength and elasticity coefficient were 37.2 MPa, and 2.4 GPa, respectively (Table 2).

2.3 Experimental method

Central axial load experiments on columns were carried out as shown in Figure 4. The control was set at a speed of 0.5 mm/min using an oil jockey with the capacity of 12,000 kN, and seats were installed on the lower and upper parts to prevent eccentricity. Axial strains were measured in the experimental section from four sides using LVDT (100 mm).

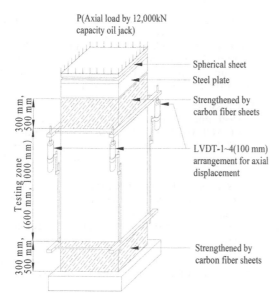

P(Axial load by 12,000kN capacity oil jack)

Spherical sheet

Steel plate

Strengthened by carbon fiber sheets

LVDT-1~4(100 mm) arrangement for axial displacement

Strengthened by carbon fiber sheets

Testing zone 300 mm, 500 mm (600 mm, 1000 mm) 300 mm, 500 mm

Figure 4. Test setting.

3 EXPERIMENTAL RESULTS

3.1 *Development of cracks and failures mode*

As a result of the central axial load experiment, cracks began to occur at 80–90% of ultimate strength in all columns, and the filled concrete popped out and cracks developed. The ultimate strength was reached when the filled concrete popped out, and the degree of failure after the experiment differed according to the arrangement type.

In the cross-ties arranged column, a loosening occurred at the 90° hooks when the ultimate strength was reached, and the buckling of the longitudinal bar structure was severe. On the other hand, V-tie arranged columns confined the core concrete without evidence of any loosening of internal V-ties or pulling-out. Buckling of the longitudinal bar structure was also worse in the cross-ties arranged column.

3.2 *Relationship between axial load and axial strain*

The axial load-axial strain relationship which was measured in the experiment is shown in Figure 5. The axial load-axial strain relationship in the column was not affected by the steel arrangement method before filled concrete popped out. In addition, it met all the ultimate strength related formulas present in ACI 318-11, regardless of the arrangement method. However, the ultimate strength decreased sharply in the cross-tie arranged columns, while their 90° hooks loosened and the filled concrete popped out.

Although V-tie arranged columns had approx. 89% of the transverse reinforcement steel of cross-tie arranged columns, the ultimate strength of V-tie arranged column was equivalent to or higher than that of the cross-tie arranged columns. In addition, the descending slope was gentle. The increase in ultimate strength created by the increase in the embedded length of V-ties was 1.02 times, which is insignificant. The descending slope was gentler in the V-tie arranged column. The reason the descending slope is sharp in the cross-tie arranged column is the loosening of the 90° hooks.

3.3 *Strain of transverse ties*

The load-strain relationship of internal transverse ties in the column was shown in Figure 6. Strain ascending sections were similar in the columns arranged with cross-ties and with V-ties. However, although the strain of internal transverse ties in the cross-tie arranged column could not reach the yield strain, the strain in the V-tie arranged column reached the yield strain. This is because the internal cross-ties could not confine the longitudinal bar, but were loosened at the 90° hooks. On the

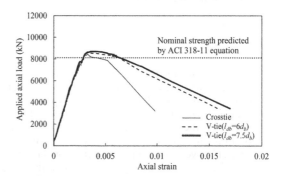

Figure 5. The axial load-axial strain relationship in the column.

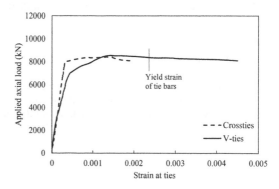

Figure 6. Strains of supplementary tie bar against applied axial load.

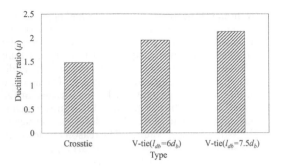

Figure 7. Axial ductility ratio of specimens.

other hand, the internal transverse ties in the V-tie arranged column successfully resisted buckling of the longitudinal bar structure up to the yield point.

3.4 *Axial ductility ratio*

Axial ductility ratio (ρ_{svh}) was evaluated as an index to determine the confining force which confines the core concrete within the steel reinforcement columns. As a model for determination, the equation of Razvi and Saactioglu was used.

$$\mu = \frac{\varepsilon_{85}}{0.004} \tag{1}$$

where, ε_{85} = means the strain is at 85% of ultimate strength, after reaching the maximum ultimate strength. The axial ductility ratio of specific columns is shown in Figure 7. V-tie arranged columns have 182% the ductility ratio of cross-tie arranged columns. In other words, the V-tie arrangement is more effective at preventing buckling of the longitudinal bar and confining core concrete than the cross-tie arrangement. While the embedded length of V-tie varied from $6d_b$ to $7.5d_b$, the axial ductility ratio increased by 1.03 times. In conclusion, the embedded length of V-tie has an insignificant effect on the axial ductility ratio.

4 CONCLUSIONS

Central axial load experiments were carried out in order to support the V-tie arrangement method for replacing cross-ties which have been used for internal transverse ties in columns. The V-tie arrangement method and cross-ties arrangement method were compared by calculating the ultimate strength and ductility from the axial load and axial strain relationship of column members that meet the amount of transverse steel specified in ACI 318-11.

1. Cross-tie columns reinforced with the transverse steel amounts required by the seismic standard of ACI 318-11 experienced a loosening of the 90° hooks at ultimate strength; however, V-tie columns experience neither loosening nor pullout
2. As a result of the central axial load experiment, V-tie arranged columns have a ultimate strength equivalent to or higher than cross-tie arranged columns, even though V-tie arranged columns have less transverse reinforcement steel

V-tie arranged columns have 180% capacity of the ductility of cross-tie arranged columns.

ACKNOWLEDGEMENT

This work was supported by the Small and Medium Business Administration (SMBA) grant funded by the Ministry of Trade, Industry & Energy, the Republic of Korea (No. S2187208).

REFERENCES

American Concrete Institute, 2011. *Building Code Requirements for Structural Concrete and Commentary (ACI 318-11)*, ACI Committee 318, U.S.A.
Korea Concrete Institute, 2012. *Concrete Design Code and Commentary*, Kimoondang Publishing Company.
Razvi, S. & Sattcioglu, M. 1992. Strength and Ductility of Confined Concrete, *Journal of Structural Engineering*, 118(6): 1590–1607.
Sim, J.I. 2008. Flexural *Behavior of Reinforced Concrete Columns Strengthened with Wire Ropes and T-shape Plates*, Master Thesis, Mokpo National University, Mokpo, South Korea.
Yang, K.H. Lee, Y.H. Kwak, N.H. & Chung, H.S. 1999. A Study on the Effectiveness of Hook Type of Cross-ties in the Flexural Behavior of Confined Concrete Columns. *Journal of the Architectural Institute of Korea*, 15(9): 63–70.

An analysis of heating energy consumption and savings in Multi-Dwelling Units in South Korea using energy performance certificates

H.K. Jung

Korea Institute of Civil Engineering and Building Technology, Goyang-Si, Korea

ABSTRACT: Using Korea's building energy performance certificates, a Multi-Dwelling Unit (MDU) evaluation model was set up for MDUs in South Korea. The insulation performance of the building for fenestration, exterior walls, roofs, and floors was modified, and the heating energy consumption and energy saving rates of MDUs were analyzed. Among the cases in the alternative design, the heating energy consumption rates in cases 3 and 16 were 29.057 GJ/year and 22.904 GJ/year, respectively; the energy saving rates were 26.73% and 44.29%, respectively. The insulation performance of the fenestration was higher in Case 3 than in Case 2. However, because of low shading coefficient performance, the heating energy consumption rate was 1.334 GJ/year, resulting in the energy savings rate 1.06% lower. For MDUs in Korea, rather than using low-e glazing, increasing both the airspace depth in the fenestration system and the heat resistance of the air layer can work better and needs to be tested and verified.

1 INTRODUCTION

Many certificate systems are in operation worldwide to improve building energy performance. South Korea adopted building Energy Performance Certificates (EPCs) in 2001. Among the EPC-targeted buildings, evaluation methods for residential use involve calculating housing units' energy consumption, then comparing results with the heating energy consumption seen in standard housing, and then finalizing the energy savings rates. These rates determine the certificate rating according to EPC rating standards.

Certificates comprise preliminary certificates that evaluate drawings before construction, as well as main certificates that evaluate final as-built drawings and construction inspections. Depending on the rating of the acquired certificate, remission criteria will be applied by the related legal regulations regarding floor area ratio, landscape area, and building height. This study uses EPCs to set up an evaluation model for MDUs in South Korea, and gradually strengthens its insulation performance according to each material's physical characteristics to analyze and compare heating energy consumption and energy savings rates of the different housing units.

2 METHODOLOGY

2.1 Certificate rating

EPCs assign energy savings ratings to buildings voluntary application. The evaluation targets are the newly built MDUs of more than 18 units, and the rating system is shown in Table 1.

2.2 Evaluation standard and method

EPCs calculate energy consumption both for each household and each unit of the target MDU in the application, in addition to that for a standard housing unit. The energy savings rates e_{au} and e_{ab} are then added to the total energy savings rates. The calculation methods are as follows:

Table 1. Building energy performance certificate rating.

Rating	Energy savings rate
1	$\geq 33.5\%$
2	23.5%–33.5%
3	13.5%–23.5%

$$e_u = \left\{ (E_s - E_a)/E_s \right\} \times 100 + e_{au} \qquad (1)$$

$$e_b = \sum (e_u \times A_u)/A_b + A_{ab} \qquad (2)$$

$$e = \sum (e_b \times A_b)/A_T \qquad (3)$$

Here,

e: energy saving rate (%) of the housing in application

e_u: energy saving rate (%) of a housing unit

e_{au}: energy saving rate (%) of the added items of a housing unit

e_{ab}: energy saving rate (%) of the added items of a housing unit

e_b: energy saving rate (%) of an MDU

E_s: heating energy consumption rate of a unit of standard housing (GJ/year)

E_a: heating energy consumption rate of a housing unit in application (GJ/year)

A_u: area for exclusive use of a housing unit (m²)

A_b: total area for exclusive use of an MDU (m²)

A_T: total area for exclusive use of a housing unit in application (m²)

2.3 Standard housing concept

Standard housing refers to housing that meets current legal regulations in energy-saving criteria. The established standards are shown in Tables 2 and 3.

The air change rates for each household includes air infiltration rates and ventilation from living. The air change rates for each household's precertification is assumed to be 0.7/hr, as defined by the code. After a building's construction, the rates will be replaced by the values achieved through blower door tests.

2.4 Evaluation model

The evaluation model for MDUs is taken from popular high-rise apartment building units. The unit summary is shown in Table 4 and in Figure 2.

The air change rate for the evaluation model is set to be 0.7/hr, as defined by the code, and the model is district heated. The fenestration to wall area ratios consists of 58% south-facing and 39% north-facing windows. The glazing system's Shading Coefficient (SC) values are set to be 0.8. The insulation values are set to be the mean values from the values taken from the EPC-awarded cases and are illustrated in Table 5.

The MDUs include enclosed verandas by the living rooms, and there are glazed sliding doors between the verandas and the living rooms. These sliding doors are categorized into household fenestration, whereas the glazed enclosures for the verandas are part of veranda fenestration.

Table 2. Setup standards of MDUs.

Item	Notes
Housing shape and size	Same as housing in application
Floor height and ceiling height	Same as housing in application
Building direction	East (N-S)
Other factors such as substructures	Same as housing in application
Energy sources and boiler types	Same as housing in application
Boiler's rated performance	80% (Total energy standard)
Boiler's seasonal load loss factor	5% (Individual heating, central heating)
Plumbing heat loss factor	5% (Central heating, district heating)

Table 3. Setup standards of standard housing units.

Item	Heating area	Non-heating area
Interior temperature	20°C	Details
Plans and floor area	Same as housing in application	
U-values of exterior walls, roofs, and floors	U-value according to the local codes	4.0 W/m²K
U-values of fenestration	3.3 W/m²K (Household) 6.6 W/m²K (Veranda)	6.6 W/m²K
Solar heat gain	Same as housing in application	Same as housing in application
Shading coefficient	0.45	
Front door types	2.1 m² (Opaque)	
Front door u-value	2.6 W/m²K	
Ventilation rate	0.7/hr	2.0/hr

Table 4. Evaluation model summary.

Location	Seoul, Korea
Area	84.99 m²
Floor height	2.8 m
Ceiling height	2.3 m
Direction	South facing

Table 5. Insulation performance (Evaluation model).

Division		U-Value [W/ m²K]
Fenestration	Household fenestration	3.3
	Veranda fenestration	3.3
	Exterior wall	0.446
	Roof	0.256
	Floor	0.316

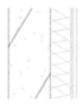

Exterior wall Section

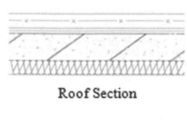

Roof Section

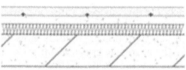

Floor Section

Figure 1. Sections of exterior walls, roofs, and floors.

The exterior walls are constructed of reinforced concrete, insulation materials, and gypsum boards, as shown in Figure 1. The roof structure includes plain concrete for waterproofing. The floor slabs are covered with foamed concrete and mortar and are finished with laminate flooring materials.

Local heating performance is calculated to be 100% in the EPC system, and 5% is the plumbing heat loss factor.

2.5 Alternative design

U-values from "the building energy-saving design standards" are provided, as illustrated in Table 6, for the insulation performance of the model's fenestration.

The thermal conductivity and depths of the exterior walls, roofs, and floors are considered, and

Figure 2. Plan (Evaluation Model).

Table 6. Insulation performance of fenestration.

No.	U-value [W/m²K]	Shading coefficient	Fenestration system
1	3.3	0.8	PVC Frame, DGU, Airspace (Argon Gas) 6 mm
2	3.0	0.8	PVC Frame, DGU, Airspace (Argon Gas) 12 mm
3	2.9	0.6	PVC Frame, Low-E DGU, Airspace (Argon Gas) 6 mm
4	2.4	0.6	PVC Frame, Low-E DGU, Airspace (Argon Gas) 12 mm
5	2.2	0.6	PVC Frame, Low-E DGU, Airspace (Argon Gas) 12 mm

Table 7. Insulation performance.

No.	U-value [W/m²K]		
	Exterior wall	Roof	Floor
A	0.446	0.256	0.316
B	0.364	0.23	0.273
C	0.343	0.216	0.259
D	0.322	0.202	0.243
E	0.307	0.201	0.237
F	0.287	0.188	0.223

their insulation performances are described and compared as U-values, as shown in Table 7.

The evaluation model described in Table 4 was given the insulation performance described in Table 5 as a base point. Then, using the data

Table 8. Alternative design cases.

| Case no. | Fenestration | | Exterior walls, roofs, and floors |
	Fenestration of unit	Fenestration of veranda	
Case 1	1	1	A
Case 2	2	2	A
Case 3	3	2	A
Case 4	4	2	A
Case 5	4	4	A
Case 6	2	2	B
Case 7	2	2	C
Case 8	2	2	D
Case 9	2	2	E
Case 10	2	2	F
Case 11	3	2	D
Case 12	4	2	D
Case 13	4	4	D
Case 14	4	2	F
Case 15	4	4	F
Case 16	5	4	F

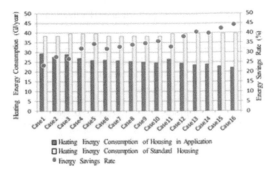

■ Heating Energy Consumption of Housing in Application
□ Heating Energy Consumption of Standard Housing
● Energy Savings Rate

Figure 3. Heating energy consumption and energy savings rates of the alternative design.

in tables 6 and 7, heating energy consumption and energy savings rates were calculated. For the alternative design, the itemized data shown in Table 8 are used to calculate the heating energy consumption and energy savings both for the housing in the application and for the standard housing. Case 1 is set as the base point of the alternative design.

3 RESULTS

In Case 1, the heating energy consumption of the housing in the application was 29.412 GJ/year, and that of standard housing was 38.377 GJ/year; the energy savings rate was 23.36%.

The heating energy consumption and energy savings rates of the alternative design are shown in Figure 3.

In Case 3, the energy saving rate was 26.73%, and the heating energy consumption was 29.057 GJ/year. In Case 16, the energy saving rate was 44.29%, and the heating energy consumption was 22.094 GJ/year.

The heating energy consumption in Case 3 was higher than the consumption seen in Case 2 by 1.344 GJ/year, and the energy savings rate was also higher by 1.06%.

Regarding household fenestration, the U-values were 2.9 W/ m²K for Case 3 and 3.0 W/ m²K for Case 2, showing higher insulation performance in Case 3 than in Case 2. However, the SC values of the low-e glazing are lower, therefore indicating less Solar Heat Gain and higher heating energy consumption.

The heating energy consumption in Case 4 was lower than that in Case 2 by 0.761 GJ/year.

In cases, 3 and 4, where low-e glazing is designed, heating energy consumption varies depending on airspace depth.

When the insulation performance conditions of the exterior walls, roofs, and floors changes from those of Case 6 to those of Case 10, heating energy consumption decreases by 1.48 GJ/year, and the energy savings rate by 3.85%.

When the insulation performance conditions of the fenestration changes from Case 8 to Case 12, the heating energy consumption decreases by 0.819 GJ/year, and the energy savings rate by 4.2%.

Since energy savings rates are calculated in accordance with rates of heating energy consumption of standard housing and the housing in application, the heating energy consumption savings can be big depending on the standard housing model, while energy savings rates remain low. On the other hand, heating energy consumption savings can be small when energy savings rates are high.

4 CONCLUSIONS

Using the EPC program and targeting MDUs in South Korea, heating energy consumption and energy savings rates were analyzed for the alternative design.

Since the EPC program evaluates heating energy, fenestration with high insulation performance can generate high heating energy consumption because of SC values. The heat resistance of the airspace can be secured by raising SC values or increasing air-space depth.

As heating energy consumption and energy savings rates vary depending on the standard housing model of the EPC system, it is necessary to review the insulation performance conditions and SC values of fenestration in evaluating heating energy consumption and energy savings rates of MDUs.

ACKNOWLEDGEMENT

This research was supported by a grant from an Evaluation Project (Project No.: 20150021–001–01) funded by the Korea Institute of Civil Engineering and Building Technology.

REFERENCES

Jung, H.K. 2011. *A Study on the Current Systems and Design Alternatives*, Incheon: Inha University.

Korea Energy Management Corporation. 2007. *Building energy performance rating certificate system operation standards, South Korea*, Korea Energy Management Corporation.

Ministry of Land, Infrastructure, and Transport. 2008. *Building energy performance rating certificates and related rules. South Korea*, Ministry of Land, Infrastructure, and Transport.

Yu, K.H. Cho, D.W. & Song, K.D. 2006. A Study on the Energy Performance Rating and Certification of Apartment Houses, *Journal of the Architectural Institute of Korea*, 22(12): 319–326.

Advances in Civil, Architectural, Structural and Constructional Engineering – Kim, Jung & Seo (Eds)
© 2016 Taylor & Francis Group, London, ISBN 978-1-138-02849-4

A construction of the heat dissipation data base for measuring the heat value of electrical machine and appliance

Y.M. Kim & J.H. Kim
Korea Institute of Civil Engineering and Building Technology, Gyeonggi, South Korea
Daewoo Institute of Construction Technology, Suwon, South Korea

G.S. Choi
Korea Institute of Civil Engineering and Building Technology, Gyeonggi, South Korea

ABSTRACT: The heat value data of electrical machine and appliances for HVAC capacity calculations is provided by the maker or is used to extract from some Code. Overdose design is occurred in HVAC capacity calculations and the heat dissipation DB on the new item is incomplete. In this study, the heat dissipation item is analyzed by building applications and new test method for the heat dissipation measurement is proposed.

1 INTRODUCTION

Upon making a request for heat value of an electronic appliance or device to the electrical design team to calculate HVAC capacity, the team gets the value and provides it to the conditioning design team. The data that the electrical design team applies is provided by the manufacturer or is extracted from existing codes (ASHRAE 2009, Bhatia 2011, 2012), but the issue is the values result in the cooler with unexpectedly large capacity selected by the conditioning design team. Also new products from the advancement of technology require heat dissipation data, but the lack of such suitable data causes difficulty with their application as the foundation (Mohammad 1999).

Therefore this study aims to propose a more efficient method to measure heat dissipation by analyzing heating appliances per purpose of building and existing method of measuring heat dissipation and research method of usage ratio, as well as to design optimal capacity for a conditioning equipment with per purpose of building by measuring major heat dissipation through establishing and operating a new type of heat dissipation measuring equipment.

2 BUILDING A NEW HEAT DISSIPATION MEASURE EQUIPMENT

Heat dissipation measuring equipment has been built with the purpose of quantitative analysis of influence from energy consumption efficiency as well as the goal of determining capacity of conditioning equipment based on conditioning load by measuring the amount of heat value generated by loss of power on any device powered by electricity, the secondary energy created by burning the primary energy sources such as coal or gas. Figure 1 shows the concept of such equipment of which size is 4560(W) × 2250(D) × 3150(H) mm with the structure to withstand the load of up to 300 kg. It can generate a maximum load of 380 VAC/200 A with the power of AC 380 V, 4 PH and 60 Hz. Major subjects of heat dissipation measuring equipment include devices used by industrial plants such as distributing board, MCC (Motor Control Center) and inverter panel, which are difficult to transport due to quite large size and custom-manufacturing system on the site, thus the heat dissipation test has to be conducted at the site. Wheels have been installed on the bottom of the measurement equipment for easier transportation.

In order to measure the heat lost to outdoor from the measurement equipment, a heat flow

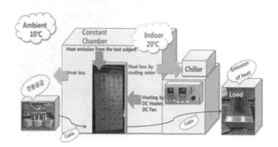

Figure 1. Heat Dissipation Measuring Equipment.

sensor has been attached to the inside, top and both sides of the equipment, while the port with direct contact with the outdoor air has been applied with insulation material to minimize the heat loss.

Basic concept of calculating heat dissipation of the test subjects is 'Heat dissipation = reduced heat' inside the constant temperature room in normal state.

Heat dissipation is the total of heat from the DC heater and DC fan inside the constant temperature room and the heat from the test subject itself, while reduced heat is the sum of the heat reduced by coolant and the heat lost to outdoor through the wall of the constant temperature room, latter of which means the heat lost due to temperature difference between inside constant temperature room and outdoor air. As the testing has been conducted in winter, this loss of heat has been included in the reduced heat. Thus the following equation is established by defining the value excluding the heat from the test subject, or the heat from the DC heater and DC fan as input heat, and defining the reduced heat as output heat.

Heat dissipation from test subject
= Output heat − Input heat (1)

In order to calculate the heat dissipation from the test subject, power consumed by the DC heater and DC fan, heat reduced by coolant and heat flow meter through the wall have been measured. Also to establish the interaction formula between the load condition of test the subject and heat dissipation, test has been conducted under the aforementioned conditions to get the correctional heat of test the subject so that influence from uncertain elements such as sensors and devices of the measuring equipment can be minimized. Correction of the equipment has been made by repeated measurements of heat from standard specimen with accurate heat dissipation, and heat can be adjusted

by controlling the voltage and current supplied to the specimen consisting of the DC heater and DC fan. Correctional heat has been measured while heat from standard specimen is controlled in consideration of measurement under different load conditions, but the result was around 55 W without much influence on heat dissipation. Figure 2 shows the standard specimen used to correct the heat dissipation measuring equipment.

3 MEASURING HEAT DISSIPATION FROM ELECTRIC APPLIANCE

Size of test the subject, the dry transformer, is $800 \times 1120 \times 2500$ mm and it can be connected to the maximum load of 64 kW (3.4 kW × 20 EA). Since the massive power of 64 kW is only available from large scale commercial buildings or plants where the subject is installed and operated, this study used actually available electricity for testing to get the estimated heat dissipation under the maximum load. Measurement has been conducted at the factory where subject was manufactured, and a heavy-duty truck was mobilized to transport the measurement equipment to the site. Figure 3 shows the process from transportation to an installation of the equipment.

Table 1 shows pre-determined cases under the load conditions to get the interaction formula by

Figure 3. The process from transportation to the installation of the equipment.

Table 1. Conditions for test.

Category	Load [kW]	Temperature inside constant temperature room [°C]	Coolant temperature [°C]
Case 1	30.6 (3.4 kW × 9 EA)	20	15
Case 2	20.4 (3.4 kW × 6 EA)		15
Case 3	10.2 (3.4 kW × 3 EA)		17

Figure 2. Standard heat dissipation specimen.

measuring the heat from the subject. Temperature set inside the constant temperature room has been set to 20°C for all cases.

Stabilization has been conducted until the temperature inside the constant temperature room is determined as back to normal from fluctuation from temperature difference due to the supply of coolant. Table 4 shows changes of temperature in this situation. After the temperature inside the constant temperature room has been stabilized at 20°C, measurement was taken for about 2 hours and the values from the last 1 hour are judged as the average heat dissipation. Coolant temperature for Case 2 was set at 15°C as with Case 1, and minute adjustment of the temperature inside the constant temperature room has been done with the DC heater inside the room. Coolant temperature for Case 3 was increased to 17°C for stabilization as heat dissipation was too little.

Table 2 shows the input heat dissipation was calculated by Formula (2).

Input heat dissipation $[W]$ = (*DC Heater current* × *DC Heater voltage*) + (*DC Fan current* × *DC Fan voltage*)   (2)

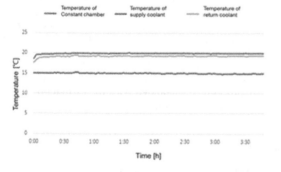

Figure 4. Changes of temperatures of coolant and inside the constant temperature room (Case 2).

Table 2. Input heat dissipation [W] per the test subject's load condition.

		Case 1	Case 2	Case 3
DC Heater	Voltage [V]	13.09	20.19	17.15
	Current [A]	7.37	11.20	9.59
	Power [W]	96.41	226.10	164.41
DC Fan	Voltage [V]	24.91	24.91	24.91
	Current [A]	3.05	3.05	3.05
	Power [W]	75.91	75.93	76.02
Input heat dissipation [W]		172.31	302.03	240.43

DC fan has been kept at the certain level in each cases to show similar heat dissipation. To compare the input heat dissipation of Case 1 and Case 2 where temperatures set for coolant and inside of constant temperature room, Case 2 maintained constant temperature by supplying more heat by 130 W than Case 1. Output heat dissipation can be calculated with Formula (3).

Output heat dissipation $[W]$
 = *Heat reduced by coolant*
 + *Heat lost through walls of*
    constant temperature room   (3)

Table 3 shows the output heat per load condition of test the subject. Ambient temperature around the constant temperature room has been steadily maintained so that the temperature difference of inside and outside of the room is kept constantly too, resulting in the judgment that heat dissipation through the wall of the room is similar regardless of load condition. Case 1 and Case 2 where coolant temperature has been kept at the same level had similar amount of heat reduced by coolant, but Case 3 turned out to lose 240 kW less heat than the other two cases.

Heat dissipation of test the subject can be calculated by applying the input/output heat calculated with above processes to Formula (4).

Heat dissipation of test the subject $[W]$
 = *Output heat* − *Input heat*   (4)

Table 4 shows the heat dissipation of test the subject according to load condition. Based on the calculated heat dissipation, the graph and interac-

Table 3. Output heat dissipation [W] per the test subject's load condition.

Item	Case 1	Case 2	Case 3
Heat through wall	118.95	114.90	116.48
Heat reduced by coolant	609.31	613.05	369.84
Output heat	728.26	727.95	486.32

Table 4. Heat dissipation of test the subject [W] per load condition.

Item	Case 1	Case 2	Case 3
Input heat [W]	172.31	302.03	240.43
Output heat [W]	728.26	727.95	486.32
Corrective heat [W]	54.34	54.34	54.34
Heat from subject [W]	610.29	480.26	300.23

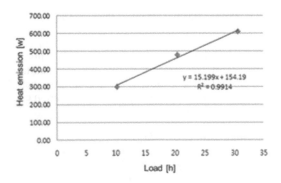

Figure 5. Graph of heat dissipation per load condition of test the subject.

Table 5. Comparison of calculated value from the test subject and tested value.

Load condition for subject		Heat from subject [W]	
Number of load [EA]	Amount of load [kW]	Calculated value [W]	Tested value [W]
1	3.4	210.24	–
2	6.8	260.46	–
3	10.2	310.67	300.89
4	13.6	360.89	–
5	17.0	411.11	–
6	20.4	461.33	480.92
7	23.8	511.55	–
8	27.2	561.76	–
9	30.6	611.98	610.95
10	34.0	662.20	–

tion Formula (5) have been established for the relation between load condition and heat dissipation of test the subject as shown in Figure 5.

Heat dissipation of test the subject $[W]$
$$= 15.199x + 154.19 \ (R^2 = 0.9914) \qquad (5)$$

In order to estimate heat dissipation of test the subject under diverse load conditions based on the interaction formula, range of load has been set to between 3.4 and 662.2 kW. As Table 5 shows, the result was similar to value calculated with the condition of 10.2 kW, 20.4 kW and 30.6 kW.

4 CONCLUSIONS

This study aims to reduce initial investment and heating and cooling energy by selecting the appropriate capacity of conditioning equipment and optimal product by establishing a database of heat dissipation of electric appliance and devices and applying the data. To achieve this goal, the existing heat dissipation calculation method to build a new calculation method and equipment. Heat dissipation of the test subject is calculated with the interaction formula of input and output heat dissipations. Also to get the most accurate heat dissipation, corrective heat dissipation has been applied to this formula to minimize influence from inaccurate elements such as construction, sensor or components of the equipment.

Results of measuring heat dissipation of a dry transformer showed that Case 1 had bigger heat dissipation (610.29 W for Case 1, 480.26 W for Case 2) by about 130 W than Case 2 despite the same temperatures of coolant and constant temperature room. Case 3 with increased coolant temperature by 2°C was 300.23 W, about 180 W less heat than Case 2. To estimate heat dissipation under the maximum load on the test subject based on these results, an interactive formula has been established for different load conditions and heat from the subject. The result was similar to the value calculated under the conditions of 10.2 kW, 20.4 kW and 30.6 kW, meaning that heat dissipation of devices with heavy loads can be calculated with the formula to get the value under maximum load.

REFERENCES

ASHRAE, 2009. ASHRAE handbook fundamentals.
Bhatia, A. *Cooling Load Calculations and Principles*, Continuing Education and Development.
Bhatia, A. 2012. *HVAC Refresher—Facilities Standard for the Building Services (Part 2)*, PDH online Course M216.
Mohammad, H. 1999. *Experimental Results for Heat Gain and Radiant/Convective Split from Equipment in Buildings*, SE-99-1-4 (RP-1055).

Water, irrigation and architectural engineering application

Advances in Civil, Architectural, Structural and Constructional Engineering – Kim, Jung & Seo (Eds)
© 2016 Taylor & Francis Group, London, ISBN 978-1-138-02849-4

The stress-strain state determination of a large diameter composite rod during pultrusion process

S.N. Grigoriev
Moscow State Technological University "STANKIN", Moscow, Russia

I.A. Kazakov & A.N. Krasnovskii
Department of Composite Materials, Moscow State Technological University "STANKIN", Moscow, Russia

ABSTRACT: This work is focused on the stress-strain state determination of a composite rod with a large diameter during pultrusion process. The understanding of the mechanism of a crack formation in the product allows to obtain high quality composite rods with a lower prime cost. A two-dimensional finite-difference analysis of cure and process induced residual stress is presented. The micro-mechanics model was used to predict the cure dependent composite mechanical properties. Thermal expansion and cure shrinkage strains were included in the analysis. The present results were validated by the comparisons with the results of the experiments from the pertinent literature. The results showed that the main crack in the composite rod occurs at some distance from the die due to circumferential stresses exceeding allowable values. The maximum stresses are concentrated on the surface of the rod above the place where the cone of uncured resin ends, not in the vicinity of the die it was considered earlier.

1 INTRODUCTION

Pultrusion is one of the most common ways to manufacture fiber-reinforced composite rod with a large diameter. Such rods are widely used as electrical insulators, structural parts of composite bridges, parts of supports of high-voltage power lines, their application area is constantly expanding.

Several researchers (Joshi 2001, Carlone et al. 2013) dedicated their work to pultrusion process simulations. These models take into account the mutual interactions between heat transfer, resin flow and cure reaction, variation in the material properties and suitable for stress determination. However, these models use a finite element method involving high-level software packages (ANSYS, ABAQUS), which are difficult to use for several reasons. In addition, the authors do not present concrete results for stress-strain state determination for large-sized pultruded rods. Obtained stresses for small diameter rods (Carlone et al. 2013) have no practical significance due to their insignificant values.

One distinctive feature of the pultrusion of large-sized composites is a low pulling speed (not exceed 70 mm/min for 80 mm rod diameter). If the pull speed is high enough (>60 mm/min), then polymerization process of the oversized product does not have time to finish before it exit from the die (Figure 1). Under these conditions, the

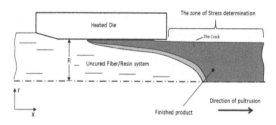

Figure 1. The process of passing the composite material through a heated die.

stresses occurred in a polymerized part of the rod may exceed the maximum allowable values for the material. It is leads to the main crack appearance (Krasnovskii & Kazakov 2012). Thermal expansion and chemical shrinkage of material affect the value of stresses.

2 STATEMENT OF THE PROBLEM

2.1 *Governing equations for stress-strain state*

The task of determining the stress-strain state in the present study is a logical continuation of our earlier works (Grigoriev, Krasnovskii & Kazakov, 2012–2014), and has more advanced and improved theory. As a new approach the first step was to define analytically an equation for the displacements. Then, this equation solved using the finite difference method. The next step is to determine the strains and stresses using known displacements.

Hooke's law for orthotropic solid in cylindrical coordinates taking into account the chemical shrinkage is given by

$$\varepsilon_x = \frac{1}{E_x}\sigma_x - \frac{v_{x\theta}}{E_\theta}\sigma_\theta - \frac{v_{xr}}{E_r}\sigma_r + \alpha_x T + \varepsilon_{\chi x}, \qquad (1)$$

$$\varepsilon_\theta = \frac{v_{\theta x}}{E_x}\sigma_x + \frac{1}{E_\theta}\sigma_\theta - \frac{v_{\theta r}}{E_r}\sigma_r + \alpha_\theta T + \varepsilon_{\chi\theta}, \qquad (2)$$

$$\varepsilon_r = \frac{v_{rx}}{E_x}\sigma_x - \frac{v_{r\theta}}{E_\theta}\sigma_\theta + \frac{1}{E_r}\sigma_r + \alpha_r T + \varepsilon_{\chi r}, \qquad (3)$$

where, σ_r, σ_x, σ_θ—radial, axial and circumferential and stress components;
ε_r, ε_x, ε_θ—radial, axial and circumferential and strain components;
α_x, α_r, α_θ—linear thermal expansion coefficients in corresponding directions;
$T = T(r,x)$—temperature function;
ε_χ, $\varepsilon_{\chi x}$, $\varepsilon_{\chi\theta}$—chemical shrinkage strains in corresponding directions.
E_r, E_x, E_θ—modulus of elasticity of material in corresponding directions;
$v_{x\theta}$, v_{xr}, $v_{\theta x}$, $v_{\theta r}$, v_{rx} and $v_{r\theta}$—Poisson ratios.

We considered the rod as a transversely isotropic cylindrical solid. For this solid

$$E_r = E_\theta = E, v_{r\theta} = v_{\theta r} = v, v_{x\theta} = v_{xr}, v_{\theta x} = v_{rx},$$
$$\alpha_r = \alpha_\theta = \alpha_{KTE}, \varepsilon_{\chi r} = \varepsilon_{\chi\theta} = \varepsilon_\chi. \qquad (4)$$

The chemical shrinkage and linear thermal expansion coefficient for the material along the fibers were assumed to be zero, i.e. $\varepsilon_{\chi x} = 0$, $\alpha_x = 0$. Furthermore, for a generalized plane strain state it was assumed that

$$\varepsilon_x = \varepsilon_0, \quad E_x = \infty. \qquad (5)$$

This model can be used due to the elastic modulus of unidirectional composite significantly higher elastic modules in the transverse direction.

The equilibrium equation for an infinitely small element in the local cylindrical coordinates for $r\theta$ plane is:

$$\frac{d\sigma_r}{dr} + \frac{\sigma_r - \sigma_\theta}{r} = 0. \qquad (6)$$

Geometrical relations

$$\varepsilon_\theta = \frac{u}{r}, \quad \varepsilon_r = \frac{\partial u}{\partial r}. \qquad (7)$$

Boundary conditions ($\sigma_r = 0$ on cylinder surface):

$$\frac{\partial u}{\partial r} + \frac{u}{R}\cdot v - (1+v)\cdot(\alpha_{KTE}T + \varepsilon_\chi) = 0, r = R. \qquad (8)$$

Solving Equations (1)–(3) by stresses, using (4)–(7) the following equation for displacements can be obtained:

$$\frac{\partial^2 u}{\partial r^2} + \left(\frac{1}{E}\frac{\partial E}{\partial r} + \frac{2v}{1-v^2}\frac{\partial v}{\partial r} + \frac{1}{r}\right)\frac{\partial u}{\partial r} +$$
$$\left(\frac{v}{E}\frac{\partial E}{\partial r} + \frac{1+v^2}{1-v^2}\frac{\partial v}{\partial r} - \frac{1}{r}\right)\frac{u}{r} =$$
$$(1+v)\left[\alpha_{KTE}\frac{\partial T}{\partial r} + \frac{\partial \varepsilon_\chi}{\partial r}\right] +$$
$$(1+v)\left(\frac{1}{E}\frac{\partial E}{\partial r} + \frac{1}{1-v}\frac{\partial v}{\partial r}\right)(\alpha_{KTE}T + \varepsilon_\chi). \qquad (9)$$

The modulus of elasticity of material was calculated as (Vasiliev, Morozov 2013):

$$E = \pi \cdot E_m \cdot r(\lambda)/2 \cdot v_f \cdot (1 - v_m^2), \qquad (10)$$

where, E_m – the elasticity modulus of resin, v_m – Poisson's ratio of resin, v_f – fiber-volume fraction, dimensionless parameter $r(\lambda)$ is calculated as

$$r(\lambda) = \frac{1}{\sqrt{1-\lambda^2}}\cdot \tan^{-1}\sqrt{\frac{1+\lambda}{1-\lambda}} - \frac{\pi}{4}, \lambda = \frac{4\cdot v_f}{\pi}. \qquad (11)$$

It is assumed that the modulus of elasticity E_m is proportional to degree of cure according to simple mixture rule (Travis 1992):

$$E_m = (1-\alpha)E_\infty /1000 + E_\infty \cdot \alpha, \qquad (12)$$

where, $E_\perp$—the elasticity modulus of completely cured resin, α—degree of cure.

Similarly it was considered that the Poisson's ratio of resin is proportional to the degree of cure:

$$v_m = (1-\alpha)\cdot 0.5 + v_\infty \cdot \alpha, \qquad (13)$$

where, $v_\perp$—the Poisson's ratio of a completely cured resin.

The equation for Poisson's ratio of the Fiber/resin system has the form.

$$v = v_f \cdot v_v + (1-v_f)\cdot v_m, \qquad (14)$$

where, v_v – the Poisson's ratio of fiber.

The linear thermal expansion coefficient is defined by (Travis 1992):

$$\alpha_{KTE} = (1+v_v)\cdot\alpha_{KTRf}\cdot v_f + (1-v_f),$$
$$(1+v_m)\cdot\alpha_{KTEm} - \left[v_v\cdot v_f + (1-v_f)\cdot v_m\right],$$
$$\frac{\alpha_{KTEf}\cdot E_f\cdot v_f + \alpha_{KTEm}\cdot E_m\cdot(1-v_f)}{E_f\cdot v_f + E_m(1-v_f)}. \tag{15}$$

where, E_f—the elasticity modulus of fibers, α_{KTEf} and α_{KTEm}—linear thermal expansion coefficients for the fibers and matrix (resin) correspondingly.

Expression to determine the chemical shrinkage strain is compiled by analogy with the model of White-Khan (White & Hahn 1992):

$$\varepsilon_\chi = CSC_3\cdot\chi_0\cdot 10^{-3.08(1-\alpha)}, \tag{16}$$

where, α is a degree of cure, CSC_3 is a coefficient taking into account the fiber-volume fraction of the composite:

$$CSC_3 = (1-v_f)\cdot(1+v_m) - \left[v_v\cdot v_f + (1-v_f)\cdot v_m\right]\cdot$$
$$\frac{E_m\cdot(1-v_f)}{E_f\cdot v_f + E_m\cdot(1-v_f)}, \tag{17}$$

χ_0—is the resin shrinkage for one of three directions:

$$\chi_0 = \sqrt[3]{\chi_V + 1} - 1 \approx \frac{1}{3}\chi_V. \tag{18}$$

where, χ_V is a full volumetric resin shrinkage.

For the stress-strain state determination we need to know the temperature and degree of cure which can be obtained for each section by solving a non-linear heat transfer and polymerization problem (Grigoriev, Krasnovskii & Kazakov, 2012–2014).

Equation (8) is applied to each rod section using the finite difference method. Then, the strains are determined by Equation (7) using known displacements. Consequently, the radial and circumferential stresses are obtained using Equations (1–3) for all rod sections. These stresses are compared with ultimate values to prevent main crack appearance.

2.2 Failure criteria

At high pull speeds the stress occurred in a polymerized part of the rod may exceed the maximum allowable values for the material. It leads to the main crack appearance. As a strength criterion the maximum stress criterion is chosen which has the following form (Vasiliev & Morozov 2013):

$$\sigma_r, \sigma_\theta \le \sigma_m\cdot\frac{r(\lambda)}{2\cdot v_f}\cdot(\pi - 4\cdot v_f), \tag{19}$$

where, σ_m—ultimate stress for the completely cured resin.

3 RESULTS AND DISCUSSION

For the current study, the following parameters were used when solving the problem. The fiberglass rod radius $R = 38$ mm, pull speed 5 and 6 cm/min. Die temperature settings (the length of straight portion of the Die is 1 m): 110–150–190–160°C. The Fiber/resin temperature at the Die inlet: 50°C. Environment temperature: 25°C. The fiber volume fraction: $v_f = 0.6$. The rest model inputs and material physical properties are presented in Table 1.

Figures 2–5 show the results of the mathematical model. For these Figures the axial distance from the die outlet in the horizontal direction is denoted in meters. Figure 2 shows the temperature (°C) and degree of cure distributions within the rod. It can be seen that the central part of the rod at the distance of 0.28–0.42 m. from the die outlet has a maximum temperature of 180°C. The external layers of the rod at the same time begin to lose temperature to the surrounding air due to cooling. The product is completely solidified at a distance of 0.36 m from the die exit. It is assumed for the finished product $\alpha \ge 0.95$.

Table 1. Material physical properties.

Property	Value
Heat of reaction, H_{tot}, J/g	373.2
Activation energy, J/mol	127200
Pre-exponential factor, c^{-1}	exp(31.4)
Equation superscript	1.8
Fiber density, g/cm³	2.56
Elasticity modulus of completely cured resin, MPa	3447
Elasticity modulus of fiber, MPa	73080
Poisson's ratio of completely cured resin	0.35
Poisson's ratio of fiber	0.22
Linear thermal expansion coefficient of fiber, 1/°C	$5.04\cdot 10^{-6}$
Linear thermal expansion coefficient of resin, 1/°C	$57.6\cdot 10^{-6}$
Full volumetric resin shrinkage, %	5
Ultimate stress for the completely cured resin, MPa	70
Ultimate stress, MPa, according to (19)	47

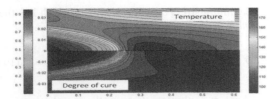

Figure 2. The temperature and degree of cure distributions within the rod (pull speed is 5 cm/min).

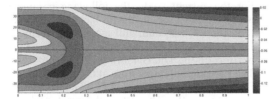

Figure 3. Displacement field within the rod, in mm. (pull speed is 5 cm/min).

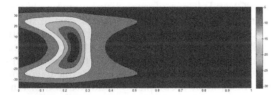

Figure 4. Radial stresses distribution within the rod, in MPa. (pull speed is 5 cm/min).

Figure 3 shows the displacement field within the rod. The displacements at the rod surface are close to zero near the die outlet. The rod diameter decreases as the cooling due to the presence of chemical shrinkage. The radial stresses at the rod surface are close to zero (Figure 4). This indicates that the boundary conditions (8) are satisfied. The results indicate the existence of large compressive stresses in the center of the rod, near the top of the cone formed by uncured resin (see Figure 2). The circumferential stresses (Figure 5) have maximum values at the surface of the rod. They do not exceed 35 MPa at a pull speed of 5 cm/min (Figure 5a) but becomes critical when the pull speed is increased to 6 cm/min (Figure 5b).

When the circumferential stresses reach a value of 47 MPa the main crack formation occurs, which is confirmed by the experimental data presented in literature for this rod (Safonov 2006).

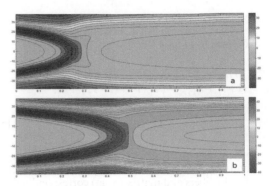

Figure 5. Circumferential stresses distribution within the rod, in MPa. (pull speed is 5 cm/min (a) and 6 cm/min (b)).

4 CONCLUSIONS

The results show that the maximum circumferential stress at the rod surface appears above the top of the cone, which is formed by uncured resin. This could be explained by the most unfavorable concentration of thermal and shrinkage stresses in some of the rod sections. The numerical model is a useful tool to determine the stress-strain state of polymer composite materials at the outlet of the Die through the selection of appropriate values for the different process control parameters. The obtained results will help to find the optimal regime of the pultrusion process (Grigoriev et al. 2011) and achieve high quality products.

ACKNOWLEDGMENT

This work is a part of the State project in the field of scientific activity which has been granted by the Education and Science Ministry of the Russian Federation.

REFERENCES

Carlone, P. Baran, I. Hattel, J.H. & Palazzo, G.S. 2013. Computational Approaches for Modeling the Multiphysics in Pultrusion Process. *Advances in Mechanical Engineering*, 2013: 1–14.

Grigoriev, S.N. Krasnovskii, A.N. Kazakov, I.A. & Kvachev, K.V. 2013. An analytic definition of the border polymerization line for axisymmetric composite rods. *Applied Composite Materials*, 20(6): 1055–1064. DOI: 10.1007/s10443–013–9317–8.

Grigoriev, S.N. Krasnovskii, A.N. & Kazakov, I.A. 2014. The friction force determination of large-sized composite rods in pultrusion. *Applied Composite Materials*, 21(4): 651–659. DOI: 10.1007/s10443–013–9360–5.

Grigoriev, S.N. Krasnovskii, A.N. & Kazakov, I.A. 2015. The Impact of Pre-heating on Pressure Behavior in Tapered Cylindrical Die in Pultrusion of Large-sized Composite Rods. *Advanced Materials Research,* 1064: 120–127. DOI: 10.4028/www.scientific.net/AMR.1064.120.

Grigoriev, S.N. Krasnovskii, A.N. & Khaziev A.R. 2011. Optimum designing of long complicatedly reinforced polymeric composite structures. *Mechanics of composite materials and structures,* 17(4): 545–554.

Joshi, S.C. & Lam, Y.C. 2001. Three-dimensional finite-element/nodal-control-volume simulation of the pultrusion process with temperature-dependent material properties including resin shrinkage. *Composites Science and Technology,* 61(11): 1539–1547.

Krasnovskii, A.N. & Kazakov, I.A. 2012. Determination of the optimal speed of pultrusion for large-sized composite rods. *Journal of Encapsulation and Adsorption Sciences.* 2(3): 21–26.

Krasnovskii, A.N. & Kazakov, I.A. 2012. The study of the stress-strain state of the material during the pultrusion process. *Plastics,* 10: 22–26.

Kutin, A.A. Krasnovskii, A.N. & Kazakov, I.A. The Fiber Orientation Angle Determination for a Composite Anisotropic Solid Rod in Pultrusion. 2014. *Advanced Materials Research,* 941–944: 262273–2278. DOI: 10.4028/www.scientific.net/AMR.941–944.2273.

Safonov, A.A. 2006. *Mathematical modeling of the technological process of glass-reinforcement elements production in pultrusion: Dissertation for the degree of Candidate,* tech. Science. Moscow.

Travis, A. Bogetti, J.W. & Gillespie, Jr. 1992. Process induced stress and deformation in thick-section thermoset composite laminates. *Journal of Composite Materials,* 26(5): 626–660.

Vasiliev, V.V. & Morozov, E.V. 2013. *Advanced mechanics of composite materials and structural elements,* Third edition– Oxford: Elsevier Science Ltd.

White, S.R. & Hahn, H.T. 1992. Process Modeling of Composite Materials: Residual Stress Development during Cure. Part I. Model Formulation. *Journal of Composite Materials,* 26(16): 2402–2422.

Advances in Civil, Architectural, Structural and Constructional Engineering – Kim, Jung & Seo (Eds)
© *2016 Taylor & Francis Group, London, ISBN 978-1-138-02849-4*

Effect of ureolytic bacteria on compressive strength and water permeability on bio-concrete

N. Othman

*Micropollutant Research Centre (MPRC), Faculty of Civil and Environmental Engineering,
Universiti Tun Hussein Onn Malaysia, Malaysia*
*Jamilus Research Centre for Sustainable Construction (JRC), Faculty of Civil and Environmental
Engineering, Universiti Tun Hussein Onn Malaysia, Malaysia*

J.M. Irwan, A. Faisal Alshalif & L.H. Anneza

*Jamilus Research Centre for Sustainable Construction (JRC), Faculty of Civil and Environmental
Engineering, Universiti Tun Hussein Onn Malaysia, Malaysia*

ABSTRACT: This paper presents on results of ureolytic bacteria effect on compressive strength and water permeability of bio-concrete. Ureolytic bacteria that was used in this study was isolated from fresh urine. Water was partially replaced through three different percentages 1%, 3%, and 5% by liquid of growth medium containing ureolytic bacteria. Tests were conducted on the 7th, 14th, and 28th days of the specimens curing. A maximum increase of compressive strength (23%) and reduction of water permeability (10.4%) is achieved by 3% ureolytic bacteria concentration. The improvement in compressive strength and water permeability was due to the deposition on the bacteria cell surface within the pores.

1 INTRODUCTION

Concrete is widely used as a construction material. The properties of concrete are ever evolving to suit the need of the industry as well as to produce a greener concrete. Therefore, many supplementary cementations materials are used in concrete to reduce cement content, improve workability, increase strength and enhance durability through hydraulic or pozzolanic activities. As a matter of fact, most researchers in concrete technology focused on exploring to a green material that resulted to a good compressive strength. An attempt is by minimization quantity of cementations materials.

It is now recognized that the strength of the concrete is not sufficient in some cases. Understanding on environmental conditions' effects should be considered at the design stage (Chahal et al. 2012).

Bio-concrete is an option of concrete technology that is more environmental friendly. Bacteria are used in concrete mix in order to enhance concrete properties as well as reduce negative impact towards the environment. Microbial mineral precipitation of calcium carbonate resulting from metabolic activities of specific microorganisms in concrete is used to improve the overall behavior of concrete. (Siddique & Chahal 2011). These bacteria are able to influence the precipitation of calcium carbonate by the production of urease enzyme. Precipitation of calcium carbonate crystals occurs by heterogeneous nucleation on bacterial cell walls once super saturation is achieved. The deposition of a calcite layer of the specimens is due to the bacterial culture and medium composition. Previous study had proved that the use of bacteria such as Sporosarcina pasteurii is able to influence the precipitation of calcium carbonate by the production of urease enzyme. This enzyme catalyzes the hydrolysis of urea to CO_2 and ammonia, resulting in an increase of the PH and carbonate concentration in the bacterial environment (Chahal 2012).

The aim of this research is to analyze effect of ureolytic bacteria in improving compressive strength and water permeability.

2 METHODS

2.1 Isolation of bacteria

Ureolytic bacteria was isolated in the environmental lab from fresh urine. The isolation process of ureolytic bacteria was conducted following the process of enrichment, serial dilution, streaking plate, strain purification and gram staining. All process and media used were autoclave at 121°C for 15 minutes for sterilization. The enrichment process of ureolytic bacteria was done under special environment to tolerate the alkaline environment within the concrete. The compositions of media for ureolytic bacteria are as follows.

Table 1. Mix proportion of bio-concrete with different percentages of ureolytic bacteria.

Ratio (%) of UB	Cement (Kg)	Water (Kg)	Fine agg (Kg)	Coarse agg (Kg)	Bacteria (Kg)
0%	420	210	685	1115	0
1%	420	207.9	685	1115	2.1
3%	420	203.7	685	1115	6.3
5%	420	199.5	685	1115	10.5

1a. Composition of ureolytic bacteria consist of 25 ml (nutrient broth) + 10 ml (urea 40%) + 1 ml (urine).
1b. Composition of control sample for ureolytic bacteria consists of 25 ml (nutrient broth) + 10 ml (urea 40%).

Bacteria can't tolerate with extreme pH value. Under acidic conditions, some bacteria cell will hydrolyze or enzyme inactivate. The pH value plays an important role in microbial life that will influence the dissociation and solubility of many molecules that indirectly influence microorganism (Ramakrishnan et al. 1998) & (Subhashree et al. 2012). Therefore, after media preparation the pH of the samples was adjusted to an alkaline condition. The pH adjustment was done using NaOH until the pH value in the range of 9–11.

Other condition that was monitored along this study is anaerobic condition of the enrichment. Nitrogen gas was purged in the enrichment flask before enrichment started. The concentration of oxygen was measured by maintained 0 ppm of oxygen in the flask. The flask was shaken at room temperature. The adjustment of pH and anaerobic condition were done every day over the enrichment period of 20 days.

Ureolytic bacteria added to the concrete mix in the form of liquid at the end of the concrete mix process.

2.2 Concrete mixture proportion

The concrete mix design calculated using DOE method and the mix proportions are given in Table 1. Sample was designed by grade 30 and 0.54 w/c to achieve 30 MPa after 28 days.

2.3 Preparation of test specimens

Each mix proportion consist different percentages namely, 1%, 3%, and 5% of bacteria. Concrete cubes were prepared with different percentage of bacteria. Therefore, control concrete cubes were cast without additional of ureolytic bacteria. All experiments were performed in triplicates.

The specimen properties were determined at age 28 days at the dry condition.

2.4 Compressive strength

In this study cubes, size of 150 mm × 150 mm × 150 mm was used to study the compressive strength of the concrete. Concrete specifications of compressive strength cubes were cast without bacteria for the control sample. Another mix of concrete with the same material properties were done with ureolytic bacteria to produce bio-concrete specimen with three different percentages of water replacement by 1%, 3%, and 5% of bacteria in different mixes. After compaction and de-molding, the specimens were left for curing. The compressive strength test is tested on the 7th, 14th, and 28th days of curing. The test was carried out using Compressive Test Machine to conduct according to BS EN 12390–3:2002.

2.5 Water permeability test

In order to determine the reduction of water permeability in concrete, cube molds size of 150 mm × 150 mm × 150 mm were prepared both with and without ureolytic bacteria the concrete specimens were cured for 7th, 14th, and 28th days. After curing, the specimens weigh before being placed in the test machine. According to BS EN 12390–8, constant pressure of 5 KPa for 72 hours was used to measure the water permeability. To measure the water permeability depth the specimen was split in half and the maximum depth of penetration of water measured in (mm) as shown in Figure 1.

Figure 1. Penetration of water in the cube.

3 RESULT AND DISCUSSION

3.1 *Properties of colony of bacteria*

The results showed that the bacteria are capable to grow in alkaline environment and anaerobic condition. Two distinct colors such as cream and yellow were observed for all colonies during the purification process. The cultured colonies formed in the isolation medium with diameters ranging from 0.2 mm to 1 mm. The study observed that colonies were smooth and they had moist surface. Table 2 shows ureolytic bacteria was in the shape of coccus with the period of enrichment of 10 and 20 days respectively. The observation also showed that colony morphology for 10 and 20 day enrichment with gram stain was positive for ureolytic bacteria. Therefore, bacteria had the thicker peptidoglycan wall and blue color as shown in Figure 2.

3.2 *Compressive strength*

Compressive strength of concrete with and without ureolytic bacteria are shown in Figure 3. It is proved that compressive strength increase with 1% and 3% of the bacteria quantity of the concrete compared to the control sample. Similarly, there was an increase of compressive strength with 5% of ureolytic bacteria, but it was lower compared to 1% and 3%. By comparing the control concrete specimen which contained (0% UB) with the specimens contained 1%, 3% and 5% of UB in 28 days curing, the percentage of increment of strength is 16.48%, 22.97%, and 15.22% respectively. Whereas, the maximum increase of compressive strength for bacteria specimens compared to control is with 3% of ureolytic bacteria that achieved the highest strength of 36 Mpa compared to 1% (35 Mpa) and 5% (34 Mpa). Throughout the compressive strength test during the 7th, 14th and 28th day of testing, the concrete with 3% ureolytic bacteria always achieved the highest compressive strength as compared to control, 1% and 5%. The observation of the enhancement in compressive strength of ureolytic bacteria is due to the deposition of $CaCO_3$ on the microorganism cell surfaces within the concrete pores, which plug the pores within the binder matrix. Similar results were reported by other researchers (Wu et al. 2012).

3.3 *Water permeability*

The water permeability measured was not immediately constant, but decrease during several days, as can be seen in Figure 4. The decrease was due

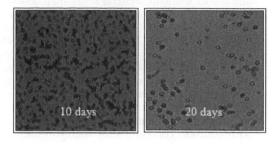

Figure 2. Bacteria picture for 10 and 20 days respectively.

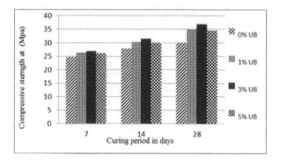

Figure 3. Compressive strength of concrete with different ratio of ureolytic bacteria at 28 days.

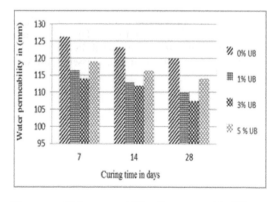

Figure 4. Water permeability of concrete with different ratio of ureolytic bacteria at 28 days.

Table 2. The characteristic of ureolytic bacteria in the period of 10 and 20 days.

Source of bacteria	Enrichment period	Colony color	Gram stain	Shape	Colony size
Ureolytic bacteria	10 days	Yellow	+ ve	Coccus	In the range of 0.2 mm
	20 days	Cream	+ ve	Cocci	to 1 mm each type

to incomplete saturation of the specimens and the unavoidable existence of air bubbles in the specimens, even though special care had been taken (Achal et al. 2009). The results demonstrated that the water permeability of concrete with 1%, 3%, and 5% of ureolytic bacteria lower than concrete with 0% of ureolytic bacteria. The decrement shows that at 28th day, the water permeability reduces by 8.2%, 10.36% and 4.94% of concrete specimens containing 1%, 3% and 5% ureolytic bacteria respectively compared to control specimens. The specimen with 5% of ureolytic bacteria has the lowest performance as compared to 1% and 3% ureolytic bacteria. On the contrary, the maximum reduction of water penetration was observed with concrete contain 3% of ureolytic bacteria. Therefore, the deposition of a layer of calcium carbonate on the surface and inside the pores of the concrete specimens resulted in a decrease of water permeability and absorption (Kim et al. 2010).

4 CONCLUSIONS

The results presented that ureolytic bacteria play a significant role in increasing compressive strength up to 22.97% and 10.36% of water permeability reduction was achieved. The maximum increase of compressive strength and water permeability was with 3% replacement of water by ureolytic bacteria growth media.

ACKNOWLEDGMENT

This study has been granted by the Ministry of Education (MOE) under The Fundamental Research Greant Schemen (FRGS Vot 1211) and supported by Universiti Tun Hussein Onn Malaysia (UTHM).

REFERENCES

Achal, V., Mukherjee, A., Basu, P.C. & Reddy, M.S. 2009. Lactose mother liquor as an alternative nutrient source for microbial concrete production by Sporosarcinapasteurii. *Journal of industrial microbiology & biotechnology*, 36(3): 433–438.

BS. 1992. *Specification for Aggregate from Natural Resources for Concrete*. British Standards Institution, London. 882.

Chahal, N., Siddique, R. & Rajor, A. 2012. Influence of bacteria on the compressive strength, water absorption and rapid chloride permeability of concrete incorporating silica fume. *Construction and Building Materials*, 37: 645–651.

EN, B. 2000. *Cement. Composition, Specifications and Conformity Criteria for Common Cements*. British Standards Institution, London. 197–1.

EN, B. 2002. *Standard Test Method for Compressive Strength of Cement Mortar*. British Standards Institution, London. 12390–3.

Kim, V.T., Nele, D.B., Willem, D.M. & Willy, V. 2010. Use of bacteria to repair cracks in concrete, *Cement and Concrete Research,* 40(1): 157–166.

Nemati, M. & Voordouw, G. 2003. Modification of porous media permeability using calcium carbonate produced enzymatically in situ. *Enzyme and Microbial Technology*, 33: 635–642.

Ramakrishnan, V., Bang, S.S. & Deo, K.S. 1998. *A novel technique for repairing cracks in high performance concrete using bacteria*, Proceeding of the International Conference on High Performance, High Strength Concrete, Perth, Australia, 597–617.

Siddique, R. & Chahal, N.K. 2011. Effect of ureolytic bacteria on concrete properties. *Construction and building materials*, 25: 3791–3801.

Subhashree, P., Ram, C.S., Kailash, C.S. & Pradeep, K.C. 2012. Isolation and Identification of Pathogenic Bacteria from Brackish Waters of Chilika Lagoon, Odisha, *India for Pharmaceutical Use*, 8(3): 197.

Wu, S., Chen, X. & Zhou, J. 2012. Tensile strength of concrete under static and intermediate strain rates: Correlated results from different testing methods. *Nuclear Engineering and Design*, 250: 173–183.

Advances in Civil, Architectural, Structural and Constructional Engineering – Kim, Jung & Seo (Eds)
© 2016 Taylor & Francis Group, London, ISBN 978-1-138-02849-4

New deals in urban growing tools, procedures and feasibility strategies

S. Miccoli & R. Murro
Department of Civil, Building and Environmental Engineering, University of Rome La Sapienza, Rome, Italy

F. Finucci
Department of Architecture, Roma Tre University, Rome, Italy

ABSTRACT: The idea of a new urban settlement on the island of Amager, following the high housing demand, is dated at the beginning of the Sixties, but since the end of the '80 s Copenhagen was suffering problems related with slow economic growth, high rate of unemployment and growing debts. Copenhagen was not a very attractive city for families and companies. Nowadays, Ørestad is considered one of the most livable and appreciated urban settlement of Europe and internationally oriented companies are set-tled in this attractive area, midway between the airport and downtown Copenhagen. The paper intend to explore the feasibility factors that made possible the implementation and the success of the project.

1 INTRODUCTION

The creation of an urban settlement on the island of Amager and within the Municipality of Copen-hagen dates back to the beginning of the Sixties, following the high housing demand. Ørestad was considered as the optimal area for urban develop-ment of Copenhagen, in that it allowed it to grow without affecting its integrity. At the same time, it was regarded as the gateway between Europe and Scandinavia and as such was highly attractive for international investors in multiple industries. Still, because of the 1972 oil crisis, Denmark had suf-fered from a very deep economic recession. This crisis went up until the early nineties, and pre-vented this new urban space from coming to life.

Today Ørestad is about to become a new, impor-tant district of Copenhagen, located on a 310-Ha area between the city centre and the protected Amager Fælled and Kalvebod Fælled areas. Øres-tad stretches over more than 5 km of a 600-m wide strip of land situated between the city centre of Copenhagen and the International Airport, along a motorway leading to the new Øresund connec-tion. The connection on the Øresund international strait, between the Kastrup peninsula (Copenha-gen, Denmark) and Limhamn (Malmö, Sweden), was based on two main factors: a) the remarkable growth in population density which characterized the Danish side (Copenhagen) and the Swedish side (Skåne) towards the end of the eighties; b) the difficulty in crossing the Øresund strait, hav-ing the effect of slowing down social, economic and cultural exchange between the two neighbour-ing countries. In this situation the Øresundsbron

project intended to factually contribute to the development of a communities on the two sides of the strait (approximately 3.5 million inhabitants), thus fostering integration. The Oresund cross bor-der area, thanks to a steady connection across the strait, stood as a European-level region featuring a remarkable potential for economic and social development, also due to its Great Belt and Lit-tle Belt link with Central Europe. The infrastruc-ture consists of a double railway and a four-lane road stretching for 16 km across the two coastlines (Miccoli et al. 2015a).

The Ørestad project was based on two strategic principles, combined synergistically, for the pur-poses of reducing system costs and facilitate the decision-making process: a) developing Copen-hagen with a new central densely built, multi-functional structure (universities and research cen-tres, cultural facilities, residences and commercial buildings, as well as high-tech manufacturing facili-ties) – outside of the existing city—capable of being an economic driver for Copenhagen and Malmö alike; b) expanding the existing public transport system by means of a new subway line effectively connecting the Amager areas surrounding Copen-hagen, with the city centre.

2 THE ØRESTAD PROJECT

In 1993, based on the provisions of the Øres-tad Act, the Ørestad Development Corporation (ØDC) was established as a development company held by the Copenhagen Municipality by 55% and the Danish State (represented by the Ministry of

Transport) by 45%. The ØDC basically had a double role: manage the creation of the new Ørestad urban development and the new subway line (ØDC 2003).

In 1994, the ØDC was asked to take charge of a masterplan for the new city of Ørestad. The masterplan had to stand as the base for the integration of the Copenhagen Regional Plan. For the purposes of guaranteeing quality solutions, an international competition was launched, in which 119 architecture firm from 17 different countries participated (Jordan 2002).

In November of the same year the winner was announced, namely, Finnish group ARKKI. Based on ARKKI's project, the ØDC defined the masterplan defining the general structure of the urban development of Ørestad; in December 1995 the City of Copenhagen published the plan. In the same time other projects were received for the new subway line, for its maintenance and control centre and for the Ørestad road network.

In 1996, at the end of the public inquiry, the Copenhagen City Council approved all projects. In 1997 the Ørestad masterplan was included in the Copenhagen Municipal Plan. The masterplan intended to divide Ørestad into four districts, to be built separately, each having its own identity. The districts were separated by large green areas and at the same time connected by the subway line and the Ørestad Boulevard; the latter thoroughfare crossed Ørestad from north to south.

The project mainly considered two aspects: the relation with the surrounding landscape and the water issue; indeed, out of 310 Ha, approximately 100 Ha were designed for landscape projects. South of the university area, by the big natural basin of Lake Grønjordssø, a 72-Ha park was planned. Moreover, the districts had to include two large urban parks. Ørestad had another outstanding element, represented by a network of navigation canals stretching over 10 km and fitted with waterfronts and spaces meant for social activities to take place.

Ørestad Nord, the first district to be built, was meant to house the Copenhagen University and the IT University, besides the university campus and the Danish Broadcasting Corporation headquarters. Between 2011 and 2013 the IT, theology, law and human sciences departments opened in the area. To date there are approximately 17,000 students and 800 staff, numbers which will certainly increase with the completion of the new buildings. The Ørestad Gymnasium was opened in 2007, which since then has become the most-applied for upper school in Denmark. In 2011, next to the Gymnasium, the first public school in Ørestad was opened, run by the City of Copenhagen, together with a new public library. In 2009 the first

large permanent park in Ørestad Nord, Grønningen, was opened. The park was divided into three zones: a urban, a park and a nature zone, all freely accessible. Inside, the park hosts a large number of small, very frequented sports facilities. Currently, Ørestad Nord is the most developed part and includes, besides the University, the Copenhagen Concert Hall and more than 1000 residential units, 500 of which are situated in the Tietgenkollegiet as university residences. The important institutions that have settled here turned this district into an international research and development centre in such sectors as culture, media and communications technologies.

Ørestad City (Figure 1), the second, central district of Ørestad, runs along the motorway and railway connecting the Copenhagen airport with the Øresund infrastructure line. Ørestad City was planned as a densely populated large urban community, complete with residences, offices, businesses, services and leisure facilities. In 2006 architect Libeskind presented a Masterplan for an area called Ørestad Downtown (total surface 187,000 m^2) including an 11-building complex, a central square and a glass-roofed gallery. Its construction began at the end of 2007 and in 2009 the first building, the 709-room Cab Inn Metro Hotel, was completed. In this area the international pharmaceutical company Ferring International A/S and a large shopping mall (146,000 m^2) are also located.

In 2005 the construction of the last two districts, namely Ørestad Syd and Amager Fælled, was approved. Ørestad Syd will house office buildings, residential areas, shops, schools and public services. Once completed, it will be the most populated district of Ørestad and will count over 10,000 inhabitants. The Ørestad Syd plan was approved in 2005 and to date the transformation plans regarding the Copenhagen Towers, the Ørestad Business Centre and the Hannemannparken complex were presented. The construction of the

Figure 1.   Ørestad City overview.

Table 1. Growing of Ørestad population.

Historical population		
Year	Pop.	±%
2004	100	—
2005	375	+275.0%
2006	1,739	+363.7%
2007	3,377	+94.2%
2008	5,410	+60.2%
2009	5,610	+3.7%
2010	6,142	+9.5%
2011	6,839	+11.3%
2012	7,445	+8.9%

Copenhagen Towers began at the end of 2007 and is still in progress today, while the Ørestad Business Centre building site opened in 2009.

The Amager Fæller district was completed only in its eastern part. The western portion will be the last part of the city of Ørestad to be completed and it is still under construction.

Basically the completion of the Ørestad project was expected to take 30/40 years; the district will be home to 20,000 people, 60,000 workers and 20,000 university students. At 2012 the registered residents were 7,445 (Table 1).

The underground line project (21 Km and 22 stations, 6 of which in Ørestad) was developed in three phases. The first phase began in 1997 and finished in the fall of 2002; it included the construction of an 11 Km network from Ørestad to the centre of Copenhagen. Phase two, ended in October 2003, involved the connection between the Centre of Copenhagen and nearby Frederisksberg. The last phase, namely the connection between Copenhagen and the airport, was concluded in October 2007.

3 ECONOMICAL, SOCIAL AND PROCEDURAL ELEMENTS

The Ørestad development is a self-funded project, where funds are provided by the urbanisation of the area, that is, by selling the building allotments. Basically, the Ørestad Act took inspiration from the principles of the British experiences of New Towns and urban renewal processes, where facilities were funded by the increase in value of the surrounding areas that resulted from the developments themselves (Miccoli et al. 2015b, c). Proceeds from the new subway help repay the investment. It has been estimated that 60% of the overall investment will be repaid by capital gains (50% from the sales of the land and 10% from property tax); 30% by revenues from the new subway; the remaining 10%

by subsidies provided by the regional bodies that joined the project without land transfer.

Under the Ørestad Act, the State of Denmark and the Municipality of Copenhagen should give ØDC the eastern area of Amager Fælled that they have owned since 1963 and some properties in the harbour. When they sold them, the market value of the Ørestad land was estimated at 122 million Danish kroner and that of the harbour properties at approximately 324 million. The land began to be sold in 1997 and ØDC expects all of them to be sold by 2035.

The first 6 billion Danish kroner's public subsidies were invested in the relocation of some institutional premises. The overall cost of the urban development project was about 1.6 million Danish kroner (about 215 million euros). The estimated cost of the subway is 12 billion Danish kroner (about 1.6 billion euros), with 6.7 billion for phase one, 3.4 billion for phase two, and 1.9 billion for phase three.

Considering the developments of the property market, the banks' interest rates, the number of passengers and the fares of the subway, the entire investment in the Ørestad project should be repaid by 2039.

Until 2007, ØDC managed the entire Ørestad redevelopment, including the subway; then, when company was dissolved, the subway-related activities were taken over by the Metro Corporation, and the redevelopment activities were managed by CPH City & Port Development.

Over the last few years, CPH City & Port Development has done such a good job that investors have started to show more interest in the completion of Ørestad. Development rights for terrace-houses in northern Ørestad were sold in 2013, and a large building zone for commercial use was sold in 2014. The areas sold in the previous years are being developed in Ørestad Syd, with most of the investments being made primarily in the so-called Arena District (CPH City & Port Development 2013).

Socially, the involvement of the community was one of the key factors that the development, as well as the management of the businesses that settled there, have been built on. An example of this are the Culture Days of Ørestad, launched in 2006 as a dance festival and then grown into a broader, much more far-ranging cultural festival. The efforts made by both residents and private companies in these events led to the establishment of a company in 2010, Ørestad Culture, which, as well as organising the festival, also manages major events throughout the year with know-how and, to a limited extent, money. Like the Ørestad Culture, the Grundejer-foreningssekretariatet plays a key role in Ørestad's city life too. In 2008, the householders' associations of the North, City and Syd districts decided to join

Ørestad's Waterworks Corporation in establishing a joint secretariat. As well as serving the local associations, the secretariat acts as a shared platform for contacts between residents, businesses etc. and organises joint events for all the quarters of Ørestad (BY & HAVN 2012). For example, in 2010 they developed a new website for comprehensive information by and for the residents of Ørestad.

4 CONCLUSIONS

So far, the Ørestad development has been going on nicely; a local development plan was adopted in northern Ørestad for the August Schade district, which also involves the building of Nordea's new headquarters. In Ørestad Syd the Copenhagen Towers are being built; in addition, many new buildings are being built, including a new day hospital, public housing, family and youth houses, etc. A local development plan was adopted in 2013 for the Arena district; a 15,000-seat roofed amphitheatre is under construction and should be completed in the first few months of 2016. Plans have also been adopted for the building of a skating rink and a primary school, alongside a local development plan for about 200,000 square metres of residential and commercial developments. CPH City & Port Development will take care of all the roads and communal areas; the project started in 2014. In partnership with the Municipality of Copenhagen, the company kept working at the design of a municipal sports centre in Ørestad City, which is expected to become the main sporting and cultural facility in Ørestad. In January 2014, the Municipality of Copenhagen approved an addition to the local development plan so that the project can be started. The sports centre should be ready for use by the end of 2015.

The Ørestad redevelopment has been widely welcomed by residents and everyday users alike since the very beginning; several surveys found that they are extremely satisfied with the quality of the housing, the vast green areas and the fast subway connections to downtown Copenhagen. At the same time, though, some people complain that the area lacks a true 'urban life' because the buildings are too tall, sparse and distant; the neighbourhood assumedly fails to convey that sense of belonging, that sense of community that is typical of a historical city.

The Ørestad project allowed the internationally oriented companies to settle in an attractive area, midway between the airport and downtown Copenhagen, near the High Speed line, the subway, the motorway and the international connection of the Øresund. In addition, the project is trying to respond to the increased demand for housing with high-quality buildings and high levels of environmental sustainability. The executive strategy of the project, the social inclusion of new residents and effective financial management, has made Copenhagen grow and turn into an international city.

REFERENCES

BY & HAVN 2012. Copenhagen Growing The Story of Ørestad, available at http://www.orestad.dk.
CPH City & Port Development 2013. Annual Report no. 30823702.
http://www.arup.com (last visited May 2015)
http://www.oecd.org (last visited May 2015)
http://www.orestad.dk (last visited May 2015)
http://www.orestadsselskabet.dk (last visited May 2015)
http://www.orestadsyd.dk (last visited May 2015)
Jordan, T. 2002. Kopenhagens Ørestad in der Kritik. In Detail, n. 4, April 2002.
Miccoli, S., Finucci, F. & Murro, R. 2015a. A Strategic Model for a Complex Infrastructure in Northern Europe. *Applied Mechanics and Materials,* 744–746: 2131–2135.
Miccoli, S., Finucci, F. & Murro, R. 2015b. A Sustainable and Integrated Approach to Urban Regeneration: Tools and Procedures for a Complex Area in London. *Applied Mechanics and Materials,* 737: 885–888.
Miccoli, S., Finucci, F. & Murro, R. 2015c. A New Generation of Urban Areas: Feasibility Elements. *In course of publication on WIT Transactions on Engineering Sciences,* paper accepted in 28.02.2015.
Ørestad Development Corporation. 2003. Annual Report, available at http://www.orestad.dk.

Advances in Civil, Architectural, Structural and Constructional Engineering – Kim, Jung & Seo (Eds)
© 2016 Taylor & Francis Group, London, ISBN 978-1-138-02849-4

Numerical simulation of multi-stage channels for a specific height-to-depth ratio

J.H. Tang
Department of Civil Engineering, Chung-Yuan Christian University, Chung-Li, Taiwan, R.O.C.

L. Du
Department of Bioenvironmental Engineering, Chung-Yuan Christian University, Chung-Li, Taiwan, R.O.C.

ABSTRACT: Artificial canals are basically sorted into irrigation, water delivery, drainage or irrigation-drainage. Owing to uplands having different elevation levels and being generally steeper than plain, the hydraulic drop design is often used to reduce the rate of water flow in the canal and dissipate energy, thus minimizing erosion to the bottom of the canal and constrained flow center. This research used ANSYS FLUENT 14.0 of Computational Fluid Dynamics (CFD) code in order to analyze the influence of energy consumption on the multi-stage channel to hydraulic drop design. Compared with the results obtained from the multi-stage vertical drop water energy dissipation hydraulic simulation test, performed by the Zhongxing engineering consultant agency, the results obtained were finely verified and also place technical base for further investigation on multi-stage drop water canal design and flood control in rivers. Finally, we conclude that the Napped flow is the best drop flow in terms of energy dissipation efficiency. Changing the canal design to form a Napped flow will reduce the water flow rate and the scouring effect on the riverbed to a maximum extent, thus dissipating energy carried by the water.

1 INTRODUCTION

1.1 Research motivation

Multi-stage hydraulic engineering is priority in cases of steeper slope, fast-flowing and erosion of riverbed, and where flood cannot be withheld by a single-stage canal.

Taking the environment into consideration, the study makes an optimal canal design, irrespective of high or low water period, ensuring that when the water flows through the canal, the friction shear force due to drop water, refluxing or votexing can fully eliminate strong scouring energy in the flow, so as to maintain the stability of the riverbed.

1.2 Research purpose

To sum up, the hydraulic drop canal is one of the most efficient engineering methods on energy dissipation. Therefore, the purposes of this study are:

1. Exploring the effect of energy dissipation on the multi-stage canal of hydraulic drop design.
2. Calculating the efficiency of the multi-stage canal of hydraulic drop design at different flow rates and water depths.

3. Comparing with the experiment of Zhongxing Engineering Consultant Agency to verify the feasibility of simulation.
4. Discussing on the development of the multi-stage canal of hydraulic design.

2 MODEL

2.1 Equations

Continuity equation is given by

$$\frac{\partial \rho}{\partial t} + \frac{\partial (\rho u)}{\partial x} + \frac{\partial (\rho v)}{\partial y} = 0 \tag{1}$$

Momentum equation is given by

$$\rho \left(\frac{\partial u}{\partial t} + u \frac{\partial u}{\partial x} + v \frac{\partial u}{\partial y} \right) = \rho F_x + \frac{\partial p_{xx}}{\partial_x} + \frac{\partial p_{xy}}{\partial_y} \tag{2}$$

$$\rho \left(\frac{\partial u}{\partial t} u \frac{\partial u}{\partial x} + v \frac{\partial u}{\partial y} \right) = \rho F_y + \frac{\partial p_{yx}}{\partial_x} + \frac{\partial p_{yy}}{\partial_y} \tag{3}$$

k-equation is given by

$$\rho \frac{dk}{dt} = \left[\left(\mu + \frac{\mu_t}{\sigma_k} \right) \frac{\partial k}{\partial x_i} \right] + G_k + G_b - \rho_\varepsilon - Y_m \tag{4}$$

ε-equation is given by

$$\rho \frac{d\varepsilon}{dt} = \frac{\partial}{\partial x_i}\left[\left(\mu + \frac{\mu_t}{\sigma_\varepsilon}\right)\frac{\partial_\varepsilon}{\partial x_i}\right] + C_{1\varepsilon}\frac{\varepsilon}{k} - C_{2\varepsilon}\rho\frac{\varepsilon^2}{k} \quad (5)$$

Turbulent viscosity coefficient is given by

$$\mu_t = \rho C_\mu \frac{k^2}{\varepsilon}$$
$$C_{1\varepsilon} = 1.44, C_{2\varepsilon} = 1.92, C_{3\varepsilon} 0.09, \sigma_k = 1, \sigma_\varepsilon = 1.3 \quad (6)$$

The numerical code adopted in this study is ANSYS FLUENT 14.0, which is based on the control volume method. The basic concept is that the following equation must be satisfied for the variable φ in the control volume V:

$$\int_v\left[\frac{d}{dxx}(\rho u\varphi) - \frac{d}{d}\left(\Gamma\frac{d\varphi}{dx}\right)\right]dv = 0 \quad (7)$$

2.2 Experimental measurements

From the report regarding the "Multi-stage vertical drop water energy dissipation hydraulic simulation test," by the Zhongxing engineering consultant agency, we found the energy dissipation effect η to be 89.6% and 63.5%, as shown in Figure 1, which is adopted from the report, for multi-stage vertical drop for five consecutive steps with a height-to-length ratio of 1/4. The energy dissipation effect is defined as the energy loss (ΔE) due to the drops to the original total energy (Eo).

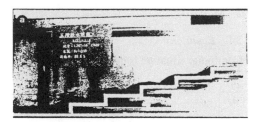

Napped Flow，Fr = 0.99，（η）= 89.6%：

Transition Flow，Fr = 1.32，（η）= 63.5%

Figure 1. The experimental results adopted from the report by Zhongxing engineering consultant agency.

2.3 Design model

Since the report only shows the figure without the detail design of the channel geometry and inlet velocity, we made the assumption of the geometry. Based on one specific case, we set the height (h) and length (l) ratio, $h/l = 1/4$, inflow discharge Q = 3.71×10^{-2} cm and Froude number F = 1.4. After several calculations for verification, we found the water depth = 0.2 m and the channel width B = 0.095 m were better and adopted in the model simulation.

Then, we applied numerical simulation for two cases with data shown in Figure 1, which is as follows:

Case 1. Napped Flow
  Based on Q = 1.763×10^{-3} cm, Fr = 1, B = 0.095 m, q = Q/B = 0.0186, we can find the water depth h.

$$Fr = \frac{Vh}{\sqrt{gh}} = \frac{V}{\sqrt{9.8 \times h}} = 1$$
$$Vh = q \Rightarrow h = \frac{q}{V}$$
$$V^2 = Fr^2 \times gh = Fr^2 \times g \times \frac{q}{V}$$
$$V^3 = Fr^2 \times g \times q = 1^2 \times 9.8 \times 0.0186$$

We found that V = 0.576 m/s and h = 0.033 m.

Case 2. Transition Flow
  Based on Q = 1.06×10^{-2} cm, Fr = 1.32, B = 0.095 m, q = Q/B = 0.112, the water depth h can be found as follows:

$$Fr = \frac{Vh}{\sqrt{gh}} = \frac{V}{\sqrt{9.8 \times h}} = 1.32$$
$$Vh = q \Rightarrow h = \frac{q}{V}$$
$$V^2 = Fr^2 \times gh = Fr^2 \times g \times \frac{q}{V}$$
$$V^3 = Fr^2 \times g \times q = 1.32^2 \times 9.8 \times 0.112$$

Thus, V = 1.24 m/s, and h = 0.09 m.

2.4 Design mesh

In the numerical simulation, as shown in Figure 2, a model of 3 m long and 0.9 m high was designed

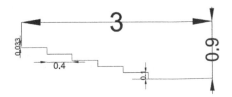

Figure 2. The designed set-up of the multi-stage vertical drop.

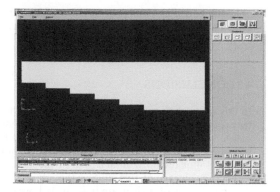

Figure 3. The designed set-up of mesh of the multistage vertical drop.

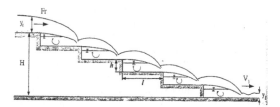

Figure 4. Napped flow schematic diagram.

with mesh size $\Delta X = 0.003$ and the number of mesh of 236,165.

The designed set-up of mesh is shown in Figure 3, and the general expression of the Napped flow in the multi-stage vertical drop is shown in Figure 4.

3 SIMULATION RESULTS

3.1 Numerical simulation results

The simulation results of water profiles such as free surface contours and velocity vectors at various time steps are shown in Figure 5.

From Figures 5 and 6, we can see that the free surface profile simulated matches very well with the Napped flow phenomena shown in Figure 4. In both figures, we can see the development of the free surface profile at different time steps in the multi-stage channel and the formation of backflow vortex at the vertical zone of the steps.

3.2 Calculation

From the numerical results, for both cases, the flow field is clearly of the Napped flow.

In each stage, reflux vortex is formed. From the last contours, the water flow rate has been declined due to the energy dissipation effect. Based on

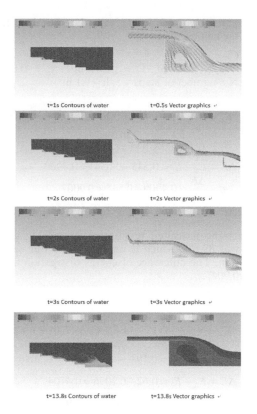

Figure 5. The contours of free surface and velocity vectors for various time steps for case 1.

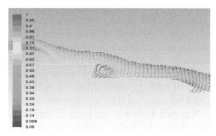

t=1.5s, Vector graphics

t=8s, contours of water.

Figure 6. The contours of free surface and velocity vectors for various time steps for case 2.

Figure 4, the following values are obtained from the numerical results:

Case 1.

$y_1 = 0.05$ m, $v_1 = 0.284$ m/s, $y_c = 0.03$ m, N = number of steps = 5, $h = 0.1$ m

$$E_0 = N \times h + \frac{3}{2} y_c = 5 \times 0.1 + 1.5 \times 0.03 = 0.545$$

$$E_1 = y_1 + \frac{V_1^2}{2g} = 0.05 + \frac{0.284^2}{2 \times 9.8} = 0.0541$$

$$\Delta E = E_0 - E_1 = 0.545 - 0.0541 = 0.4909$$

$$\eta = \frac{\Delta E}{E_o} = 0.4909 \div 0.545 = 0.9007$$

$$\Delta E = 0.4909, \ \eta = 90.07\%$$

The measured result from Zhongxing agency was 89.6%, which is very closed to our calculation. The difference of energy dissipation is only 0.0047.

Case 2.

$$E_0 = N \times h + \frac{3}{2} y_c = 5 \times 0.1 + 1.5 \times 0.08 = 0.62$$

$$E_1 = y_1 + \frac{V_1^2}{2g} = 0.19 + \frac{(1.24 \times 0.5)^2}{2 \times 9.8} = 0.2096$$

$$\Delta E = E_0 - E_1 = 0.62 - 0.2096 = 0.4104$$

$$\eta = \frac{\Delta E}{E_o} = 0.4104 \div 0.60 = 0.6619$$

The measured result from Zhongxing agency was 63.5%, which is very close to our calculation. The difference in energy dissipation is only 0.0027.

Based on the same procedures, we checked the other measured case and found that the energy dissipation is matched well to the simulation results. Therefore, the assumption for the designed geometry is verified. Moreover, to improve the efficiency of energy dissipation of hydraulic fall, it should be ensured that the flow of water produces the Napped flow on the ladder, the canal is fully bonded and the reflux vortex is formed. Ideal status produces backflow vortex at the vertical zone of the steps, and then the water flows horizontally along the canal.

We also found that the Napped flow is formed under the following condition: when the water flows through the ladder, the drop location should be 1/4 of the ladder field. This allows the water to fully swirl and be dragged by the horizontal shear force.

This simulation was subjected to verification testing, but not completely the same as in the natural environment, so this simulation can be taken as a reference only, just to provide a support for the theoretical data

or hydraulic design. For a project, more sophisticated experiments are required. This simulation can be used as the basis of screening data plan.

4 CONCLUSIONS

In this study, the numerical simulation was verified by experimental measurements. The energy dissipation of the multi-stage channel was proved to have a better effect when the free fall drop location was the 1/4 of the horizontal length of the ladder.

The significance of this study is that after water flow conditions of a river are investigated, we can take the advantage of this methodology, design all kinds of fall hydraulic models and make simulation calculation pointing to the same water flow, and then find out several models having the best energy dissipation efficiency.

This study can be an enormous help to the project schedule. Verified by the experiments, this study has proved its authenticity. It is possible to modify the canal model, change any properties of water flow and exclude various irrelevant factors.

Through this study, allowing the water flow to form Napped flow is the best strategy to increase energy dissipation efficiency in drop hydraulic engineering.

In the future study, we will determine how to design a water fall canal by making simulations pointing to different water flow conditions so as to maximize its advantage. For water flow with a big flow rate and depth, we may extend the length of each step, thus enlarging the ladder with a fixed height-to-length ratio.

REFERENCES

Chanson, H. 1994. *Hydraulic design of stepped cascades, channels, weirs and spillways*, The University of Queensland.

Chow, V.T. 1973. *Open-channel hydraulics*, McGraw-Hill Book Company.

French, R.H. 1987. *Open-channel hydraulics*, McGraw-Hill Book Company.

Richardson, E.V., Harrison, L.J. & Davis, S.R. 1991. *Evaluating Seoiat at Bridges*, U.S. Department of Transportation.

USBR, 1973. *Design of Small Dams (2nd ed.)*, Denver, Water Resources Technology.

Veronese, V. 1937. *Erosion De Fond En Aval D'une Decharge*, IAHR Meeting for Hydraulic Works, Berlin.

Zhongxing engineering consultant agency, 2001. *Multi-stage vertical drop water energy dissipation hydraulic simulation test.* 189–258.

Advances in Civil, Architectural, Structural and Constructional Engineering – Kim, Jung & Seo (Eds)
© *2016 Taylor & Francis Group, London, ISBN 978-1-138-02849-4*

An observation technique and GPS buoy processing strategy for ocean surface monitoring

A. Mohd Salleh & M.E. Daud
Faculty of Civil and Environmental Engineering, Universiti Tun Hussein Onn Malaysia,
Batu Pahat, Johore, Malaysia

ABSTRACT: This paper presents a kinematic positioning approach that uses a Global Positioning System (GPS) buoy for precise ocean surface monitoring. A field test was conducted to validate the observation technique and the processing strategy applied in this study; observation was performed at three stations located at different distances from the control station. Then, the observation technique was applied during the field observation off the Senggarang Coast. The GPS buoy data obtained from this observation were processed through a precise, medium-range differential kinematic technique. These data were collected over a period of more than 24 hours from a nearby coastal site at a high rate (1 Hz), along with the measurements from neighboring tidal stations to verify the estimated sea surface heights. The kinematic coordinates of the GPS buoy were estimated via epoch-wise pre-elimination and the backward substitution algorithm. Test results show that centimeter-level accuracy in sea surface height determination can successfully be achieved with the proposed technique. The centimeter-level agreement between the two methods also suggests the possibility of enhancing (or even replacing) current tidal gauge stations with this inexpensive and more flexible GPS buoy equipment.

1 INTRODUCTION

1.1 *Wave observation*

Offshore waves and storm surges must be detected before arrival at the coast to prevent coastal disaster (Gonzalez et al. 1999). A decade ago, coastal tide stations were the sole means to determine ocean and storm surge profiles, and these stations complicated offshore observation (Kato et al. 1998). Seabed-installed coastal wave gauges have recently identified various ocean profiles successfully based on continuous data acquisition (Kato et al. 2005).

Nevertheless, such wave gauges are installed in a limited area with a water depth of less than 50 m as per maintenance requirements. In addition, buoy-type wave gauges with acceleration sensors cannot detect long-period ocean and storm surge profiles because acceleration is low in long-term fluctuations. Therefore, a new offshore observation system must be established.

To this end, the authors of the current study have recently developed a novel system that uses a Global Positioning System (GPS) buoy. This system does not necessitate manual seabed maintenance and can be installed in any sea area without water depth limitation. A field test is conducted to validate the observation method and the data-processing strategy used, and the obtained results are applied to the field observation at sea. This paper explains these findings and the outcomes of the observation at sea.

1.2 *Field test*

The field test was conducted on August 25, 2014 at three stations situated at different distances from the control station (0.5, 10, and 20 km) to determine whether distances among stations exert a technical effect on the observation technique and on the data-processing strategy employed in this study. Observations were completed in a single day, and the travel time among observations was shortened to reduce atmospheric and tropospheric influences (Kelecy et al. 1994). The observation period at each station was limited to two hours.

The test was performed using a slider machine (Figure 1) that can replicate wave movement consistently. This machine was transported to each station for an observation period of two hours, and the data collection rate was 1 Hz. The position of the GPS buoy was estimated according to a fixed reference GPS receiver located on the roof of the Faculty of Civil and Environmental Engineering building at the Universiti Tun Hussein Onn Malaysia (UTHM) (Figure 2). The GPS determined a 3D position relative to the reference station instantaneously.

The local vertical component of the GPS solution is one measure of the GPS receiver on the slider machine. This component was compared with the "truth," that is, the standard value of the slider machine wave. By contrast, no comparative "truth" was available for contrasting with the horizontal components of the GPS buoy.

1.3 *Field observation at sea*

The field observation at sea was conducted off the shore of Senggarang, Batu Pahat, Johor on February 17, 2014. A specially designed buoy equipped with a GPS antenna was deployed 20 km from the UTHM control station and approximately 3 km off the Senggarang Coast (Figure 3).

The GPS buoy was constructed from a rubbermaid float to determine buoyant volume. A GPS antenna and a solar charger controller were installed in the GPS buoy; moreover, a steel pipe measuring 0.9 m with approximately 7.5 kg of ballast was fixed to the bottom of this buoy for sta-

(a)   (b)

Figure 1. Slider machine that consistently shifts the antennas from positions (a) to (b) throughout the field observation.

(a)   (b)

Figure 2. (a) Slider machine at a field test station; (b) setup of the reference station located at UTHM.

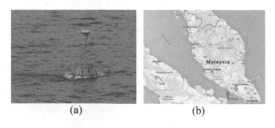

(a)   (b)

Figure 3. (a) GPS buoy; (b) field observation site.

bility and for minimizing its motion due to ocean waves. Prior to this fieldwork, the fully assembled GPS buoy was tested in the calm waters of the UTHM lake, and a calibration measurement was taken to determine the height of the GPS antenna above the water line. Two Leica GPS receivers were used in the experiment.

To generate results for the GPS buoy, a "truth" comparison was conducted at sea level as measured by a digital tide gauge at a nearby buoy location. Data were obtained at 50-second intervals from the tide gauge and referenced against a benchmark. The GPS buoy position solution also provided a measure of sea level that was relative to this benchmark. In sum, two independent methods of measuring sea level were established for comparison.

2   KINEMATIC PROCESSING

The data obtained from the fieldwork were processed by post-processing kinematic GPS positioning software based on the proposed methodology developed at the Astronomical Institute, University of Bern, Switzerland. Dual-frequency data must be obtained for linear ionosphere-free (Lc) resolution. Given that the program was designed for both post—and real-time data reduction, the GPS precise ephemeris technique was utilized by default to process data. The satellite elevation cutoff angle was set to 5°. Once phase ambiguities were successfully resolved in the initialization procedure, L1 and L2 phase measurements were used to conduct continuous epoch-by-epoch kinematic positioning. When the number of continuously tracked satellites dropped below four, the program automatically reverts to the stage of ambiguity integer identification to determine the correct phase ambiguities (Colombo, 1996 & Tsuji et al. 1997).

Throughout the entire field campaign, the buoy constantly measured its location. The Cartesian coordinates (x, y, z) of the antenna reference point of the GPS buoy were readily converted into their corresponding geodetic quantities (ϕ, λ, h). The instantaneous water level was obtained after reducing the height component h from the antenna reference point to water level according to known antenna height information.

A digital tide gauge provided water-level data every 50 seconds. The observed tidal heights were regarded as "ground truth," and they were compared with a GPS running average obtained over 150 seconds (to reduce the effect of waves). The ellipsoidal height of the buoy was determined and corrected for the Earth's body tide (but not for ocean loading).

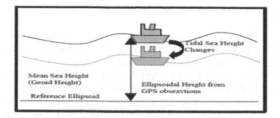

Figure 4. Schematic image to explain geoidal height, tidal height changes, and height estimation with the kinematic GPS technique.

3 GEODETIC SEA-HEIGHT MODEL

A simple reference model was prepared to compare the estimated sea surface heights. Such a model was based on a model of geoidal height distribution derived from the reference ellipsoid of WGS-84 and from changes in tidal sea surface height. The super-position of tidal height changes and geoidal height provided a simple model of sea surface height (Figure 4).

This model did not consider the oceanographic effect, atmospheric pressure changes, and sea wave changes, among others.

The typical frequency related to sea waves is significantly higher than those of other effects; furthermore, vertical movement may by estimated and removed based on the data collected by a dynamic motion sensor on a board survey vessel (Born et al. 1994). In the current study, we eliminated high-frequency changes with the moving average (boxcar window).

4 DATA ANALYSIS AND RESULTS

A field test was conducted at three sites situated at different distances from the UTHM control station. Figure 5 shows the change in the position (z axis) of the antenna at these stations. The "wave" heights at all stations ranged from 0.31 m to 0.35 m, unlike the standard slider machine value of 0.32 m. The wave patterns were uniform in all locations, and only the data in the third station were unclear in comparison with those in the first observation station. These differences may be attributed to the atmospheric and tropospheric variations during observation. Moreover, the root mean square (r.m.s.) values for the three stations were 2.2, 2.9, and 4.5 cm.

On the basis of the favorable results obtained from the field test, observation techniques and processing strategies were applied to conduct a fieldwork at sea. The positions of the GPS buoy were calculated for distant sites every second. Figure 6 depicts the mean water level that is relative

to the UTHM reference station, which clearly indicates that the variation in low and high tides was approximately 2.6 m and that the r.m.s was 1.7 cm. The maximum wave height recorded during field observation was 0.4 m, which was true for this area, unlike the historical wave data derived from the Malaysian Meteorological Department. The tidal record from the Kukup tide gauge was compared with the change in buoy height given the 1 Hz solution generated after using the running average obtained over a period of 150 seconds to eliminate the wave effect.

Figure 7 displays the ocean tide based on the Kukup Pier gauge. At this point, the actual tide has

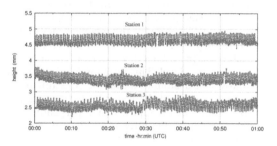

Figure 5. Wave heights at the three stations as observed with the slider machine.

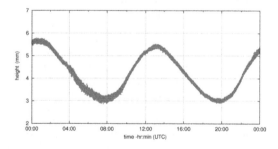

Figure 6. Waves and tides during the test conducted at Kukup as observed with GPS on a buoy. The medium-baseline differential solution is applied.

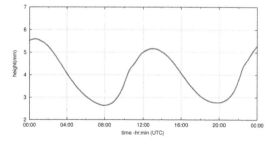

Figure 7. Change in sea height during the test conducted at a local tidal station.

just crested and has started to drop. The tidal variation was significant during the field observation period; by contrast, the mean GPS-determined height that was corrected for the solid earth tide exhibited a small variation in the low and high tides. The comparison between the wave observations of the two methods differed by 20 cm at the first low tide.

5 CONCLUSIONS

The slight differences in the field test results may be ascribed to the variation in the atmospheric and tropospheric factors between stations 1 and 2. Hence, the observations of the two stations should be conducted in the same weather condition.

A possible cause for the inconsistency between GPS buoy and tide data is the imprecise height reduction from the GPS antenna reference point to the water level. This antenna height reduction follows the assumption that the GPS antenna reference point and the water-level reference point are located at the same normal perpendicular to the surface of the reference ellipsoid. However, this assumption is not necessarily true because the ocean waves surrounding the GPS buoy can alter buoy attitude constantly. The instantaneous attitude should be monitored in applications that require highly accurate post-processing for height determination. This process can be performed with an on-board GPS antenna array that consists of at least three GPS antennas with known relative geometric relations. In conclusion, this experiment demonstrates that sea surface height can be determined precisely with the post-processing kinematic GPS technique. The centimeter-level agreement between the results obtained by the two methods for ocean surface monitoring also suggests the eventual possibility of improving on (or even replacing) tide gauge stations with this inexpensive and more flexible GPS buoy.

ACKNOWLEDGMENT

This research was conducted with financial assistance from the Ministry of High Education: Fundamental Research Grant Scheme (Vot 1229: Research on the Application of GPS for Natural Hazard, i.e., coastal erosion, earthquake, and tsunami). The authors obtained the data on the coastal tides at Kukup Port from the Department of Survey and Mapping Malaysia (Jabatan Ukur Dan Pemetaan Malaysia). The authors also express their sincere gratitude to these concerned persons.

REFERENCES

Born, G.H. Parke, M.E. Axelrad, P. Gold, K.L. Johnson, J. Key, K.W. Kubitschek, D.G. & Christensen, E.J. 1994. Calibration of the TOPEX altimeter using a GPS buoy, *Journal of Geophysical Research*, 99: 24517–24526.

Colombo, O. 1996. *Long-distance kinematic GPS, Chapter 13, in GPS for Geodesy (2nd edition)*, Springer, 537–567.

Gonzalez, F.I. Bernard, E.N. Milburn, H.B. & Mofjeld, H.O. 1999. *Early detection and real-time reporting of deep ocean tsunamis*, Abstracts of International Union of Geodesy and Geophysics 99 in Birmingham, B.127.

Kato, T. Terada, Y. Ito, K. Hattori, R. Abe, T. Miyake, T. Koshimura, S. & Nagai, T. 2005. Tsunami due to the 2004 September 5th off the Kii peninsula earthquake, Japan, recorded by a new GPS buoy, *Earth Planets Space*, 57: 297–301.

Kato, T. Terada, Y. Kinoshita, M. Isshiki, H. & Yokoyama, A. 1998. GPS tsunami—system development, monthly meetings, *Special*, 15: 38–42.

Kelecy, T.M. Born, G.H. Parke, M.E. & Rocken, C. 1994. Precise mean sealevel measurements using the Global Posi-tioning System, *Journal of Geophysical Research*, 99: 7951–7959.

Tsujii, T. Harigae, M. & Murata, M. 1997. The development of kinematic GPS software, KINGS, and its application to observations of the crustal movements in the Izu-islands area, *Journal of the Geodetic Society of Japan*, 43(2): 91–105.

Advances in Civil, Architectural, Structural and Constructional Engineering – Kim, Jung & Seo (Eds)
© 2016 Taylor & Francis Group, London, ISBN 978-1-138-02849-4

The community project in a public urban space

V. Šafářová
VŠB-TUO, Ostrava, The Czech Republic

ABSTRACT: This article describes the architectural and social research method of urban activism, which manifests itself in the form of community projects, namely Urban Interventions Ostrava in 2013. This project is expanded in two countries of Central Europe, namely the Czech and Slovak Republics, where 15 million people live. This article deals with the architectural views of the public and responses of Ostrava City Council, which then started again, entertaining the urban center of Ostrava, e.g. in public debates. Furthermore, a new community club (Town for People) and an old club (Beautification Committee for Beautiful Ostrava) are also included in the topic. This article aims to inform the expert public about the so-called Urban Activism in architecture, whose events are organized by activists including volunteers from architects, sociologists, artists, engineers, conservationists, students and other (mostly financially supported by local authorities). It is a specific way of revitalized urban space and environment using the principle "from the bottom up", which means the public addressing representatives of state administration and local government, that need to be investigated.

1 URBAN ACTIVISM

Join a vibrant community that cares about civic participation and contribute to one of NYC's most beloved public spaces. High Line, 2015, (http://www.thehighline.org/)

In 1889, Camillo Sitte, the Viennese architect and urban planner, as an expression of disagreement with the urban concept Ringstrasse, wrote a book titled "Building cities according to artistic principles" (Der Städtebau nach seinen künstlerischen Grundsatzen), which describes various examples of proper formation of cities, spatial composition and its public space, criticizes urban schematism and admires the principles of composition organically grown medieval towns. He favored the "irregular and organically grown native urban compositions medieval cities (Greece, Italy) for their scenic beauty and human scale" (Camillo Sitte, 1889). Now these are the topics of urban activism and community planning deals. In the world of urban activism, the term is more often used for movement and actions that are more revolutionary, critical, and destructive (in the positive sense). Here such an activism that is creative, positive and cooperative is meant. The activity is initiated from the public to the authorities and representatives of the city, political power, and the mechanism is called "from the bottom up." The problem is that the government perceiving these voices are few and are not interested to know what the public needs. The smaller the town is, the more provincial view persists that the politicians them-

selves know what to improve and where to invest the money, and then dominated by an atmosphere of distrust and suspicion stemming from the past.

Optimists believe, however, that even the climate is changing, and it is due to the natural criticism from city residents and activists who work in the field as the first sensors (Jan Gehl, 2013). This situation is in the former Eastern bloc in Central Europe, particularly in both Czech and Slovak Republics, as well as in Poland, which is relatively new.

The successful community project is "The High Line in New York" (Figure 1). In 1999, there

Figure 1. High Line in New York (photo: Jitka Jelínková).

was an initiative of two local citizens, non-profit organization Friends of the High Line, which fought for the preservation of monuments within the city layering as a document of its time and the fact that Manhattan is not always financial, fashionable and artistic center. The High Line is a public park built on a historic freight rail line elevated above the streets on Manhattan's West Side. It runs from Gansevoort Street in the Meatpacking District to West 34th Street, between 10th and 12th Avenues. Since 2006, it has been realized, still developed and cultivated. The High Line design is a collaboration between James Corner Field Operations, Diller Scofidio + Renfro, and Piet Oudolf.

1.1 Urban Interventions project and its method

Urban Interventions project was devised by the architects Matúš Vallo and Oliver Sadovský in the Slovak capital of Bratislava in 2008, at the young architectural studio Vallo + Sadovský Architects. Since then, the project has taken place in 13 cities within the Czech and Slovak Republics, and created more than 800 projects—interventions (Urban Interventions 2015, p. 4)

Ostrava is the third largest city in Bohemia, with about 300 000 inhabitants. In Ostrava Urban Interventions project version Ostrava in 2013, the first point was the announcement of the project, the establishment of websites, public presentation of the project to an internationally recognized platform Pecha Kucha http://www.pechakucha.org/, Ostrava formation Pecha Kucha Night in Ostrava (2013), presentation Project to Ostrava Universities (OU Faculty—Department of Sociology, Department of Philosophy, Faculty of Fine Arts, Faculty of Education OU), collecting proposals "interference" from the public (60 proposals of which 2 excluded). The results of the project were three exhibitions in Ostrava: multi-genre center Cooltour, Ostrava City Hall, and a shopping mall OC Forum Karolina, followed by three public presentation of proposals including an invitation key figures (founder—Matúš Vallo, urban activists, architects Yvette Vašourková and all participants' suggestions). Subsequently, it was accompanied by a catalog of the exhibition Urban Interventions Ostrava 2013 (published 2014) and made the implementation of two hits in the Ostrava public space (2014), and the third intervention is in preparation.

1.2 Statistics of project participants UI

In Bratislava Urban Interventions Publication 2015, Matúš Vallo, founder of the project, reports that the analyses presented in this book

are mostly carried out on participants who discussed the idea/concept/manifesto (28%), the activation of the city (26%), street (12%), park (10%), interconnection (7%) and transport solutions (7%), square (3.5%), sports (3.5%) and parking (1.7%) (City Interventions 2015, p. 8–12).

Less than 30% of projects were in the range of 101–500 Euros; it was the biggest part. The average age of the authors was 26–30 (at 41%), which was broadest representation of age followed by 31–35, then 20–25 and 36–40.

In a publication from 2011, it stated that 255 architects donated city Bratislava, Prague and Brno work worth 186,900 Euro and spent on it over 9800 hours.

The publication Urban Interventions Ostrava 2013 presented the main themes: idea/concept (33%), the activation of the city (32%), revitalization (21%), interconnection (8%), transport solutions (2%) and other (4%).

The average age of the authors was 25–34 (53%), which was the broadest representation of age followed by 24, then 35–44 and 45–64.

It follows that participating in both Ostrava and Bratislava, like dealing with inventing concepts and visions for the city (Figure 2), has become popular topics without specific place, continuing to feel the motivation and need to tell the city what it sees as a problem in specific public areas and corners (Figure 3), re-energize, renew or associate road surface sidewalks, minimize distractions that separate people from each other (Eliška Macková) and cast out parking spaces and creating one square (Tomáš Čech), bedecked bridges and Square pots of flowers (Radim Václavík) revitalized subway (Aleš Vojtasík, Vendula Šafářová, David Kotek).

Figure 2. High Line in New York (photo: Jitka Jelínková).

Participants´ motivation

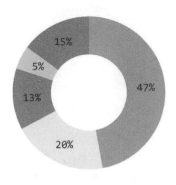

- ■ do something for Ostrava
- ■ present work publicly
- ■ create freely without assignment
- ■ elaborate an interesting topic
- ■ other

Figure 3. Graph participants of project Urban Interventions Ostrava 2013, details of participants' motivation (analysis: Soňa Frídlová).

2 ACTIVISM IN OTHER CITIES OF CENTRAL EUROPE

2.1 Bratislava—Process as a project

Subsequently, the studio Vallo + Sadovský architect founded other civic associations. We have city and Aliancia Old Market that continue a more professional version of community activity. Also, their projects have increased the scale and value. Enforced at the Town Hall renovation project object dilapidated Old Market in downtown Bratislava, they achieved that city marketplace leased them for a token price, for it is an association committed to monthly market that will invest in a certain amount of euros they searched investors, and today is the renovated Market and functional, transformed into a multifunctional building, where he organized exhibitions, concerts, conferences, theater performances for 653 visitors, below the Grocery U Paulika, three cafes with entrances to the street, one of them working homeless, besides there space leased Cooking Jem and other market further investments. It is officially named City's Cultural Center. Every Saturday at the marketplace place are held food markets and bazaar

books and clothes, as well Visegrad Food Fest, as a local product is produced there sodas.

Another project is WhatCity? - We are looking for good neighbors, which is similar Baugruppe respectively. Co-housing (in terms of building a community whose aim is the implementation of the concept of urban living, which is based on the initiative of the future residents of the house and is based on a community way of life).

2.2 Activism in Ostrava

Since the days of totalitarianism divided us has passed a quarter of a century, but also freer and more responsible relationship to public space we were looking for a long time. Only in recent years, we have begun to understand that its appearance and life depends mainly on us. We are witnessing an explosion of interest in an almost urban environment, public space, local activism, neighborhood, short life on the streets. Urbanism "from the bottom up", is transforming the city into a better place to live, initiated or implemented directly by the inhabitants, just do not resonate with us, it was also one of the themes of last year's Architecture Biennale in Venice, say the organizers of the festival Day of Architecture, 2014. http://www.denarchitektury.cz/ (2015).

Cooltour as a multicultural center is a platform that aims to change the cultural mindset of the public, seeks and supports community projects, consisting of volunteers working for free in an effort to improve something. They want to change either their neighborhood or society's view on some ingrained thing, such as the use of public space, to cooperate in the Department of Sociology FF OU (Mgr. Petra Šobáňová, PhD. and Nikol Horáková, PhD.). Their initiative came civil association for people City (director Soňa Frídlová). Recently, the research center revival, whose results are still being processed. Their chests project will be held for the second year and it is to rent a flower box and one in the season to worry and water, the project has mainly an educational dimension and a widely used school. Festival Different City Experience "is a nationwide celebration of life in public space" is happening in Prague, Hradec Králové, and this year will be held for the third time, will be held as well at three locations in Ostrava (city center, Poruba, Zábřeh). This year I will use the space next to the Church of St. Wenceslas (the oldest building in Ostrava, 13th century) that was selected based on the intervention of Little Copenhagen, where the main idea was to revitalize the neglected area, including the involvement of local tenants, so that pedestrianized and picturesque was set for relaxation and fun. The Prague event is organized by the initiative Auto * Mat. "We are trying to transform the space around us and improve the quality of life in the city. We support public, pedestrian and bicycle traffic "Project Restaurant Day is dedicated to local products that people bring by themselves to boil and taste, it brings people into space and thus enliven it and settle. The last project is the revitalization of the area behind Cooltour in a community garden in the center, where a member of its own planting plants and cares for the whole year.

Another association's Beautification Committee for beautiful Ostrava (printed bulletin issued since 2008), initiated by the workers antiquarian Fiducia, has more than 15 years of tradition and is an important facilitator of cultural events in Ostrava. The association leads Ilona Rozehnalová and helping her is a historian of art and architecture Martin Strakoš. Currently, so-called organized. Soundboard, which is inspired by the eponymous Prague initiative. Its contents are the public debate on the thorny issue of the recovery center and the construction of a new research library (i.e. Black Cube).

There also organizes Day of Architecture, held on the occasion of the World Day of Architecture (6/10) in Bohemia, the Prague-based association o.s. KRUH. It is a fact that architecture is a day dedicated to the celebration of architecture, and architects, historians and theorists of architecture events are being organized for the general public. The event is now already supported by the Ministry of Culture, which is reflected in the quality and the program menu. Program (around 108 shares) offered guided tours, conferences, trips, bike rides, workshops for children and others, a total of 55 Czech and Slovak cities. Educational activities o.s. KRUH is continuing in secondary, primary and nursery schools, a series of programs in architecture schools.

3 CONCLUSION—THE CONTRIBUTION OF URBAN ACTIVISM

Resignation is a disease of our latitudes and generation (Matúš Vallo, 2011) (Figure 4).

The aim is to regenerate urban activism and transform the city to a living organism; the initiative stems from the general public. Sometimes it can act unprofessionally and spontaneously, but it is important to accept it, because even this phenomenon is visibly changing public urban space. And it is welcomed and supported by the city both politically and financially, because for one thing motivates people to actively participate in public life (bringing community projects). Second, it is financially beneficial (Urban interventions have produced a lot of work for free) and, third, it has educational results (Chests).

The aim of community planning is to enter into urban planning and to transform with respect to

Main topics

- idea / concept
- activation of the city
- revitalization (square, building, park etc.)
- connection
- traffic solutions
- other

Figure 4. Graph participants of project Urban Interventions Ostrava 2013, details of the main topics (analysis: Soňa Frídlová).

the specific requirements of inhabitants themselves smaller residential units, mostly small-scale internal parterre and the like (M. Teo, M. Loosemore, 2014). It is not a fight for political power or aggravates offices' work, but disability negative nuances backstage thinking that office by their nature can reflect.

The aim of the projects organized by the center and its volunteers Cooltour is, according to a project manager Soňa Frídlová first "change of thinking". This is happening through organizing community projects such as Urban Interventions, Chests, Different City Experience, Restaurant Day or implementation of community gardens in the city.

These little tiny steps civic societies are changing the face of the city and people's behavior creates a pleasant synergy of residents, students, municipal administration in restoring public urban space.

This article aims to inform and support urban activism, community planning, and evidence on some examples (High Line New York, Urban Interventions and The Day of Architecture), and demonstrate that the activity is beneficial to the public and civil service.

ACKNOWLEDGMENT

The author thanks Mr Soňa Frídlová, MBA, for the information and materials.

REFERENCES

Gehl, J. & Svarre B. 2013. *How to study public life*. Washington: Island Press.

http://raumlabor.net/

http://staratrznica.sk/tag/my-sme-mesto/

http://www.denarchitektury.cz/

http://www.era21.cz/qx27a/era21_1_2012.htm

http://www.liptov.zasahy.sk/

http://www.mestskezasahy.cz/

http://www.mestskezasahyostrava.cz/index.php/aktuality

http://www.thehighline.org/

http://www.uia.archi/en/exercer/nouvelles/8866#.VXFOGs_tmko

http://www.uzemneplany.sk/vranokonovo

http://www.zaopavu.cz/

http://www.zaopavu.cz/search.php?rsvelikost=sab&rstext = all-phpRS-all&rstema = 38

http://zazitmestojinak.cz/

https://www.facebook.com/events/168246079891539/

Safarova, V. & Frídlová, S. 2014. The City Interventions Ostrava 2013 Ostrava: VŠB.

Sitte, C. 2012. The construction of cities according to artistic principles. Brno Institute for Spatial Development.

Teo, M., Loosemore, M. 2014. *Getting to the heart of community action against construction projects*. Proceedings 30th Annual ARCOM 2014, 855–864.

Vallo, M. & Sadovský, O. 2011. Municipal intervention. Bratislava: Slovart.

Vallo, M. & Sadovský, O. 2015. The City Interventions 2015. The take- sla -va OZ We 've city.

Advances in Civil, Architectural, Structural and Constructional Engineering – Kim, Jung & Seo (Eds)
© *2016 Taylor & Francis Group, London, ISBN 978-1-138-02849-4*

Digital diagrams in architecture

K. Soliman
Dessau Institute of Architecture, Dessau, Germany

M.A. Mogan
"Ion Mincu" University of Architecture and Urbanism, Bucharest, Romania

ABSTRACT: The present study captures the importance and recent evolution of diagram use in the architectural discourse, seen as a mediation between intrinsic abstract ideas and communication of ideas. Besides the main characteristic of the brief explanation of the process and the end result, the diagrams can turn their role to a more creative direction as a core medium in digital design process. This duality is expressed through a case study envisioned in our Maser Thesis.

1 INTRODUCTION

1.1 *Meaning*

Architectural diagrams are methods of developing, configuring an idea, concept, space, through various graphical means. Departing from the etymological sense, the diagram or 'diagramma' composes of the two terms: dia—across, between two and gramma-a figure, mark, line that is made, and represents an abstract concept which describes a transfer of meaning between two parts by graphical elements. The information transmitted reflects one or more parameters and specific features of the design process, many times being a simplified sketch or a complex drawing with data clusters. In fact the diagram is the abstract space where the information-data is embedded and evolves into the architectural realm. This area of transformation and configuration has a wide potential for the design process. Here is the moment where the concept is generated, where meanings are sorted, analyzed, folded and elucidated. It is the space where various knowledge streams and the subject converge to the point where the story is created and the spaces are articulated.

1.2 *Medium*

Two main diagram components can be highlighted: once is the message, content, information, concept which holds the entire construct together and secondly, the varied graphical language that enables multiple readings and expands the architectural innovation potential. In other words it reduces to answering the questions: which is the story and how is it drawn. In this sense, when speaking about the intentionality, we could observe the main feature in the case of the diagram as reductive or additive. The reducing process doesn't encounter a mode of eliminating the valuable or the meaningful characteristics of a construct, but filters the wide range of properties which defines it. Almost like a dissecting process which variously takes apart particular features, from the main concept to other significant characteristics of the project, like materiality, functionality or energy. Looking at the creative approach, many times the ideas crystallized through diagrams are hidden within the process, pending to stand in front with accuracy and clearness at the end. Further, when being brought to the light, the diagrams need to be designed at their time, in order to express the design story, such as a scenario would do for the play.

2 THEORETICAL BACKGROUND

2.1 *Diagram territories*

From a broader debate and philosophical perspective, Gilles Deleuze, examining Foucault major topics of knowledge, power, subjectivity, defines the diagram as "a supple set of relationships between forces" (Deleuze 1986/88:36) and further more highlights that "it is no longer an auditory or visual archive, but a map, a cartography that is coextensive with the whole social field. It is an abstract machine" (Deleuze 1986/88:34). This key concept refers at an infinite state of existence, where the diagram unfolds in a process of grasping all the interconnectivities between anteriority, interiority and exteriority. Those terms formulated by Eisenman in the same article, round up the creative space of the diagram. Firstly is the anteriority as it 'summation' of architectural history, and then comes the interiority which goes to the subtle ideas of project's possible options and in the end the exteriority represented as an outer source of meaning, such as the specific site, the program or the history. The abstract machine comes along with some key concepts such as field and force, and substitutes a rigid approach based on types and relations. The terms express another vision towards objects, mediums and open up wider understanding on the design process. In this case any entity is composed out of forces that configure the shape and the program and ultimately this forces act as fields of influence between subparts. At a first glance many times this apparent excess of graphics, connectivity, data with multiple inputs and outputs such as a complex machine, might seem as a diffuse, vague, even messy condition, but this is a matter of the multiplicity and in the end it calls for finding the optimal solutions. In consequence the challenge remains to create this exploration territory where all elements, forces and fields dynamically interact and lead to multiple scenarios. The hierarchies are no longer fixed and it is a matter of finding the sequences of understanding the importance of the items involved in the design process.

2.2 Digital platform

Within the context of new digital tools and advanced technological fabrication, the design process is enriched by the variability, flexibility and multitude of methods which come along with the new paradigm. This new design perspective opens new fields of understanding and experimentation where the complex approach to geometries, program, and structure leaves a lot of space for the unexpected situations and innovation. From the digital point of view the diagrammatic process could reveal this openness towards design.

2.3 New type of representation

Opposed to any typical diagram, the digital approach spans the possibilities for high complex geometries and for reading and articulating huge amount of data, by processing them as complex drawings. Expanding the discussion to the formation process, the digital diagram is a generative drawing where multilayered information gets interwoven as a variable interchangeable network, afterwards are transformed into 2d maps or 3D models and further more into architectural space. Primary, the input data is assimilated and ready for updates along the process according to the needs and usage. Afterwards, the specific drawing scenario and its performance with open multiple

combinations, unfolds the speculative towards finding the optimal layout, geometry and organization of the space. This syntax process envisions the zone where architecture is being generated, as Patrick Schumacher is referring to as "proto representations" (Schumacher 2010).

2.4 Data processing

At the very bottom of information age lays the concept of data which is embedded in any possible medium such as society, landscape, urbanity etc. The main challenge of contemporary architecture is to decode and to highlight all the hidden integrated knowledge. With the new digital tools and algorithms, multiple information levels are taken in consideration, from a normative view to a progressive perspective where it unveils unprecedented facts and reveals new understanding, which will lead to more innovative, creative, specific context related architectural spaces. The architectural evolution makes a step forward from the object based design strategy, long time in action, towards a performance data driven process. We are now more than ever, able to consider influential factors, facts, information embedded in the surrounding context.

Hence if we manage to analyze, confront, synthesize this livable cloud data, with a specific close-up to the scale, we might find out surprising situations, never thought before. Overall the actual challenge is about performance through design, and how we manage to create space with a higher sensibility to the users, context of material or immaterial sort. In the same time by grasping carefully data, the design process leads to sustainable results. Following the theory and ideas of Winy Mass (MVRDV) related to the Metacity, we can develop a comprehensive and extended study of urbanity and territories. Such datascapes demonstrate the high capacity and complexity of abstract maps as diagrams of our cities, to envision multiple criteria which together propose a more sensitive and flexible construct of the space.

3 CASE STUDY

3.1 Mission

As a case study and experiment for digital diagrams, we will look on the project developed as our final master thesis, a proposal for the new headquarter of United Nations Organization. The brief in discussion appears to be an ideal condition for an experiment which encounters a huge set of parameters and represents a global challenge from political, social and cultural point of view. The preliminary questions related to the performance, evolutions, complexity and richness of all nations, should lead to the theoretical and representational reevaluation of the future nations house. The project intends to explore the data patterns and the possibilities of transferring all their meanings with the use of diagrams to express a statement envisioned in one coherent architectural proposal.

3.2 *Evolution traces*

The features, values and performance tracks of all nations combined together create the real image of United Nations, which will be reflected in the new design. In this sense the strategy was to undertake a study on various data clusters around the topic of human development and growth, based on 192 nations. The forms of expression resulted through series of digitally generated diagrams, more specifically 2d and 3d graphs (scatter plot, historigrams, and network diagrams) describe with a scientific precision a cut out in the human development evolution.

3.3 *Analysis*

If we had to analyze the diagrammatic approach, according to Ben van Berkel's theory in a text dedicated to the influential role of the diagram in design, we could consider splitting the process in three stages: "selection, application and operation"(van Berkel, Ben 1999). Taking them apart, in separate rational phases almost like an experimental procedure, this action compose the motherboard for the creativity which comes along. The first stage represented the collection of annual databases lists for each nation over the last half century, concerning four main factors: Gross Domestic Product per capita (GDP), life expectancy at birth, literacy ratio, Gross Enrolment Ratio (GER) which are taken in consideration to calculate the Human Development Index (HDI). In the second stage we developed code definitions that transpose the accumulated data into comprehensive diagrams, tracking down the nation's development patterns (Figure 1).

This application phase refers to the analytical approach, which reveals the evolutions of nations, observing an increasing trend among the medium developed countries. The development rhythms vary between the four groups very high, high, medium, and low developed.

3.4 *Analysis conclusions*

The top countries proceed with a more or less constant ratio, adding small steps, meanwhile the medium and low ones in general are performing with accelerated rhythms, having huge challenges and important steps to achieve till a better position, but in the same time with many risk of instability. The overall conclusion points out the substantial

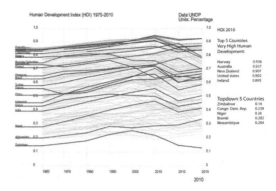

Figure 1. Human Development Index HDI 1960–2010/ © MirceaMogan, Karim Soliman/Master Thesis: United Nations Headqurters/Dessau Institute of Architecture, 2011.

progress and evolution of nations, of human kind in the last century, due to groundbreaking improvements in technological, medical care system, emergent economies, all together with an increase of the population of almost seven times bigger today than the beginning of the last century.

3.5 *Diagram-design process*

And we reach the last process phase, one that requires the transition from the analysis results towards the architectural proposal. The operation reflects the space generation, considering scenarios where particular features or reading of the data processed in the diagram deliver substantial tectonics, materiality, hierarchy, and structural conditions. And here is raised a critical point of whether these constructs are literal or abstract forms generated from the diagram. There might be cases where the diagram traces are more visible in the final model, but as long as the concept and intrinsic values stand as a powerful construct, which enables surprising spaces, innovative program and enriched materiality, this will turn out to release architectural qualities within the data driven design.

In the case study elaborated above, the diagram transforms into the real space tectonic, described by the multilayered intersecting paths of development for each nation transforming into merged slabs. The operation implies a prediction towards 2050 by extending the paths towards the highest point of development, as a message and humanity's goal to be reached in the near future (Figure 2). The final diagram represents a 3D graph implementing the data on a xyz system with time paths and HDI values of UN members from the 1960 to 2050 (Figure 3). The diagram's potential translated within a system of intersecting multi strata brings two important ideas the discussion: one is

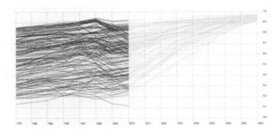

Figure 2. Human development index HDI prediction 1960–2050/© MirceaMogan, Karim Soliman/Master Thesis: United Nations Headqurters/Dessau Institute of Architecture, 2011.

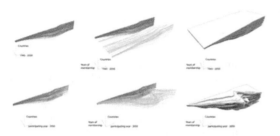

Figure 3. Human Development Index HDI prediction 1960–2050/© MirceaMogan, Karim Soliman/Master Thesis: United Nations Headqurters/Dessau Institute of Architecture, 2011.

Figure 4. 3D HDI diagrams formation/© MirceaMogan, Karim Soliman/Master Thesis: United Nations Headqurters/Dessau Institute of Architecture, 2011.

the expression of the long term strategy towards a highly developed world, and on the other side the manifold of intersections between nations paths, suggesting an architecture system where floors are softly merging into each—other, resulting in a continuous flowing space where all members and UN nation representatives will collaborate, debate, vote for the prosperity of the future world (Figure 3). In opposition with a standard packed overlapped floor system, the continuous slab construct softly ramping from floor to floor allows for a seamless communication and circulation for an office layout, creating multiple spaces for meetings, both informal and formal. The same diagrammatic layout incorporates the complete program including the general assembly and other conference rooms, according to proportions between the intersecting tracks of development (Figure 4).

Following the example of this particular project, data transforms itself through a sequence of applications and operations into built form, making it possible to construct a space that performs and represents the values of a certain topic, in this case United Nations organization headquarters. The project triggers to raise the awareness to the unprecedented growth and development of human kind in the last century, and in the same time to highlight the gaps which still need to be solved, and ultimately to propose a new architectural set up that envisions the diversity of all nations in one common and open platform, human development city.

4 CONCLUSIONS

In order to condense the ideas behind the digital diagramming, there can be emphasized the main features for merging abstract data, proposing hybrid scenarios, transformations and explorations of architectural features like tectonics, performance systems, organizational systems and experimental programme. Whether perceived as analytical or generative tools, digital diagrams contribute to pushing new boundaries of scanning our environment and formulating an improved design of the surrounding medium. Combined with the high potential of computational methods, the capacity of informing architecture with tremendous data sets enriches the design's variability, precision and complexity. In the same time the geometrical language deployed within the multiple configurations of the diagram emerges in a state of innovation and further exploration towards non linear dimensions. Digital Diagrams are the territories where geometry, data, intuition and creativity converge to envision the abstract space described in flexible patterns featured by precision and speculative strategies.

REFERENCES

Deleuze, G. 1986/88. *Foucault,* University of Minnesota Press, Minneapolis.
Eisenman, P. 1999. *Diagram Diaries*, Universe Publishing, New York.
Garcia, M. 2010. *The Diagrams of architecture*, Ed. Wiley & Sons, London.
Gausa, M. et al. 2000. *Metapolis dictionary of advanced architecture*, Ed. Actar, Barcelona.
Lefebvre, H. 1991. *Production of Space*, Blackwell publishing, XXXXX.
Polhill, R.M. 1982. *Crotalaria in Africa and Madagascar.* Rotterdam: Balkema.
Schumacher, P. 2010. *"Parametric Diagrams" in: Mark Garcia*, ed. *The Diagrams of Architecture*, AD Reader, John Wiley & Sons, London, 260–269.
Van Berkel B. & Bos, C. 1999. *Move,* Goose Press, the Netherlands.

Advances in Civil, Architectural, Structural and Constructional Engineering – Kim, Jung & Seo (Eds)
© 2016 Taylor & Francis Group, London, ISBN 978-1-138-02849-4

Offshore platform decommissioning cost index: A solution to decommissioning planning

N.A.W.A. Zawawi, A.B. Ahmed & M.S. Liew
Offshore Engineering Centre, Department of Civil and Environmental Engineering, University Technology PETRONAS, Tronoh, Perak, Malaysia

ABSTRACT: Offshore platforms decommissioning has become a central issue in the oil and gas or hydrocarbon industry in recent times. The burden and liabilities of decommissioning earlier platforms which did not incorporate the costs of decommissioning in their concession agreements with operators falls squarely on national governments. However, in recent concession contracts, operators are now mandated to set aside an annual amount into a special account created specifically to cater for decommissioning at the end of the concession or economic life of the platform. However, a common challenge facing the industry is determining accurate decommissioning costs for offshore platforms. Further challenge is the fact that no two platforms are the same, thereby introducing further element of uncertainty into current cost estimates. Hence taking this into cognizance, a cost index for offshore platform decommissioning was established to enumerate a range rather than an exact cost estimate. Therefore, this article attempts to use Laspeyres indices approach to generate a cost index for updating the cost of offshore platform decommissioning in Malaysia. Using historical cost data from earlier decommissioning carried out in the shallow waters in Malaysia, the index was able to predict future costs for similar scope of work with an accuracy range of plus or minus 20%. Consequently, given the degree of accuracy the index was able to achieve on a consistent basis, it can serve as a planning tool to assist governments, policy makers, senior management and concessionaires in estimating the annual amounts required to be set aside for the purpose of platforms decommissioning at their end of economic life. The index can also be used to carry out bid assessments among various contractors bidding for the same decommissioning contract.

1 INTRODUCTION

Energy supply serves as the backbone in modern societies and consolidates economic growth and sustainable development (Brookes 2000). There has been consistent trend showing an increase in energy consumption globally. It has been reported that about 53% of the total global energy consumption is expected to rise by the year 2030, but with the demand of about 70% coming from developing countries (Islam et al. 2009). This is due to economic growth, new innovations and sophistication in technology coupled with growing increase in global population which is manifesting in third world countries.

Of all the alternative sources of energy, the importance of oil and natural gas cannot be overemphasised, because about 80% of global energy will come from fossil fuels (Shafiee & Topal 2009, Shaari et al. 2012). While some nations have vast deposit of natural resources others depend largely on importing them. For example, South Korea is importing about 97% of its total energy demand because of the inefficient reserves of natural resources. In the year 2013, South Korea was ranked as the second-largest importer of Liquefied Natural Gas (LNG). Also, this East Asian country was the fourth importer of coal as well as fifth importer of total petroleum and other chemicals in liquid form (Weiss 2014). In another example, in Malaysia, oil and natural gas consumption in 2011 was estimated to be 76% of the total alternative source of energy (IEA 2013) as shown in Figure 1:

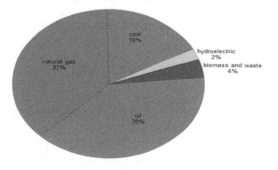

Figure 1. Malaysia's primary energy consumption, 2011 (IEA 2013).

Having acknowledged and appreciated the importance of oil and natural gas as a driving force for nation's development, there is need to explore the other side of it. Most of these oil and natural gas reservoirs are found offshore which require the fabrication and installations of offshore platforms to develop, extract, process and transport the extracted hydrocarbon to the shore for further treatment and uses.

As such, oil and gas industry has increased its exploration and drilling activities with the construction of newer, bigger, and more complex platforms to reach these hydrocarbons deposit in these offshore reservoirs.

However, design life span of these offshore platforms is usually 25 to 30 years, at the end of these economic period, these platforms need to be remove and decommissioned in line with international conventions which requires all abandoned platforms be removed in a way that will not cause harm to the environment. However, this represents the end of the production life cycle of oil and gas installations (Kaisera & Byrd 2007).

Figure 2 shows a typical life cycle of an offshore oil and gas platform:

Usually, offshore platform decommissioning can be achieved by complete removal or partial removal. Once removed, the platforms can be turned into other uses such as Remote Reefing, Conversion to a Weather Forecast Center (can be in-situ) and even conversion to a Tourist Attractions Center (can be in-situ). However, tourist attraction center and weather forecast zone have been considered uneconomical and hence unrealistic because the cost of maintenance is usually high (Mallat et al. 2014). Due to international conventions on ocean safety and prevention of marine pollution that gives more emphasis on complete removal, owners of dilapidated platforms are under pressure to seek compliance which will cost the oil and gas industries in excess of US$40 billion (Salcido 2005). In addition to that, the most widely used contractual regime is the Production Sharing Contract (PSC) equally referred to as Production Sharing Agreement (PSA) also neglect to consider

decommissioning of these platforms in sufficient details and hence the burden of the liability heavily falls on the platforms owners (Hamzah 2003). The omission of decommissioning liability in the early contractual regime have troubled many National Oil companies (NOCs) with gigantic liabilities as more than US$155 million is required to decommission a Harmony platform in Pacific Outer Continental shelf in California region in the U.S.A. (Anthony et al. 2000).

Establishing an accurate cost estimate for offshore platform decommissioning at the planning stage does not only assist in proper planning and executing the work, but also provide an accurate benchmark for bargaining with the potential contractors and hence provide basis for competitive bidding evaluation. It also assists to determine funding requirements and financial liabilities.

For complete removal, there are ten (10) steps to the process, these are: Engineering, Project Management and Planning; Permitting and Regulatory Compliance; Platform Preparation; Well Plugging and Abandonment; Conductor Removal; Mobilization and Demobilization of Derrick Barges; Platform Removal; Pipeline and Power Cable Decommissioning; Materials Disposal; and Site Clearance (Proserve 2010).

2 DECOMMISSIONING OVERVIEW

2.1 Some international conventions

It has earlier been established that the word decommissioning has not appeared in the 1982 United Nations Conventions on Law of Sea (UNCLOS). Moreover, the word is as well missing in the 1952 Geneva Convention on the Continental Shelf. It is also pertinent to note that the International Maritime Organization (IMO) guidelines and standard have not defined the word decommissioning (Hamzah 2003). In spite of the fact that decommissioning is not defined by the aforementioned treaties, they all mentioned and stressed the need to remove all abandoned offshore installations. Obviously the word "offshore platform decommissioning" has a recent origin. It attracts the attention of international oil and gas industries following the Brent Spar controversy of 1995 (Osmundsen & Tveteras 200, Lofstedt & Renn 1997).

2.2 Decommissioning phases

The following are the three main phases for offshore platform decommissioning:

– Pre-decommissioning Activities: Also known as the planning stage, at this stage a decommissioning plan is developed in detail and the programme of work is formulated;

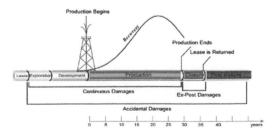

Figure 2. Typical life cycle of an offshore oil and gas structure (Ferreira et al. 2004).

- Decommissioning Activities: This is the main decommissioning stage as it involves removal and re-use, recycling, leaving in-situ, or disposal of all, or part, of the installation as the case, may be; and
- Post-decommissioning Activities: site survey, site clearance and post-decommissioning inspection.

2.3 *Global decommissioning market*

There are more than 7,500 offshore oil and gas installations globally, thousands of these offshore structures (about 85%) are approaching the end of design life and will require decommissioning within the next decade (Fowler et al. 2014). Significant percentages of such have been in service for more than 16 years, many have more than 20 years of operation, and some are abandoned and due for decommissioning (Hamzah 2003).

It is important however to note that, huge amount of money is needed to decommission an offshore oil and gas platforms. In Malaysia for example, it was estimated to cost PETRONAS, Malaysia's NOC, about US$2 billion to remove over 200 obsolete offshore platforms (Twomey 2010). A similar study carried out in Norway which shows that about US$27 billion is required to remove about 500 platforms (Bjerkelund 2002, Namdal 2010). The cost escalation in Norway is as a result of rough sea condition, depth of the installations and enormous steel jacket network. In the Gulf of Mexico (GOM) the decommissioning marker reaches about US$2.4 billion (Kaiser & Liu 2014). Most of the experiences related to decommissioning offshore installations are found in the Gulf of Mexico, in which about 1,600 structures have so far been decommissioned. Most of these installations were relatively small in size, with less than 10% of the total contributed to the United State Artificial Reef programme. In the North Sea, majority of offshore oil and gas installations decommissioned to date are either small steel platforms, floating or subsea installations (Ayoade 2002). This shows that to date, the removal and disposal or larger and gigantic network of steel structures remained new and need to be explored.

3 THE NEED FOR A COST INDEX FOR DECOMMISSIONING

Malaysia has grown in the energy market from a net importer to a major industry player over the last couple of decades. The oil and gas sector contributes about 40% of Malaysia's total revenue and 17% of its Gross Domestic Product (GDP) (Lintzer & Salomon 2013). However, many of Malaysia's platforms constructed in seventies

or earlier are either due for decommissioning or about to be due and need to be decommissioned (Kurian & Ganapathy 2009). Of the 300 offshore platforms in Malaysia, many of which are in shallow waters (50–70 m depth), about 60% have exceeded their design life (Zawawi & Liew 2013). Although Malaysia's long experience in offshore oil and gas is fully appreciated by other ASEAN countries and has been relied on by Vietnam in the development of its own infrastructure and regulation, it does not have any specific legislation on decommissioning (Lyons 2013). Moreover, because the earlier concessions did not originally include decommissioning, the task of removing oil platforms has become a liability for the government which operates through PETRONAS. The 1974 Petroleum Development Act (PDA) and the Production Sharing Contract (PSC) documents entitled PETRONAS as the legal owner of all offshore platforms in Malaysia. These two provisions make PETRONAS the sole concessionaire of petroleum resources and ownership of upstream facilities respectively; hence its liability for decommissioning and its residual liability (Fewings 2005).

Furthermore, in Malaysia, decommissioning plans will have to comply with at least eight (8) other local laws (Ibanez 2011). However, it is important for PETRONAS to work in line with global best practices in this area to develop cost indices or models for determining probable estimates for decommissioning to aid negotiations with contractors. A similar practice has been adopted by the U.S Mineral Management Services (MMS) by ascertaining the cost of decommissioning at a point in time and updating the cost every five years to reflect the impact of market, technology, inflationary, and regulatory policy changes on costs. One of the easiest ways to achieve this is to develop a decommissioning cost index. Although indices have been criticised highly subjective, vague and crude, they, however play a vital roles in establishing variations in a given attribute for making logical and systematic comparisons (Minogue 2005). This is why they are the most favoured means of showing improvements overtime with indices such as Resource governance Index, Global Competitiveness index etc being adopted globally.

In the Malaysian housing sub-sector for instance, the Malaysian House Price Index launched in 1997 can be used to monitor the trend of house prices in Malaysia and assist policymakers in formulating national economic policies with respect to housing and property development (Zahari & Nasir 2002). A similar decommissioning price index can also be used to monitor shifts in the cost of offshore platform removal in order to assist in updating the decommissioning accounts appropriately. In the U.S., a general bond covers decommissioning while a supplemental bond is used to update the cost over time. The purpose of the supplemental bond is to protect the U.S. Government from incurring financial losses by ensuring sufficient funds are set aside to cover the full cost of decommissioning by another party in the event the current operator/lessee becomes financially insolvent and is unable to carry out its contractual obligations under the lease (Proserv 2010).

Every platform is unique in design and complexity, however, this uniqueness is limited to design and weight (size) while the major features remain the same. Where differences exist between platforms, they do so simply on the basis of size or proportion; hence, a factor can be calculated to cater to such proportional differences pro-rata. It has also been found that early concession in a negotiations depended on the point a negotiator intends to stand within their 'Zone of Possible Agreement' (ZOPA). A promotion focused party gained an upper hand in negotiations if the prevention focused party conceptualised their goals within the lower range of their ZOPA (Trotschel et al. 2013). The implication of this in negotiations is that the client needs to understand the needs of the contractor before stating their own position. However, the ZOPA of the client tends to often be lower given the fact that they are promotion-focused, hence want the contract executed

4 METHODOLOGY

4.1 *Nature of data for computing Offshore Platform Decommissioning Cost Index (OFDCI)*

Secondary sources of data are generally suitable for constructing OFDCI. However, the challenges of developing a good and quality OFDCI can be outlined as follows:

i. Offshore platforms are basically heterogeneous in nature. That is to say, no two platforms are similar or identical.
ii. Decommissioning cost often varies upon the location, water depth, sea condition etc. The cost of an offshore platform decommissioning is not fixed and can vary from one period to another.
iii. Offshore platform decommissioning itself is infrequent. In many countries, very few or none of the platforms are decommissioned annually

4.2 *Mathematical formulation*

Suppose in a given time t there are Q^t_{jk} item of cost marked j in area or zone k, with the total cost of the item i in cluster j in zone k is given as X^t_{jk} as

$i = 1,\ldots\ldots, Q^t_{jk}$ $(j = 1, 2, \ldots, M^{tk})$ is known as the structure in zone k at a given period t. The mean price P^{tk}_j for a cluster j in zone k at a given time t, can be written as:

$$P^{tk}_j = \sum_{i=1}^{X^{tk}_{jk}} X^{tk}_{ij}/Q^t_{jk} \tag{1}$$

The geometric mean of the Laspeyre's as well as the Paasche's indices is referred to as the Fisher index. It can be expressed as: Let $P^t \equiv [P^t_1, \ldots\ldots, P^t_j]$ and $Q^t \equiv [Q^t_1, \ldots\ldots, Q^t_j]$ is the time t vectors quantities and cell prices. Laspeyres cost index, P^{0t}_L, from the base period 0 to a point at time t can be computed as:

$$P^{0t}_L(P^0, P^t, Q^0) \equiv \frac{\sum_{i=1}^{M^{tk}} P^{tk}_j Q^t_{jk}}{\sum_{i=1}^{M^{0k}} P^{0k}_j Q^0_{jk}} \tag{2}$$

Equation (2) can be written as cell price index $P^{0t}_L = P^t_j/P^0_j$ and value share $W^0_j = P^0_j Q^0_j / \sum_{j=1}^{M} P^0_j Q^0_j$. While Paasche price index is given as:

$$P^{0t}_L(P^0, P^t, Q^t) \equiv \frac{\sum_{i=1}^{M^{tk}} P^{tk}_j Q^t_{jk}}{\sum_{i=1}^{M^{tk}} P^{0k}_j Q^t_{jk}} \tag{3}$$

Therefore the Fisher cost index for time t in relation to the base year 0 P^{0t}_F, is established by the geometric mean of Equation (2) and Equation (3) above and could be written as:

$$P^{0t}_{KF}(P^0, P^t, Q^0, Q^t)$$
$$\equiv \sqrt{\left(\frac{\sum_{i=1}^{M^{tk}} P^{tk}_j Q^t_{jk}}{\sum_{i=1}^{M^{ok}} P^{ok}_j Q^o_{jk}}\right)\left(\frac{\sum_{i=1}^{M^{tk}} P^{tk}_j Q^t_{jk}}{\sum_{i=1}^{M^{tk}} P^{ok}_j Q^t_{jk}}\right)} \tag{4}$$

Given that at time t from cluster k there are Q^t_k platforms decommissioned, with the given cost as i in cluster k and is the same as X^{tk}_i for $i = 1, \ldots\ldots, Q^t_k$ i The median cost P^{*t}_k, and overall cost V^t_k, for all platforms decommissioned in zone k at time t, are given by:

$$P^{*t}_k = \text{Median}\left\{X^{tk}_i, i = 1, 2, \ldots\ldots, Q^t_k\right\} \tag{5}$$

And

$$V^t_k = \sum_{i=1}^{Q^t_k} X^{tk}_i \tag{6}$$

This means that the quantity Q^{*t}_k of the total cost of all platforms types in cluster k at given period of time t is:

$$Q_k^{*t} = \frac{V_k^t}{P_k^{*t}} \quad (7)$$

When time t share all costs of all platforms types in cluster k, S_k^t, (if $k = 1, 2, \ldots, j$) can be written as:

$$S_k^t = \frac{V_k^t}{\sum_{r=1}^{j} V_r^t} \quad (8)$$

The Total Törnqvist-Theil index P_{kT}^t, is given by

$$P_{kT}^t = \left[\left\{ \left[\frac{P_k^{*t}}{P_k^{*0}} \right] \right\} \right]^{\left\{ \frac{1}{2} \left[S_k^0 + S_k^t \right] \right\}}$$

The total weighted average or mean of all the platform gives an overall Törnqvist Theil index for Malaysian waters P_{kT}^t, and it can be written as:

$$P_{kT}^t = \exp \left\{ \sum_{k=1}^{6} \left\{ \frac{1}{2} [S_k^0 + S_k^t] \right\} \log_e \left\{ \frac{P_k^{*t}}{P_k^{*0}} \right\} \right\}$$

4.3 Laspeyres index calculation

Based on the data available, year 2011 served as the base year. Formally, the calculation is written as shown below;

$$L_t = \frac{\sum_{j}^{n} p_{jt} q_{j0}}{\sum_{j}^{n} P_{j0} q_{j0}}$$

where, subscript "$j0$" and "jt" refers to the base year value and current year value of the index, while "p" and "q" represent the price and quantities respectively.

4.4 Rate of inflation

$$\text{Inflation Rate}_{F-1} = \frac{F - I}{I} \times 100\%$$

where, F = value of index in the succeeding year
I = value of index in the preceding year

4.5 General assumptions

The following are the general assumptions considered in this research:

- Onshore normal hour/day = 9
- Offshore normal hour/day = 12
- Offshore effective hour/day = 8.5
- Man-hours efficiency @ onshore = 75%

- Man-hours efficiency @ offshore = 65%
- Efficiency during harsh weather = 40%
- Costs are estimated in 2014 and in Malaysian Ringgit (RM)
- Reverse installation techniques will be use in removing platforms
- Derrick Barges' mobilized from Asia (Malaysia)
- Complete removal of all platforms is considered.
- Explosives will not be used during the decommissioning process.
- No salvage or resale value has been considered for the structures, pipelines or power cables that are removed
- One DB mobilization/demobilization cost is included for the whole project
- The round-trip mobilization/demobilization times for Derrick Barge (DB) is: 120 days for a DB having a 2,000 ton maximum lift capability (DB 2000) mobilized from southeast Asia;
- The weather contingency downtimes for demolition operation is assumed to be: 15%
- No downtime is assumed due to the presence of whales or marine mammals.
- A general contingency (provisional work) of 20% is applied to all phases of the decommissioning

5 RESULTS AND DISCUSSION

Table 1 below shows a summary of the unit cost index and the inflationary trend.

5.1 Discussion on results

Table 1 shows the indices calculated, it can be deduced that there is a stable increase in the cost of offshore platform decommissioning in Malaysia as seen from the table. Year 2011 has been considered and earmarked as the base year and the index should always equal to 100. The subsequent years show a stable and gradual increase to the indices. In the year 2012, there was an increase 18.3% from the preceding year 2011. Similarly, Malaysia witnessed another 0.18 or 18% increase in the year 2013. However, the average annual index value was found to be 0.181 indicating the figure as a multiplayer for the average annual increase to the

Table 1. Index value and inflation rate.

Year	Index value	Inflation rate (%)
2011	100.0	18.30
2012	118.3	14.96
2013	136.0	–
Average	118.1	16.63

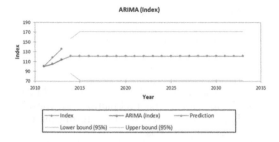

Figure 3. Prediction graph (ARIMA).

decommissioning cost. It serves as the required multiplier in updating the cost of offshore platform decommissioning. This value can be forecasted for future planning using Autoregressive Integrated Moving Average also known as ARIMA.

5.2 Autoregressive Integrated Moving Average (ARIMA)

However, due to the paucity of data at hand, the ARIMA analysis presented hereunder is the product of time series data of only three inputs thus the graph shows a stationary movement after the third year. It needs to be updated as soon as the decommissioning market boom in Malaysia.

5.3 Rate of inflation

From the index results obtained, the rate of inflation related to offshore platform decommissioning in Malaysia has been established. The inflation rate calculated shows a deflationary trend between year 2011 and 2012.

6 CONCLUSIONS

Offshore platform decommissioning presents a great liability to the government. Due to international conventions that laid more emphasis on complete removal, countries with abandoned offshore platforms in their territorial waters are under pressure to comply with those conventions. As a result of the heterogeneous nature of offshore platforms, it is difficult to establish cost estimate for offshore platform decommissioning.

However, due to huge sums of money incurred in decommissioning an offshore installation, the Malaysian government for example has incorporated from the year 2001 in its PSC/PSA, a clause that stipulates regular remittance of certain amount of money during the economic life of a platform in order to cater and provide sufficient fund required to decommission the platforms at the end of its

economic life. This is commonly known as Cessation Fund for decommissioning.

This article, however, attempt to use Laspeyres technique to establish a cost index for offshore platform decommissioning for Malaysian waters. These indices were applied to an earlier decommissioned platform and were able to achieve the predetermined benchmark of plus or minus 20% of the actual cost. Moreover, this study helps in reducing the wide margin of cost overruns which is seen more often than not in almost all decommissioning costs estimates. Therefore, cost index will assist policy makers at the planning stage to efficiently plan for offshore platform decommissioning at the end of its economic life by periodic updating of the decommissioning cost.

ACKNOWLEDGEMENT

Foremost, I would like to express my deepest gratitude to my supervisor, Dr. Noor Amila Wan Abdullah Zawawi, for her excellent guidance, patience, constructive recommendations and providing me with an excellent atmosphere for conducting this research. My special appreciation also goes to my co-supervisor, Associate Professor Ir Dr Mohd Shahir Liew, for his supervision and constant support.

REFERENCES

Anthony, N., Ronalds, B. & Fakas, E. 2000. *Platform decommissioning trends*, in SPE Asia Pacific Oil and Gas Conference and Exhibition.

Ayoade, M.A. 2002. *Disused Offshore Installations and Pipelines: Towards*, Sustainable Decommissioning, 17: Kluwer Law International.

Bjerkelund, M.H. 2002. *Decommissioning of Offshore Installations—Experience Related to Safety and Environment and the Philosopy: "How Clean is Clean Enough?"*, in ASME 2002 Engineering Technology Conference on Energy, 483–488.

Brookes, L. 2000. Energy efficiency fallacies revisited, *Energy Policy*, 28: 355–366.

Cresswll, J.W. 2002. *Research Design: Qualitative*, Quantitative and Mixed Methods Approaches: SAGE Publications, Inc.

Ferreira, D., Suslicka, S., Farleyc, J., Costanzac, R. & Krivovc, S. 2004. A decision model for financial assurance instruments in the upstream petroleum Sector, *Energy Policy*, 32: 1173–1184.

Fewings, P. 2005. *Construction Project Management: An Integrated Approach*. Abingdon: Taylor & Francis, 2005.

Fowler, A., Macreadie, P., Jones, D. & Booth, D. 2014. A multi-criteria decision approach to decommissioning of offshore oil and gas infrastructure, *Ocean & Coastal Management*, 87: 20–29.

Hamzah, B. 2003. International rules on decommissioning of offshore installations: some observations, *Marine Policy*, 27: 339–348.

Ibanez, M.F.A. 2011. *Towards the sustainable decommissioning of Offshore installations: A regulatory Challenge for ASEAN States,* in Law in a Sustainable Asia: 8th Asian Law Institute Conference-Thursday and Friday, 26 and 27 May 2011, Kyushu, Japan, Singapore, 1–11.

IEA, 2013. Countries/Malaysia, in International Energy Agency, ed.

Islam, M.R., Saidur, R., Rahim, N.A. & Solangi, K.H. 2009. Renewable Energy Research in Malaysia, *Engineering e-Transaction*, 4(2): 69–72.

Kaiser, M.J. & Liu, M. 2014. Decommissioning cost estimation in the deepwater U.S. Gulf of Mexico – Fixed platforms and compliant towers, *Marine Structures*, 37: 1–32.

Kaisera, D. & Byrd, C. 2007. *Current Practices of Decommissioning.* United States: Gulp Publishing Company.

Kurian, V.J. & Ganapathy, C. 2009. *Decommissioning of Offshore Platforms,* in 2nd Construction Industry Research Achievement International Conference (CIRAIC).

Lintzer, M. & Salomon, M. 2013. *Greater Transparency and Accountability in Managing Malaysia's Oil Wealth Urgently Needed,* in Revenue Watch institute, ed, 2013.

Löfstedt, R.E. & Renn, O. 1997. The Brent Spar controversy: an example of risk communication gone wrong, *Risk Analysis*, 17: 131–136.

Lyons, Y. 2013. *Abandoned Offshore Installations In Southeast Asia And The Opportunity For Rigs-To-Reefs,* Centre for International Law, National University of Singapore Singapore.

Mallat, C., Corbett, A., Harris, G. & Lefranc, M. 2014. *Marine Growth on North Sea Fixed Steel Platforms: Insights From the Decommissioning Industry,* in ASME 2014 33rd International Conference on Ocean, Offshore and Arctic Engineering, V01AT01A021-V01AT01A021.

Minogue, M. 2005. Apples and Oranges: Problems in the Analysis of Comparative Regulatory Governance, *The Quarterly Review of Economics and Finance*, 45: 195–214.

Nåmdal, S. 2010. *Decommissioning of offshore installations,* Helsfyr, Oslo2010.

Osmundsen, P. & Tveterås, R. 2003. Decommissioning of petroleum installations—major policy issues, *Energy policy*, 31: 1579–1588.

ProServe Offshore. 2010. *Decommissioning Cost Update for Removing Pacific OCS Region Offshore Oil and Gas Facilities,* U.S. Department of the Interior, Minerals Management Service, Houston.

Salcido, R.E. 2005. Enduring optimism: Examining the rig-to-reef bargain, *Ecology LQ*, 32: 863.

Shaari, M.S., Hussain, N.E. & Ismail, M.S. Relationship between energy consumption and economic growth: empirical evidence for Malaysia, *Business Systems Review*, 2: 17–28.

Shafiee, S. & Topal, E. 2009. When will fossil fuel reserves be diminished?, *Energy Policy*, 37: 181–189.

Trötschel, R. Bündgens, S. Hüffmeier, J. & Loschelder, D.D. 2013. Promoting prevention success at the bargaining table: Regulatory focus in distributive negotiations, *Journal of Economic Psychology*, 38: 26–39.

Twomey, B. 2010. Study assesses Asia-Pacific offshore decommissioning costs, *Oil and Gas Journal*, 15.

Wan Zahari, W.Y. & Nasir, M.D. 2002. *House Price Dynamics: Evidence from a Malaysia Case Study,* in International Real Estate Research Symposium, Kuala Lumpur, 297–310.

Weiss, D.J. 2014. *Trade Implication of US Energy Policy and the Export of Liquefied Natural Gas (LNG),* Center for American Progress.

Zawawi, N.A.W.A. & Liew, M.S. 2013. *Rig to reef scenario in Malaysia,* in Rigs-to-Reefs: Prospects in Southeast Asia, Singapore.

Advances in Civil, Architectural, Structural and Constructional Engineering – Kim, Jung & Seo (Eds)
© 2016 Taylor & Francis Group, London, ISBN 978-1-138-02849-4

Flood simulation using rainfall-runoff for Segamat River Basin

M.S. Adnan, E. Yuliarahmadila, C.A. Norfathiah & H. Kasmin
Micro Pollutant Research Centre, University Tun Hussein Onn Malaysia, Parit Raja, Johor, Malaysia

N. Rosly
Centre for Energy and Industrial Environment Studies, University Tun Hussein Onn Malaysia,
Parit Raja, Johor, Malaysia

ABSTRACT: Malaysia experienced two types of monsoon, which is North-East monsoon and South-West monsoon. During these monsoon seasons, Malaysia received abundant of rainfall and this has resulted in flood in certain areas, especially east coast and southern part of Peninsula Malaysia. This disaster has resulted in big losses either in term of facilities, infrastructures, money or life. Thus, a flood mitigation project must be carried out to reduce the losses. One of the most effective ways in managing and identifying the flood magnitude is by using modeling. Thus, this study was conducted with the aim to identify the flood area by using the rainfall-runoff model. InfoWorks RS was utilized as the software engine to model the flood. The required data such as rainfall, water level, flow rate and maps were retrieved from respective authorities and field works. The duration of 2 years to 200 years of return period were used to calculate and generate the rainfall pattern. The result clearly indicates that the study area is susceptible to flood if the ARI is used more than 25 years. The maximum water level is 11.94 m with the flow rate of 1325.68 m³/s. Calibration process was conducted to verify the result by comparing the real data from flood event in 2007. The difference between the measured and calculated results is 0.02%. It can be concluded that the result obtained can be acceptable.

1 INTRODUCTION

Flooding is a natural phenomenon that occurs almost every year in Malaysia, and this phenomenon is often triggered by the high intensity of rainfall. There are several factors that contribute to the flood occurrences such as natural and man-made factors. For the natural factor, it includes longer duration of time with high intensity of rainfall, the elevation or terrain type is flat and located adjacent to the river. In contrast, for the man-made factor, under-design capacity of hydraulic structure, blockage of the drainage system and uncontrolled land use plan have worsened the scenario. The study by Adib et al. (2011) has demonstrated that the uncontrolled land use activities lead to serious flood problems in Kota Tinggi.

The collection of hydrology data always becomes a major problem in Malaysia due to unavoided actions such as gauging instrument fault and improper data recording system. This has led to a major problem when involving a prediction of hydrological variables such as precipitation and river stages. Hydrological modeling becomes one of the important tasks for planning any water resources project since the hydrological phenomena are extremely complex (Solomatine & Khada 2003).

Since the precipitation is the main variable that is used to model the runoff, the discharge behavior along the river was calculated. This value was used to perform the analysis. The sequential data generation was utilized to estimate the required data. By using the rainfall-rainoff data, the flood modeling was performed. The specific objective of this study is to simulate the flood by using the rainfall-runoff model and to analyze the factor that contributes to this event.

2 METHODOLOGY

2.1 Study area

The study area is located at the southern part of Peninsula Malaysia. Segamat River Basin is located in the state of Johor with the catchment area size of approximately 777 km². Segamat River is one of the tributaries of the Muar River, which originate from Gunung Besar with 23 km long, 14 m wide, 14 m above mean sea level, and flows through the Segamat town (Liew & Teo 2014). Figure 1 shows the location of the study area.

2.2 Data collection

Prior to the modeling and simulation work, numerous data input is needed as the input data for the model. Rainfall and river flow data were obtained from the Department of Irrigation and Drainage Malaysia. All the hydrological data from the year 2001 to 2012 were used to derive the Intensity-Duration-Frequency (IDF) and the peak flow (Q_p) for the return period of 2, 5, 10, 20, 50 and 100 years. For the river cross-section, manual measurement was conducted combined with the data from the consultant.

For the land use, soil and elevation maps, data were obtained from Malaysia Centre for Geospatial Data Infrastructure (MaCGDI). ArcGIS Software is used to form digital topographic maps that are used for basic layer, developing TIN and analyzing the hydraulic data. Digital Elevation Model (DEM) and Triangular Irregular Networks (TIN) are created to present the ground surface topography or terrain in a digital or virtual platform. DEM is present in TIN to produce a 3D flood model for Segamat Town catchment (Figure 1).

2.3 Rainfall-runoff analysis

In the rainfall-runoff process, kinematic wave methods have been applied to describe the flow over planes. The characteristic equations can be solved analytically to simulate the outflow hydrograph in response to rainfall of a specified duration (Chow 1988). The kinematic wave model of the rainfall-runoff process offers the advantage over the UH methods, which is the solution of physical equation governing the surface flow in 1-D (Bell et al. 2001).

Rainfall analysis is used to obtain the intensity-duration-frequency (IDF) or depth-duration-frequency (DDF) for the return period (T_R) of time to produce rainfall patterns for the Segamat

telemetry station from 2001 to 2012. This analysis was performed by referring to the standard documented by DID (2012). The annual maximum precipitation for giving duration is selected by applying the methods outlined in a year for each year of historical record (Chaw 1988).

InfoWorks RS used the soil conservation service (US SCS boundary) for runoff analysis. Prior to simulation, runoff coefficient (CN value), rainfall pattern, rainfall depth, time of concentration, peak of time, lag time and initial value data should be calculated. Equations (1)–(3) give the SCS formula obtained at a time of concentration (t_c):

$$T_L = \frac{L^{0.8}\left[\left(\dfrac{1000}{CN}\right)-1\right]^{0.7}}{1900\, S^{0.5}} \tag{1}$$

$$T_p = 0.67 T_C \tag{2}$$

$$T_C = 1.667\, T_L \tag{3}$$

2.4 InfoWorks RS

A one-dimensional (1-D) steady and non-steady hydrodynamic model is applied to the hydrological and hydraulic modeling to the Segamat River by utilizing the 1-D hydrodynamic model that is widely used for flood simulation. These are based on conservation of the mass and momentum laws for shallow water to calculate free surface flow for steady and unsteady flows in open channels. The Saint Venant equation is used to describe the flow and water level variations in dimensional models (Chaw 1988). This software is selected due to numerous numbers of successful stories in modeling the flood (Goodarzi 2010).

3 RESULTS AND DISCUSSION

3.1 Hydrological analysis

In order to calculate the runoff, the SCS method is applied, as this method is one of the methods that is available in the Info Work RS database. Table 1 provides the summary of the data used.

For the value of Intensity-Duration-Frequency, the values for all the regions in Malaysia have been provided in the Malaysia Stormwater Manual (MSMA). Table 2 presents the result of IDF for various return periods and storm durations. In order to determine the rainfall pattern, the design rainfall depth is obtained from the IDF, and the value is shown in Figure 2. Stream flow data are calculated based on various return periods,

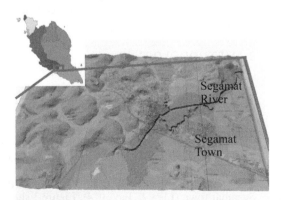

Figure 1. Study area located at the Segamat River.

as presented in Table 1. The calculated data are then imported into the InfoWorks RS database to estimate the flow magnitude as well the unit hydrograph graph. Hydrological data inclusive of river section, water level and rainfall data are also imported into InfoWorks RS for the calibration and verification purposes.

The temporal distribution of rainfall within the design storm is an important factor that affects the runoff volume, and the magnitude and timing of the peak discharge. Design rainfall temporal patterns are used to represent the typical variation of rainfall intensities during a typical storm burst. Standardization of temporal patterns allows standard design procedures to be adopted in flow calculation. Design rainfall pattern for 24 hours is used to determine the maximum and minimum of rainfall depth (Figure 2). Temporal patterns are used as the main data for the identification of rainfall depth and intended for use in hydrograph generation design storm. The form of the temporal pattern and runoff computation is closely interlinked (DID, 2012).

As presented in Table 2, the flow rate varies based on the ARI values. For the 100-year ARI, the lowest value of flow rate is 105.074 m^3/s, while the highest flow rate is 1152.479 m^3/s. To validate the model results, the observed data and calculated

data are compared. Based on the model results, the maximum of stream flow during the flood event is 1183.9 m^3/s.

Simulations are grouped into single networks by utilizing event data sets. Direct runoff hydrograph derived from the rating curve is used as the boundary input. Figure 3 shows the longitudinal cross section and vertical cross section for three selected chainage points that represent the upstream, middle and downstream cross section of the Segamat River. Based on Figure 3, the difference in elevation is identified, with the highest elevation being 5 m while the lowest elevation being 2 m respectively. In addition, this figure also shows that the right and left banks are flooded during the event where the water elevation is higher than the river bank. This result can identify and recognize the vulnerable areas or flood-prone area that can provide the important information to the local authority for flood mitigation projects.

Based on Figure 3, all the areas adjacent to the Segamat River in the township area are susceptible to flood since the water from the river flows over the bank to the nearby locations. This finding is in agreement with the previous study by Gasim et al., (2012), which indicated that the highest water level along the Segamat River had passed the river bank and floods the adjacent areas.

To verify the quality of produced results, the calibration processes are carried out by comparing the model result with the observed data. The observed data in 2007 are used to compare with the calculated result. Figure 4 shows the differences between the calculated and observed data. Based on this figure, the difference between the calculated and observed results is 2%. This result reflects the accuracy of the calculated result. This result can be accepted as the difference is small, as mentioned by Liew et al. (2009). The maximum tolerance of difference between the model and observed result is not more than 10%.

Table 1. Summary of the data used in this analysis.

Catchment	Area (acre)	Avg slope (%)	CN value	T_L (hr)	T_c (hr)	Δt	T_p (hr)
1	291.63	6.88	63.31	0.84	1.40	0.19	0.93
2	170.98	3.51	70.13	0.76	1.27	0.17	0.84
3	109.39	4.32	79.92	0.42	0.70	0.09	0.46
4	290.71	3.13	66.77	1.14	1.89	0.25	1.26
5	168.68	0.11	83.66	2.83	4.71	0.63	3.14
Average		3.59	72.76	1.20	1.99	0.27	1.33

Table 2. Maximum stream flow for various return periods and storm durations.

Return period	Storm duration (minutes)											Q_T (m^3/s)
	15	30	60	90	120	180	360	540	720	900	1440	
2	95.82	60.64	38.38	29.37	24.29	18.59	11.76	9.00	7.44	6.42	4.71	105.074
5	131.62	83.82	53.38	41.00	34.00	26.11	16.63	12.77	10.59	9.16	6.74	350.402
10	147.64	93.76	59.54	45.66	37.82	29.00	18.41	14.12	11.69	10.10	7.43	535.985
20	179.07	114.52	73.23	56.38	46.83	36.05	23.06	17.75	14.74	12.77	9.43	721.568
50	209.33	134.14	85.96	66.26	55.09	42.46	27.21	20.97	17.44	15.11	11.17	966.896
100	231.50	148.46	95.20	73.41	61.05	47.08	30.19	23.28	19.36	16.78	12.41	1152.479
200	253.67	162.79	104.46	80.59	67.03	51.71	33.18	25.60	21.30	18.46	13.67	1338.062

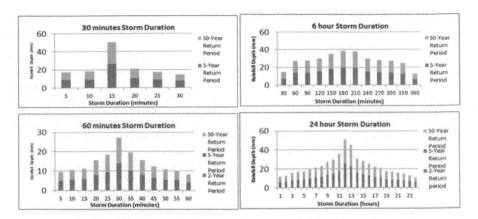

Figure 2. Rainfall pattern for various storm durations and return periods (MSMA).

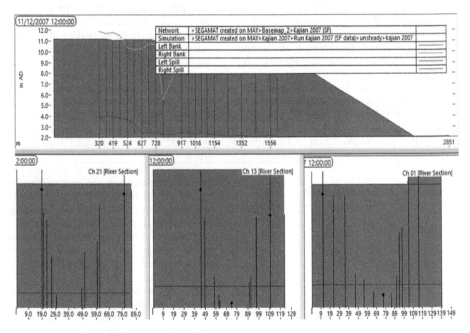

Figure 3. Water elevation at several chainage points.

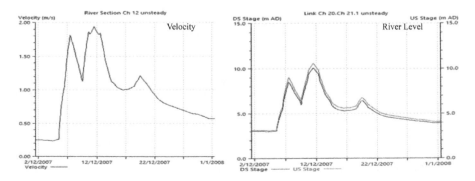

Figure 4. Comparison between the calculated and observed results.

4 CONCLUSIONS

In this study, the rainfall-runoff analysis was conducted to determine the peak discharge. The temporal pattern was then performed to determine the design hydrograph value. All the values were used in the main data input. Water elevation and magnitude of the flow rate were calculated. Based on the model results, the maximum flow rate was 1152 m³/s. While the maximum observed flow rate was 1183.9 m³/s. These values were used to verify the quality of the model. It was observed that the difference between the calculated and observed results was 2%. Furthermore, the flood area could be identified, which could be an important additional information for local authority to design the plan for the mitigation work in this area.

ACKNOWLEDGMENT

The authors would like to express the gratitude to the Ministry of Education Malaysia and Universiti Tun Hussein Onn Malaysia for funding this research under research grant vot RAGS R017.

REFERENCES

Adib, M.R.M., Saifullizan, M.B., Wardah, T., Rokiah, D. & Junaidah, A. 2011. Flood Inundation Modeling for Kota Tinggi Catchment by Combination of 2D Hydrodynamic Model and Flood Mapping Approach. *Lowland Technology International*, 13(1): 27–35.

Bell, V.A., Carrington, D.S. & Moore, R.J. 2001. *Comparison of rainfall-runoff models for flood forecasting. Part 2: Calibration and evaluation of models.* Bristol, UK, Environment Agency, 239. (R&D Technical Report W242, CEH Project Number: C00260).

Chow, V.T., David, R.M. & Larry, W.M. 1988. *Applied Hydrology*. McGRAW-Hill International Editions.

Gasim, M.B., Surif, S.M.M., Toriman, M.E.A., Rahim, S. & Chong, H.E. 2010. Flood Analysis December 2006: Interest in Urban Areas Segamat, *Johor, Sains Malaysia*, 39(3): 253–361.

Goodarzi, M.S. 2010. *Flood Mapping of Sungai Skudai using InfoWorks RS*. Universiti Teknologi Malaysia: Thesis Master of Engineering.

Liew, Y.S. & Zalilah S. 2009. *The development of Two Dimensional Flood Risk Map for Muar River*. Publisher NAHRIM—River Research Center.

Solomatine, D.P & Khada, N.D. 2003. Model Trees as an Alternative to Neural Networks in Rainfall-Runoff Modeling. *Hydrological Science Journal*. 48(3): 399–411.

Advances in Civil, Architectural, Structural and Constructional Engineering – Kim, Jung & Seo (Eds)
© *2016 Taylor & Francis Group, London, ISBN 978-1-138-02849-4*

Analysis on the effect of Landsat NDVI by atmospheric correction methods

M.H. Lee, S.B. Lee & Y.D. Eo
Department of Advanced Technology Fusion, Konkuk University, Seoul, South Korea

M.W. Pyeon & K.I. Moon
Department of Civil Engineering, Konkuk University, Seoul, South Korea

S.H. Han
Department of Geoinformatics Engineering, Kyungil University, Gyeongsan, South Korea

ABSTRACT: The atmospheric correction types can be divided into physical-based and image-based. In the case of physical-based, biophysical analysis can be difficult because the atmospheric conditions of the exact time when the image is acquired are also needed. This paper shows NDVI results from Landsat images that were corrected by various atmospheric correction methods. The calculated NDVIs in each of the atmospheric correction methods were compared and analyzed. Experimental results 6S indicated the smallest standard deviation of 0.16. The result from DOS is similar to the 6S. The DOS method can be applied to future utilization of the time-series NDVI mapping.

1 INTRODUCTION

For climate change and environmental monitoring, we used precise sensors in a local area and the satellite images in landuse/landcover and vegetation data analysis of large areas. The VI (Vegetation Index) is the ratios of bands designed to show properties of vegetation. Among more than 150 VIs, the NDVI (Normalized Difference Vegetation Index) is most popular for many vegetation applications. Especially, NDVI is used in the urbanization index, urban green, and agriculture surveys. NDVI time-series analysis and multisensory integration are essential to improve diverse atmospheric science models related to the continuous monitoring of information related to forests (Seong 2000). Landsat images with medium resolution have been provided since 1972 and have a high result reliability because the images have been utilized within a variety of research fields. This paper analyzed the effect of Landsat NDVI by atmospheric methods. The experiment included the seasonal and temporal Landsat NDVI. The experimental result may be applied for future applicability of atmospheric correction as preprocessing.

2 DATA AND METHODOLOGY

2.1 Study site and image data

Recently, Suwon city has been developed with a view toward economy and housing. The area around Suwon was selected as a study site to determine the landuse/landcover changes from 1994 to 2003. Landsat-5 TM and Landsat-7 ETM+ with level 1T and a small amount of cloud cover were acquired from the USGS (http://earthexplorer.usgs.gov/). The details of the satellite image used in this study are provided in Table 1.

2.2 Methodology

The pixel data of an image contains information about soil and vegetation. In the case of information about vegetation, NDVI is widely used, as shown in Equation (1). NDVI infers vegetation vitality by using the difference in energy reflected from the vegetation in red and NIR wavelength (Jensen 2000):

$$NDVI = \frac{NIR - Red}{NIR + Red} \tag{1}$$

Table 1. Satellite image for the experiment.

Date	Satellite	Sensor	Cloud cover (%)
April, 1996	Landsat 5	TM	0.0
Sep., 1994			0.0
April, 2003	Landsat 7	ETM+	0.0
Sep., 2001			1.0

where *NIR* is the near-Infrared wavelength band and *Red* is the Red wavelength band.

Before reaching the sensor, the energy reflected from the surface is attenuated by atmospheric materials such as gas and aerosol (Lee et al. 2015). Atmospheric correction is needed to guarantee consistency in results by correcting the attenuated energy. The TOA (Top of Atmosphere), FLAASH (Fast Line-of-sight Atmospheric Analysis of Spectral Hypercubes), DOS (Dark Object Subtraction) and 6S (Second Simulation of the Satellite Signal in the Solar Spectrum) were considered for the Landsat images. TOA reflectance is the reflectance measured by space-based sensors above the earth's atmosphere. TOA reflectance does not take into account the energy attenuated from the atmosphere. FLAASH and 6S are modules that correct the energy attenuated from the atmosphere by using a physics equation.

FLAASH is based on the MODTRAN4 (MODerate resolution atmospheric TRANsmission) and used from ENVI software (FLAASH 2007). 6S is the module composed of the Fortran code and uses the SOS (Successive Order of Scattering) algorithm (Vermote et al. 1997). Exact atmospheric values are needed in the case of the atmospheric module with the physics equation and acquired from the same time with a satellite image. The exact atmospheric values with the same acquisition time of satellite image are required. For this reason, a DOS-based image on the information of the pixel in the image is proposed (Chavez 1996). In this study, a DOS tool called 'Dark Subtraction' in ENVI was used.

3 EXPERIMENTAL RESULTS

Landsat images were corrected by using the FLAASH, 6S and DOS as atmospheric correction methods, and NDVI was extracted from the corrected images. After that, the difference of each NDVI in the images was compared. Each result image is shown in Figure 1, and the average of NDVI in each image is presented in Table 2.

Based on the results of the comparison with the NDVI extracted from the TOA reflectance, the highest improved NDVI image is the DOS image on 6 April 2003, with differences of 0.198. In the case of the NDVI image utilizing the 6S method, the best reliability was 0.16 of standard deviation. All of the average NDVIs applied to atmospheric correction are higher than those not applied. There was no significant difference in NDVI results extracted from the atmospheric corrected images. Figure 1 shows the difference in NDVI according to the seasonal changes and the atmospheric correction method.

There were little wider areas of vegetation in the image during September as opposed to the images of April. Because the change in NDVI was low in the 1990s and 2000s, urban development took

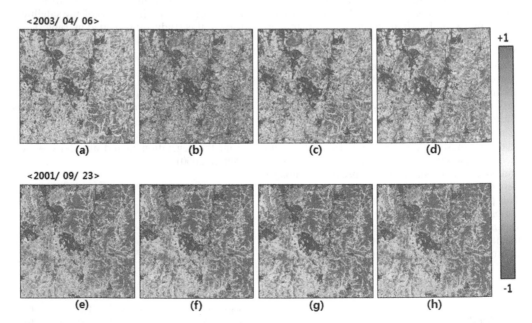

Figure 1. NDVI images applied the various atmospheric correction methods. (a), (e) TOA; (b), (f) FLAASH; (c), (g) 6S; and (d), (h) DOS.

Table 2. NDVI of various atmospheric correction methods.

Satellite	Date	Atmospheric correction methods			
		TOA	FLAASH	6S	DOS
L-5	1996-04-10	0.172	0.239	0.238	0.352
L-7	2003-04-06	0.162	0.280	0.213	0.523
L-5	1994-09-28	0.523	0.627	0.660	0.360
L-7	2001-09-23	0.524	0.614	0.658	0.626

place in existing city areas, not in forest or vegetation areas.

4 CONCLUSIONS

This study compared changes in NDVI by using atmospheric correction methods and seasonal differences in the study area from the 1990s and the 2000s. The result indicated that atmospheric correction affects NDVI computation. NDVI utilizing the 6S method gives the smallest standard deviation of 0.16 than that utilizing the other methods. None of the three methods made a big difference. Therefore, when extracting the NDVI, DOS also shows the possibility for use in future time-series mapping.

ACKNOWLEDGMENT

This work was financially supported by Korea Ministry of Land, Infrastructure and Transport (MOLIT) as "U-City Master and Doctor Course Grant Program".

REFERENCES

Chavez, P.S. 1996. Image-based atmospheric corrections-revisited and improved. *Photogrammetric engineering and remote sensing,* 62(9): 1025–1035.
FLAASH. 2007. *Atmospheric correction module: QUAC and FLAASH user's guide,* version 4.7. ITT visual Information Solutions Inc., Boulder, Co.
Jensen, J.R. 2000. *Remote Sensing Environment,* Prentice hall.
Lee, S.B., La, P.H., Eo, Y.D. & Pyeon, M.W. 2015. Generation of simulated image from atmospheric corrected Landsat TM images. *Journal of the Korean Society of Surveying, Geodesy, Photogrammetry and Cartography,* 33(1): 1–9.
Seong, J.C. 2000. Characteristics and application of large-area multi-temporal remote sensing data. *Korean Journal of Remote Sensing,* 16(1): 1–11.
Vermote, E.F., Tanré, D., Deuzé, J.L., Herman, M. & Morcette, J.J. 1997. Second simulation of the satellite signal in the solar Spectrum (6S): An overview. *Geoscience and Remote Sensing. IEEE Transactions on,* 35(3): 675–686.

Advances in Civil, Architectural, Structural and Constructional Engineering – Kim, Jung & Seo (Eds)
© 2016 Taylor & Francis Group, London, ISBN 978-1-138-02849-4

Water–to–binder ratios for the equivalent compressive strength of concrete with Fly Ash

H.S. Yoon & K.H. Yang
Department of Plant and Architectural Engineering, Kyonggi University, Suwon-si, Gyeonggi-do, South Korea

ABSTRACT: This study proposed a k-value as an index to determine the water–to–binder ratio (W/B) of concrete with Fly Ash (FA) with regard to the equivalent value to the compressive strength of cement concrete. The k-value was derived from the nonlinear exponential relationship between W/B and compressive strength of concrete. The experimental constants in the relationships were obtained from the regression analysis using 6160 concrete mixes compiled from the different laboratory and plant tests. The k-value for FA concrete commonly decreased as the weight ratio of FA as a partial replacement of cement decreased. This indicates that a lower W/B is required for FA concrete to obtain the equivalent compressive strength of cement concrete.

1 INTODUCTION

The reduction of Greenhouse Gas (GHS) emissions is one of the hottest issues all over the world. It is generally estimated that the amount of GHS emitted from the worldwide production of Ordinary Portland Cement (OPC) corresponds to approximately 7% of the total GHS emissions into the Earth's atmosphere (Malhotra 2002). For these reasons, a large proportion of the concrete industry has made a great effort to reduce the consumption of OPC in concrete production. Fly Ash (FA) and Ground Granulated Blast-Furnace Slag (GGBS) are practically useful as a partial replacement of OPC due to their sound environmental performance such as the recycling of by-product materials, low carbon dioxide emissions and low energy consumption. As a result, the replacement level of OPC by the addition of FA and GGBS have gradually increased in concrete production.

It has been also generally recognized (Lee et al. 2013) that the addition of FA is unfavorable to compressive strength development of concrete, especially at an early age, due to a pozzolanic reaction with calcium hydroxide. This implies that a lower water water–to–binder ratio (W/B) would be required for FA concrete to obtain an equivalent value to the compressive strength of OPC concrete. CEN/TR 16339 provision (2014) suggests utilizing k-values to determine the W/B of FA concrete with regard to the equivalent strength to OPC concrete. However, the provision assumes the relationship of compressive strength of concrete and W/B to be a linearity. This results in an unreality solution for k-value as W/B exceeds 80%. Furthermore, the k-values in the provision is determined using very limited concrete mixes with the OPC replacement level by FA (R_F) not exceeding 0.3. Therefore, k-value for FA concrete needs to be examined further for different the OPC replacement levels using extensive test data.

The present study proposed k-values to determine the W/B of FA concrete with regard to equivalent value to the compressive strength of OPC concrete. The equation for k-values were derived from the nonlinear exponential relationship between W/B and compressive strength of concrete. The experimental constants in the relationships for the concrete with each R_F were obtained from the regression analysis using 6160 concrete mixes compiled from the different laboratory and plant tests. Ultimately, a simple equation was formulated to straightforwardly obtain the k-value for a given R_F.

2 FORMULATION OF K-VALUE

The most critical parameter influencing the compressive strength of concrete is W/B. Thus, from the relationship between W/B and compressive strength of concrete, the k-value to obtain the compressive strength of FA concrete equivalent to OPC concrete can be expressed as follows (CEN/TR 16639):

$$\omega_o = w_a/(c_a + ka) \tag{1}$$

$$k = (w_a/\omega_o - c_a)/a \tag{2}$$

where, ω_o is the water–to–cement ratio of OPC concrete, w_a is the unit water weight of FA concrete, and c_a and a are the unit weights of cement and FA, respectively.

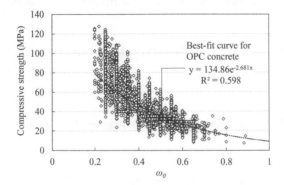

Figure 1. Effect of W/C on compressive strength of OPC concrete.

It is widely accepted (Han et al. 2004) that compressive strength of concrete decreased with increasing in W/B. The decreasing rate in compressive strength is close to nonlinearity rather than linearity, as shown in Figure 1. Hence, the relationship of W/B and compressive strength of OPC concrete ($f'_{c(o)}$) can be formulated as follows:

$$f'_{c(o)} = A_o \exp(B_o w_o) \tag{3}$$

where, A_o and B_o are the experimental constants to account for the decreasing rate in compressive strength. Substituting Equation (1) into Equation (3), the following relationship of W/B and compressive strength of FA concrete ($f'_{c(a)}$) can be obtained:

$$f'_{c(a)} = A_a \exp\left(B_a \frac{w_a}{c_a + ka}\right) \tag{4}$$

where, A_a and B_a are experimental constants. According to the equivalent strength definition, $f'_{c(a)}$ should be equal to $f'_{c(o)}$, regardless of R_F. Thus, from Equations (3) and (4), k-value can be formulated as follows:

$$k = \frac{(c_a + a)(\ln A_o - \ln A_a + B_o \omega_o)}{B_a \omega_o a} - \frac{c_a}{a} \tag{5}$$

3 EXPERIMENTAL CONSTANTS IN EQUATION (5)

3.1 Database

To determine the experimental constants (A_o, B_o, A_a, and B_a) in Equation (5), an comprehensive database including 6160 concrete mixes were analyzed using the generalized least square approach. The experimental data were compiled from different sources including laboratory tests and ready-mixed concrete plant tests. In the database, the range of primarily parameters is as follows (Table 1): compressive strength between 7.67 and 136 MPa, W/B between 13.8 and 88.9%, and R_F between 2 and 66%. The compressive strength of concrete tended to decrease with the increase in R_F.

3.2 $f'_{c(o)}$–W/C Relationship (for OPC Concrete)

Figure 1 shows $f'_{c(o)}$–W/C relationship for OPC concrete. On the same figure, the best fit curve determined from the test data is also plotted. Because of the wide variety of data sources and limited available data such as the size and grading of aggregates, and loading rate, the scatter of data against the fitting line is yielded. However, the typical trend on $f'_{c(o)}$–W/C relationship for OPC concrete can be obtained as follows:

$$f'_{c(o)} = 134.9 \exp(-2.68 \omega_o) \tag{6}$$

Overall, the experimental constants A_0 and B_0 are determined to be 134.9 and -2.68, respectively.

3.3 $f'_{c(a)}$–W/B Relationship (for FA Concrete)

Figure 2 shows $f'_{c(a)}$–W/B relationship for FA concrete with different R_F. Considering the distribution of test data and their reliability, R_F is selected to be between 20 and 50%. The experimental constants A_a and B_a determined from the best-fit curves are summarized in Table 2. When RF increases from 20% to 50%, the value of A_a increased from 136 to 142, whereas B_a value decreased from -3.2 to -3.8.

Table 1. Incidence of various parameter values in 6160 concrete mixes.

Type of binder	Range of f'_c (MPa)			
	3–30	30–60	60–100	100–170
OPC	1278	1636	711	119
OPC + FA	1769	539	108	–
Type of binder	Range of W/B (%)			
	13–20	20–40	40–60	60–90
OPC	20	1671	1745	308
OPC + FA	2	581	1538	295
Type of binder	Range of R_F (%)			
	2–20	20–40	40–60	60–90
OPC + FA	678	1625	24	6

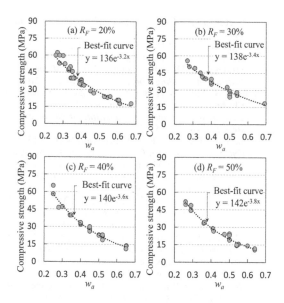

Figure 2. Effect of W/B on compressive strength of concrete with different R_F.

Table 2. Determination of experimental constants A_a and B_a for FA concrete.

Experimental constants	Range of R_F (%)			
	20	30	40	50
A_a	136	138	140	142
B_a	−3.2	−3.4	−3.6	−3.8

4 DETERMINATION OF K-VALUE

The experimental constants (A_o, B_o, A_a, and B_a) obtained in the previous section were inputted into Equation (5) and then k-values for FA concrete were calculated (Table 3). The calculated k-values ranged between 0 and 1. This indicates that W/B of FA concrete varied from 0 to 100%. The k-value for FA concrete tended to decrease as the W/B increased, as shown in Figure 3. When R_F increased from 20% to 50%, the k-values for W/B of 0.2 increased from 0.295 to 0.530 and those for W/B of 0.7 increased from 0.213 to 0.451. The increasing rate in k-values against R_F was insignificantly affected by W/B. At the same W/B, the k-value increased with the increase in R_F. This indicates that a lower W/B for FA concrete is required with the increase in R_F to obtain the equivalent strength to the OPC concrete.

Figure 3 demonstrates that the main parameters on the k-value for FA concrete are R_F and ω_o. To straightforwardly determine the k-values for FA concrete, rational equation was formulated using

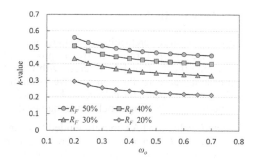

Figure 3. Effect of W/B on the k-value of FA concrete.

Table 3. k-values determined from concrete mixes with different R_F.

ω_o	k-value			
	$R_F = 20\%$	$R_F = 30\%$	$R_F = 40\%$	$R_F = 50\%$
0.2	0.295	0.433	0.510	0.560
0.25	0.272	0.405	0.479	0.530
0.3	0.257	0.385	0.459	0.510
0.35	0.246	0.372	0.445	0.495
0.4	0.238	0.361	0.434	0.484
0.45	0.231	0.353	0.425	0.476
0.5	0.226	0.347	0.418	0.469
0.55	0.222	0.342	0.413	0.463
0.6	0.218	0.337	0.408	0.459
0.65	0.215	0.334	0.404	0.455
0.7	0.213	0.330	0.401	0.451

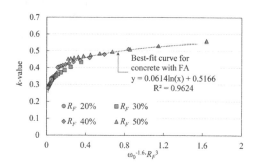

Figure 4. Modeling of k-value for FA concrete.

non-linear multiple regression analysis of datasets given in Table 3.

To establish basic models, the two parameters were combined and tuned repeatedly by trial-and-error approach until a relatively acceptable correlation coefficient (R^2) was obtained. In summary, a simple equation for k-value of FA concrete can be obtained as follows (Figure 4):

$$k = 0.06 \ln(\omega_o^{-1.6} R_F^3) + 0.5 \qquad (7)$$

5 CONCLUSION

Based on 6160 concrete mixes, k-values for FA concrete were simply formulated to straightforwardly determine the W/B of FA concrete with regard to equivalent value to the compressive strength of OPC concrete. From the calculated k-values for FA concrete, the following conclusions may be drawn.

1. The k-value for FA concrete tended to decrease as the W/B increased. The increasing rate in k-values against R_F was insignificantly affected by W/B.
2. At the same W/B, the k-value increased with the increase in R_F.
3. The k-value for FA concrete was simply formulated as a function of the two parameters R_F and ω_o.

ACKNOWLEDGEMENTS

This research was supported by a grant (12CCTI-C063722-01) from Construction Technology Innovation Program (CTIP) funded by Ministry of Land, Transport and Maritime Affairs of Korean government.

REFERENCES

CEN/TR 16639. 2014. *Use of k-value Concept*, Equivalent Concrete Performance Concept and Equivalent Performance of Combinations Concept. CEN/TC104.

Han, C.G., Hwang, Y.S., Lee, S.H. & Kim, G.D. 2004. Properties of Strength Development of Concrete at Early Age with Water Cement Ratio and Cement Factor. *Journal of Architectural Institute of Korea*. 20(4): 72–79.

Jaung, J.D., Cho, H.D. & Park, S.W. 2012. Properties of Hydration of High-Strength Concrete and Reduction Strategy for Heat Production. *Journal of the Korea Institute of Building Construction*. 12(2): 203–210.

Jung, Y.B., Yang, K.H. & Choi, D.U. 2014. Influence of Fly Ash on Life-Cycle Environmental Impact of Concrete. *Journal of the Korea Institute of Building Construction*. 14(6): 515–522.

Kim, T.H., Tae, S.H., Roh, S.J. & Kim, N.H. 2013. Life Cycle Assessment for Carbon Emission Impact Analysis of Concrete Mixing Ground Granulated Blast-furnace Slag (GGBS). *Journal of the Architectural Institute of Korea*. 29(10): 75–82.

Lee, J.H., Kim, Y.R., Park, J.H. & Jeong, Y. 2013. Study on the Mineral Admixture Replacement Ratio for Field Application of Concrete with High Volume Mineral Admixture. *Journal of the Korean Recycled Construction Resources Institute*. 1(2): 93–100.

Lee, S.S., Song, H.Y. & Lee, S.M. 2009. An Experimental Study on the Influence of High Fineness Fly Ash and Water-Binder Ratio on Properties of Concrete. *Journal of the Korea Concrete Institute*. 21(1): 29–35.

Lee, Y.J., Shin, S.Y. & Kim. Y.S. 2013. A Study of Compressive Strength Property of Mortar with Fly Ash using Water Eluted from Recycled Coarse Aggregates. *Journal of the Architectural Institute of Korea*. 29(3): 89–96.

Malhotra, V.M. 2002. Introduction: sustainable development and concrete technology. *Concrete International*. 24(22).

Moon, J.H. & Lee, S.S. 2013. Dynamic and Durability Properties of the Low-carbon Concrete using the High Volume Slag. *Journal of the Korea Institute of Building Construction* 13(4): 351–359.

Yang, K.H., Seo, E.A., Jung, Y.B. & Tae, S.H. 2014. Effect of Ground Granulated Blast-Furnace Slag on Life-Cycle Environmental impact of Concrete. *Journal of the Korea Concrete Institute*. 26(3): 13–21.

Advances in Civil, Architectural, Structural and Constructional Engineering – Kim, Jung & Seo (Eds)
© 2016 Taylor & Francis Group, London, ISBN 978-1-138-02849-4

Design of Information-Measuring and Control Systems for intelligent buildings: Areas for further development

I.Yu. Petrova, V.M. Zaripova, Yu.A. Lezhnina & V.Ya. Svintsov
Astrakhan Civil Engineering Institute, Astrakhan, Russia

ABSTRACT: The article describes the current requirements for integrated control systems for smart-houses. The hierarchical classification of building automation system is proposed. The classification shows different levels of information transfer. The development trends of Information-Measuring and Control Systems for intelligent buildings were reviewed. The Complex structure of information-measuring and control subsystems (IMCS) for the smart home elaborated. Authors used Energy-informational model of the knowledge base on physical and technical effects to develop a computer-aided conceptual design system to devise elements of building information-measuring and control systems. Such approach allows to expand the expert knowledge base in dozens of times and to reduce the designing period in two or three times by choosing the more efficient options and built-in calculation of the significant characteristics of the design conceptual models, which greatly reduces the amount of prototyping and field testing.

1 BASIC DEFINITIONS

The concept of "intelligent building" was formulated by the Institute of intelligent buildings in Washington in the 70-ies of the last century as "the building, ensuring productive and efficient use of space thanks to the optimization of its four basic elements: structure, systems, services and management, as well as the relationship between them."

In intelligent buildings additional facilities for residents are created through an integrated information-measuring and control systems (BMS—building management system), in which we can distinguish 3 main components: variety of sensors and actuators; telecommunication network, transmitting this information to the center; control system, processing the information and forming solutions.

"Smart (intelligent) home environment" is defined as the physical infrastructure (sensors, actuators and network), that allow to operate in ambient intelligence (Nakashima et al. 2010, Badica et al. 2013). In addition, now the term "domotics" is often using. It is the mix of the Latin word "domus" (house) and modern terms: informatics, telematics and robotics i.e. DOMus infOrmaTICS. In the book (Frances 203) this term is defined as "the Integration of technologies and services for better quality of life" and 5 levels of home automation is chosen. There are:

Homes contain intelligent objects—homes contain one or more objects operating with intelligence features.
Homes contain intelligent objects, communicating with each other—homes contain technical objects and devices which function intelligently and can transmit information each other that extend their functionality.

The connected homes—a home have an internal and external network that allows create interactive and remote control of building systems and provides access to information services both within the house and outside.

Learning homes—recurring management scenarios (patterns) in homes are recorded and stored in knowledge bases to manage the BMS, and then the accumulated data are used to predict the needs of users and monitoring the situation.

Adaptable (attentive) homes—the activity and location of people and objects in such homes are constantly registered, and this information is used to control of the home subsystems anticipating the needs of the residents.

This classification of home automation levels allows allocate different levels of information transmission within and outside the home as well as self-learning and adaptive building management systems. The classification is hierarchical: each level has wider functionality and richer opportunities for residents to delegate some management functions to an automated system. This transfer of control increases the potential of a smart home functionality. The BMS itself can solve an increasing range of problems relating to the comfort of residents, their convenience, safety, and entertainment.

The diversity of needs of the resident's safety, comfortable and clean atmosphere, efficient use of energy and water resources, the availability of

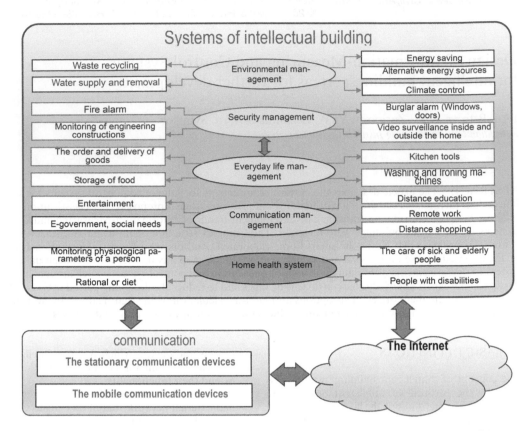

Figure 1. Complex structure of Information-Measuring and Control Subsystems (IMCS) for the smart home.

means of communication and entertainment, and ease of management of this complex mix, lead to a complex structure of home Information-Measuring and Control Systems (IMCS) shown in Figure 1. Such system may use different technologies with different communication protocols in different layers.

The importance of implementation of building automation is stressed by the following factors. According to the Association for the Efficient and Environmentally Friendly Use of Energy (ASUE) 41% of the total worldwide primary energy is consumed by buildings. About 85% of that is used for heating or cooling of rooms and 15% for lighting.

According to research of consulting firm Markets and Markets (USA) the global market of equipment for intelligent buildings in the period 2015 to 2020 will increase annually by 17% (20,38 billion in 2014 to 58,68 in 2020) (Marketsandmarkets 2015).

The high cost of automation can be reduced by taking the modular structure of subsystems of a smart home, which will allow people to choose necessary functions (safety, multiroom, lighting, etc.). Market research on smart homes

in the USA showed that development is possible by analogy with the automotive industry, where the buyer are formed additional to the basic functions of the car based on ready-made modules. An important requirement is that the system should be flexible, easily upgradable to extend their functionality and meet the needs of residents.

2 PROSPECTS OF DEVELOPMENT OF HOME SYSTEMS

Modern technologies of microelectronics, telecommunications and information systems make a significant contribution to the development of intelligent home systems. Immediate prospects include:

– Design based on open automation protocols, including software for easy project management, automatic configuration of the system.
– Integrated communication capabilities of all devices in the home including all possible network and data transfer environment, the necessary gateways and converters.
– Reliable, durable and affordable touch device (for example, to detect gases, odors, tempera-

ture, humidity, pressure, flow, etc.) and actuators (e.g., switches, actuators, valves) for testing the response of the system and decentralized energy supply of all its components.

- Simple, sustainable, distributed operating system relevant opportunities enter information and data visualization (keyboard input, touch screen, voice input/output information, mental control, etc.).
- A variety of compact intelligent device connected to the home network (e.g., smart electric socket or door washing machine, home robots or pocket computer with satellite Navigator) with the possibility of communication with the home network and global communications systems.
- Software development for distributed applications (for example, Jini Network Technology), tool enabling communication of any electronic device in the house with the Internet.
- Currently, there are three main types of user interaction with the home network (Home Area Network—HAN):
- Stationary communication device (e.g., touch screens, keyboards, and button-type devices)
- Mobile communication devices (smartphone, tablet, etc.)
- Wearable devices (smart bracelets, computerized watches, smart glasses, smart clothing).

The prospect of the near future—the implantable bioelectronics, which will allow for individual diagnostics of human health, to control the metabolism and to transmit data to the relatives or to the doctors.

3 PROSPECTS FOR THE DEVELOPMENT OF SENSORS FOR SMART BUILDINGS

Intelligent building must include a large number of information points (sensors), which receives data on the condition of the equipment and the environment. According to the norms of the US and the EU, information points should be at least 15,000. The development of the vital areas of "Internet of Things" (IoT) leads to a further increase in the need for a variety of sensors. According to the research of IDC firm, the global IoT market will grow from $655.8 billion in 2014 to $1.7 trillion in 2020. The IDC predicts that the number of "IoT endpoints," connected devices such as cars, refrigerators and everything in between, will grow from 10.3 billion in 2014 to more than 29.5 billion in 2020 (Norton 2015).

Thus, modern buildings require many types of sensors. Home monitoring requires multiple sensors, including temperature sensor, humidity sensor, water-level or dust count sensor, air velocity detector, cooking alarm, physical property sensors, gas detector, oven flame sensor, etc. Home security requires temperature sensors, heat radiation sensors, smoke alarm, fire gas detector,

seismic sensors, watts transducer, overheat sensor, vibration sensor, doors and windows ultrasound or optic sensors, biometric facial and fingerprint recognition systems and voice identification system Energy control requires sensors to measure the power, voltage, current, gas or water flow, temperature, water level, solar radiation, and lighting. Remote monitoring and diagnostic systems require the measurement and transmission via communication channels many types of signals for remote control of various home systems and the implementation of telemedicine functionality and care of the disabled or elderly.

In (Gassmann & Meixner 2001, Petrova et al. 2014) the classification of physical effects and phenomena (mechanical, thermal, electrical, magnetic, optical and chemical) is shown that can be used to create systems of management of intelligent buildings. The sensors, installed in control systems for smart buildings must meet the following requirements:

1. Performance specifications of sensors, such as accuracy, linearity, high reliability and long life cycle, low energy consumption, and the absence of service requirements. New smell—or voice-sensors and speech synthesis devices give opportunities for intelligent household appliances with a remote control.
2. Smart sensors enable the integration of primary conversion of the signal and its further process. Intelligent sensor is a sensor with integrated electronics (analog-to-digital Converter, a microprocessor, a digital signal processor, system on chip, etc.), as well as the implementation of a digital interface and a network communication protocols. Smart sensor has the possibility of harmonizing the measuring path with the signal source sensitivity, dynamic range, and selectivity and noise rejection. It adapts its parameters to the external factors and conditions. Microprocessor and sensor form a single integrated device can significantly improve the speed of processing sensor signal. In addition, the integration of sensors and electronic circuits processing reduces the cost of such devices.
3. Multi-sensor systems allow to combine multiple primary converters in a single technology platform with common software. This allows automatically perform many different measurements, to process and obtain a comprehensive result. For example, multi-sensor fire detectors make decision on alarm situation on base of complex factors such as smoke, carbon dioxide, the flame and the temperature rises.
4. Miniaturization of sensors. The requirements of high-tech sectors, such as aerospace engineering, medicine, and entertainment industry, led to a rapid reduction of the dimensions and

weight of the sensors. Typical dimensions of the individual elements are in the range from 1 micrometer to 100 micrometers. The use of nanotechnology in the near future will open access to an even smaller size.

5. Microelectromechanical System (MEMS) is a device that combines microelectronic and micromechanical components. Microminiaturization sensors contributed to the further integration of sensors and control micromachines (actuators) on a common substrate. Modern MEMS technology of integrated circuits are formed on a single substrate sensor, actuators, and control circuits with the dimensions of the elements from a few hundred to a few microns.

6. Standardization of sensor interfaces. A sharp decline in the cost of microminiature sensors and actuators (actuators), associated with its mass production led to the emergence of so-called sensor networks (Growth 2009). Standardized interfaces are required to transmit information from complex sensors via the telecommunications system to the upper level systems to control the house. Standardization helps to make it easier to process information from various sensors. A growing is importance to standards for wireless technologies Bluetooth, WiFi and ZigBee.

Increased market demands are forcing manufacturers to intensively expand the range of commercially available sensors. However, even the leading firms in the field of production of sensitive elements using no more than 27–30% of the available Fund of physical phenomena that potentially could significantly expand the range of these devices (the data of the firm Endress & Hauser, industrial equipment (Endress 2004)).

Thus, creation of the automated systems conceptual design of sensor equipment, that would allow to unify and centralize the development process and provide the engineer with a powerful theoretical framework. Such systems allow to reduce the time and complexity of creating a new product, to quickly adapt to market changes, to involve the end user in the design process, many times to increase the amount of actively used knowledge in the training of engineers.

4 CONCEPTUAL DESIGN OF ELEMENTS OF THE INFORMATION-MEASURING AND CONTROL SYSTEMS ON THE BASIS OF ENERGY-INFORMATION METHOD

(Petrova et al. 2014) discusses a new technology of conceptual design of building management system elements. The technology is characterized by a higher productivity of designers and the increasing level of automation of the design invention process. In (Zaripova & Petrova 2014, Zaripova & Petrova 2015) it is proposed to use the energy-informational model of chains of different physical nature (EIMC) on the basis of the phenomenological equations of non-equilibrium thermodynamics. The model allows synthesis of sensors or sensing elements on basis of knowledge about physical effects of a certain nature (mechanical, thermal, electrical, etc.) and their structural decisions. From the point of view of the ontological approach any Physical Effect (PTE), connecting two chains of the i-th and j-th physical nature or parameter of the circuit of the i-th physical nature can be represented as tuples of type:

$$P = \{H_P, B_{i\,in}, B_{i\,out}, K, K_O, KM_P, \\ D_{i\,in}, D_{i\,out}, EX_{(n|1)N}\} \quad (1)$$

$$\Pi = \{H_\Pi, B_{i\,in}, B_{i\,out}, \Pi, \Pi_O, \Pi M_\Pi, \\ D_{i\,in}, D_{i\,out}, EX_{(n|1)N}\} \quad (2)$$

Tuples can be divided into two groups where the first group represents the description of the physico-technical effect: H_P, H_Π—the name PTE or parameter text value, $B_{i\,in}$—type of the input variable of i - the physical nature, $B_{j\,out}$—type of the output variable of j—the physical nature, K—ratio of PTE that reflects the dependence of the output values from input values (the simplest case is a linear relationship $B_{j\,out} = K_{ij} \cdot B_{i\,in}$), Π—parameter of circuit of the i-th physical nature (for example, $B_{i\,out} = \Pi_i \cdot B_{i\,in}$), K_o or Π_o—a text value that represents a textual description of the coefficient K_{ij} or parameter and the formula in the function of well-known physical constants of the material parameters and geometrical dimensions, KM_P—a mathematical model of the PTE, which specifies the factors influencing the functional relationship of physical values of input and output, for example the influence of the fields (takes the value 1 or 0), ΠM_Π—the mathematical model of the parameter (taking the value 1 or 0), $D_{i\,in}$, $D_{j\,out}$—the range of change of input and output values, for ensuring efficiency of chain requires observance of the rules of crossing the ranges of values of the output of each of the previous effect and input of each subsequent effect in the chain:

$$D_{i\,in(\nu)} \cap D_{i\,out(\nu-1)} \neq 0 \text{ and } D_{j\,out(\nu)} \cap D_{j\,in(\nu+1)} \neq 0 \quad (3)$$

The second group represents the set of operational characteristics (from 0 to 10), the set of these characteristics and their average numerical values are determined by a group of experts in a given subject area: $EX_{(n|1)N}$—variables for the calculation of the operational characteristics of the synthesized physical principle. If at least one kind of operating characteristics known to all PTE have been included in the synthesized chain of the physical principle of the device, it is possible to calculate these performance specifications for synthetic devices in General.

Necessary and sufficient condition for the synthesis of the principle of operation of the technical device is a complete coincidence of the output value of the previous effect with an input value of the next effect in the chain:

$$T = (P_{i1j1}, P_{i2j2}, ..., P_{iNjN} \mid P_{ikjk} \in DB \wedge j_k$$
$$= i_{k+1} \wedge B_{out\,jk} = B_{in\,jk+1}) \qquad (4)$$

The technical unit will be operable if the ranges of the corresponding values intersect.

Operational characteristics of the device are computable—if the performance characteristics are calculated for each effect in the chain:

$$E_{nT} = f(E_{nT}, (\forall P \in T)(\exists E_{nP})) \qquad (5)$$

On the basis of the logical model of PTE and the above expressions it is possible to calculate the minimum information that necessary for a successful procedure for the synthesis of a new sensor:

$$P = \{H_P, B_{iin}, B_{jout}, 1, 0, 0, (-\infty, +\infty),$$
$$(-\infty, +\infty), \{0\}, 0, 0, 0, 0, 0\} \qquad (6)$$

In the computer-aided system for the synthesis of new technical solutions, developed on the basis of the energy-information method of modeling are implemented dynamic graphical representation of the principle of PTE and dynamic graphical synthesis design on morphological characteristics. In addition, passports of PTE have a set of 10 performance characteristics (sensitivity, reliability, etc.) to evaluate the performance properties of the synthesized technical solutions (Zaripova & Petrova 2014, Zaripova & Petrova 2015).

5 CONCLUSIONS

The hierarchical classification of home automation levels is proposed. It allows allocate different levels of information transfer within and outside the home with level-by-level increasing in functionality and opportunities for the residents to delegate some management functions to an automated system. This increases the potential functionality of a smart home in services related to the comfort of residents, their convenience, security,

and entertainment. Tendencies of development of information and measuring and control systems for intelligent building are:

- Gradual transition to self-learning adaptive robotic systems and complexes;
- Transition to wireless mobile management systems that can be used both in newly constructed buildings and in old buildings without the high cost of reconstruction.

Energy-informational model of the knowledge base of physical and technical effects described in the article enables developing a system for the automated support of the stage of conceptual design of elements of information-measuring and control systems. Using this knowledge base allows to expand the amount of actively used expert knowledge in dozens of times and to reduce the designing cycle in two or three times by choosing the more efficient options calculation of their significant characteristics, which greatly reduces the amount of prototyping and field testing.

REFERENCES

Nakashima, H., Aghajan, H. & Augusto, J.C. 2010. *Handbook of Ambient Intelligence and Smart Environments.* New York: Springer.

Badica, C., Brezovan, M. & Badica, A. 2013. *An Overview of Smart Home Environments: Architectures, Technologies and Applications.* In Christos K. Georgiadis (ed.), Local Proceedings of the Sixth Balkan Conference in Informatics, Proc. Conf., Thessaloniki, Greece 19–21 September 2013: 78–86. New York: The Association for Computing Machinery.

Frances K.A. 2003. *Smart Homes: Past Present, and Future. In Richard Harper (ed.) Inside the Smart Home*: 17–39. London: Springer.

Markets and markets. 2015. Smart Homes Market by Product, Protocol and Technology, Service, and Geography—Trend and Forecast to 2020. http://www.marketsandmarkets.com/Market-Reports/smart-homes-and-assisted-living-advanced-technologie-and-global-market-121.html (accessed July 31, 2015).

Norton S. 2015. Internet of Things Market to Reach $1.7 Trillion by 2020: IDC. http://blogs.wsj.com/cio/2015/06/02/internet-of-things-market-to-reach-1-7-trillion-by-2020-idc/ (accessed July 31, 2015).

Gassmann, O. & Meixner, H. 2001. *Sensors in Intelligent Buildings.* Weinheim: Wiley-VCH.

Petrova, I. Zaripova, V. & Lezhnina, Yu. 2014. *Sensors for information-measuring and control systems for hi-tech building, In Michal Mokrys & Stefan Badura (ed.)*, Proceedings in Advanced Research in Scientific Areas, Proc. Intern. Conf., Zhilina, 1–4 Desember 2014: 336–342. Zhilina: EDIS—Publishing Institution of the University of Zilina.

Endress, K. 2004. The state and prospects of development of instrumentation for the process. *Industrial automation and controllers,* 1: 45–48.

Zaripova, V. & Petrova, I. 2014. *System of Conceptual Design Based on Energy-Informational Model in Progress in systems engineering. In Selvaraj H. at all (ed.)*, Progress in Systems Engineering, Proc. Intern. Conf., Las Vegas,19–21 August 2014: 365–373. Switzerland: Springer International Publishing.

Zaripova, V.M. & Petrova I.Yu. 2015. *Ontological Knowledge Base of Physical and Technical Effects for Conceptual Design of Sensors. In Kravets A. at all (ad.)* Creativity in Intelligent Technologies and Data Science, Proc. Intern. Conf. Volgograd, 15–17 September 2014: 224–237. Switzerland: Springer International Publishing.

Advances in Civil, Architectural, Structural and Constructional Engineering – Kim, Jung & Seo (Eds)
© 2016 Taylor & Francis Group, London, ISBN 978-1-138-02849-4

Study on measuring the thermal conductivity of EVA insulation material used under floors in residential buildings

Young-Sun Jeong, Kyoung-Woo Kim & Hae-Kwon Jung
Korea Institute of Civil Engineering and Building Technology, Goyang-Si, Korea

ABSTRACT: In Korea, a radiant floor heating system (Ondol) is generally used in residential buildings. Thermal insulation materials installed under floors can minimize heat loss and reduce noise and impact from the upper layers, and are therefore highly important building materials. In this study, the density and thermal conductivity of insulation materials made from 30 Ethylene Vinyl Acetate (EVA) materials used under flooring in residential buildings were measured. The heat flow measurement method was used in this study. The bottom surface of the insulation materials had protrusions to provide cushioning characteristics that reduce floor impact noise. We chose three type specimens of protruding shapes as well as a flat plate specimen for this experiment. Thermal conductivities for the different external shapes were measured. We also compared the thermal conductivity between single-layer and double-layered specimens. The measurement results showed that thermal conductivity increased proportionally as the density of EVA insulation materials was increased. The thermal conductivity of insulation materials with dot-shaped protrusions was the lowest at 0.0464 W/(m·K). The overlapping specimen with flat plate and dot-shaped-protrusion specimen had the lowest effective thermal conductivity.

1 INTRODUCTION

Efforts to improve the environmental performance of buildings by employing sustainable building technologies and building materials have been accelerating. Resources and energy consumed during the building life cycle are significantly large. Accordingly, it is very important to select materials that can minimize the adverse effect of buildings on the environment. Using appropriate insulation materials for buildings is the most basic method that can maximize energy reduction in buildings.

The climate in the Korean Peninsula has four distinctive seasons among which winter is dry and cold while summer is humid and hot. Residential buildings in Korea have the characteristic Ondol structure system under bedrooms to cope with the cold winter weather. Ondol is a radiant floor heating system with hot water pipe. Thus, there are many energy uses for space heating in residential buildings during winters.

Many people live in apartments in major cities in Korea. The typical apartment buildings have 5–30 stories and the residential area per household is between 50 and 150 m². The radiant floor heating system has been used conventionally in Korea (Seo et al. 2011, Lee et al. 2015). Hence, an insulator used for flooring is one of the important building materials to reduce heating energy. Insulation materials used in concrete slabs not only prevent heat loss but also absorb impact and thus function

as resilient materials for reducing noise and vibration against floor impact (Jeong et al. 2009, Kim et al. 2009).

Figure 1 shows the floor structure of residential buildings in Korea. The floor largely consists of a structure layer, floor heating layer, and finish layer. The structure layer is a 210-mm-thick reinforced concrete slab. The floor heating layer is laid around hot water pipes and consists of a 20-mm-thick insu-

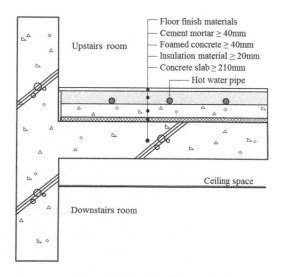

Figure 1.   Section of the dwelling floor structure.

lation layer, hot water pipes, 40-mm-thick light-weight foamed concrete, and 40-mm-thick mortar finish. The final layer is flooring finish, which can be vinyl, timber, or carpet.

The flooring structure should comply with the insulation standards to prevent heat loss. Insulation materials installed under the floor prevent heat from hot water in hot water pipes used for indoor heating from being lost into space below the floor. It can also reduce noise and vibrations due to impact on the upper layer. Because of this, dynamic stiffness of insulation materials was must be less than 40 MN/m².

This study measured the thermal conductivity of insulation materials used in residential buildings' floorings and compared the effective thermal conductivity corresponding to different external shapes of insulation materials.

2 MEASUREMENT METHOD

2.1 *Measurement subjects*

An insulation used in buildings is a material that blocks heat flow due to conduction, convection, or radiation. Its thermal conductivity should generally be below 0.058 W/(m·K). Table 1 shows the grade and range of thermal conductivity for insulation materials recommended in 'the Building Energy Saving Design Standards' in Korea (MOCT 2014).

Test specimens were Ethylene Vinyl Acetate (EVA) specimens that can satisfy resilient performance to reduce vibrations due to floor impact and also provide insulation performance. Insulation materials used in residential flooring mostly have a flat plate shape. To enhance the resilience performance, insulation materials have the unevenness of the bottom.

In this study, the thermal conductivity and density of 30 EVA specimens were measured. The thickness of the measured specimens was 10–30 mm and the size was 300 mm × 300 mm. All specimens were rectangular with a flat surface.

Four single-layered specimens and three double-layered specimens composed of two single-layered specimens were used to measure thermal conductivity. The four single-layered specimens were a flat plate specimen, a specimen with one directional protrusion, dot-shaped protrusions, and multi-directional protrusions. The three double-layered specimens composed of two single-layered specimens were made by laying a specimen with surface protrusions over the flat plate specimen. Thus, an air space was formed in the middle of the double-layered specimen. The entire double-layered specimens were covered with a thin vinyl to make an air layer due to the unevenness. Figure 2 shows the test specimens with the unevenness of bottom.

Table 1. Classification of insulation material grade and range of thermal conductivity.

Type	Thermal conductivity W/(m·K)	Insulation materials
A	< 0.034	Extruded polystyrene board Rigid urethane foam board
B	0.035–0.040	Expanded polystyrene board Type 1, 2, and 3 Rock wool and glass wool boards
C	0.041–0.046	Expanded polystyrene board Type 4 Other insulation materials
D	0.047–0.051	Other insulation materials Rock wool and glass boards

2.2 *Test procedure*

Thermal conductivity is a physical property of homogeneous materials and is defined as the calories that flow through a 1 m² surface area when the temperature difference between the two sides of a 1 m-thick homogeneous material is 1°C. Most insulation materials contain an embedded layer of small-sized foams. The thermal characteristics of insulation materials are determined by the insulation performance of the foamed layer.

The test method used in this study to measure thermal conductivity was KS L 9106 (heat flow method, KSA 2010) and ASTM C 518. The equipment used to measure the thermal conductivity was HFM 436 Lambda of Netzsch. Figure 3 and Table 2 show details of the test equipment used to measure thermal conductivity.

For pre-treatment, the test specimens were stored in a test chamber maintained at the standard conditions specified in KS A 0006 (dry bulb temperature of 23 ± 2°C and 50 ± 5% Relative Humidity (R.H.)).

The high-temperature side was 30°C and low-temperature side was 10°C for test condition. Respectively, the mean temperature was 20°C. The measurements were acquired from both the high and low temperature sides to measure the surface temperature of the specimen and the heat flow meter measured the calories that passed through the specimen. When all measured data were reached at the normal stay-state, the measured data were recorded and effective thermal conductivity was calculated. The thickness of the insulation materials was measured using a digital vernier caliper.

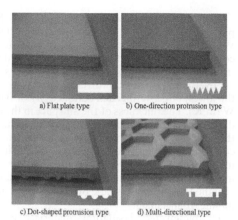

a) Flat plate type   b) One-direction protrusion type

c) Dot-shaped protrusion type   d) Multi-directional type

Figure 2. Photograph of test specimens.

Figure 3. Photograph of test equipment.

Table 2. Test equipment.

Maker	Netzsch
Model	HFM 436 Lambda series
Sample geometry	Square
Test range	0.005 to 0.50 W/(m·K)
Repeatability	0.5%
Accuracy	± 1 to 3%

3 RESULTS AND DISCUSSION

In general, the thermal conductivity of insulation materials has a high correlation with its density. According to a study by Jeong (2009), the thermal conductivity of manufactured insulation materials made from EVA and Polyethylene (PE) increased with density. The dynamic stiffness decreased as the thickness of the measured specimen increased. Furthermore, if insulation materials were configured such that specimens with high dynamic stiffness were laid over specimens with low dynamic stiffness, the dynamic stiffness of the layered specimen was comparable with that of specimens with low dynamic stiffness (Kim et al. 2009).

According to a study by Choi (2007), when the thickness of the materials was increased from 20 mm to 80 mm, approximately 4 to 8% of thermal conductivity was changed because of heat loss from the side of specimens. However, they did not observe any significant effects of composite structures on thermal conductivity using the heat flow method, which is the same method used in this study.

Figure 4 shows the measured results for flat plate type insulation materials. The measured correlation between thermal conductivity and density was similar to that of a previous study (Jeong et al. 2009). The minimum and maximum values of density among the measured specimens were 60.5 kg/m³ and 193.8 kg/m³, respectively. The minimum and maximum values of thermal conductivity were 0.0388 W/(m·K) and 0.0741 W/(m·K), respectively. An increasing trend in thermal conductivity was clearly observed as the density of the measured specimens increased. When the density of the specimens increased three times, the thermal conductivity increased approximately 1.8 times.

The density values of the measured specimens were concentrated between 60 kg/m³ and 90 kg/m³ and thermal conductivity were concentrated between 0.039 and 0.046 W/(m·K).

Table 3 shows the measured thermal conductivity for single-layered specimens. The Type 1 flat plat shaped specimen had a thermal conductivity of 0.0495 W/(m·K). The highest thermal conductivity of 0.0782 W/(m·K) was observed for the one direction protruded specimens. The thermal conductivity of dot-shape protruded specimens was the lowest at 0.0464 W/(m·K). The multi-direction protruded insulator had a thermal conductivity of 0.0526 W/(m·K).

Table 4 shows the measured effective thermal conductivity of double-layered specimens. The specimen (Type 5) in which the one direction protruded specimen was laid over the flat plate

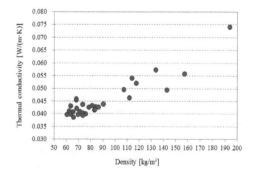

Figure 4. Thermal conductivity and density of flat plate EVA insulation materials.

Table 3. Types of insulation materials and thermal conductivity (single-layer).

Type	Shape	Thickness mm	Thermal conductivity W/(m·K)
1	Plate	24.5	0.0495
2	One direction	24.1	0.0782
3	Dot-shape	29.9	0.0464
4	Multi-direction	30.6	0.0526

Table 4. Types of measured insulation materials and thermal conductivity (double-layer).

Type	Shape	Thickness mm	Thermal conductivity W/(m·K)
5	1type+2type	47.8	0.0614
6	1type+3type	53.0	0.0466
7	1type+4type	54.9	0.0502

specimen had a thermal conductivity of 0.0614 W/(m·K), which is higher than that of the flat plate specimen (Type 1). The air space thickness formed in the Type 5 specimen was 5 mm. The thermal conductivity of the Type 5 specimen was close to the mean value of thermal conductivity of Types 1 and 2.

The thermal conductivity of the Type 6 specimen in which a Type 3 specimen was laid over a Type 1 specimen was 0.0466 W/(m·K); this value is similar to that of the Type 3 single layered specimen. The air space thickness in the Type 6 specimen was 10 mm. The thermal conductivity of the Type 7 specimen is lower than that of the Type 4 specimen. The air space thickness formed in the middle of the Type 7 specimen was 15 mm. The thermal conductivity was measured differently depending on the formation of air space and the thermal transfer characteristics due to the protrusion shape.

4 CONCLUSIONS

Insulation materials are used in the inside of flooring structures of residential buildings to prevent heat loss and reduce floor impact noise. Thermal conductivity of EVA insulation materials used in the flooring of residential buildings was measured, and the effective thermal conductivity of three types of double-layered specimens was also measured.

The thermal conductivity and density of 30 specimens, flat plate type EVA insulation materials were also measured. It was found form the result that the thermal conductivity of EVA insulation materials increases proportionally with its density.

The thermal conductivity of the dot-shape protruding specimen was the lowest at 0.0464 W/(m·K). According to the measured thermal conductivity values of the flat and overlapped protruding shape insulation materials, specimens with an overlaid dot-shape protruding specimen had the lowest effective thermal conductivity of 0.0466 W/(m·K). The thickness of the air space in the middle of the double-layered specimen was 10 mm, and the air space was dot shaped. The effective thermal conductivity of insulation materials consisting of multiple layers was affected by not only the thermal conductivity of each constituent material but also the thickness and shape of the air space in the middle of specimen.

Double-layered insulation materials with a dot-shaped air space, whose thickness is around 10 mm, are expected to have low effective thermal conductivity.

ACKNOWLEDGMENTS

This research was supported by a grant from a Strategic Research Project (A study on noise reduction solution for adjacency household in apartment house) funded by the Korea Institute of Civil Engineering and Building Technology.

REFERENCES

Choi, G.S. et al. 2007. An experimental study on thermal properties of composite insulation. *Thermochimica Acta,* 455(1–2): 75–79.

Jeong, Y.S. et al. 2009. A study on the thermal conductivity of resilient materials. *Thermochimica Acta,* 490: 47–50.

Jeong, Y.S. & Yu, K.H. 2014. Experimental study of thermal conductivity of insulation materials made of expanded polypropylene, ethylene-vinyl acetate co-polymer, and polyethylene. *Advanced Materials Research,* 831: 40–43.

Kim, K.W. Jeong, G.C. Yang, K.S. & Sohn, J.Y. 2009. Correlation between dynamic stiffness of resilient materials and heavyweight impact sound reduction level. *Building and Environment,* 44: 1589–1600.

Korea Ministry Construction & Transportation (MOCT). 2014. *Korea's building energy code.* Korean government.

Korea Standards Association (KSA). 2010. *KS L 9016*: Test methods for thermal transmission properties of thermal insulations. Korea.

Lee, S. Joo, J. & Kim, S. 2015. Life cycle energy and cost analysis of thin flooring panels with enhanced thermal efficiency. *Journal of Asian Architecture and Building Engineering,* 14(1): 167–173.

Seo, J. Jeon J. Lee, J.H. & Kim, S. 2011. Thermal performance analysis according to wood flooring structure for energy conservation in radiant floor heating systems. *Energy and Buildings,* 43: 2039–2042.

Advances in Civil, Architectural, Structural and Constructional Engineering – Kim, Jung & Seo (Eds)
© *2016 Taylor & Francis Group, London, ISBN 978-1-138-02849-4*

Application of SWAT and GIS to simulate the river flow of the Sembrong River, Johor, Malaysia

M.S. Adnan, S.N. Rahmat & L.W. Tan
Micro Pollutant Research Centre, Universiti Tun Hussein Onn Malaysia, Parit Raja, Johor, Malaysia

N. Rosly
Centre for Energy & Industrial Environment Studies, Universiti Tun Hussein Onn Malaysia, Johor, Malaysia

Y. Shimatani
Watershed Management Laboratory, Faculty of Engineering, Kyushu University, Fukuoka, Japan

ABSTRACT: The power of Geographic Information System (GIS) has been demonstrated in many previous studies in various fields of study. In this study GIS was integrated with a watershed model known as Soil and Water Assessment Tools (SWAT) to model the stream flow of the Sembrong River, Malaysia. In this study the main objective of this study to test the applicability of SWAT in modeling the stream flow in selected study area. Several required parameters are needed to run the model which is meteorological data, land use, soil type, and topography maps. Through the processing the sub-basin was generated that consist of unique value to increase the performance of results. Then, the model was calibrated and validated to ensure that the generated result is acceptable. The calibration and validation based on the r^2 were used as the main reference to validate the model. Based on the r^2 value, the value is more than 90% which is 0.9161 that conclude that the result is acceptable and reliable. Overall, reviewing the results it seems that overestimate and underestimate values are within the acceptable range. As a conclusion, the simulated stream flow was within the acceptable for the Sembrong watershed and this model is suitable to be used as a platform to predict the future changes.

1 INTRODUCTION

Recently, due to uncontrolled human activities that pursue the modernization, many changes in the watershed have been reported. Many of the changes have led to watershed degradation either from the quality or quantity perspective point of view. To monitor the changes involved a large area, one of the promising methods that can be applied is by using Geographic Information System (GIS). There are several functions of GIS as discussed by other researcher such as data capture, database management, geographic Analysis, Manual Digitalization, and Scanning System. Moreover, the advantage of GIS is this application can be integrated with other model through the addition of extension. This function significantly enhances the functionality and capability of GIS in watershed modeling.

A distributed watershed model is growing rapidly with so many models and software was developed. The results from this model normally being used in making and backing decisions in management strategy. In this study, the Soil and Water Assessment Tool (SWAT) embedded with GIS was fully utilized. SWAT is a river basin or watershed scale model which has the capableness to simulate both the spatial heterogeneity and the physical processes which has separated the large area of watershed into several smaller modeling units known as Hyrologic Response Units (HRU's) for the sustainable planning and management of surface water resources of rivers (Abbaspour et al. 2007).

SWAT model has been applied in several countries around the world in specific study such as management of water quality, sediment yield, pollution loading, river flow, estimation of base flow and groundwater recharge and prediction of climate change. Abbaspour et al. (2007) has used the SWAT model to simulate all related process affecting water quantity, sediment and nutrient loads in the watershed in Switzerland. Zhu et al. (2015) have discussed the application of SWAT to simulate the streamflow in the Haifa River basin by integrating precipitation data. While Rahman et al. (2014) has demonstrated the application of SWAT in measuring the impact of subsurface drainage on streamflow in the Red River on the North Basin.

A dam was constructed with the intention to serve several functions such as water supply, flood protection, water reservoir and meet the agricultural demand. However, the construction of this structure has trigger negative effect on the watershed itself. Study by Adnan et al. (2013) has shown that the quantity and quality of water flowing over the dam have changed. The Sembrong Dam has been constructed since 1980 and since then no comprehensive study has been conducted to measure the change on the downstream of this dam. In this paper, the application of GIS and SWAT was utilized in determining the applicability of the model to predict the changes of streamflow in the study area.

2 METHODOLOGY

2.1 Study area

The study area located at the Southern part of Peninsula Malaysia. The Sembrong River basin is located in the state of Johor with the catchment area size approximately 273 km² with the total length approximately 22.3 km. This river is a sub-catchment of the Batu Pahat River. It originates from the Sembrong Dam and flows through south eastern part of Johor and afterward flow into the Bekok River and the Simpang Kanan River (Figure 1).

2.2 Data collection

Prior the modeling and simulation works can be carried out, numerous data input is needed as the

Figure 1. Location of the study area.

input data for the model. Rainfall and river flow data were obtained from the Department of Irrigation and Drainage Malaysia. All the hydrological data from the year 2012 to 2014 were used to derive the Intensity-Duration-Frequency (IDF) and the peak flow (Q_p) for the return period of 2, 5, 10, 20, 50 and 100 years. For the river cross-section, manual measurement was conducted combined with the data from the consultant.

While for the landuse, soil and elevation maps the data were obtained from Malaysia Centre for Geospatial Data Infrastructure (MaCGDI). ArcGIS Software is used in formed of digital topographic maps that used to basic layer, developing TIN and analyse the hydraulic data. Digital Elevation Model (DEM) and Triangular Irregular Networks (TIN) have been created to present the ground surface topography or terrain in a digital or virtual platform.

2.3 Streamflow analysis

Once the landuse, soil and slope data layers have been overlaid, the distribution of Hydrologic Response Units (HRUs) within the watershed will be determined. This process will divide to catchment into sub-basin. Subdividing the watershed into areas having unique land use and soil combinations enable the model to reflect differences in evaporation and other hydrologic conditions for different land cover/crops and soils. Runoff is predicted separately for each HRU and routed to obtain the total runoff for the watershed. This increases the accuracy of load predictions and provides a much better physical description of the water balance (Neitsch et al. 2005).

Calculating stream flow involved solving an equation that examines the relationship among several variables including stream cross-sectional area, stream length and water velocity. Equation 1 was integrated into SWAT to predict the flow;

$$Q = \frac{ALC}{T} \tag{1}$$

where
  A = Average cross-sectional area of the stream
  L = Length of the stream reach measured
  C = A coefficient or correction factor
  T = time.

3 RESULTS AND DISCUSSION

The rainfall data for three years was used as the main input data for the analysis. The data were created in dbf file in access as a database since the swat only recognized the .dbf file. Figure 2 shows the relationship between rainfall and stream flow.

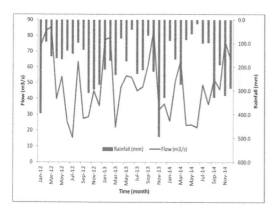

Figure 2. The relationship between flow and rainfall between 2012–2014.

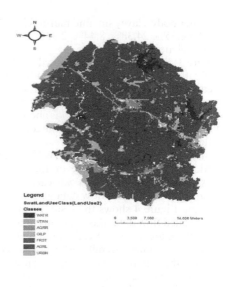

Based on this graph is clearly a show that the rainfall for the monthly range between 90 mm to 400 mm. Normally, during the wet season from October to March the rainfall is higher compared to other month in a year. While for the annual rainfall, for 2012, 2013 and 2014 the rainfall are 2357.9 mm, 2463.4 mm and 2018.1 mm respectively. While for the value of stream flow is ranging between 15 m³/s to 90 m³/s.

Each watershed is first divided into sub basin and then in Hydrologic Response Unit (HRUs) based on the land use and soil distributions. The HRU created in ArcSWAT geodatabase that stores values for SWAT HRUs input parameters. Landuse, soil and slope characterization for a watershed is performed using commands from the HRU analysis menu on the ArcSWAT. All the mentioned data were loaded into the current project, evaluate slope characteristics and determine the land use/soil/slope combinations and distribution for the delineated watersheds. By performed the HRUs analysis, the landuse, soil and slope characteristic for each sub-basin were generated.

Land use in this watershed was classified into seven (7) types of landuse which are water, road, agriculture, oil palm estate, forest, rural agricultural and urban (WATR, UTRN, AGRR, OILP, FRST, AGRL, URBN). Based on the watershed delineation using SWAT, it shows that the grasses area is covered of 2.19%, agricultural land generic is consisted of 92.79%, residential area of 3.25% and water of 3.96% of the total area (Figure 3).

Figure 3 shows the soil map after the delineation process. Based on this map, five (5) categories of soil types have been identified. The soil type in this area consists of peat, clay, clayey silt, marine clay and water body. Based on the percentage, soft soil the dominant soil type in the watershed, which clay contributes the largest percentage followed by

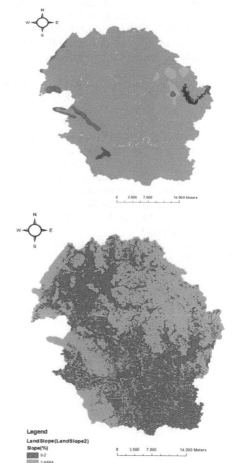

Figure 3. The landuse, soils and slope map for the Sembrong Catchment.

peat, water body, clayey silt, and marine clay which covered 64.35%, 24.38%, 8.44%, 2.32% and 0.51% of the total area respectively.

Figure 3 shows the slope map based on the new classification which separated the landform into two categories which is less than 2% of slope and more than 2% of slope. Based on this classification, the area of slope less than 2% is about 17.45% and more than 2% is 82.55% from the total area.

After the successful of SWAT simulation, all the simulation results saved in the SWAT database. In order to verify and validate of the simulated results, the calibration process was carried out by comparing the simulated flow data with the measured data. Figure 4 shows the result of measured and simulated flow data. Based on this figure, even though there are some overestimation and underestimation of values, overall performed is acceptable since the difference is still intolerance range. Figure 5 shows the result of calibration process based on the R^2 value. Based on this figure, the agreement between measured and simulated is more than 90% which is 0.9161. Legendre and Fortin (2010) have mentioned that the R^2 can be used to evaluate the result in showing the agreement and similarity.

Based on the analysis conducted, the minimum, maximum, mean and standard deviation for both measured and simulated stream flow over the modeling period were calculated. The simulated minimum was 26 m³/s and the minimum measured was only 15.5 m³/s Moreover, the simulated maximum was 94 m³/s, but the maximum measured was only 85.5 m³/s. The standard deviation for simulated flow was lower compared to measured flow. On average, the difference between simulated and measured is 14.19% and this value can regard as not much different.

Nevertheless, a lack of information regarding the aquifer systems, both deep and shallow, can

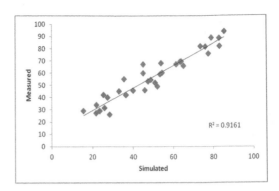

Figure 5. Calibration between measured and simulated stream flow.

impact on the base flow modeling. Indeed, temporal variations in the origin and constitution of the, hydrologic, recharged water and human factors may result in periodic varies in groundwater mechanism. These changes can be attributed with nature phenomena or human activities (Adnan et al. 2013).

4 CONCLUSIONS

The application of GIS and SWAT was used to simulate the stream flow in the Sembrong River, Johor located in the southern part of Peninsular Malaysia. The differences between simulated and measured stream flow is still in the acceptable range. The average difference between simulated and measured is 14.19%. To validate this result, the value of r^2 was used. Based on the r^2 the value is 0.9161 indicating a good result. It can be concluded that, the application of SWAT in modeling the stream flow is successful and this could be used as a based study to predict the future stream flow for this watershed.

ACKNOWLEDGEMENT

The Author would like to express the gratitude to the Ministry of Education Malaysia and Universiti Tun Hussein Onn Malaysia for funding this research under research grant FRGS 1232.

REFERENCES

Abbaspour, K.C., Yang, J., Maximov, I., Siber, R., Bogner, K., Mieleitner, J., Zobrist, J. & Srinivasan, R. 2007. Modeling hydrology and water quality in the pre-alpine/alpine Thur watershed using SWAT. *Journal Hydrology*, 333: 413–430.

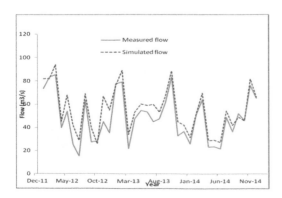

Figure 4. Monthly measured and simulated stream flow of the Sembrong watershed.

Adnan, M.S., Heru, H. & Doni, P.E. 2013. Groundwater Model as a Tool For Sustainable Groundwater Management. *International Journal of Integrated Engineering*, 5(1): 46–57.

Legendre P. & Fortin M.J. 2010. Comparison of the Mantel test and Alternative Approaches for Detecting Complex Multivariate relationships in the Spatial Analysis of Genetic Data. *Molecular Ecology Resources*. 10: 831–844.

Neitsch, S.L., Arnold, J.G., Kiniry, J.R. & Williams, J.R. 2005. *Soil and Water Assessment Tool Theoretical Documentation version 2005 USDA*. Grass Land, Soil and Water Research Laboratory, Agricultural Research Service 808 East Blackland Road, Temple, Texas.

Partha, P.S. & Ketema, Z. 2014. Modeling Streamflow Response to Climate Change for the Kyeamba Creek Catchment of South Eastern Australia. *International Journal of Water*. 8(3): 241–258.

Rahman, M.M., Lin, Z., Jia, X., Steele, D.D. & DeSutter, T.M. 2014. Impacts of Subsurface drainage on Streamflows in the Red river of the North Basin. *Journal of Hydrology*. 511: 474–483.

Zhang, X.S., Hao, F.H., Cheng, H.G. & Li, D.F. 2003. Application of SWAT model in the Upstream Watershed of the Luohe River. *Chinese Geographical Science*, 13(4): 334–339.

Zhu, H., Li, Y., Liu, Z., Shi, X., Fu, B. & Xing, Z. 2015. Using SWAT to simulate streamflow in Huifa River Basin with ground and Fengyun Precipitation Data. *Journal of Hydroinformatics*. doi:10.2166/hydro.2015.104.

Advances in Civil, Architectural, Structural and Constructional Engineering – Kim, Jung & Seo (Eds)
© 2016 Taylor & Francis Group, London, ISBN 978-1-138-02849-4

Flow visualization in a malodor absorption system with porous baffles and a rotating disk drum via a PIV technique

J.H. Lee, K.W. Kim, H. Ali & C.W. Park
School of Mechanical Engineering, Kyungpook National University, Buk-gu, Daegu, Korea

ABSTRACT: Eco-friendly malodor absorption systems can be employed to remove the enhanced odor-producing gases from the surroundings to restore the health of an ecosystem. Malodor gas is introduced into a water basin that contains effective microorganisms for water absorption. In the present study, the Particle Image Velocimetry (PIV) technique was used to experimentally investigate the effects of a rotating disk drum and porous baffles on the fluid properties of a laboratory-scale odor absorption system. The disk drum was tested at various rotational speeds to determine their effects on the velocity field, streamlines, and spatial Reynolds number distribution. Results showed that a rotating disk drum and cantilever porous baffles can extend gas residence time and improve liquid circulation.

1 INTRODUCTION

Effective Microorganisms (EMs) are anaerobic organisms used in secondary water treatment processes to absorb malodor gas emissions from sewage systems. These malodor-producing gases harm the surrounding human population; therefore, EMs are employed to restore the health of an ecosystem (Shalaby 2011, Ting et al. 2013). EM odor absorption systems are an innovative industrial technology that uses EMs to treat malodor gases from sewage systems and public washrooms. Malodor gas from surrounding areas enters the water basin of an EM absorption system through gas nozzles at the bottom of the basin. Then, EMs interact with the malodor gas and decomposes it into CO_2 and nutrients that are consumed in the EM growth process.

Gas bubbles tend to increase in number and accumulate in the water basin due to buoyancy force. The accumulation of gas in the basin results in circulatory liquid motion. The interaction of gas with EM cells increases as a result of the increased circulatory motion of liquid; consequently, the absorption of malodor gas from the surroundings is enhanced (Rubio et al. 1999). Improved liquid circulation increases the working rate of the EM system by dissolving increased amounts of malodor gas. Therefore, continuous mixing is required for the effective interaction of EM with gas (Razzak et al. 2013, Lananan et al. 2014). However, the high rotational speeds of mixing devices can generate a turbulent flow that may damage the mechanical cell structure of EMs (Papoutsakis 1991). Moreover, the geometrical features of water

basins also affect the odor removal process by influencing the flows of the water and gas within. Couvert et al. (1999) studied the effect of flow-deflecting baffles on the gas dissolution process in a chemical reactor. However, little experimental research has been conducted on the influence of baffles and mechanical mixing in EM absorption systems.

The present study aims to experimentally investigate the effects of porous baffles and of a rotating disk drum on fluid dynamic characteristics using the Particle Image Velocimetry (PIV) technique. In this work, various cantilever porous baffles are employed to control the residence time of the gas in a water basin to enhance its interaction with EM cells. A rotating disk drum structure with many fins is also installed to improve liquid circulation and gas absorption in the upper part of the water basin. The flow structure is measured at three different locations to observe the effects of porous baffles and of the rotating disk. Various rotational speeds of the rotating disk drum are studied as well to determine the effects of properties that enhance mechanical flow mixing [i.e., velocity field, streamlines, and Reynolds number (Re) distributions].

2 EXPERIMENTAL SETUP AND METHOD

2.1 *Water basin*

An acrylic laboratory-scale EM water basin with a Length (L) of 0.6 m, Width (W) of 0.4 m, and Height (H) of 0.9 m was used for the PIV experiment (Figure 1). This study utilized nine porous flow directing baffles (0.17 m × 0.25 m × 0.008 m)

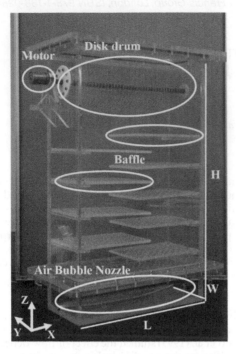

Figure 1. Experiment model in a water basin with baffle plates and rotating disks.

to increase the residence time of the gas in the basin. The porosity of baffles was set at a value of 0.019 to generate bubbles and enhance flow mixing. A rotating disk drum (acrylic) with a diameter of 0.15 m was installed in the basin to enhance flow mixing; this drum consists of 40 disk sheets that were positioned at a distance of 10 mm from one another. A brushless DC motor was employed to rotate the disks via a connected shaft. This motor was powered by an analog controller that adjusted rpm speed. A high rotational speed may damage the EM cell walls; therefore, the rotational speed to enhance flow mixing was set in the range of 0 rpm to 15 rpm (Papoutsakis 1991). Gas was introduced through gas holes (diameter = 1 mm) in the bottom of the water basin with the aid of an air compressor (pressure = 80 kPa). This study determined turbulent flow in the water basin based on Reynolds number (Re).

2.2 PIV apparatus

This research used a two-dimensional PIV system to measure the velocity fields in the water basin.

The PIV system consisted of an Nd:YAG dual-head laser source, a lens set, Charge-Coupled Device (CCD) camera, synchronizer, image frame grabber, and computer. The tracer particles used

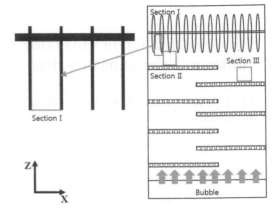

Figure 2. Locations of Section–I, Section–II, and Section–III in the water basin (ZX-plane view).

for flow visualization in the basin are fluorescent polymer microspheres with a diameter of 15 μm and a density of 1.1 g/cm³. The CCD camera obtained an image of the tracing particles at a spatial resolution of 2 K pixels × 2 K pixels. A low-pass filter and a Q-switch with a 532 nm wavelength were connected to the CCD camera. The camera captures two frames in double-exposure mode using an image grabber with a delay time of 8 ms. The shooting distance was set to five pixels/shot to obtain enhanced images of the tracing particles. The effects of the porous baffle and of the rotating disk drum on the fluid field in the basin were observed by measuring the fluid velocity in three selected sections (Figure 2). Section–I was located between the disks to observe the disk effect. Section–II and Section–III were positioned near baffles to determine the influence of the rotating disk on gas bubbles. All the experiments were conducted under room temperature conditions.

3 RESULT AND DISCUSSION

3.1 Velocity field and streamlines

Figure 3 shows the velocity field in Section–I (XY plane), which is located between the disks of the rotating drum. Initially, the flow field was measured without rotating the disk drum, that is, at a rotational speed of 0 rpm (Figure 3a). Subsequently, the disk drum was rotated at a speed of 15 rpm, and the effects of this rotation on flow field were determined (Figure 3b). The flow particles failed to follow the geometrical shape of the rotating disk drum, and recirculation zones were formed although the rotating disk did not move. These zones were generated by the movement of gas bubbles in the water and were enlarged by an

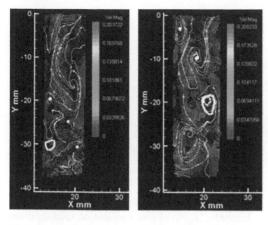

Figure 3. Velocity contour field at Section–I (a) rotating disk speed of 0 rpm and (b) rotating disk speed of 15 rpm (ZX-plane view).

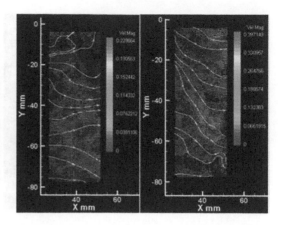

Figure 4. Velocity contour field at Section–II (a) rotating disk speed of 0 rpm and (b) rotating disk speed of 15 rpm (ZX-plane view).

increase in the rpm of rotating disk. This enlargement improves the gas dissolution process. The gas absorption process is strengthened near the rotating disk drum as a result of improved liquid recirculation (Razzak et al. 2013, Lananan et al. 2014).

Figure 4 presents the velocity field in terms of streamlines and vector plots for Section–II (XY plane). The velocity gradient increased significantly, and the number of recirculation zones was reduced near the baffle plate by the smooth path followed by the fluid particles (Figure 4a). The increasing velocity enhanced the turbulent flow, which in turn limited the working efficiency of EMs by deteriorating their cell walls (Papoutsakis 1991). Velocity magnitude increased significantly when the disk drum was rotated at a speed of 15 rpm (Figure 4b). The results suggest that gas dissolution will be ineffective near the baffle areas because of the limited liquid recirculation (Lananan et al. 2014).

The velocity field in Section–III (XY plane) is depicted in Figure 5. The results were computed at rotating disk drum speeds of 0 and 15 rpm. Velocity magnitude was reduced because the effect of the rotating disk on this region was weakened. The recirculation process was short when the rotating disk was stationary (Figure 5a). The recirculation zones were particularly prominent when the rotational speed was 15 rpm (Figure 5b). Velocity magnitude also increased with rotational speed. This outcome implies that the bubbly flow is more prominent in Section–I than in other regions. Therefore, more gas is dissolved near the disk drum than on the surface of the baffles. Significant gas bubbles production limited the liquid circulation in Section–II and Section–III. Therefore, low porosity baffles are recommended for the EM odor systems because a large amount of gas bubbles can restrict

the odor removal due to the extended residence time and the increased mixing power required for gas dissolution.

3.2 Turbulent flow

Razzak et al. (2013) suggested that a mechanical mixing device is necessary to mix gas with all microorganisms. However, high mixing rates may damage the cell structures of these microorganisms through the high shear force of fluid in the basin. Therefore, the present study determines the intensity of the turbulence in the water basin under the effect of different disk drum rotational speeds through Re. Figure 6 shows the local Re values in Section–I, Section–II, and Section–III at different rotational speeds. Re is a function of rotational speeds, and the value is lower in Section–I than in the other two sections because the influence of the gas bubbles is weakened. Turbulent flow increases in Section–II and Section–III as a result of the gas bubble turbulence caused by increased number of gas (Rubio et al. 1999).

The magnitude of turbulent flow also increased in Section–II due to the increased amount of gas bubbles. Re value was maximized when the disk drum rotated at a speed of 15 rpm. The cell membrane of EM was threatened at high rotational speeds by high turbulent flow; consequently, odor removal is limited (Papoutsakis 1991). The EM in Section–II is placed under continuous threat of cell damage because of high turbulence, and the effectiveness of the odor removal process is reduced. The results suggest that the gas dissolution process is more efficient in Section–I than in the other two sections due to low turbulent flow, and high rotational speeds do not affect the EM

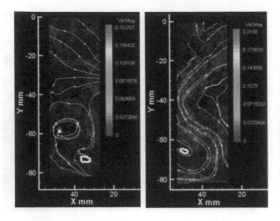

Figure 5. Velocity contour field at Section–III (a) rotating disk speed of 0 rpm and (b) rotating disk speed of 15 rpm (ZX-plane view).

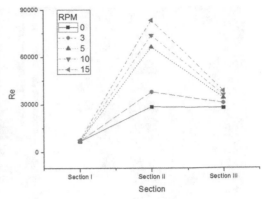

Figure 6. Effect of rpm variation on spatially different Reynolds number distributions in Section-I, Section-II, and Section-III.

cell structure. Therefore, odor removal is more effective in this section than in the other two as a result of the low turbulence and the improved liquid circulation (Lananan et al. 2014).

4 CONCLUSION

In the present study, the PIV technique was employed to for an experimental investigation into the effects of porous baffles and of a rotating disk drum on the fluid dynamic characteristics within a water basin for a malodor removal system. As per the basic platform designs, cantilever porous baffles and a rotating disk drum were installed to increase gas residence time and flow mixing in the water basin. This drum was tested at various rotational to determine their effects on the velocity field, streamlines, and Re distributions.

The entrainment flow around the rotating disk drum can effectively improve the gas absorption process. Gas is actively dissolved near the rotating disk drum as a result of the improved recirculating flow structure. Stationary dead zones were formed between the disks. Furthermore, porous baffles were installed as another enhancing structure at the lower part of the water basin. These baffles can facilitate a smooth and long flow passage for effective malodor absorption. Nonetheless, rapid, bubbly oncoming flow can reduce gas absorption rate around the surface of baffles due to shortened residence time.

Much malodor gas is absorbed in Section-I because of the limited influence of gas bubbles. Turbulent flow increases with rotational speed as a result increased Re value. The EM cells in Section-II are at risk because of high turbulent flow. Thus, odor removal is more effective in Section-I than in the other sections because of the low Re and the enhanced liquid circulation.

ACKNOWLEDGEMENTS

This work was supported by the National Research Foundation (NRF) of Korea, by a grant funded by the Korean government (MEST) (No. 2012R1 A2 A2 A01046099), and by a grant from the Priority Research Centers Program through the NRF as funded by MEST (No. 2010-0020089).

REFERENCES

Couvert, A., Roustan, M. & Chatellier, P. 1999. Two-phase hydrodynamic study of a rectangular air-lift loop reactor with an internal baffle. *Chemical Engineering Science*, 54(21): 5245–5252.

Lananan, F., Abdul Hamid, S.H., Din, W.N.S., Ali, N., Khatoon, H., Jusoh, A. & Endut, A. 2014. Symbiotic bioremediation of aquaculture wastewater in reducing ammonia and phosphorus utilizing Effective Microorganism (EM-1) and microalgae (Chlorella sp.). *International Biodeterioration & Biodegradation*, 95: 127–134.

Papoutsakis, E. 1991. Fluid-mechanical damage of animal cells in bioreactors. *Trends in Biotechnology*, 9(1): 427–437.

Razzak, S.A., Hossain, M.M., Lucky, R.A., Bassi, A.S. & de Lasa, H. 2013. Integrated CO2 capture, wastewater treatment and biofuel production by microalgae culturing—A review. *Renewable and Sustainable Energy Reviews*, 27: 622–653.

Rubio, F.C., et al. 1999. Steady-state axial profiles of dissolved oxygen in tall bubble column bio-reactors. *Chemical Engineering Science*, 54(11), pp. 1711–1723.

Shalaby, E.A. 2011. Prospects of effective microorganisms technology in wastes treatment in Egypt. *Asian Pacific Journal of Tropical Biomedicine*, 1(3): 243–248.

Ting, A.S.Y., et al. 2013. Investigating metal removal potential by Effective Microorganisms (EM) in alginate-immobilized and free-cell forms. *Bioresource Technology*, 147: 636–639.

Advances in Civil, Architectural, Structural and Constructional Engineering – Kim, Jung & Seo (Eds)
© 2016 Taylor & Francis Group, London, ISBN 978-1-138-02849-4

Numerical study to evaluate the human thermal comfort and the buildings energy demand inside the urban canyon

A. Vallati

DIAEE Department of Sapienza University, Rome, Italy

ABSTRACT: This study shows the results of a thermal comfort analysis which was carried out in order to investigate the parameters that influence thermal comfort conditions and buildings energy demand in urban canyon environment. For the creation of the model have been compared various configurations of urban canyon varying geometric parameters, such as the aspect ratio (High-Width), using the common methods for mitigating the ur-ban heat island, such as vegetation and the use of surfaces for the roadbed low-reflective, with the objective to determine the optimal configuration climate and the thermal comfort inside this area and reduce the energy consumption. The results, for the various comparisons, show that the use of a tree lined at the center of the road allows a reduction of the air temperature, soil temperature and mean radiant temperature. Actual improvements, in addition, are also obtained using a unitary ratio between the height of buildings and road width and using low reflective surfaces and vegetation.

1 INTRODUCTION

In recent years several studies have been carried out on thermal comfort outdoors (Bottillo, de Lieto Vollaro, Galli, Vallati (2013 and 2014), since it is an issue strongly related to health and well-being (Galli, Vallati, Recchiuti, de Lieto Vollaro, Botta (2013). Also comfortable urban conditions are related to human activity out-doors.

The thermal comfort parameters that need to be analyzed are: air temperature, soil temperature, the mean radiant temperature and velocity of the wind. An urban canyon is defined as an arc road relatively narrow, laterally delimited, in its longitudinal development, by two continuous rows of buildings and it is the basic geometric units of the urban agglomeration. Generally, an urban canyon can be defined as a function of geometrical parameters such as the aver-age height H of the buildings facing the roadway and the width W and length L of this canyon that affect the thermal exchanges in buildings and affect the comfort thermo hygrometric (de Lieto Vollaro, De Simone, Romagnoli, Vallati, Botillo (2014); also the geometrical parameters of the canyon can affect the exchange of mass of air favoring concentration of pollutants emitted from the human activities as vehicular traffic and heating homes in the ends of the canyon. The main objective of this paper, is to study the thermo-hygrometric parameters in the urban can-yon and develop a microclimate optimal model ratios H/W and L/H of the canyon in relation for opti-

mizing the thermal comfort and an optimal configuration studying materials of the ground and the vegetation inside the canyon (Kroener et al. 2014), to obtain an urban network with low energy consumption and thermally comfortable for the cities of the south Europe and Mediterranean area (de Lieto Vollaro, Vallati, Bottillo (2013) and Grignaffini Vallati (2007). It is studied especially the thermal behavior of the canyon in the summer season because the climatic conditions in-fluence the human activities much more than the winter season.

2 THERMAL ANALYSIS AND CALCULATION MODEL

The model is made by the ENVI-met and was al-ready validated by experimental measures (Bottillo et al. 2013). The ENVI-met software is able to simulate and repro-duce the behavior of micro-climate and physical ur-ban and rural areas. The calculation model studies the interactions between buildings, surfaces, vegetation, air flow and energy of the urban area simulated by the climatic conditions of the geographical con-text. In this study through Envimet it is modeled an urban canyon and has carried out an analysis which it is possible to extrapolate all the microclimatic parameters (temperature, velocity of air, Mean Radiant Temperature, calculated on nine points inside the canyon (Figure 2).

3 RESULTS

The model of the canyon studied in this paper follow the classic urban configuration of an European cities where there are structures on the height of 25 meters with average width roadways of 15–20 meters.

In a first step, the configuration described is modeled varying the ratios H/W and the orientation north-south and then east-west (Figure 1) to analyze the thermal conditions and parameters (temperature of the air, temperature mean radiant, temperature of the soil and velocity of the air) in the summer season, considering some significant days with worst climatic conditions (Bottillo et al. 2014) in Rome.

Below the lists of the main characteristics of the urban canyon considered for the study: day of simulations is 26/06/2012, Duration of the simulations is 72 h, Building height is 22 m, width buildings is 14 m, road width is 18 m, wind speed at 10 m is 3 m/s, wind direction is est, relative humidity is 63%, Atmospheric temperature potential is 299 K, indoor temperature is 293 K, albedo asphalt is 01, albedo walls is 0.2.

In the model, nine receptors were placed inside the urban canyon, along three transverse axes (north, center and south) representing nine ideals pedestrians.

In the first phase of the study we have modeled with software Envimet three canyon's geometry with three different ratios H/W (interior height of the edifice W width of the road that separates the two buildings) to analyze the geometric configuration more comfortable for the man who is inside the canyon.

The three cases analyzed reports have H/W (Case A H/W = 1,2 – Case B H/W = 0,5 and Case C H/W = 3) chosen by analyzing the structural characteristics of European cities.

The Simulation models about these three configurations, considering the data of a city like Rome in the summer season, and we have analyzed four fundamental microclimate parameters as temperature of air, mean radiant temperature, temperature of the ground and air velocity inside the canyon shown in the nine receptors in Figure 2, in the day and the

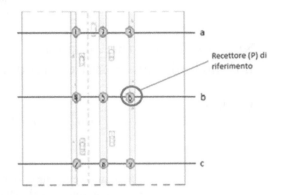

Figure 2. Position of nine receptors inside the canyon.

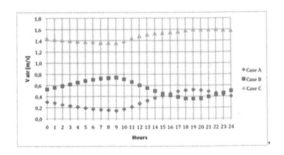

Figure 3. Hours plot of air velocity for the three configurations.

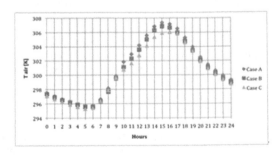

Figure 4. Hours plot of air temperature for the three configurations.

time for the hottest summer in the city of Rome, according to the climatic data libraries for Envimet.

In addition, we chose a point P inside the canyon and we compared the four climate parameters values in the point P within 24 hours of the hottest day for the three canyons (Case A, B, C) and the Case C has parameters of comfort significantly better than other two, with the air temperature, the mean radiant temperature and the temperature of the ground significantly lower in the nine receptors at 14 p.m. and at point P for 24 hours and an air flow velocity higher. Figures (3–6).

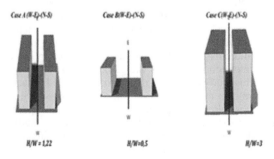

Figure 1. The various model configurations of the canyon.

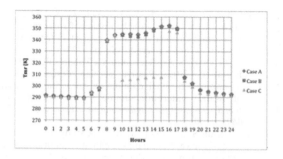

Figure 5. Hours plot of the radiant temperature for the three configurations.

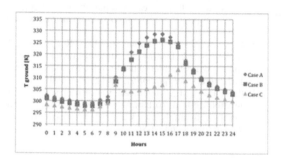

Figure 6. Hours plot of ground temperature for the three configurations.

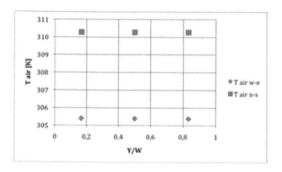

Figure 7. The plot of air temperature varyng the canyon orientation.

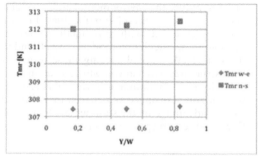

Figure 8. The plot of radiant temperature varyng the canyon orientation.

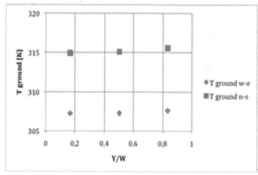

Figure 9. The plot of ground temperature varyng the canyon orientation.

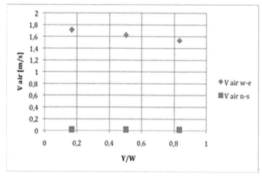

Figure 10. The plot of air velocity varyng the canyon orientation.

Then it is studied the four parameters: air temperature (Ta °C), mean radiant temperature (Tmr °C), ground temperature (Tg °C), air velocity (Va m/s) for the three cases varying the orientation, principal direction, of the canyon in the first NS and then WE, always considering the weather data of hottest summer day for the city of Rome; it is considered the warmer time of the day (14 p.m.) and the impact of the global solar radiation in that hour (14 hours); therefore it is analyzed the point P for 24 hours in order to evaluate the optimal orientation (Figures 7–10).

4 CONCLUSIONS

The study resulted in the development of some design concepts for new urban areas with the aim of optimizing the thermo hygrometric parameters, in order to increase the comfort of the man who lives in urban areas. Indirectly, the goal is to reduce the contribution of the thermal energy plant in order to reduce the environmental impact of active systems at the service of our facilities. From the studies developed with this methodology and through these calculation models it was found that the ratio H/W optimal, among those tested, is 3 with structures with canyon whose structures have a development predominantly vertical H respect to the horizontal distance W between them, in order to take advantage of the summer shading and the flow of air entering the canyon above.

ACKNOWLEDGEMENT

This material is based upon work funded by Zhejiang Provincial Natural Science Foundation of China under Grant No. LQ12E09002; Project (51308497) supported by National Natural Science Foundation of China.

REFERENCES

Bottillo, S., Vollaro, A. de. L., Galli, G. & Vallati, A. 2013. Fluid dynamic and heat transfer parameters in an urban canyon. *Solar Energy*, 99: 1–10.

Bottillo, S., Vollaro, A. de. L., Galli, G. & Vallati, A. 2014. CFD modeling of the impact of solar radiation in a tridimensional urban canyon at different wind conditions. *Solar Energy*, 102: 212–222.

Galli, G., Vallati, A., Recchiuti, C., Vollaro, R. de. L. & Botta, F. 2013. Passive cooling design options to improve thermal comfort in an Urban District of Rome, under hot summer conditions. *International Journal of Engineering & Technology*, 5(5): 4495–4500.

Grignaffini, S. & Vallati, A. 2007. A study of the influence of the vegetation on the climatic conditions in an urban environment. *WIT Transactions on Ecology and the Environment*, 102: 175–185.

Kroener, E., Vallati, A. & Bittelli, M. 2014. Numerical simulation of coupled heat, liquid water and water vapor in soils for heat dissipation of underground electrical power cables. *Applied Thermal Engineering*, 70 (1): 510–523.

Vollaro, A. de. L., Simone, G. De., Romagnoli, R., Vallati, A. & Botillo, S. 2014. Numerical study of urban canyon microclimate related to geometrical parameters, *Sustainability*, 6(11): 7894–7905.

Vollaro, R. de. L., Vallati, A. & Bottillo, S. 2013. Differents Methods to Estimate the Mean Radiant Temperature in an Urban Canyon. *Advanced Materials Research*, 650: 647–651.

Author index

Printed and bound by CPI Group (UK) Ltd, Croydon, CR0 4YY

24/10/2024

01778295-0009